THE CORRESPONDENCE OF ISAAC NEWTON

VOLUME IV
1694–1709

EAST END OF CEILING IN THE NATIONAL MARITIME MUSEUM, GREENWICH

The illustration shows the figures of John Flamsteed and Thomas Weston, together with the 'great Mural Arch and ¡Tube'. On the balustrade is a paper which gives Flamsteed's prediction of the eclipse of the Sun, 22 April 1715.

THE CORRESPONDENCE OF
ISAAC NEWTON

VOLUME IV
1694–1709

EDITED BY

J. F. SCOTT, D.Sc.

CAMBRIDGE
PUBLISHED FOR THE ROYAL SOCIETY
AT THE UNIVERSITY PRESS
1967

CAMBRIDGE UNIVERSITY PRESS
Cambridge, New York, Melbourne, Madrid, Cape Town, Singapore, São Paulo, Delhi

Cambridge University Press
The Edinburgh Building, Cambridge CB2 8RU, UK

Published in the United States of America by Cambridge University Press, New York

www.cambridge.org
Information on this title: www.cambridge.org/9780521058155

First published 1967
This digitally printed version 2008

A catalogue record for this publication is available from the British Library

Library of Congress Catalogue Card Number: 59-65134

ISBN 978-0-521-05815-5 hardback
ISBN 978-0-521-08589-2 paperback

CONTENTS

THE CORRESPONDENCE

CONTENTS

LIST OF PLATES

xv

PREFACE

Volume IV of *The Correspondence of Isaac Newton* covers the period 1694–1709, a period which was probably the most varied of Newton's whole career. The *Principia*, published seven years earlier, had already established Newton as the world's foremost mathematician and natural philosopher. In spite of the abstruse nature of the mathematical treatment adopted in its pages, the first edition was rapidly exhausted and, within a very few years, Newton was being urged to consider the preparation of a second edition. This was to contain, *inter alia*, his further researches upon the motion of the Moon, the solar system, and the behaviour of the comets. Not until 1694, however, did his thoughts upon this project assume definite shape. To carry out his plan, he had need of the most accurate observations available, and for these he turned to the Observatory at Greenwich, where John Flamsteed had been installed, in March 1674/5, as King's Astronomer, his duties being to 'rectify the tables of the motions of the heavens and the places of the fixed stars, so as to find out the much-desired longitude of places, for the purpose of perfecting the art of navigation' (Cudworth, p. 40). So came about that close association between the two men which was to last for many years, though not without frequent interruptions. Newton visited Flamsteed at the Observatory on 1 September 1694. On this occasion Flamsteed showed him 'about fifty positions of the Moon reduced to a synopsis' and offered to show him a further hundred. In acknowledging these, Newton gave an assurance that he would not disclose them without Flamsteed's consent: 'Sept. 1. 1694. Received then of Mr Flamsteed two sheets of MS of ye places of the Moon observed & calculated for years 89 90 & part of 91 wch I promise not to communicate Without his consent.' An account of their meeting is given by David Gregory in Memoranda (Number 468). Newton made known his further requests to Flamsteed two months later in the words: 'I desire only such Observations as tend to perfecting the Theory of the Planets in order to a second edition of my book' (Letter 478).

It may well be thought that the association of two such men, each pre-eminent in his own field, could not fail to produce results of the utmost importance. This indeed proved to be the case, despite the fact that they were of widely different temperaments. Newton, in his eagerness 'to set right the Moons Theory this winter' (Letter 473), became more and more impatient, and frequently accused Flamsteed of being needlessly dilatory. 'Pray let me have your Lunar Observations as soon as you can', he wrote on 20 December 1694 (Letter 485). This however was not the tone adopted six months later:

'I received your solary Tables and I thank you for them', he wrote. 'But these & almost all your communications will be useless to me unless you can propose *some practicable way or other of supplying me wth Observations*...I want not your calculations but your Observations only...or else let me know plainly that I must be content to lose all the time & pains I have hitherto taken about the Moons Theory & about the Table of refractions.' There was abundant reason for Newton's impatience: nevertheless, it is not a matter of wonder that requests couched in such terms could only irritate Flamsteed, who, not unnaturally, rebelled against Newton's assumption that it was his duty, as a public servant, to make his observations available merely upon demand. Yet Flamsteed was still disposed to be helpful, for on receiving this letter, he assured Newton of his willingness 'to accomodate you with what is necessary for cleareing the Motion of the Moon & how small a returne I desire'. A month later (Letter 522) he wrote: 'When ever you let me know that it lies in my power to serve you, I shall doe it freely.'

It was, however, not only with the object of perfecting the lunar theory that Newton had need of Flamsteed's observations. The comets, particularly that of 1680/1 had been the object of close study both by Newton and by Halley. Flamsteed was strongly of the opinion that the comet which had been already observed in November reappeared after passing the Sun in February and March following. Newton at first dissented, however he came round to accept Flamsteed's view (Letter 528, note (2)). Many of Newton's speculations on the motion of the comet were modified in the second edition of the *Principia*, no doubt as a result of his correspondence with Halley on this subject (Letters 533–6, and 538).

The tables of the motions of the celestial bodies in use at this time were far from accurate, and Flamsteed resolved to undertake the colossal task of presenting to the world tables which were free from error. Nevertheless, he realized that to accomplish such a task unaided would be beyond his powers. It must be remembered that Flamsteed suffered from prolonged periods of ill-health, and was rarely free from pain. A much more serious obstacle, however, was the parsimony of the Government. When, therefore, towards the end of 1704, Prince George of Denmark, Queen Anne's consort, who had been newly elected to a Fellowship of the Royal Society, generously undertook to bear the expense of the printing of Flamsteed's work, the *Historia Cœlestis Britannica*, it might well be thought that Flamsteed's troubles were over. This, however, was far from being the case. The Committee of Referees, consisting of Fellows of the Royal Society with Newton at the head, which had been appointed to see that the Prince's money was well spent, rarely consulted Flamsteed; not only that, they did little to expedite the printing of the work, or to execute it in the way Flamsteed desired. Yet he had no alternative but to accept the terms dictated by the referees. 'Sir Isaac Newton has at last forced me to enter into articles for printing my works with a bookseller very disadvantageously to myself', he complained to his assistant, Abraham Sharp, on 20 November 1705 (Cudworth, p. 89). The frustrations which he experienced from those he had hitherto looked upon as his friends aroused feelings of resentment which found expression in his letters to Sharp. These letters, which have been made available through the courtesy of Mr F. S. E. Bardsley-Powell, a descendant of Sharp, constitute a continuous record of the delays and difficulties which beset Flamsteed in the production of his great work. It should however be recognized that, apart from these letters, there is little evidence to support the view that Newton and the referees were responsible for the repeated delays in the production of the *Historia*.

The hostility which had been gradually building up between Flamsteed and Newton was in no way diminished by the latter's close friendship with Halley, concerning whom Flamsteed had developed an intense distrust and suspicion, even going so far as to brand him 'a malitious thief'. 'S.I.N. has put our R.S. into great disorder by his partiality for E.H.', he complained on 14 July 1711. David Gregory, another of Newton's close friends, also became the object of Flamsteed's dislike, and not without reason. His 'officious flattery' did much to exacerbate the Flamsteed–Newton relations.

The present volume, like its predecessors, contains many letters of which Newton was neither the sender nor the recipient. To have excluded these would have left serious gaps in the narrative. Among such are the letters which passed between Wallis and Flamsteed, with relation to the latter's alleged discovery of a parallax of the pole-star, leading to Newton's extraordinary outburst on 6 January 1698/9 (Letter 601). Flamsteed's letter to Wren (Letter 747), in which he recounted the rapid deterioration in his relations with Newton, and in which he vigorously repudiated the charge that he was responsible for the repeated delays in the production of the *Historia*, makes it abundantly clear that henceforth no quarter could be sought or given between the two men. Though the letter was addressed to Wren, there is little doubt it was intended to fall into Newton's hands, as indeed it appears to have done.

In March 1696 Newton was appointed Warden of the Mint. Although the post was offered to him for his greater ease, as Montague's letter suggests (Letter 545), yet he entered into the duties of this office with such zeal that there was an almost complete cessation of his philosophical studies. The great recoinage scheme being practically complete, nevertheless Newton found other tasks to absorb his energies. Among these were his efforts to stamp out currency offences, which had now grown to such proportions as to be a grave embarrassment to the Government. Moreover, troubles of a serious nature broke out at some of the country Mints, notably that of Chester, where Halley had been installed as Deputy Comptroller. Throughout his many years at the Mint, Newton carried out his duties so faithfully and so conscientiously as to set a standard for integrity in the public service in an age when corruption was by no means uncommon.

In April 1705 Newton was knighted. The reasons why this honour should have been conferred upon him have been the subject of conflicting opinions. Trevelyan (*Ramilies and the Union with Scotland*, 1932, p. 28) states that 'the Queen knighted Isaac Newton, her Master of the Mint', but the citation in the *London Gazette*, where he is described as 'formerly Mathematic Professor' suggests that the Queen took the unusual step of recognizing outstanding scientific achievement. This opinion is strengthened by Conduitt; see the article by E. N. da C. Andrade in *Newton Tercentenary Celebrations* (1946), p. 16 (quoted in Letter 692, note (1)).

The letters in this volume show clearly that Newton was not only a man of superlative genius but also a man of the highest principles. Yet it would be doing science no good service, as Professor Andrade has shown clearly in the article cited above, to pretend that his character was free from blemish. He was easily irritated, as a man who has wearied himself with prodigious and concentrated effort might well be. An occasional reluctance to pay tribute to the achievements of others, coupled with a certain misgiving that his own work was not always appraised at its true value, was his greatest fault. 'He is a nice man to deal with, and a little apt to raise in himself suspicions where there is no ground'; such was the judgment of Locke (King, *Life of John Locke*, vol. II, p. 38), after referring to his 'wonderful skill in mathematics and divinity'. Nevertheless, as we read through the pages of the *Correspondence* we cannot fail to notice his generosity towards younger men and his readiness to help many who were intellectually his inferiors. In any case, as Professor Andrade continues: 'Such imperfections of character are not inconsistent with high performance, even in spiritual matters, as anyone who has studied the Church Fathers must admit.'

This, however, does not explain, still less excuse, his uncompromising attitude towards Flamsteed, to whose labours he owed much. Yet on his side, Flamsteed does not escape censure. It is true that he was conscientious in his beliefs and jealous for the reputation of the office he held. It was, however, an act of gross imprudence on his part to impute to Newton any unworthy motive for his actions, for such an imputation would inevitably arouse bitter resentment. It will be recalled that the criticisms of Linus regarding Newton's investigations into the nature of light and colour drew from the great innovator the resolve to 'bid adew to it [philosophy] eternally' (see vol. II, pp. 183, 189–92, 254–60). Flamsteed's criticisms were of such a nature that they could not be easily dismissed. For not only did he declare that Newton's Treatise of Light and Colours contained many errors (Letter 678), he went on to disparage Newton in the most intemperate language. 'Plainly, his design was to get the honour of all my pains to himself, as he had done formerly', he wrote to Sharp on 4 May 1704, 'and to leave me to answer for such faults as should be committed through his mismanagement; but having known him formerly, and his sole regard to his own interests, I was careful to give him no encouragement to expect I should give him anything gratis, as I had done formerly' (Cudworth, p. 80). Two years later he wrote: 'I have allways hated such low practices... he [Newton] thruste himself into ye business purposely to be revenged of me because I found fault both of his Opticks & Corrections of my Lunar Numbers. Sir Is: carrys himselfe very cunningly. I deal plainly wth him & doubt not but God will let me see a good effect of it' (Letter to Sharp, 2 March 1705/6). Newton could hardly fail to be aware of these attacks, nor could he remain indifferent to them, since they concerned what he prized most highly, his integrity. This Flamsteed must have realized. It will always remain a matter of the deepest regret that the cordial relations which had existed between these two men in the early days of their association did not persist. It was a melancholy fate which ordained that the best years of these two distinguished men should have been distracted and embittered by personal animosity and opposing principles.

The Act of Union with Scotland, which came into force on 1 May 1707, brought additional responsibilities to Newton. The Scottish Mint, situated in Edinburgh, was now required to conform to the English method of coinage. The currency was called in, melted down into ingots and recoined to English standards. 'For setting on foot the coinage in the Mint in Scotland with expedition' Newton made certain recommendations to the Treasury (Letter 724). These included the sending of his friend, David Gregory, with some of the London Mint officials, to advise and supervise on Newton's behalf the staff of the Edinburgh Mint. The difficulties resulting from the different traditions in the two mints are brought to light in the correspondence which follows. By the end of 1707 Gregory returned to London and he reported that 'he saw the Methods of the Mint in the Tower well understood, and exactly practised by all concerned, and the Recoynage advanced so, that they coyned Six Thousand Pounds a week' (Letter 732).

In the preparation of this volume, the Editor has been fortunate in obtaining the assistance and counsel of many scholars, each of whom is expert in his own particular field. Without their guidance and encouragement, the preparation of a work such as this would hardly have been possible. The Chairman and the other members of the Newton Letters Committee have spared no efforts to render the work worthy of the Royal Society who have sponsored it. Professor Andrade has ungrudgingly placed his wide knowledge of the history of science of the period at the disposal of the Editor, and has made many suggestions of great value. Sir John Craig, the well-known authority on matters connected with the Mint, has made most helpful contributions. His many references to Newton's activities at the Mint will undoubtedly be read with interest. Mr H. G. Stride, a welcome new member of the Committee, has read the proofs with care and attention and has also made many significant suggestions. Dr Esmond S. de Beer's painstaking and expert examination of the historical background has rendered a service the value of which cannot be over estimated. Dr Robert Schlapp, who gave much expert assistance in the preparation of the earlier volumes, has been no less assiduous in his contributions to this volume, and his helpful and penetrating criticisms have been a source of encouragement to the Editor. In a work which contains so many Latin passages, the guidance of an expert Latinist soon became essential; and here the Committee has been very fortunate in having the willing help of one of its members, Professor W. H. Semple, of the University of Manchester, who has not only checked most of the translations, but has also clarified many passages which had long remained obscure. Tribute must also be paid to Mr P. S. Laurie, the archivist of the Royal Greenwich Observatory, who has freely placed his profound knowledge of the early history of the Observatory at the disposal of the Editor. Others who have made suggestions of value are Professor Douglas McKie and Dr Armitage, of University College London, Dr D. T. Whiteside, of Cambridge, Dr E. G. Forbes, of Edinburgh University, Professor I. Bernard Cohen, of Harvard, Mr D. R. Cooper, Mr D. Starck and Mr G. P. Dyer, of the Royal Mint, Lt.-Commander D. W. Waters and Mr Westby Percival-Prescott, of the National Maritime Museum.

In the collection of material, and its preparation for the Press, Dr E. W. J. Neave has proved himself a most helpful colleague, and one to whom no labour seems to have been too much.

Finally, grateful thanks are due to Mr I. Kaye, the Librarian of the Royal Society, and to his staff, who have never wearied in meeting the demands of the Editor.

The Cambridge University Press has shown its customary skill and care in the production of this volume.

J. F. SCOTT

ACKNOWLEDGMENTS

In addition to those who are named in volume I, the President and Council of the Royal Society of London gratefully acknowledge the help given by the undermentioned, many of whom have supplied photographic reproductions of letters in their possession, and have allowed copies to be made. All have readily agreed to the publication of the documents in their possession:

The Governor and Company of the Bank of England; the Bolling Hall Museum, Bradford (Abraham Sharp Exhibition, 1963); the Clerk of Christ's Hospital; the Astronomer Royal, Royal Greenwich Observatory; the Somerset Herald, College of Heralds; Professor R. S. Westfall of Grinnell University, Iowa.

Thanks are due to Messrs Sotheby and Company for permission to quote from their *Catalogue of the Newton Papers* (1936).

They also express their gratitude to the Librarians of the following institutions who, with their staffs, have been generous in providing help in many ways:

The British Museum; Corpus Christi College, Oxford; the University of Cambridge; the Guildhall, London; the Institute of Historical Research, London; the House of Commons; the National Maritime Museum, Greenwich; the Royal College of Physicians, London; the Public Record Office, London; the Scottish Record Office, Edinburgh; City of Westminster Public Library.

The frontispiece is reproduced by permission of the Admiral President, Royal Naval College, Greenwich; Plates I, II, V by permission of the Trustees of the National Portrait Gallery and Plate III by permission of the Deputy Master of the Royal Mint.

SHORT TITLES AND ABBREVIATIONS
FOR PUBLISHED WORKS

Aubrey, *Brief Lives*	Aubrey's *Brief Lives*, edited by Oliver Lawson Dick (1949).
Baily	*An Account of the Revd John Flamsteed.* By Francis Baily. London, 1835.
Birch	*The History of the Royal Society of London.* By Thomas Birch. 4 vols., London, 1756–7.
Brewster	*Memoirs of the Life, Writings, and Discoveries of Sir Isaac Newton.* By Sir David Brewster. 2 vols., Edinburgh, 1855.
Burke's *Peerage*	*Burke's Peerage, Baronetage and Knightage,* London, 1963.
Cajori, *Principia*	*Sir Isaac Newton's Mathematical Principles of Natural Philosophy and His System of the World.* By Florian Cajori. Berkeley, 1946.
C.E.	*Commercium Epistolicum D. Johannis Collins, et aliorum de Analysi Promota.* London, 1712.
Child	*Early Mathematical Manuscripts of Leibniz.* By J. M. Child. London and Chicago, 1920.
Cooper, *Annals*	*Annals of Cambridge.* By C. H. Cooper. 5 vols., Cambridge, 1842–53.
Craig, *Mint*	*The Mint. A History of the London Mint from A.D. 287 to 1948.* By Sir John Craig. Cambridge, 1953.
Craig, *Newton*	*Newton at the Mint.* By Sir John Craig. Cambridge, 1946.
Cudworth	*Life and Correspondence of Abraham Sharp.* Edited by William Cudworth. London, 1889.
D.N.B.	*Dictionary of National Biography.*
De Analysi	*Analysis Per Quantitatum Series, Fluxiones, ac Differentias: cum Enumeratione Linearum.* Edited by W. Jones. London, 1711.
Dreyer, *Tycho Brahe*	*Tycho Brahe, A Picture of Scientific Life and Work in the Sixteenth Century.* By J. L. E. Dreyer. Edinburgh, 1890.
Edleston	*The Correspondence of Sir Isaac Newton and Professor Cotes.* By J. Edleston. London, 1880.
Foster	*Alumni Oxonienses,* compiled by Joseph Foster.
G.M.V.	*James Gregory Tercentenary Memorial Volume.* Edited by H. W. Turnbull. London, 1939.
Hooke, *Diary*	*The Diary of Robert Hooke, 1672–80.* Edited by H. W. Robinson and W. Adams. London, 1935.
Horsefield, *B.M.E.*	*British Monetary Experiments, 1650–1710.* By J. Keith Horsefield. Harvard University Press, Cambridge, Mass., 1960.
Horsley	*Isaaci Newtoni Opera quæ exstant omnia.* By Samuel Horsley. 5 vols., London, 1779–85.
Hutton	*A Philosophical and Mathematical Dictionary.* By Charles Hutton. 2 vols., London, 1815.

Huygens, *Œuvres*	*Œuvres Complètes de Christiaan Huygens.* 22 vols., The Hague, 1888–1950.
Luttrell	*A Brief Historical Relation of State Affairs from September 1678 to April 1714.* By Narcissus Luttrell. 6 vols., Oxford, 1857.
MacPike, *Halley*	*Correspondence and Papers of Edmund Halley.* By E. F. MacPike. Oxford, 1932.
MacPike, *Hevelius, Flamsteed and Halley*	*Hevelius, Flamsteed and Halley. Three Contemporary Astronomers and their mutual Relations.* By E. F. MacPike. London, 1937.
Montucla	*Histoires des Mathématiques:* (Nouvelle edition). By J. F. Montucla. Paris, 1799–1802.
More	*Isaac Newton.* By Louis Trenchard More. New York and London, 1934.
O.E.D.	*Oxford English Dictionary.* 1933.
Rigaud	*Correspondence of Scientific Men of the Seventeenth Century.* Edited by Stephen Peter Rigaud. 2 vols., Oxford, 1841.
Rigaud, *Essay*	*Historical Essay on the First Publication of Sir Isaac Newton's Principia.* By Stephen Peter Rigaud. Oxford, 1838.
Rouse Ball, *Essay*	*An Essay on Newton's Principia.* By W. W. Rouse Ball. London, 1893.
Ruding	*Annals of the Coinage of Britain and its Dependencies, from the Earliest Period of Authentick History.* By the Rev. Rogers Ruding. 1817–9; 3rd ed. 1840.
Scott, *Descartes*	*The Scientific Work of René Descartes, 1596–1650.* By J. F. Scott London, 1952.
Sotheby	*Catalogue of the Newton Papers, published by Messrs Sotheby for the Sale in 1936.*
Spencer Jones	*General Astronomy.* By Sir Harold Spencer Jones. 3rd ed. London, 1951.
Taylor, *Mathematical Practitioners*	*The Mathematical Practitioners of Tudor and Stuart England.* By E. G. R. Taylor. Cambridge, 1954.
Turnor, *Grantham*	*Collections for the History of the Town and Soke of Grantham.* By Edmund Turnor. 1806.
Venn	*Alumni Cantabrigienses.* Compiled by J. and J. A. Venn.
de Villamil	*Newton: The Man.* By R. de Villamil. London, n.d.
Wallis, *Opera*	*Johannis Wallis S.T.D. Opera Mathematica.* 3 vols., Oxford. Vol. I, 1695; vol. II, 1693; vol. III, 1699.
Ward	*Lives of the Professors of Gresham College.* By J. Ward. London, 1740.

INTRODUCTORY NOTE ON
THE LUNAR THEORY

The following notes aim at giving the accurate and relevant data which will enable the reader to judge what Newton achieved over and above the traditional theory, a theory which did account for observational data to a remarkable extent, but which was incomparably improved by Newton's relating it to one single idea, namely, gravitation, and then adding still further refinements theoretically.

The angular position of the Moon is conveniently given by longitude θ and latitude λ, measured respectively in and perpendicularly to the plane of the ecliptic, θ being reckoned from ♈, the first point of Aries as origin, in the west–east sense, and λ, which is numerically small, varying north and south of the ecliptic. Since the Moon has a nearly uniform angular velocity in longitude, its true motion never differs much from its mean motion p, equal to one orbital revolution per lunar month, which is given by $\theta = pt$, t being the time of describing the angle θ.

The relation $\theta = pt$ would be true if the Moon's orbit were circular and in the plane of the ecliptic, with the Earth at the centre of this circular orbit. But

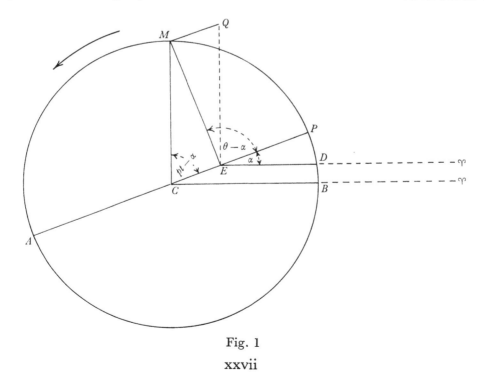

Fig. 1

the motions of the Moon are so irregular and subject to so many corrections (equations) that many years elapsed before even the most prominent of them could be expressed mathematically. Hipparchus (140 B.C.), noting that neither the Sun nor the Moon had a constant angular velocity about the Earth, accounted for this by supposing that the Sun and the Moon each described a circular orbit uniformly, but that the Earth was no longer at the centre of this orbit. He called such a circle the *eccentric*, and the distance of its centre C from E, the Earth, the *eccentricity*. The orientation (in the plane of the orbit) of the diameter upon which E lay with respect to the direction of ♈ on the celestial sphere (i.e. the angle α) was determined by observing the points P (*perigee*) and A (*apogee*) at which the movement in the orbit, as viewed from E, appeared fastest and slowest respectively; equivalently, P is the nearest, and A is the furthest point of the orbit viewed from E. Denote the longitude of P, i.e. the angle DEP by α, and that of the Moon, i.e. the angle DEM by θ, then since CM has a constant angular velocity p, the angle PCM is $pt-\alpha$, from which it follows that the angle CME, or M, is $\theta-pt$. From the triangle CME we have

$$CE:CM = \sin M : \sin(\theta-\alpha),$$

or, denoting the ratio $CE:CM$ by e

$$\sin M = e \sin(\theta-\alpha)$$

or $\qquad\qquad M = \sin^{-1}\{e \sin(\theta-\alpha)\}.$

But $M = ♈EM - ♈EQ = \theta-pt$, therefore $\theta = pt+M$

$$= pt + e \sin(\theta-\alpha)$$

if we neglect e^3, and from this M may be determined. (Fig. 1).

If we complete the parallelogram $ECMQ$ (Fig. 2), then as M describes the

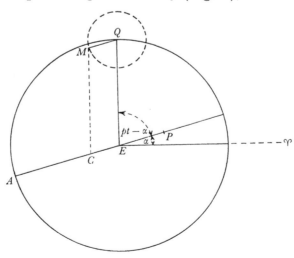

Fig. 2

circle MAP of Fig. 1, with centre C, Q will describe an equal circle with centre E. Since QM is equal and parallel to EC, the Moon's position may be obtained, as Hipparchus showed, by alternately considering M to be carried on a small circle of radius EC, or MQ, an *epicycle*, whose centre Q describes the original circle uniformly about E as its centre. He found in the case of the Moon the ratio $CE:EQ$ (or $CE:CM$) to be equal to $5° 1'$; in effect his eccentricity was given by $e = CE/EQ = \sin 5° 1' = 1/12$, nearly. This small angle $5° 1'$ is the maximum value of the angle M of the triangle CEM, and it satisfies the relation $\theta = pt + M$ since $\Upsilon EM = \theta$ and $\Upsilon CM = pt$.

Hipparchus also found that the point A, the apogee, advanced progressively about $3°$ in each lunar revolution, as if the eccentric circle AMP (Fig. 1) was movable, with its centre C rotating uniformly around E, and carrying the apsidal line AP with it, or equally well, as if each of the (parallel) rays QM and EA (Fig. 2) rotated through about $3°$ while EQ performed one revolution. This implied that the longitude α of perigee was variable. Ptolemy (A.D. 150) confirmed this, and discovered further that the progressive advance of the apsidal line was affected by a periodic disturbance, which he interpreted by using one movable eccentric and one movable epicycle. This device satisfied the requirements for lunar longitude, but failed to account for the correct angular size, and therefore the distance of the Moon correctly. Eventually, Copernicus (A.D. 1543) accounted for both longitude and distance by using two epicycles (Fig. 3), one with centre Q, as in Fig. 2, and radius QR, parallel

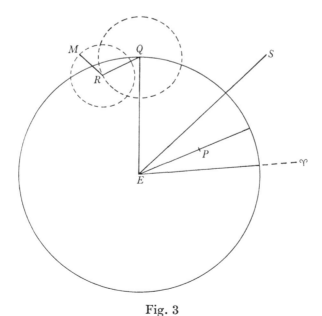

Fig. 3

to EP, and a second carrying the Moon on its circumference, with centre R, the radii QR, RM being chosen so that

$$\frac{QR-RM}{EQ}=\sin 5° 1' \quad \text{and} \quad \frac{QR+RM}{EQ}=\sin 7° 40'$$

and the angle QRM being twice the angle SEQ, S being the position of the Sun. The lunar longitude is now

$$\theta = \Upsilon EM = \Upsilon EQ + QER + REM$$
$$= pt + u_1 + u_2$$

say, where pt, the first term, is the longitude of Q, as in para. 2 above, u_1 the Hipparchian correction (the QEM of Fig. 1), whilst u_2 is a further Ptolemaic correction (later called the evection), which places the Moon on the circumference of the second epicycle. Each term u_1 and u_2 is a periodic circular function, precisely defined by the corresponding epicycle, whose radius is equal to the amplitude of the periodic oscillations.

When Tycho Brahe (A.D. 1580) extended the range of his observations he detected another inequality. Having computed the positions of the Moon for different parts of her orbit, he noted that she was always in advance of her computed place from syzygy to quadrature, and behind it from quadrature to syzygy, the maximum of this *variation* occurring at the octants, or the points equidistant from syzygy and quadrature. Tycho fixed the maximum value of this inequality at 40′ 30″. The value which results from modern observations is 39′ 20″.

Tycho also discovered a fourth inequality, called the *annual equation*. Having calculated the position of the Moon corresponding to any given time, he found that the observed place was behind her computed one while the Sun moved from perigee to apogee, and before it in the other half year.

So far we have considered longitude only, as if the Moon's orbit ΥNmn (Fig. 4) lay in the plane of the ecliptic; actually it lies in a plane NMn inclined at about 5° to the ecliptic, as Hipparchus discovered. These two planes meet in the nodal line nN through the two points, the *nodes*, common to the two orbits, and the difference between the arc NM in the lunar orbit, and its longitude Nm ($=\theta$) in the ecliptic, is a fifth inequality, the *reduction*. Here Mm is the latitude of M. Hipparchus had found that the nodes had a retrograde motion in the ecliptic (in contrast to the progressive motion of the lunar apogee). Tycho further discovered that this retrogression of the nodes was accompanied by an oscillatory motion. Each of these inequalities has an effect on the longitude θ, which can be represented by an epicycle of suitable radius, and equivalently, by a new periodic term added to the series

$$\theta = pt + u_1 + u_2 + \dots.$$

In Fig. 3 there would be three further epicycles, each centred on its predecessor, with M fixed on the fifth. Each of the five radii was determined numerically from observations at the maximum of each inequality. Compared with the radius of the lunar orbit each of the five was small.

Before Newton all these inequalities were calculated empirically, and this was possible because the maxima of the angles QER, RER_1, ...R_3EM occur at distinct positions of longitude, octants for variation, quadratures for evection, syzygies for u_1, etc.

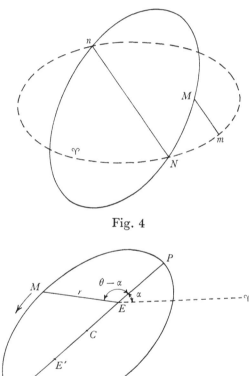

Fig. 4

Fig. 5

By a careful and elaborate study of the motions of the planet Mars, Kepler (1571–1630) had been led to the discovery of his laws of planetary motion, which proved that the idea of circular motion, with its attendant epicycles, must be discarded for the much simpler one of motion in an ellipse, the Sun being located in one focus.

Kepler's discovery of the elliptic orbit led Horrocks (1638) to apply the new theory to the Moon. Instead of Fig. 1 we now take Fig. 5, in which M, (r, θ)

describes the ellipse $l/r = 1 + e \cos(\theta - \alpha)$ about a focus E (the Earth), where $r = EM$ and $\theta - \alpha = PEM$; AP is the major axis, $AC = CP = a$, l is the semi-latus rectum $= b^2/a$, e is the eccentricity, and $b^2 = a^2(1 - e^2)$.

Newton was the first to account for these inequalities in the Moon's motion by gravitation. This is discussed in the *Principia*, Book I, Sections 8 and 9 and in Proposition 66 and its corollaries. He was the first to achieve a real breakthrough in the lunar theory, which up to his time had been purely empirical, hardly a theory in the real sense of the word. Pre-Newtonian astronomers concentrated their efforts on finding empirical formulae to represent the difference between the observed position θ, and the position pt, on the assumption of uniform motion in a circle with the Earth as centre. Newton clearly did envisage the problem as a dynamical perturbation problem based on the inverse square law of attraction between bodies. 'For I have been...about getting a generall notion of all the equations on wch her motions depend', he wrote to Flamsteed (Letter 480, p. 47), and this can only mean that he is seeking the general form of the correcting terms, and later on, he tries to evaluate the various coefficients by comparison with observation.

For a more exhaustive account of the theory of the Moon's motion, the reader is referred to H. Godfray, *Treatise on the Lunar Theory* (1853), and W. M. Smart, *Text-Book on Spherical Astronomy* (1949).

THE CORRESPONDENCE

467   A MANUSCRIPT BY NEWTON

From the original in the University Library, Cambridge

THEORIA LUNÆ[1]

Motus Planetarum Astronomi Veteres per circulos excentricos in orbibus solidis exhibere conati sunt. Tycho[2] per observationes Cometarum deprehendit cælos non esse solidos, sed motus circulares tamen retinuit, causas motuum minimè expendens. Keplerus tandem motus Planetarum c[i]rcum Solem in Ellipsibus fieri[3] ex observationibus Astronomicis primus deprehendit. Et Horroxius noster, ut natura sibi consona esset Lunam quoque in Ellipsi circum Terram revolvi voluit. Et quemadmodum Keplerus Planetas radio ad Solem in inferiore orbium umbilico communi constitutum [ducto] areas tempori proportionales describere docuit:[3] sic Horroxius[4] Lunam radio ad Terram in inferiore Orbis umbilico constitutam ducto, aream tempori proportionalem describere supposuit; excentricitatem vero non eandem manere ut in Planetis primariis sed per vices augeri ac diminui pro positione Apogæi Lunæ ad Solem. Et Halleius ex observationibus astronomicis collegit superiorem Orbis Lunaris umbilicum æquabili motu in circumferentia circuli circumferri cujus radius est semidifferentia excentricitatis maximæ et minimæ et cujus centrum uniformi motu circum Terram ad distantiam excentricitati mediocri æqualem revolvitur. Nos in præcedentibus applicando Theoriam gravitatis ad phænomena cælestia non solum causas motuum investigare conati sumus sed et quantitates motuum ex causis suis computare. Et ejus rei exempla quædam dedimus in motibus etiam Lunæ. Hin[c] enim sequitur revolutiones Planetarum omnium a gravitate regi et & non solum orbes solidos in materiam fluidam resolvendos esse sed etiam hanc materiam rejiciendam ne motus cælestes a gravitate pendentes impediat, ac perturbet.

Motus autem cælestes a legibus gravitatis[5] deducendo deprehendimus insuper quod Æquatio annua medij motus Lunaris[6] quam Keplerus & Horroxius cum æquatione temporis[7] composuerunt, Flams[t]edius vero separatim edidit, oriatur a varia dilatatione orbis Lunaris per vim Solis, juxta Corol. 6, Prop. LXVI Lib I.[8] Hæc vis in perihelio Terræ major est, in aphelio Terræ minor est, et orbem illum contrahi permittit. In orbe dilatato Luna tardius revolvitur, in contracto citius. Et hæc inæqualitas compensatur per æquationem quæ in Aphelio et perihelio Terræ nulla est, in mediocri Terræ a Sole distantia ad 11′ 55″ circiter ascendit et additur medio motui Lunæ ubi Terra pergit a perihelio suo versus aphelium, et in opposita Orbis magni parte subducitur.

Deprehendimus etiam quod in perihelio Terræ propter majorem vim Solis apogæum et nodi Lunæ celerius moventur quam in Terræ aphelio, & quod hæ inæqualitates compensantur per æquationes quæ in mediocri Solis distantia ad 20′ 9″ in motu Aphelij et ad 9′ 34″ in motu nodi ascendunt.

Hæ tres æquationes annuatim augentur ac diminuuntur in eadem ratione cum æquatione quæ dicitur centri Solis.[9] Et ex hypothesi quod excentricitas Orbis magni sit ad ejus Radium ut $16\frac{11}{12}$ ad 1000 computavi earum quantitates maximas 11′ 55″, 20′ 9″ & 9′.34″. Si excentricitas illa statuatur major vel minor, augendæ vel minuendæ erunt hæ æquationes maximæ in triplicata ratione excentricitatis.

Ex Theoria gravitatis constitit etiam quod actio Solis in Lunam paulo major sit ubi major Orbis Lunaris diameter producta transit per Solem quam cum ad rectos est angulos cum linea Terram et Solem jungente, et propterea Orbis Lunaris paulo major est in priore casu quam in posteriore. Et hinc oritur alia æquatio motus medij Lunaris pendens a situ Apogæi Lunæ ad Solem, quæ quidem maxima est cum Apogæum Lunæ versatur in Octante cum Sole,[10] et nulla cum illud ad quadraturas vel Syzygias pervenit, et motui medio additur in transitu Apogæi Lunæ a Solis quadrato ad syzygiam & subducitur in transitu Apogæi a syzygia ad quadraturam. Hæc æquatio quam semestrem[11] vocabo, quantum ex phænomenis colligere potui in Octantibus quando maxima est ascendit ad 3′ 45″ circiter in mediocri Solis distantia. Augetur vero ac diminuitur in triplicata ratione distantiæ Solis inverse, adeoque in maxima Solis distantia est 3′ 34″, in minima 3′ 55″ quam proxime. Et cum Apogæum Lunæ extra Octantes versatur evadit minor estque ad æquationem maximam (posita eadem distantia Terræ et Solis ab invicem) ut sinus duplæ distantiæ Lunaris[12] apogæi a proxima syzygia vel quadratura ad Radium.

Ex eadem gravitatis Theoria innotuit etiam quod actio Solis in Lunam paulo major sit ubi linea Nodorum Lunæ transit per Solem quam ubi linea illa ad rectos est angulos cum linea Solem ac Terram jungente. Et inde oritur alia medij motus æquatio quam semestrem secundam vocabo, quæque maxima est ubi Nodi in Solis Octantibus versantur, et evanescit in eorum syzygiis et quadraturis et in aliis Nodorum positionibus proportionalis est sinui duplæ distantiæ Nodi a proxima syzygia aut quadratura: additur vero motui medio Lunæ dum Nodi transeunt a Solis syzygiis ad proximas quadraturas, & subducitur in eorum transitu a quadraturis ad syzygias. Et in Octantibus cum maxima est ascendit ad 47″ circiter in mediocri Solis distantia uti ex Theoria gravitatis colligo. In aliis Solis distantiis hæc æquatio maxima est reciproce ut cubus distantiæ Solis a Terra, ideoque in aphelio Terræ ad 45″ in perihelio ejus ad 49″ circiter ascendit.

Ex eadem gravitatis theoria didici præterea quod Sol fortius agat in Lunam

2

singulis annis ubi apogæum Lunæ et perigæum Solis conjunguntur quam ubi opponuntur. Et inde oriuntur æquationes duæ periodicæ; una medij motus Lunæ, altera motus Apogæi ejus; quæ quidem æquationes nullæ sunt ubi Apogæum Lunæ vel conjungitur cum Perigæo Solis vel eidem opponitur, & in aliis apogæorum positionibus datam habent proportionem ad invicem. Summa harum æquationum ubi maximæ sunt est minutorum primorum[13] 19 vel 20 circiter, uti ex eclipsibus Lunæ didici: differentia per eclipses Solis investigari debet.

Translation

A THEORY OF THE MOON[1]

The old astronomers tried to exhibit the motions of the planets by means of eccentric circles in solid spheres. By observations of the comets Tycho[2] discovered that the heavens are not solid, nevertheless he retained circular motions, as he paid very little heed to the causes of the motions. At length Kepler first discovered from astronomical observations that the motions of the planets take place in ellipses round the Sun.[3] Our countryman Horrocks, in order to avoid any inconsistency in nature, would have it that the Moon too revolves round the Earth in an ellipse. Just as Kepler showed that the planets describe areas proportional to the time by a radius drawn [from the planet] to the Sun, situated in the common lower focus of the orbits,[3] so too, Horrocks[4] supposed the Moon to describe, by a radius drawn from it to the Earth, situated in the lower focus of her orbit, an area proportional to the time; the eccentricity of the orbit, however, does not remain the same as is the case with the primary planets, but increases and diminishes in turn according to the position of the Moon's apogee with regard to the Sun. Halley, too, inferred from his astronomical observations that the upper focus of the Moon's orbit is borne in a uniform motion on the circumference of a circle whose radius is half the difference between the greatest and the least eccentricity, and whose centre revolves round the Earth with uniform motion at a distance equal to the mean eccentricity. In our previous discussions, we have attempted, by applying the theory of gravity to celestial phenomena, not only to investigate the causes of the motions, but also to compute the magnitude of these motions from their causes. And we supplied some examples of this in the motions of the Moon also. For it follows from this that the revolutions of all the planets are ruled by gravity, and that not only are solid spheres to be resolved into a fluid medium, but even this medium is to be rejected lest it hinder or disturb the celestial motions that depend upon gravity.

Moreover in deducing celestial motions from the laws of gravity[5] we also discovered that the annual equation of the Moon's mean motion[6] which Kepler and Horrocks coupled with the equation of time,[7] but Flamsteed published separately, arises from the varying expansion of the Moon's orbit by the force of the Sun, in accordance with Corollary 6 to Proposition 66 in Book I.[8] At the Earth's perihelion this force is greater: at its aphelion it is less, and allows that orbit to contract. In the expanded orbit the Moon revolves more slowly, and more quickly in the contracted orbit, and this inequality is taken account of by an equation which is zero at the Earth's aphelion and perihelion, but at the Earth's mean distance from the Sun amounts to about 11′ 55″; this is added to the Moon's mean motion when the Earth is travelling from its perihelion towards its aphelion, and subtracted in the opposite part of its great orbit.

We also discovered that owing to the Sun's greater pull at the Earth's perihelion, the Moon's apogee and nodes move more quickly than at the Earth's aphelion, and that these inequalities are taken account of by equations which at the Sun's mean distance amount to 20′ 9″ in the motion of the aphelion and to 9′ 34″ in the motion of a node.

These three equations increase and diminish annually in the same ratio as the equation which is called the equation of the Sun's centre,[9] and on the supposition that the eccentricity of the great orbit is to its radius as $16\frac{11}{12}$ to 1000, I calculated the greatest magnitude of these equations to be 11′ 55″, 20′ 9″, and 9′ 34″. If that eccentricity be shown to be greater or less, these maximum equations will have to be increased or diminished in the triplicate ratio of the eccentricity.

It was also established by the theory of gravity that the action of the Sun on the Moon is somewhat greater when the major diameter of the Moon's orbit passes, when produced, through the Sun, than when it is at right angles to the line joining the Earth and Sun, and consequently the Moon's orbit is somewhat bigger in the former case than in the latter. And from this arises another equation of the Moon's mean motion which depends upon the position of the Moon's apogee with regard to the Sun, which, in fact, is at its maximum when the Moon's apogee is at 45° from its conjunction[10] with the Sun, and zero when the apogee reaches the Moon's quadratures or syzygies. It is added to the mean motion when the Moon's apogee is passing from quadrature with the Sun to syzygy and subtracted when it is passing from syzygy to quadrature. This equation, which I shall call 'semestrial',[11] amounts, as far as I have been able to infer from the phenomena, to about 3′ 45″ at the octants when it is at its maximum at the Sun's mean distance. But it increases and diminishes inversely as the cube of the distance from the Sun, and so it is 3′ 34″ at the Sun's greatest distance, and 3′ 55″ at its least distance, as nearly as possible. When the Moon's apogee is situated outside the octants, the equation turns out to be smaller in value, and is to its maximum value (the distance between the Sun and the Earth being assumed to be unchanged) as the sine of the doubled distance[12] of the Moon's apogee from the nearest syzygy or quadrature is to the radius.

In consequence of the same theory of gravity, it was also ascertained that the action of the Sun on the Moon is somewhat greater when the line of the Moon's nodes passes through the Sun, than is the case when it is at right angles to the line joining the Sun and the Earth. And from this there arises another equation for the mean motion which I shall call the second semestrial equation: it is at its maximum when the nodes are situated in the octants of the Sun, and it vanishes at their syzygies and quadratures, and at other positions of the nodes it is proportional to the sine of the doubled distance of the node from the nearest syzygy or quadrature: but it is added to the Moon's mean motion

4

so long as the nodes pass from the syzygy of the Sun to the nearest quadrature and is subtracted in their passage from the quadratures to the syzygies. Further, at the octants, where it is a maximum, it amounts to about 47″ at the Sun's mean distance, as I infer from the theory of gravity. At other distances of the Sun, this equation is at its maximum reciprocally as the cube of the distance of the Sun from the Earth and so amounts to 45″ at the Earth's aphelion and to about 49″ at its perihelion.

I have learned furthermore from the same theory of gravity that the Sun acts upon the Moon more strongly in the individual years when the Moon's apogee and the Sun's perigee are in conjunction than when they are in opposition. From this there arise two periodic equations, one for the Moon's mean motion, the other for the motion of her apogee. These equations are nil when the Moon's apogee is either in conjunction with the Sun's perigee, or in opposition to it, and in other positions of the apogee they have a given proportion to each other. The sum of these equations, when they are at their maximum, is about 19 or 20 minutes,[13] as I have learned from the eclipses of the Moon: the difference [of these equations] is to be investigated by means of the eclipses of the Sun.

NOTES

(1) There is no clue as to the date of this hitherto unpublished manuscript (Add. 3966, fo. 86). It was probably written some time prior to *A Theory of the Moon*, printed in David Gregory's *Astronomiæ Physicæ et Geometricæ Elementa*, which appeared in 1702, and which is also printed in Horsley, III, 245. What Newton wrote here appears almost verbatim in the Scholium, Book III, Prop. 35 of the second edition of the *Principia*. There is nothing corresponding to it in the first edition, which suggests that this manuscript is a draft of an addition which he prepared for a subsequent edition.

(2) Aristotle accepted the homocentric spheres of Eudoxus, regarding the spheres as solid bodies to which the planets were rigidly attached. The system suggested by Tycho was a compromise between the geocentric system of Ptolemy and the heliocentric system of Copernicus. According to Tycho, the five known planets, Saturn, Jupiter, Mars, Venus and Mercury, revolved about the Sun, whilst the Sun revolved annually about the Earth, supposed fixed at the centre of the Universe. The Moon was supposed also to perform a monthly revolution about the Earth. See J. L. E. Dreyer, *Tycho Brahe* (1890). By showing that the comets were celestial bodies he finally put an end to the idea of solid spheres.

(3) Kepler, *Astronomia Nova* (1609). This contains the enunciation of his law of elliptical orbits and his law of equal areas.

(4) No effective attempt to clear up the difficulties of the Moon's motion was made until 1638 when Horrocks suggested the first improvement by considering the Moon to move in an ellipse with varying eccentricity and in such a way that the apsidal line had a libratory motion. See Letter 474, note (5), p. 31.

(5) See Letter 495 and Letter 576, penultimate paragraph.

(6) The 'annual equation of the Moon's mean motion' is a small inequality with a period of an anomalistic year which is due to the variation of the distance between the Earth and the Sun. The Earth and the Moon are both attracted by the Sun. At perihelion, when the Earth-

Moon system is nearest the Sun, the residual effect of the solar attraction is greater than at aphelion when the system is at its greatest distance from the Sun. At perihelion, therefore, there is in the mean a greater force arising from the Sun's attraction tending to draw the Earth and the Moon apart than there is at aphelion. The result is that for six months of the year around perihelion (1 October to 1 April), the mean radius of the lunar orbit is greater, and the Moon's angular velocity in its orbit is less than their annual mean values, whilst for the remaining six months, around aphelion, the mean radius is less, and the angular velocity greater than the average. See Spencer Jones, *General Astronomy*, pp. 119–20.

(7) The non-uniformity of apparent solar time causes a varying difference between the apparent and mean times. The *equation of time* is the correction which must be applied to mean time to give apparent (or sun-dial) time. See Spencer Jones, *op. cit.* p. 45.

In Newton's day, and in fact until recently, the equation of time was the correction which must be applied to *apparent* time to give *mean* time.

(8) 'Quoniam vis centripeta corporis centralis *S*, qua corpus *P* retinetur in Orbe suo, augetur in quadraturis per additionem vis *LM*, ac diminuitur in Syzygiis per ablationem vis *KL*, & ob magnitudinem vis *KL*, magis diminuitur quam augeatur.' (Since the centripetal force of the central body *S*, by which the body *P* is retained in its orbit, is increased at the quadratures by the addition of the force *LM*, and diminished at the syzygies by the subtraction of the force *KL*, and on account of the magnitude of the force *KL*, [compared with the force *LM*] the diminution is greater than the increase.)

The Sun's disturbing force, averaged over a lunation, tends to diminish the Earth's attraction on the Moon; this effect is greater the smaller the distance between the Earth and the Sun. In the figure, *S* is the Earth, *P* the Moon, *Q* the Sun. $QL/QK = QK^2/QP^2$, QK being the mean distance of *P* and *Q*. See *Principia*, Book I, pp. 173–4.

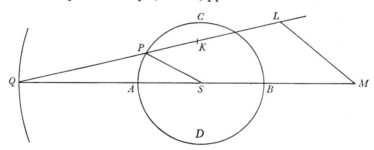

(9) The equation of the Sun's centre was defined by Hipparchus as the excess of the actual longitude of the Sun over the longitude it would have if it moved uniformly.

(10) 'In octante cum Sole.' This includes 45° from opposition as well as conjunction, in fact all the octants, 1, 3, 5, 7.

(11) *Semestrem* is derived from *sex-mensis*, or *semi-mensis*, and can therefore mean six-monthly, or fortnightly. As it is a question of orientation of the Moon's apse line with respect to the Sun, the meaning is probably fortnightly.

(12) In Newton's day, and for a long time after, the word *sine* was used to denote a length, not a ratio. With the modern definition the radius would be unity; the Latin *radius* was used as equivalent to *sinus totus*.

(13) 'first minutes', i.e. minutes. Seconds are 'second minutes'.

468   MEMORANDA[1] BY DAVID GREGORY
1 SEPTEMBER 1694
From the originals in the Library of the Royal Society of London

D Newtonus primo Septembris die 1694 Grenovici Flamstedium adiit,[2] ubi locutus est de nova editione suorum Principiorum. Credit Theoriam lunæ esse in potestate. ad illius locum inveniendum opus erit 5 vel 6 equationibus.[3] Unam Fl: aperuit quæ in quadraturis est Maxima. Fl: 50 circiter loca ☾ in synopsin reducta[4] illi exhibuit, ubi prope quadraturas erant, locum habebat æquatio Neutoniana. ubi major est vel minor ab aliis causis physicis pendet. Observationes non sufficiunt ad ☾ theoriam perficiendam. Causæ physicæ considerandæ. Flamst: alia 100 loca ☾[4] illi est exhibiturus. Opus esse causarum physicarum consideratione ad ♃ et ♄ orbitas cum coelis conciliandas; harum apsides motu libratorio agitantur.

Translation

Mr. Newton visited Flamsteed at Greenwich[2] on 1 September 1694, when he spoke about the new edition of his *Principia*. He believes that the theory of the Moon is within his grasp. To find her position he will need five or six equations.[3] Flamsteed disclosed one, which is the greatest in quadratures: he showed him about fifty positions[4] of the Moon reduced to a synopsis. Newton's equation indicated the correct position where they were near quadratures. Whether it is greater or smaller depends on other physical causes. Observations are not sufficient to complete the theory of the Moon. Physical causes must be considered. Flamsteed is about to show him another hundred[4] positions of the Moon. A consideration of physical causes is needed to reconcile the orbits of Jupiter and Saturn with the heavens. Their apses are disturbed by an oscillatory motion.

NOTES

(1) R.S. Greg. MS. fo. 26. These memoranda, hitherto unpublished, are written towards the end of a sheet of errata and emendations copied from Newton at Cambridge in May 1694 and indexed as C 58 (*Correctiones quas ad Calcem exemplaris ipsius D. Newtoni scriptas inveni* 10 *Maij* 1694 *Cantabrigiæ*). They were evidently written hurriedly and without great care, for they contain not a few grammatical errors. The handwriting confirms that they were written at a later date, presumably in September. Gregory had visited Newton in May 1694. (See vol. III, p. 382, note (1).)

(2) Newton, Halley and Flamsteed were all at this time intent on studying the irregularities of the Moon's motion. The two former were anxious to obtain accurate observations of the Moon since they were necessary to establish her theory and also to verify the equations which Newton had deduced from the theory of gravity. It was with the object of obtaining these observations, which only Flamsteed was in a position to supply, that Newton paid the visit to the Observatory which is recorded in the above Memoranda. Flamsteed's own account of the visit is as follows:

7

'1694: Mr (now Sr. I.N.) made me a visit esteeming him an obliged freind I shewed him about 150 places of ye Moon derived from my observations, by my self & Servants *hired at my own expense* with the places of ye ☽ derived to ye same times & the differences or errors in 3 large sheets of paper in order to correct the Theorys of her motions: On his earnest entreaty I lent them to him & allowed him to take copys of them. but upon this condition that what ever emendations of ye Theory he derived from them should be imparted to me before any other. Hereby I hoped to gaine leasure to begin my Catalogue of the fixed stars for which I was now furnished with a stock of observations sufficient for a begining. In order to which in ye following year I made New Tables for finding ye ☉s true place

'But I found my selfe soon deceaved for instead of saveing me labor this brought more upon me Mr Newton frequently called upon me for New observations of ye Moon. whilest some of his Creatures in town cried up his success in correcting the lunar Theory but said not one Word of his Obligations or debt to ye Royall observatory & one of them [Halley] publickly gave out that all my paines would be well employed to serve him. when I demanded therefore the performances of his promise I was put of with Excuses & delays. & sometimes even with Injurys

'1695 Nevertheless I continued to supply his demands as my other imployment of observeing (yt I might enlarge my stock for Carrying on ye Catalogue) would permit. At ye same time I had the restitution of ye Suns motions besides my night work on my hands, & was frequently ill all this year till Michaelmas when this distemper ended in a fit of ye stone yt was carried of by proper remedies & my continuall headach with it...' (R.G.O. vol. 35, fo. 156). There is no date to Flamsteed's account. But as Newton did not become 'Sr. I.N.' until 16 April 1705, the date would be subsequent to this.

Flamsteed lost no opportunity of reminding Newton of his breach of faith. He complained that Newton had imparted both to Gregory and Halley *contra datam fidem* what he had derived from the observations which Flamsteed had supplied. But when Halley visited Flamsteed (Letter 474) the latter not only showed Halley the same observations he had shown Newton, he actually allowed him to make notes of them. Flamsteed's complaint, therefore, that 'the best part of that stock of lunar observations whereof he [Halley] boasts are mine' (Letter 493) exonerates Newton particularly as the latter, in his letter of 16 February 1694/5, had repeated his promise not to communicate in print Flamsteed's results without his consent.

It should be noted, however, that Newton did communicate to David Gregory a brief note of his conclusions about the lunar theory, but it is possible that after the lapse of so long a period he felt himself no longer bound by the condition imposed by Flamsteed, particularly as he had in the meantime provided Flamsteed with some extremely valuable conclusions about the menstrual parallax. Newton, it should be noted, had assured Flamsteed (Letter 480) that he (Flamsteed) would be the first to whom he would communicate his theory once he was satisfied as to its accuracy, and he may well have thought that this was sufficient to absolve him from his obligation. See Baily, pp. 61–2, and Newton's letter to Flamsteed (Letter 494) and Flamsteed's reply (Letter 495).

(3) *Æquatio* (equation), the numerical adjustment to be added to, or subtracted from, an observed measurement in order to correct a specified type of error. See, for example, MS. 467, note (7), p. 6.

(4) In the correspondence between Newton and Flamsteed which follows there is mention of three synopses: (i) of 52 lunar observations made in 1689 and 1690, (ii) of 50 more until April 1691, and (iii) 55 more until April 1692. What Gregory writes here agrees with these figures. See Memorandum by Flamsteed (Number 516).

469   LEIBNIZ TO HUYGENS
4 SEPTEMBER 1694
From *Œuvres Complètes de Christiaan Huygens*, x, 675–82

Hanover, ce 4/14 de Septembre 1694.

MONSIEUR

Je commence par vous remercier de la communication de l'extrait de l'ouurage de Mr. Wallis touchant M. Newton.[1] Je voy que son calcul s'accorde avec le mien, mais ie pense que la consideration des differences et des sommes, est plus propre à eclairer l'esprit; ayant encor lieu dans les series ordinaires, des nombres, et repondant en quelque façon aux puissances et aux racines. Il me semble que M. Wallis parle assez froidement de M. Newton et comme s'il estoit aisé de tirer ces methodes des leçons de Mr. Barrow.[2] Quand les choses sont faites il est aisé de dire: *et nos hoc poteramus.*[3] Les choses composées ne sçauroient estre si bien démelées par l'esprit humain sans aide de caracteres. Je suis bien aise aussi de voir enfin le dechifrement des enigmes contenus dans la lettre de M. Newton à feu Mons. Oldenbourg.[4] Mais je suis faché de n'y point trouuer les nouuelles Lumieres que je me promettois pour l'inverse des Tangentes.[5] Car ce n'est qu'une methode d'exprimer la valeur de l'ordonnée de la courbe demandée *per seriem infinitam,*[6] dont je sçavois le fonds dés ce temps là, comme je témoignay alors à Mons. Oldenbourg. Et j'en ay donné le moyen depuis quelque temps dans les Actes de Leipzig, d'une maniere assez aisée et tres universelle.

···   ···   ···

Vostre explication de la pesanteur[7] paroist jusqu'ici la plus plausible. Il seroit seulement à desirer qu'on pût rendre raison pourquoy celle qui paroist dans les Astres est en raison doublée reciproque des distances. Comme je vous disois un jour à Paris qu'on avoit de la peine à connoistre le veritable sujet du Mouuement, vous me répondîtes, que cela se pouuoit par le moyen du mouuement circulaire, cela m'arresta; et je m'en souuins en lisant à peu près la même chose dans le liure de Mons. Newton; mais ce fut lorsque je croyais déja voir que le Mouuement circulaire n'a point de privilege en cela. Et je voy que vous estes dans le meme sentiment. Je tiens donc que toutes les hypotheses sont equivalentes et lors que j'assigne certains mouuemens à certains corps, je n'en ay ny puis avoir d'autre raison, que la simplicité de l'Hypothese croyant qu'on peut tenir la plus simple (tout consideré) pour la veritable. Ainsi n'en ayant point d'autre marque je crois que la difference entre nous, n'est que dans la maniere de parler, que je tache d'accommoder à l'usage commun, autant que je puis, *salva veritate....* Cependant si vous estes

9

dans ces sentimens sur la realité du mouuement, je m'imagine que vous
deuriés en avoir sur la nature du corps de differens de ceux qu'on a coustume
d'avoir....

Il y a dejà du temps que j'ay envoyé à Leipzig mes reflexions sur l'Isochrone
du Professeur Bernoulli.

...   .

N'a-t-on point des nouuelles de la restitution entiere de Mr. Newton?[8] Je
la souhaitte fort.

...   .

Que jugés vous Monsieur de l'Hypothese de Monsieur Halley, sur le noyau
mobile contenu dans le globe de la terre, pour expliquer la variation de
l'aimant?[9] Et sur ce que Mr. Newton croit avoir rendu raison du flus et
reflus de la mer.[10]

NOTES

(1) Huygens had received this extract from David Gregory (see Letter 418, vol. III, p. 275),
and on 14 (24) August 1694 sent a copy of it to Leibniz. For the passage from Wallis, see
Letters 393 and 394, vol. III, pp. 220 and 222, and p. 228, note (1).

(2) See the reference to Barrow, vol. III, p. 229, note (9), where Wallis observed that Newton's method was akin both to the differential method of Leibniz and to the earlier work of
Barrow, as expounded in the *Lectiones Geometricæ*, and that the importance of Barrow's work
was recognized by the writer (Jacques Bernoulli) of an article in the *Acta Eruditorum*, January
1691 (*Specimen Calculi Differentialis*). The following is a translation of a passage therefrom, upon
the connexion between Barrow's exposition and that of Leibniz:
'As I have inferred from the recent *Transactions* that the analysis of the celebrated Mr. Leibniz,
based upon his differential calculus, of a problem propounded by him [about the isochronous
curve solved by Bernoulli] has proved most acceptable, I have entertained the hope that the
following example of his work will also be well received, and I publish it for the gratification
of our readers who have derived pleasure from the discussion of this calculus; so that, if
perchance these very ingenious persons have not sufficiently grasped his meaning from what
he published regarding this discovery of his in the *Acta* of 1684, they may be able at least to
learn the method of applying it from this [essay]. Yet, to speak frankly, whoever has understood Barrow's calculus (which he outlined ten years earlier in his *Lectiones Geometricæ*; and of
which the whole of that medley of the propositions contained in it constitutes examples), will
hardly fail to know the other discoveries of Mr. Leibniz, considering that they were based on that
earlier discovery, and do not differ from it, except perhaps in the notation of the differentials
and in some abridgement of the operation of it.'

(3) 'We too could have done this.'

(4) See vol. II, p. 153, note (25).

(5) 'Attamen ne nimium dixisse videar, inversa de tangentibus Problemata sunt in potestate,
aliaque illis difficiliora: ad quæ solvenda usus sum duplici methodo' ('Epistola Posterior',
vol. II, p. 129).

(6) Leibniz may well have been led to expect some further light on a solution, in *finite* terms, of problems of integration, from the hints that Newton gave in the 'Epistola Posterior' (Letter 188, vol. II) where he indicated that he had found out much more than what he there communicated in the 'First Theorem' (vol. II, p. 211, note (3)). What came next (Manuscript 192, vol. II, p. 171) was not seen by Leibniz.

The general use of infinite series for squaring curves was still very much of a novelty. Thus Jacques Bernoulli had recently worked out the integrals $\int_0^1 x^2(1-x^4)^{-\frac{1}{2}}dx$ and $\int_0^1 (1-x^4)^{-\frac{1}{2}}dx$ by expansions (*Acta Eruditorum*, 1694) which drew from Leibniz the remark: 'Il prend les series de pag. 274 pour nouuelles, mais Mons. Newton et moy, nous les avons employées il y a long temps' (Leibniz to Huygens, 17 (27) July 1694; Huygens, *Œuvres*, x, 661).

(7) Leibniz is replying to a remark by Huygens (in a letter of 14 (24) August 1694) upon Fatio's theory of gravitation (see vol. III, pp. 69, 70, note (1)), in which he expresses agreement with Newton rather than with Leibniz on circular motion. See *Œuvres*, x, 669–70: 'Pour ce qui est du mouvement absolu et relatif, j'ay admirè vostre memoire, de ce que vous vous estes souvenu, qu'autrefois j'etois du sentiment de Mr. Newton, en ce qui regarde le mouvement circulaire.' See also Letter 456, vol. III, pp. 371–3.

(8) A reference to Newton's illness. See Letters 420 and 426, vol. III, p. 279 and p. 284.

(9) See article by Halley in *Phil. Trans.* **17** (1692), 563: *An Account of the cause of the Change of the Variation of the Magnetical Needle with an Hypothesis of the Structure of the Internal parts of the Earth: as it was proposed to the* Royal Society *in one of their late Meetings. By* Edm. Halley. Halley accounted for the change in the variation (declination) of the magnetic needle by supposing a mobile kernel (*le noyau mobile*) within the terrestrial globe, consisting of three concentric shells rotating about the same axis, but with different angular velocities. His hypothesis is described in the Journal Book of the Royal Society of London (25 November 1691).

(10) Leibniz in a letter to Huygens (Letter 356, vol. III, p. 81) asked: 'Que jugés vous, Monsieur, de l'explication du flus et reflus de Monsieur Newton?' The reference is to Proposition 24 of Book III: *Fluxum & refluxum Maris ab actionibus Solis ac Lunæ oriri debere.* (The flux and reflux of the sea must arise from the actions of the Sun and the Moon.) In Propositions 36 and 37 these actions are investigated. To this Huygens replied: 'Pour ce qui est de la Cause du Reflus que donne Mr. Newton, je ne m'en contente nullement, ni de toutes ses autres Theories qu'il bastit sur son Principe d'attraction, qui me paroit absurde, ainsi que je l'ay desia temoignè dans l'Addition au Discours de la Pesanteur.' See also *ibid.* pp. 81–2, note (8).

470   FLAMSTEED TO NEWTON
7 SEPTEMBER 1694
From the original in the University Library, Cambridge.[1]
For answer see Letter 473

The Observatory Sept 7. 1694

Sr.

Yesterday the included receipt was sent me by Mr Stanhop[2] whose sister Makes use of it as he tells me with good effect & wishes yours may find the same from it

At the same time the two synopses of lunar observations compared with my Tables & the French Memoires were brought me by ye peny post I have a greater Number of them lyeing by me Which I can soon collect from my book of calculations & shall willingly doe when ever you shall signifie that you have occasion for ym.

Of late I have forborne to calculate the Moons place from ye observations & compare ym with ye tables as soon as taken which I used to doe formerly that I might employ my whole time in ordering ye fixed stars[3] without whose places more accurately determined I found those of the Moon would not be so exact as they might be & I desired them.

But hereafter I intend to cause my Man to calculate them both from ye observations & tables as soon as observed whereby it will be soon evident whether ye heavens will allow those new æquations you introduce & if they will how they are to be limited.

At present I have set my selfe to enquire what Refractions will be given by ye distances of ♀ from ye ☉[4] taken wth ye sextant 4 or 5 evenings successively in Aprill 1681 & to make New tables of them. For I thinke it neither fit nor safe to rely on ye French the best of whose Instruments (I speake it without boasting) are not better yn ye worst of mine

The Calculations necessary to derive ye refractions from ye observations are long indeed & something troublesome but the method is much more exact then yt by ye ☉s observed height which both ye French & I have sometimes used. their Quadrants being onely 3 foot Radius mine $4\frac{1}{6}$: but my sextant near seven.[5]

I was forced to desert ye way by observeing the distances of fixed stars nearly verticall to each other because I found yt when one of them approached the horizon within 2 or 3 degrees ye vapors deprived me of ye sight of it especially in setled weather.[6] In clear weather after Raine I could see them nearer. but such opportunitys rarely happen & when they have of late I made use of

12

them to get the low fixed stars observed under ye Meridian. Such are those in the tayle of ye Scorpion, ye Wolf, Dove, & Southern fysh

What I gather from my observations shall be freely imparted to you, & I shall never refuse to impart either ye observations themselves or my deductions from them to any persons that will receave ym with the same Candor yt you doe. If I desire to have them witheld from others[7] who make it their business to pick faults in them to Censure them & asperse me no less unjustly then ingratefully you will not blame me for so doeing. When Mr H.[8] shews himselfe as Candid as other men I shall be as free to him as I was the first seven years of our Acquaintance. when I refused him nothing yt he desired.

I am told by a freind of his that he is very busy Calculating ye Moons places on a sudden. perhaps some hints he has got from you have set him at Worke anew but except you have been as plane with him as you was wth me I am satisfied he will never be able to find out the parallactick æquation[9] nor limit it without a bigger store of observations then he is possest of tho he have a many of mine yt I made betwixt 1675 & 82.

Since you went hence I examined the observations I employed for determineing ye greatest equation of ye Earths orbit[10] & considering the Moons places at the times of each I find that (if as you intimate the Earth inclines on that side ye Moon then is) you may abate about 20″ from it, so yt it may be onely 1°.56′.00″ [11]

If you please at your leasure to acquaint me how those observed places of ye Moon I imparted to you[12] agree with your Conceptions you will much oblige me & if you want me to compare. Please to give me a little notice of it aforehand they shall be sent you. for you shall ever freely command

Your sincerely affectionate freind & servant

JOHN FLAMSTEED

I feare I mistooke your Notion about ye Correction of ye Suns place by reason of ye common center of gravity of ye Moone & earth.[13] I shall write to you about it hereafter if you save me not ye labour by your owne explication of it J: F.

To
Mr Isaack Newton
at Trinity Colledge in
Cambridge these
present.

NOTES

(1) Add. 3979, fo. 18. This hitherto unpublished letter is the first, so far found, in the resumed correspondence between Newton and Flamsteed that followed their meeting at Greenwich in September 1694 (see Memoranda 468). This correspondence continued almost without intermission until September 1695, but the friendly spirit in which it began did not continue, and in less than a year all pretence at civility between the two men was abandoned.

(2) Possibly George Stanhope (1660–1728) of Lewisham in Kent. He eventually became Dean of Canterbury.

(3) See Letter 386, vol. III, p. 199. Newton had suggested to Flamsteed, as early as August 1691, the advisability of publishing the places of a few of the fixed stars before completing the whole catalogue. Flamsteed wisely refused to be deflected from his purpose. 'After all', he wrote to Newton (Letter 477), 'nothing certeine can be determined by observations of ye ☽ without the true places of ye Fixed Stars first stated.' Had Flamsteed yielded to Newton's request at the time he would have been no better off than Halley, who, having determined the mutual distances of the southern stars by means of the sextant only, was obliged to depend upon Tycho's observations for his fundamental points, so that the catalogue which he has provided is of little use to the practical astronomer. It was reserved for Abraham Sharp, Flamsteed's assistant, to perfect what Halley had been unable to perform (Baily, p. xxxi). See Letter 477, and especially Letter 527, where Flamsteed declared, 'The fixed stars will be my whole worke when I return.'

(4) See Flamsteed's Table of Refractions (*Synopsis Refractionum ab observatis Veneris a Sole distantiis deductarum*) sent with Letter 477. The conjunctions of Venus and the Sun gave useful data for measuring the effects of refraction.

(5) This sextant had been presented to Flamsteed on his appointment as Astronomer Royal, by Sir Jonas Moore. See Letter 488, note (2), p. 73.

(6) For the table of refractions near the horizon, Newton was particularly grateful. See Letter 475, note (3), p. 35.

(7) See Memoranda 468, note (2), p. 7.

(8) Edmond Halley. In the preface to his *Doctrine of the Sphere* (1680) Flamsteed referred to his 'singular kind Friends, the admirably Ingenious Sir *Christopher Wren*, and our Southern *Tycho*, Mr. *Edmond Halley*'. Flamsteed seems to have been anxious to maintain the cordial relations which had hitherto existed between the two men, for in a letter to Halley, dated 17 February 1680/1 (Letter 250, vol. II), he declared: 'I shall willingly answer your desires', and he subscribed himself 'Your most obliged & reall freind'. These facts have induced Dr J. L. E. Dreyer to conclude that the first rift between Flamsteed and Halley occurred after this time (see *The Observatory*, XLV, 293). But the quarrel must have had an earlier beginning, for an entry in the Hooke *Diary*, under the date Friday, 7 January 1675/6 (p. 209), reads: 'Flamsteed and Halley fallen out.' Whatever the origin of the quarrel, it is plain that by 1684 Flamsteed had become intensely suspicious of Halley: 'About E.H. . . . I have never found anything so considerable in him as his craft and forehead, his art of filching from other people, and making their works his own. . . . I had rather be without his acquaintance than to purchase it with the loss of an honest reputation' (R.G.O. vol. LXII.E). See *Notes and Queries*, vol. 168, p. 434; vol. 169, p. 122; Brewster, II, 166.

(9) See Glossary and vol. II, p. 466, note (9).

(10) That is, for determining the greatest value of the correction in the longitude of the Sun, when the Earth's supposedly circular orbit is replaced by an elliptical orbit of small eccentricity. Concerning this correction (the annual equation), Newton wrote: 'The Annual equation of the Suns motion arises from ye excentricity of his Orb which is $16\frac{11}{12}$ supposing the radius of that Orb to be 1000. Tis thence called Æquatio centri & when greatest amounts to 1 gr. 56′ 20″' (see Manuscript 622, p. 322). A short statement of these facts is given in the *Principia*, Book I, Prop. 66, Cor. 6, and Book III, Prop. 22: *Motus omnes Lunares, omnesque motuum inæqualitates ex allatis Principiis consequi* (all the motions of the Moon and all the inequalities of these motions follow from the principles which we have laid down).

(11) In the second edition of the *Principia* the figures 1° 56′ 26″ are quoted.

(12) See Letter 386, vol. III, p. 202.

(13) See Letter 475.

471   MEMORANDA[1] BY DAVID GREGORY

7 SEPTEMBER 1694
From the original in the Library of the Royal Society of London

Describenda et Chartis Consignanda
Mense Septembri MDCXCIV[2]

1. Methodus fluxionum Newtoni prout in Wallisii operibus[3] continetur plenius describenda et illustranda, deinde quid ille per Fluxiones late sumptas intelligat.

2. Calculus differentialis Libnitij ibidem explicandus prout ab ipso auctore in Actis Lipsiæ Octobri MDCLXXXIV traditur[4] utque harum duarum Methodorum Congruentia per omnia ostendatur reducendo scilicet Calculum differentialem non demonstratum ad Methodum fluxionum demonstratum,[5] et quando et ob quam rationem signa ambigua Calculum differentialem ingrediuntur.

D.T. Methodus Tangentium Dec. 1682 et Meth. de Max: et Minim: Martio 1683 huc referendæ. vid. Junium 1691 pag. 290.[6]

3. Barrovij Methodus tangentium X Lect: Geom: pag: 81 et seqq[7] proferenda, et utramque Methodum præcedentem huic superstructam esse fuse ostendendum quod et in Actis Lipsiæ mense Januaris MDCXCI pag: 14 agnitum est. Sed ut verum fatear videtur Barrovius ad solas tangentes ducendas ibi respexisse, quamvis negari nequeat illum virum ductarum tangentium usus optime cognovisse quod ex prædictis lectionibus liquet. videndum an hoc ex Fermatij methodo[8] consequatur.

4. Post explicatas hasce tres Methodus illarumque congruentiam sive identitatem patefactas harum cujusvis usus exemplis est ostendendus. primo tangentium vulgatas Curvas ductu deinde spatiorum rectis et curva vulgata comprehensorum quadratum. Atque hoc facto tam *nostri Canones* pro quadraturis quam Slusij Methodus tangentium eruenda. horum ultimum ad mentem Newtoni Cap: XCV pag 393 Wallisii.[9] Horum exempla quamplurima congerenda sunt, nec illorum delectus faciendus nec verbis parcendum.

Exempla Quadraturis et figuræ tangentium in fine Lect. X Barrovij speciatim consideranda, una cum aliis ejusdem generis exemplis[10]

5. Exempla postea particularia adducenda in quibus vel difficultas aliqua vel speciale quid occurrit, primo illa quæ Libnitius adducit Octobre MDCLXXXIV, quæ Jac. Bernoulli adducit mense Janrij 1691,[11] quæ idem adducit mense Junij 1691, atque hoc ordine et Classe distinguenda ut de tangentibus, de spatiis tam in plana superficie quam in Curva, de longitudine Curvarum tam in plana quam in Curva de Maximis et Minimis ubi Newtoni problema de Cono cui minime resistitur inserendum. punctum flexus Contrarij: item curvarum Quadratura. In hisce omnibus Exempla illustriora colligenda sunt quæque hactenus Geometrarum curam subiere.

De Curvatura Curvarum sive circulo maximo inscriptibili vel minimo circumscriptibili.[12]

6. De Curvarum Evolutionibus proxime ubi C. Hugenio citandum item Tschirnhausij curva per intersectiones radiorum reflexorum producta. Item Catenaria consideranda aliaque quæ nuper a Geometris considerata.

7. Methodus tangentium inversa etiam attentanda ex Newtoni Epistola.[13] ac inter reliqua Curva ex tracto super horizontali plano gravi determinanda, sed quæ Leibnitius noviter in Actis hac de re edidit consideranda.

8. Newtoni Quadraturæ prosequendæ quatenus in Wallisij Algebra continentur: speciatim comparatio non quadribilium cum Conisectionibus.

9. Quæcunque oc[c]urrunt problemata difficiliora huc referenda.

10. Omnia quæ possunt Newtono illustrando inservire hinc deducenda.

11. Figura optima munitionis figura plana rectilinea dat: basis et altitudinis. item figura plana minime resistentia optimus veli situs gubernaculi a rudder.[14]

VII *Septembris*

MDCXCIV

Translation

To be written out and committed to paper

September 1694[2]

1. Newton's method of fluxions, as it is contained in Wallis's works[3] is to be written out and illustrated more fully, followed by what he understands by Fluxions, in its broadest sense.

2. The differential calculus of Leibniz is to be explained at the same time just as it is presented by its author in the Leipsic Acts for October 1684[4] and in such a way that the agreement between the two methods may be shown throughout, namely by reducing the differential calculus, as yet unproved, to the method of fluxions, which has already been proved:[5] also when, and for what reason, uncertain symbols enter into the differential calculus.

The Method of Tangents of D.T. December 1682 and the Method of Maxima and Minima March 1683 are to be referred to this section. See June 1691, p. 290.[6]

3. Barrow's Method of Tangents[7] (*X Lect. Geom.* p. 81 et seqq.) must be quoted, and it must be shown at length that each of the methods mentioned above is founded upon this one, as is recognized in the Leipsic Acts for the month of January, 1691, page 14. Nevertheless, if one is to speak truly it seems that Barrow here had in mind only the drawing of tangents, though it cannot be denied that he well knew the uses which tangents might serve when drawn, as is clear from the above-mentioned passages. It remains to be seen whether he attains this result by Fermat's method.[8]

4. After these three methods have been explained, and the similarity or identity between them made clear, the application of any one of them must be shown by means of examples. First by the drawing of tangents to common curves, then by the determination of the area of spaces bounded by straight lines and a common curve; in this way our rules for squarings as well as Sluses's method of tangents are to be brought to light. The last of these accords with Newton's ideas, Wallis, Chapter 95, p. 393.[9] Very many examples are to be collected together; there must be no random selection, no economy of words. Examples for squarings, and the figures of tangents at the end of Barrow's Lect. X are to be specially considered, as well as other examples of the same kind.[10]

5. Afterwards, particular examples are to be adduced, in which either some difficulty or some special feature occurs, first those which Leibniz brings out in October 1684; those which Jacques Bernoulli adduces in January 1691, and those which he adduces in June 1961,[11] and are they to be kept distinct, in this order and classification, namely as regards tangents, areas in plane and curved surfaces, as well as lengths of curves, both in a plane and in curved surfaces about maxima and minima, in which Newton's problem of the cone of least resistance is to be inserted, as well as the determination of the point of inflexion and the squaring of curves. In all these, more illuminating examples as well as those which have hitherto occupied the attention of geometers are to receive attention.

Concerning the curvature of curves, or the greatest circle that can be inscribed, or the smallest that can be circumscribed.[12]

6. Next as regards the evolution of curves where the method is to be quoted from Huygens as well as from the works of Tschirnhaus, the curves determined by the points of intersection of reflected rays. Also catenaries are to be considered, as well as other matters lately considered by geometers.

7. The inverse method of tangents is to be attempted from Newton's Letter,[13] and among other things, the curve determined by drawing a heavy body over a horizontal plane, but what Leibniz has recently published upon this matter in the *Acta* must be considered.

8. Newton's quadratures are to be followed up as far as they are contained in Wallis's *Algebra*, especially a comparison with the conic sections of those areas which cannot be squared.

9. Any more difficult problems that occur are to be referred to this.

10. Everything that can be of service to elucidate Newton is to be deduced from this.

11. A plane rectilinear figure gives the figure of best construction of base and height; likewise a plane figure with the least resistance the best position of a sail, likewise a rudder.[14]

7 *September* 1694

NOTES

(1) These Memoranda (R.S. Greg. MS. fo. 64), begin with the note:

Mr Ns Graces

1. Good lord bless what is here offered us for use and us for thy service
2. For this, and all other mercies, the lord be blest through His Son J:C:

The handwriting suggests that it was written by David Gregory. Gregory visited Newton at Cambridge in May 1694 (see vol. III, p. 320, note (1)), and these memoranda may well be a record of what passed between them at that meeting. It is not unlikely that Gregory paid another visit some time between May and September of that year.

(2) What follows is described in Gregory's index to Folio C, U.L.E. MS. '79 Adumbratio nostræ de fluxionibus methodi' (A sketch of our method of fluxions). It would appear to be a summary, written some time in 1694, relating to, and compiled prior to, an unpublished treatise entitled 'Isaaci Newtoni Methodus Fluxionum, Ubi Calculus differentialis Libnitij, et methodus tangentium Barrovij explicantur, et exemplis quamplurimis omnis generis illustrantur' (Isaac Newton's Method of Fluxions, in which the differential calculus of Leibniz, and Barrow's method of tangents are explained, and illustrated by very many examples of all kinds). The autograph manuscript of this is preserved in the Library of the University of St Andrews. All the particulars mentioned in this sketch are found in the treatise, including the discussion of the solid of least resistance.

(3) See vol. III, p. 228, note (1). In the *Algebra*, which appeared in vol. II of the *Opera*, Wallis had included Newton's binomial theorem and many extracts from the 'Epistola Prior'. See also vol. III, p. 220, note (4), and Letter 393.

(4) *Nova Methodus pro Maximis et Minimis, itemque tangentibus, quæ nec fractas, nec irrationales quantitates moratur, & singulare pro illis calculi genus* (A new method for maxima and minima, as well as for tangents, which makes no obstacle of fractions nor irrational quantities and an unusual kind of calculus for this), *Acta Eruditorum* (October 1684), p. 467.

18

(5) No proof was given in the 'Epistola Posterior' (Letter 188, vol. II), but see Letter 209, vol. II.

(6) *Nova Methodus Tangentes curvarum expedité determinandi, per D.T., Acta Eruditorum* (December 1682). D.T. = Ehrenfried Walther von Tschirnhaus (see vol. I, p. 355, note (5)). This passage is written in the margin.

(7) Barrow's *Lectiones Geometricæ* is replete with methods of determining tangents to curves, and the areas under curves, many of which closely resemble the methods afterwards employed by Newton. See J. M. Child, *The Geometrical Lectures of Isaac Barrow* (1916). In the preface to this work, its author declares that 'Isaac Barrow was the first inventor of the Infinitesimal Calculus; Newton got the main idea of it from Barrow by personal communication'.

(8) For Fermat's method of drawing a tangent to a parabola, see J. F. Scott, *History of Mathematics*, pp. 140–1.

(9) *Hujus Exempla variis casibus accommodata.* Wallis, *Opera*, II, 383, etc.

(10) This passage is written in the margin. The following is a paraphrase of a manuscript by Barrow, entitled *Compendium pro tangentibus determinandis*.

Let AP, PM be two straight lines given in position, NM an indefinitely small arc of the curve proposed. NQ is parallel to AP, and NR to MP. NQ and QM are related by the equation of the curve as is clearly shown by the following example:

The Circle. The line AP lies along the diameter, and PM is perpendicular to it. Let $NQ = e$, $QM = a$, $AP = q$, $PM = m$, and call the diameter d. Then from the figure, $AR(d - AR) = RN^2$,

i.e.
$$d(q - e) - (q - e)^2 = (m - a)^2.$$

Since $AP(d - AP) = PM^2$, or $(d - q)q = m^2$, this becomes
$$-de + 2qe - e^2 = -2ma + a^2.$$

Now reject terms containing powers of a or e, or the product of these,
$$2qe - de = -2ma,$$
$$e/a = 2m/(d - 2q).$$

This is the ratio of NQ to MQ and this ratio persists when M and N move into coincidence and MT becomes the tangent at M.

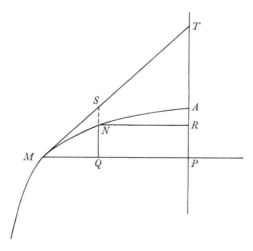

(11) *Specimen Calculi Differentialis in dimensione Parabolæ helicoidis, ubi de flexuris curvarum in genere, earundem evolutionibus aliisque.* J.B.

(12) This passage is written in the margin.

(13) The 'Epistola Posterior', Letter 188, vol. II, p. 146. Like so many of David Gregory's *Memoranda*, this example appears to have been drawn up in haste, and it is difficult to translate it without assuming that its author made a number of grammatical errors, e.g. *methodum fluxionum demonstratum* (para. 2), where one would expect *demonstratam*. Similarly, one would expect the subjunctive *ingrediantur* (cf. *intelligat*, in para. 1) as in an indirect question. *Cognovisse* (para. 3) seems the most likely rendering of a word which is almost indecipherable, as it occurs in a fold in the manuscript. The word might well be *cognobilis* (= that can be understood); this interpretation, however, would require the infinitive *esse*. In para. 4, *Figuræ* may be dative singular and refer to Fig. 115, of Barrow, *Lect.* X, or it may be plural, referring to the seven figures 115–21.

(14) The last two lines are indecipherable and have had to be surmised.

472   DAVID GREGORY TO NEWTON
24 SEPTEMBER 1694
From the original in the University Library, Cambridge.
In reply to Letter 460

Oxon 24 *Septr* 1694

Sir

I receaved yours of the 14 July[1] for which I most heartily thank you. What you wer therin pleased to communicate to me, did me both great service and pleasure. I am glade that you have done the Analysis of the Lemma in order to determining the præcession of the Æquinoxes another way.[2] this if any thing wanted your own hand. and when you shall think fitt to communicate it, it will be infinitly acceptable. But Sr what I must now beg of you is that you would at some idle or spare moment look over the approach to determining the place of a planet contained in the end of pag: 113 and all pag: 114.[3] for besides the typographical errors which I mended from your own Copy, it wants explication and demonstration mightily.[4] if you will be pleased to allow me your thoughts on this place you will add very much to the former kindnesses of this sort. I am glade to understand that you have been at London, and that your design of the new edition[5] of your book is going on and in forwardness. The Gentlemen my friends who deliver this letter have ane earnest desire to wait on you having gone from hence to see Cambridge. I am in all duty

Sir

Your most humble and most obliged servant

D. GREGORY

NOTES

(1) See Memoranda by David Gregory, vol. III, pp. 384–9. Gregory had visited Newton some time during May 1694, and had asked for further information about the lemma.

(2) *Principia*, Book III, Prop. 39, pp. 470–3. *Invenire Præcessionem Æquinoctiorum* (To find the precession of the equinoxes). In subsequent editions, pp. 472 and 473 were almost completely deleted, and a new lemma, Lemma 2, was introduced, the original Lemma 2 becoming Lemma 3, and so on. See vol. III, p. 383, note (4).

(3) *Principia*, Book I, Prop. 31, Problem 23, pp. 107–8. *Corporis in data Trajectoria Elliptica moventis invenire locum ad tempus assignatum* (To find the position of a body moving in a given elliptical trajectory at any assigned moment). This is Kepler's problem. Gregory is probably alluding to the annotated interleaved copy of the first edition of the *Principia*, now in the University Library, Cambridge (Add. b. 39. 1). Among otherwise minor corrections by Newton is the enunciation on pp. 113–14 of the formula which concludes a long discussion on Kepler's problem (pp. 107–14), namely, that the angular position, ϕ, of a planet in its orbit, referred to the *second* focus at a time t is given approximately by

$$\phi = T + V + X,$$

where T, which is proportional to the time t in which the arc BP is described, or is equal to the *mean motion*, as it is called; V is a first equation (correction) to T, and X is a second equation;

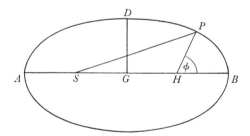

in fact, V is proportional to $\sin 2T$ and X to $\sin^3 T$. In the first edition, on page 114, X was stated to be proportional to $\frac{1}{2}$ versin $2T$, which is equivalent to $\frac{1}{2}(1 - \cos 2T)$, or $\sin^2 T$. This was set right in later editions. Newton considered the two cases $T \pm V + X$ separately, but it is better to let V change sign with $\sin 2T$. Newton expresses this in the Scholium to Prop. 31, Prob. 23, thus:

Supposing AO, OB, OD be the semi-axes of the ellipse, and L its latus rectum; D is the difference between the semi-minor axis OD, and $\frac{1}{2}L$, half of the latus rectum. Let an angle Y be found whose sine may be to the radius as the rectangle under that difference D, and $AO + OD$, the semi-sum of the axes, to the square on the major axis, AB. Find also an angle Z, whose sine may be to the radius as the double rectangle under the distance between the foci SH, and that difference D, to triple the square of the semi-major axis AO. These angles being once found, the place of the body may be thus determined. Take the angle T proportional to the time in which the arc BP was described, or equal to the mean motion, and take an angle V, the first equation of the mean motion, to the angle Y, the greatest first equation, as the sine of double the angle T is to the radius; and take an angle X, the second equation, to the angle Z, the second greatest equation, as the cube of the sine of the angle T is to the cube of the radius. Then take the angle BHP, the mean equated motion either equal to $T+X+V$, the sum of the angles T, V, X if the angle T is less than a right angle, or equal to $T+X-V$, the difference of the same if the angle T is greater than one right angle and less than two, and if HP meets the ellipse in P, draw SP, and it will cut off the area BSP nearly proportional to the time (Cajori, *Principia*, pp. 115–16).

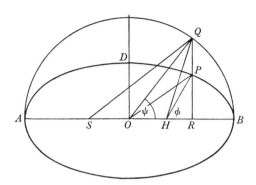

The strictly analytical equivalents of the above are:

$$V=\sin^{-1}\left\{\tfrac{1}{4}e^2 \sqrt{(1-e^2)}\right\} \sin 2T$$

and

$$X=\sin^{-1}\left\{\tfrac{4}{3}e \sqrt{(1-e^2)}-1+e^2\right\} \sin^3 T,$$

where e is the eccentricity. These become, if e is so small that e^4 and higher powers may be neglected:

$$V=\tfrac{1}{4}e^2 \sin 2T,$$

$$X=\tfrac{2}{3}e^3 \sin^3 T.$$

(4) No trace of Newton's reply to this request has been found. A proof however may be supplied.

Let the eccentric angle $BOQ=\psi$. From the early part of the above Scholium we may, in suitable terms, take the mean anomaly (the time divided by the arc BP) as

$$T=\psi+e\sin\psi. \tag{i}$$

Now if $AO=a$, $SO=OH=ae$, $DO=a \sqrt{(1-e^2)}$, then $\tfrac{1}{2}L$ is $a(1-e^2)$ and, as above, D is the difference between the semi minor axis and $\tfrac{1}{2}L$, and so, from the above Scholium,

$$\sin Y=\frac{(OD-\tfrac{1}{2}L)\,(AO+OD)}{AB^2}$$

$$=\tfrac{1}{4}e^2+O(e^4)=Y.$$

Also from the above Scholium $$\sin Z=\frac{2SH(OD-\tfrac{1}{2}L)}{3AO^2}$$

$$=\tfrac{2}{3}e^3+O(e^4)=Z.$$

From the figure, $$\tan\phi=\frac{PR}{HR}=\frac{PR}{OR-OH}$$

$$=\frac{a \sqrt{(1-e^2)} \sin\psi}{a\cos\psi-ae};$$

$$\therefore \quad \sin\phi=\frac{\sqrt{(1-e^2)} \sin\psi}{1-e\cos\psi}. \tag{ii}$$

22

Newton's problem was to eliminate ψ between (i) and (ii). Expanding (ii),

$$\sin \phi = \sin \psi \, (1 - \tfrac{1}{2}e^2 \ldots) \, (1 + e \cos \psi + e^2 \cos^2 \psi + e^3 \cos^3 \psi + \ldots)$$
$$= \sin \psi + e \sin \psi \cos \psi + \tfrac{1}{2}e^2 \, (2 \cos^2 \psi - 1) \sin \psi$$
$$+ \tfrac{1}{2}e^3 \cos \psi \sin \psi \, (2 \cos^2 \psi - 1) + O(e^4). \qquad \text{(iii)}$$

Now expanding the left-hand side as a Taylor series in variable e,

$$\sin \phi = \sin [\psi + e \sin \psi + \lambda e^2 + \mu e^3 + O(e^4)]$$
$$= \sin \psi + (e \sin \psi + \lambda e^2 + \mu e^3) \cos \psi$$
$$- \tfrac{1}{2} \sin \psi \, (e^2 \sin^2 \psi + 2\lambda e^3 \sin \psi)$$
$$- \tfrac{1}{6} \cos \psi . e^3 \sin^3 \lambda \ldots$$
$$= \sin \psi + e \sin \psi \cos \psi + (\lambda \cos \psi - \tfrac{1}{2} \sin^3 \psi) \, e^2$$
$$+ (\mu \cos \psi - \lambda \sin^2 \psi - \tfrac{1}{6} \sin^3 \psi \cos \psi) \, e^3. \qquad \text{(iv)}$$

Identifying (iii) and (iv),

$$\sin \phi = \sin \psi + e \sin \psi \cos \psi + \tfrac{1}{2}e^2 \sin \psi \, (2 \cos^2 \psi - 1)$$
$$+ \tfrac{1}{2}e^3 \sin \psi \, (2 \cos^2 \psi - 1) \cos \psi,$$

whence $\qquad \lambda = \tfrac{1}{2} \sin \psi \cos \psi = \tfrac{1}{4} \sin 2\psi$

and $\qquad \mu = \tfrac{1}{2} \sin \psi - \tfrac{1}{3} \sin^3 \psi$

and therefore $\qquad \phi = \psi + e \sin \psi + \tfrac{1}{4}e^2 \sin 2\psi + \tfrac{1}{6}e^3 \, (3 \sin \psi - 2 \sin^3 \psi),$

i.e. since $\qquad \psi = T - e \sin \psi$

and so $\qquad \sin 2\psi = \sin 2T - 2e \sin T \cos 2T + \ldots$

then $\qquad \phi = T + \tfrac{1}{4}e^2 \sin 2T + \tfrac{2}{3}e^3 \sin^3 T + O(e^4),$

which is Newton's result.

For a geometrical treatment of the problem, see Horsley, II, 135–41, where, on p. 141, a table is given of the maximum values of the corrections V and X, for the Earth and the other planets.

A clue as to how Newton obtained his result is provided at p. 114 of the *Principia* by the reference to Seth Ward (*Invento autem angulo motus medii æquati BHP, angulus veri motus HSP & distantia SP in promptu sunt per methodum notissimam Dris. Sethi Wardi Episcopi Salisburiensis mihi plurimum colendi*). For Kepler had associated three angles with the position of a point on an elliptic orbit under an attractive force to its first focus S,

    (i) the mean anomaly (Ptolemy's mean motion), T,

    (ii) the eccentric anomaly (the eccentric angle), ψ,

    (iii) the true or equated anomaly, θ.

Kepler's problem was this: Given T, to find θ, which may be done by first finding ψ in terms of T, and then θ. But Ward pointed out (Keill, *An Introduction to the True Astronomy* (1721), Lect. XXII, pp. 287, 296, Ward's Ellipticbe Hypothesis Explained) that instead of T it was better to use a fourth angle, Ward's mean anomaly, ϕ, for θ is readily found from the value of ϕ, and θ is *equal* to T if e^2 is negligible. ($T = \psi + e \sin \psi$, and after simplification (ii) yields $\phi = T + O(e^2)$, which is the analytical equivalent of Ward's result.) Newton's expressions (i) and (ii) constitute a refinement of Ward's result when e^2 and e^3 are too large to be neglected. See also J. C. Adams, 'On Newton's Solution of Kepler's Problem', *Monthly Notice of Royal Astronomical Society*, vol. 43 (1882).

The above enunciation of Kepler's problem is certainly true for Kepler himself. But it should be noted that Newton in the present proposition has taken the *complete* problem to be to find the eccentric angle ψ, given T; in other words, to resolve Kepler's problem, given T and eccentricity e, from $T = \psi + e \sin \psi$, to find ϕ.

(5) Newton was constantly being urged to publish a second edition of the *Principia*. See the letter from Fatio to Huygens, of 18 December 1691 (Huygens, *Œuvres*, x, 213), also the Journal Book of the Royal Society of London, 31 October 1694: 'A lre from Mr Leibnits to Mr Bridges was produced and read, wherin he recommends to the Society to use their endeavours to induce Mr Newton to publish his farther thoughts and emprovements on the subject of his late book Principia Philosophiæ Mathematica, and his other Physicall & Mathematicall discoverys, least by his death they should happen to be lost.'

473   NEWTON TO FLAMSTEED
7 OCTOBER 1694
From the original in Corpus Christi College Library, Oxford[1].
In reply to Letter 470; for answer see Letters 474 and 476

Sr

Since my return hither[2] I have been comparing your observations[3] wth my Theory, & *now I have satisfied my self that by both together the Moons Theory may be reduced to a good degree of exactness perhaps to ye exactness of two or three minutes.*[4] I forbore writing to you a few days, till I had considered your Observations yt I might be able to acquaint you what further Observations are requisite. And besides those fifty wch you tell me you have ready calculated,[5] & those I have already[6] your observations of this winter will be very material & therefore I am very glad you have ordered your servant to calculate them. There are requisite also your observations for the last six or seven years made in the months of March, June, September & December when ye Moons perigee or apogee is in ye syzygies or quadratures or within 5 or 6 degrees of those cardinal points & the Moon in the Quadratures or Opposition & in an eclips of the Sun. When the Moon in these cases is in the Quadratures or Opposition it will be requisite[7] to have two Observations, one a few hours before the quadrature or opposition & the other a few hours after, there being a day between the observations. If in the lunation[8] of this present month you can get two or three observations about the first quadrature pray will you endeavour to get as many opposite to them about ye last Quadrature. For observations opposite to one another when the moons apoge is in the Octants are of great moment.[9] By such a set of Observations I belive I could set right the Moons Theory this winter, only it would be requisite to have about 50 of them such as I should select, set right by the new places of ye fixt stars. The observations in March June, Sept. & December, above mentioned, will not be many. I thank you heartily for your receipt. At present *I beg your Observations of* ♃ & ♄ .[10] And what you send by the penny post, direct for Mr William Martin a Cambridge Carrier at the Bull in Bishopsgate Street & order it to be delivered there before two of ye clock on munday least he be gone. For he goes every Munday at two a clock from London towards Cambridge. I am

Yours to serve you

Is. NEWTON.

Cambridge Octob. 7th 1694.
For Mr John Flamsteed at
the Observatory at
Greenwich neare
London

NOTES

(1) C. 361, no. 39. At the top of the sheet Flamsteed has added: ' ♃ [Thursday] Sept 01 ♄ [Saturday] 1694 Mr Newton came to see me I imparted my lunar observations which occasioned the following letters' 'Oct: 7. 1694'. See Memoranda 468.

(2) That is from his visit to Flamsteed, 1 September 1694.

(3) Certainly the first synopsis, and perhaps also the second. See Memoranda 468, note (4), p. 8.

(4) Words and phrases in italics were underlined, probably by Flamsteed.

(5) The third synopsis. It was sent on 29 October (Letter 477).

(6) A note by Flamsteed in the University Library, Cambridge (Add. 3979: No. 39), says: 'Places of ye planets to be calculated from Observations taken with ye Sextant betwixt 1675 & 1689

of ye Moon	about	400
of Saturn		140
Jupiter		140
Venus & Mercury		220
Mars		100
		1000

besides Calculations of places to [be] deduced from observations made with ye Murall Arch from 1689 to 1704 compleate of which onely a third parte have passed under ye hand of my servants & hired Assistance.

J.F.'

Flamsteed had repeatedly said that all his observations were available to Newton.

(7) Newton wrote 'requite'.

(8) Lunation, the period in which the Moon returns to the same position with regard to the Sun. It is about 29½ days.

25

(9) In order that he might successfully develop his lunar theory Newton was entirely dependent upon the most accurate observations available of the successive positions of the Moon. The object of all these requests for observational data at carefully chosen positions is to determine lunar inequalities more accurately. Apart from the two uniform angular motions arising from the combination of the precession with the motion (i) of the apses (perigee and apogee), and (ii) of the nodes, each inequality is periodic, and of the form $A \sin \chi$, where the argument χ depends linearly upon one or both the longitudes of the Moon and the apse. The quadratures, syzygies and octants are places where either the value of A is easily given (as when $\chi = \frac{1}{2}\pi$) or the term $A \sin \chi$ vanishes.

(10) See note (6) above.

474　FLAMSTEED TO NEWTON
11 OCTOBER 1694

From the original in the University Library, Cambridge.[1]
In reply to Letter 473; for answer see Letter 475

The Observatory October 11. 1694

Sr.

I have yours of ye 7th Instant. before it arrived I had prepared a letter to you which I sent not because I was too late for ye post I shall give you contents of it & then answer yt I receaved last night.

After you were gone hence Mr Hally applied himselfe to me & desired I would allow him to see the lunar observations I had imparted to you I tould him I should not be unwilling provided that he in like manner would impart what he had talked so much of to ye Society. his amendmts of ye Lunar Theory. wee had some discourse of it and he tould me that there was an equation of about 9′ necessary in ye Quadratures that this was begun & ended in ye line of the Suzigies & occasiond ye variation in ye octants to be 7 or 8 minutes greater or less then ye tables make it.[2] this I perceived was your æquation[3] & told him so. he was silent.

Soon after he came to Greenwich wth one freind only in his company. I was surprised at it & took the occasion of mindeing him of his disingenious behavior in severall particulars which he bore because he could not excuse it. afterwards I shewed him the Synopses & suffered him to take a very few notes of the greatest differences of ye observations from ye tables, and affirmed the æquations of ye tables Generally to[o] small by reason ye excentricity[4] was too little. In Mr Horrox his Systeme[5] the doubled distance of ye Sun from ye Moons apoge being numbred from F in ye periphery of ye little libratory Circle to I. a perpendicular let fall from this point I on ye Syzygiacall line ZS

26

where it cuts it in *x* makes ye present excentricity *Cx*. but he[6] affirmes that not *Cx* but *CI* is the excentricity in this & all other cases [Fig. 1].

Mr Street[7] changed ye diameter of Mr Horrox his libratory Circle a very little so as ye angle *ICx* was always ye equation of ye Apoge: so altering ye diameter of ye Circle all ye present equations will be encreased & diminished wth it except ye variation.[8]

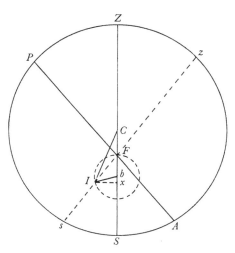

Fig. 1

To make the æquations bigger in winter yn in Summer[9] it will be requisite to make the diameter of this libratory Circle bigger in Winter yn Summer. which on your principles I affirmed & he assented to. but in what proportions he sayd not. So I perceave he is still in the darke in this point. & wants to know your determination. Hee mentioned an inæquality depending on ye moons distance from ye Node of which as I remember you gave me an hint in discourse.[10] & yesterday at London desired me to helpe him to observations made in

1687 from ye Quarter to ye full.

1688 about ye last quarter in March.

87 November about ye first Quarter

1692 Encreaseing in March decreaseing in December.

Your letter makes demands something like these & pointing at the same things I shall endeavor to satisfie you within a weeks time. but hence I gather that you have given him some hints where ye greatest errors lie & he is in pursuite of ym. Whilest you were in London I began to examine my observations of ye distances of ye declineing Sun from ♀ in order to find what the refractions were & in what proportions the[y] encreased as he descendes.[11] I employd for this purpose her distances from him observed Feb 23. Aprill ye 11. 21. 23. 25. & 26, 1681, the Calculations Were long & troublesome. the result I have drawn up in a synopsis too large to be transcribed into a letter. of which yet I may give you a copy when I have better leasure. at present I send you an empeiricall small table of ye difference of ye Refractions of the Sun & ♀ in height. when ye Sun set ♀ was in all of them above 30 & not more yn 37 degrees high. where ye French & my old tables make her refraction about $1\frac{1}{2}'$. add so much to ye differences it gives the intire refraction of ye Sun.

The Observations of Feb 26 make these differences bigger then those of Aprill 11. 21. 23. 25 by almost a minute. those of April 26 are ye least & about $1\frac{1}{2}$ less yn Februarys.

Yet all agree to make this difference when ye point observed is truely 1 degree high to be $19\frac{1}{2}$. which shews yt ye Refractions are [not] so irregular near ye horizon as they are commonly esteemed.

⊙ distanti visa a vertice	χ Refr ⊙ & ♀	⊙s refr simplex	The Refractions at	Mr Cassini[12] [°] ′ ″	La Hire[13] ′ ″	Boucher[14] ′ ″	Ye Obsr ′ ″
° ′	′ ″	′ ″					
				90:32.20	32.00	30.00	33.00
				89.27.56	26.25	25.12	23.00
77.00	2.00	3.30		88 21.04	20.43	18.24	19.30
80.	3.40	5.10		87.16.06	15.44	13.55	14.00
81.	4.00	5.30					
82.	5.00						
83.	6.00						
84.	7.00						
85.	8.00						
86.	10.00						
87.00	12.30	14.00					
87.30	13.30						
88.00	16.00						
88.25	17.25						
88.35	18.45						
88.40	19.30						
88.52	20.50						
89.00	21.30						
11	23.20						
20	24.20						
27	25.30						
30	26.30						
38	27.10						
44	28.30						
49	29.20						
51	30.00						
55	31.00						
90.02	32.00	33.30					

(left margin: dist vera ° ′ 89.00)

Mr Boucher is an English gentleman now in Jamaica if liveing who formed his Table on Tychoes observations & ye Cartesian Theory

But you see all the Theorys erre in this yt they make ye refractions to decrease but about 5 minutes betwix 89 & 90 whereas betwixt 88 & 89 they decrease above 6. on the Contrary the observations make $5\frac{1}{2}$ betwixt 88 & 89 but 10 betwixt 89. & ye horizon.

What may be the occasion of this I have not leasure at present to enquire. it seemes onely the medium in which the refractions are made is not equable as supposed by those who build their tables on theorys. this subject deserves your consideration I desire your thoughts of it at your leasure ye observations & what I deduce from them is incontestable.

Whilest I was on this subject of refractions I received from France the *Voyages Astronomique[s]*, but the title is *Recu[e]il d'Observations* &c in Folio. it conteines what Mr Richer[15] did at Cayenne Mr Picarts[16] Voyage to Uraniburg. his Cassinis & La Hires to the seaports of France with their Longitudes & latitudes determined by Observation the Voyage to Goree is not omitted but

the best part of it & greatest is Cassinis new tables of ♃s Satellits. wherein he
has corrected ye motions of ye first. he sets his Radixes to ye oppositions of ye
Sun & ♃ (very inartificially) to cover ye equation of light[17] (ariseing from ye
motion of the Earth in its orbe) which he makes when greatest 14 minutes &
allways by this means has it additionall. that other parte of it which arises
from ye change of ♃s distances from ye Sun he omitts. Againe whereas the
first Satellit moves about 2°.00′ in 14 minutes of time. he divides this in ye
proportion of versed sines & makes a table of it which he applys to all ye
satellits without any reason yt I can perceave but because it helps to salve
2 eclipses of the 2d which may be salved perhaps without it. by applying this
to the 3d he renders its Motions worse then they were for to me it seemes
equable & to need no æquation at all. he gives no examples of calculateing
ye places of ye 3d or 4th. I am apt to thinke because he found this device
would not agree in them. the Motions of ye 4th are the same with those of his
old tables.

I tould you that my observations would allow their greatest elongations
bigger then I had stated them it seemes his doe the same & that he allows their
distances to be in sesquialter proportion to ye periods of their Revolutions.[18]
tho to be thought a good catholick he says nothing of it but conceales it as he
does his allowing ye æquation of light for he makes their distances from ♃ in
semidiameters & sexagesimall partes to be

	sd ′			sd cents
of ye first.	5 . 40	or in semidiamts & Cents		5 . 66
2.	9 . 00		2.	9 . 00
3.	14 . 23		3.	14 . 38
4.	25 . 18		4.	25 . 30

I give you them thus to prevent your mistakeing them as my freind Mr Town-
ley[19] did

He still supposes ym to have one plain of their orbits but I am apt to beleive
the orbit of ye 2d lies out of ye plaine of ye orbits of the rest whose Inclination
to ♃s he now makes 2°.55′ whereas formerly he allowed it but 2°.40′. had he
known how to Calculate the length of ye line of ye passage through an ovall
shadow as well as through a Circular he needed not to have enlarged it so
much for he does it onely to make ye durations of ye Eclipses shorter as the
observations require & ye ovall shadow renders them.

You askt me once[20] when with you what were the Diameters of the satellits.
tis impossible to determine them exactly, but as well as I could when I made
my tables I stated the Angles their semidiameters subtend at ♃.

of ye 1st   28   or in partes   46⎫
    2d . 15   - - - - - -   40⎬ such as ♃s semid is 1000
    3. . 12   - - - - -   53⎭
    4.   4   - - - - -   25

I must adde that whereas you told me that ye parallactick æquation proveing double to what you esteemed it before you saw my observations argued the earth to be bigger yn you used formerly. It seemes to me the contrary that her flyeing of farther yn you thought from ye common center of gravity betwixt her & the Moon argues She should be lesse. I desire to be better informed in this particular at your leasure.

Sr I am Yours to serve you

JOHN FLAMSTEED

I shall write to you againe as soone as I can get another Synopsis transcribed. at present I am very busy about some other papers I am to send to a philosophicall Freind JF:
Mr Halley is busy about ye Moon has promised Me his corrections. intends to print something about her Systeme ere long & affirmes the meane Motion different in ye time of Albatani[21] from what it is now: J: F:

To Mr Isaack Newton
Fellow of Trinity
Colledg in Cambridg
there. these
present.
Cambridge.

NOTES

(1) Add. 3979, fo. 19.

(2) See note (3) below.

(3) See the letter of Newton to Halley, 13 February 1686/7 (vol. II, p. 464), in which Newton claims *inter alia* to have 'ye solution of your Problem about ye Suns Parallax', and also note (9) of the same letter, p. 466, where these lunar equations (corrections) which Newton called the first and second variations are specifically mentioned. The first, which was discovered and studied observationally by Tycho Brahe, appears in the mathematical gravitational theory as a second-order approximation, whereas the second, the parallactic inequality, which was discovered by Newton, is a third-order approximation.

In Fig. 2, *ACBD* is the lunar orbit, *S* being the Earth, *Q* the Sun. *P* is the position of the Moon at any instant, the points *C* and *D* representing the quadratures, *C* the first and *D* the last quarter, whilst *A* the new, and *B* the full Moon are the syzygies. In order to compare the places of the Moon in her orbit found by mathematical principles with those found by observation Newton was relying upon Flamsteed's tables, as he states in the short Scholium upon the lunar theory which follows Proposition 35 of Book III in the first edition of the *Principia* (p. 462). The tables were first printed as an appendix to *Jeremiæ Horroccij Opera Posthuma* (1673), pp. 475–88; they were printed in a more correct and enlarged form in Flamsteed's *Doctrine of the Sphere* (1680).

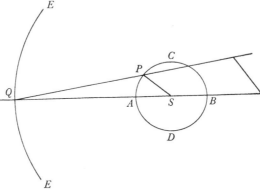

Fig. 2

Analytically, the first variation took the form $\lambda \sin 2\psi$, and the second (the parallactic inequality), $\mu \sin \psi$, where ψ is the angle QSP, the difference in longitude between Q the Sun and P the Moon. Since at $A, C, B, D, \psi = 0, \frac{1}{2}\pi, \pi, \frac{3}{2}\pi$ respectively, the first variation vanishes, so that the second, $-\mu$ at C, $+\mu$ at D, is alone apparent. Evidently Halley took $\mu = 9'$. At the syzygies $\psi = 0$ or π and the second correction is 'begun and ended there'. But at the octants $\psi = \pm\frac{1}{4}\pi, \pm\frac{3}{4}\pi$, the first variation is numerically a maximum, λ, to which the second therefore contributes $\mu/\sqrt 2$ $(=0\cdot 6363)$, that is roughly '7 or 8 minutes greater or less then ye tables make it'.

The variation (here called the first variation) is an inequality depending upon the alternate acceleration and retardation of the Moon due to the varying pull of the Sun at every quarter of a revolution (see Glossary). Tables of the variation were common; see Flamsteed, *Doctrine of the Sphere*, p. 98.

(4) According to ancient usage the excentricity (or eccentricity) is the distance from C, the centre of an orbit, to S, the position of the Earth about which the orbit is described. Hipparchus (150 B.C.) supposed that the Moon travels in a *circular* orbit ABP (Fig. 3) with uniform angular velocity about the centre C, and that the distance SC, the eccentricity, was $\sin 5° 1'$, that is, $SC:CB = \sin 5° 1'$. Ptolemy (A.D. 140) had found the same value, namely $\frac{1}{12}$.

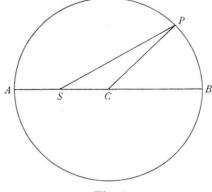

Fig. 3

(5) The motions of the Moon are so irregular and subject to so many inequalities that it was a long time before the more prominent of these irregularities could be accounted for. The earliest satisfactory attempt to account for them was due to Horrocks who gave a sketch of his

theory in letters to his friend Crabtree in 1638. In these the variation in the eccentricity is not referred to. But Crabtree in a letter to Gascoigne on 21 July 1642 gives Horrocks's rule for it. The lunar theory of Horrocks is based on an elliptic orbit *ABP* (Fig. 3) of small but variable eccentricity and having a libratory motion of the apses. This eccentricity he believed to have a maximum value of 0·06686 and a minimum value 0·04362. See note (6) below. One focus *S* was the Earth, the 'upper focus' being *H*. Hipparchus had discovered the precession of the apogee *B*, namely that the line *SB* had a small angular velocity of about 3° per lunar month, so that whilst *B* describes the orbit of 360° in the sense *BPAB*, *B* advances 3° in the same sense, and all modern observations confirm this result. Ptolemy had found that this uniform advance of *B* was accompanied by a to-and-fro disturbance along its path. Copernicus (A.D. 1543) improved on Ptolemy's treatment of this disturbance, and eventually Horrocks, in 1638, using delicate observations on the apparent size of the Moon when at apogee, formulated further improvements. They comprise two of the seven rules of his lunar theory (*Opera Posthuma*, p. 469). See article by S. B. Gaythorpe in the *Monthly Notices of the Royal Astronomical Society* (1925), pp. 858-65: 'On Horrocks's Treatment of the Evection and the Equation of the Centre, with a Note on the Elliptical Hypothesis of Albert Curtz, and its Correction by Boulliau and Newton.'

(6) Halley, presumably. In the letter to Gascoigne, mentioned above, Crabtree states the mean eccentricity to be 5524, the radius being 100,000, and a rule is given for finding the correct eccentricity in all cases, but no demonstration is given. Flamsteed, however, in his *Epilogus ad Tabulas* (*Doctrine of the Sphere*, pp. 489 and 491) has given a geometrical construction, which when translated is as follows:

'Let *ASsPZz* (Fig. 1) be the orbit of the Moon, which is carried about the Earth. *C* is the centre, *A* the apogee, *P* the perigee. Let *Cb* be the mean eccentricity. Upon *b* as centre describe a small circle whose radius *bF* shall be equal to half the difference of the greatest and least eccentricities. Through the point *F*, of least eccentricity, draw the synodical line *sIz*, cutting the small circle in *I*, from which point draw the radius *Ib* also the line *Ix* perpendicular to the axis *SZ*. Then *Ibx* will be equal to twice *bFI*, the distance of the apogee from the Sun, and *Cb* + *bx* (or *Cb* − *bx*) = *Cx*, will be the true eccentricity.'

Newton adopted a similar explanation, considering however *CI* and not *Cx* as representing the variable eccentricity. This slight correction was imparted by Halley to Newton as a secret (Baily, p. 683 note). See Letter 475.

(7) Thomas Streete (1621–89), a weaver who became an astronomer of repute and practical skill; a close friend of Robert Anderson (see vol. I, p. 312, note (1)) . He published his *Astronomia Carolina, A New Theorie of the Cælestial Motions* (1661), later re-edited by Halley. The planetary tables of Streete were founded on the hypothesis of Bullialdus (*Astronomiæ Philolaicæ Fundamenta*, 1657).

Streete was one of the six men chosen to re-survey London after the Great Fire of 1666. See E. G. R. Taylor, *Mathematical Practitioners* (1954), pp. 225–6, and Sherburne, *The Sphere of Manilius* (1675), p. 104.

(8) In the second order the variation is an additional correction to θ, amounting to $\frac{11}{8}m^2 \sin 2$ ($\mathbb{D}$ − $\odot$), which is independent of the eccentricity, and therefore unaffected here.

$\mathbb{D}$ = Moon's mean longitude at a time *t*.

$\odot$ = Sun's mean longitude at a time *t*.

m = Ratio of lunar month to a year (= 1/13).

(9) See Memoranda 468.

(10) Probably on 1 September 1694 (see Memoranda 468). The reference seems to be connected with Propositions 33 and 35 of the *Principia*: *Invenire motum verum Nodorum Lunæ* (To find the true motion of the Moon's nodes, Prop. 33, p. 455); *Dato tempore invenire Inclinationem Orbis Lunaris ad Planum Eclipticæ* (To find the inclination of the lunar orbit to the plane of the ecliptic, the time being given, Prop. 35, p. 460).

(11) See the table above. In the draft from which Baily has printed this letter (p. 135), only the right-hand side of entries is given.

(12) Giovanni Domenico Cassini (1625–1712) was an outstanding astronomer (see vol. I, p. 25, note (1)). He discovered and measured the rotation of Jupiter and Mars, and he tabulated the movements of the first satellite of Jupiter. See *Phil. Trans.* **18** (1694), 237.

(13) Philippe de la Hire (1640–1718). See vol. II, p. 299, note (12); also Huygens, *Œuvres*, x, 322–3, for the correspondence between de la Hire and Huygens relative to atmospheric refraction which is mentioned at the beginning of chapter IV of the *Traité de la Lumière*.

(14) Possibly Charles Bucher, Halley's friend and companion at Oxford.

(15) Jean Richer (1630–96), an astronomer of note. In 1672 he undertook a scientific expedition to Cayenne in French Guiana. He found that a pendulum of given length beat more slowly at Cayenne than at Paris whence he concluded that the intensity of gravity was less near the Equator than at higher latitudes, which indicated a deviation in the shape of the Earth from spherical. These results were confirmed by Halley during his expedition to St Helena. Richer was a member of the Académie Royale des Sciences; at the request of that body he wrote: *Observations Astronomiques et Physiques faites en l'Isle de Caïenne* (Paris, 1679), which is the work referred to here.

(16) Jean Picard (1620–82), physicist and astronomer, see vol. I, p. 246, note (5). He made accurate measurements of the length of a degree along a meridian, *La Mesure de la Terre*, Paris (1671). It was at his suggestion that Richer undertook his expedition to Cayenne (note (15) above). He collaborated with Römer in his observations on the satellites of Jupiter which led to the determination of the velocity of light. See *G.M.V.* pp. 181–2.

(17) In 1675 Römer had noted that the eclipse of the outermost satellite of Jupiter (Io) showed an inequality in the time of its occurrence, and this he ascribed to the time taken by light in its passage across the Earth's orbit. Most astronomers rejected this view. A note in Newton's hand in the University Library, Cambridge (Add. 3965.17, fo. 645), reads: 'According to Cassini's observations light moves from ye Sun to us in 7 minutes of time, thô Cassini ascribes ye phenom. to another cause.'

(18) Kepler's third law of planetary motions.

(19) Richard Towneley (*fl.* 1660–1705), an active mathematician of wide interests. He made a special study of the satellites of Jupiter. See vol. I, p. 78, note (1).

(20) Newton had written to Flamsteed on 30 December 1685, thanking him for information about Jupiter's satellites (Edleston, p. xxx). This may be a reference to an earlier letter.

(21) See Letter 476, note (4), p. 38.

475   NEWTON TO FLAMSTEED

24 OCTOBER 1694

From the original in Corpus Christi College Library, Oxford.[1]
In reply to Letter 474; for answer see Letter 477

Sr

I return my hearty *thanks* to you for your communications in your last &
particularly *for your table of refractions neare the horizon.*[2] The reason of the
different refractions neare ye horizon in ye same altitude, I take to be the
different heat of ye air in the lower region. For when ye air is rarefied by heat
it refracts less, when condensed by cold it refracts more. And this difference
must be most sensible when ye rays run along in the lower region of ye air for a
great many miles together; because tis this region only wch is rarefied &
condensed by heat & cold, the middle & upper regions of the air being always
cold. I am of opinion also that the refraction in all greater altitudes is varied a
little by the different weight of the air discovered by the Baroscope. For when
ye Air is heavier & by consequence denser it must refract somthing more then
when tis lighter & rarer. I could wish therefore that in all your Observations
where the refraction is to be allowed for, you would set down the height of the
Baroscope & heat of ye Air that the variation of ye refraction by ye weight &
heat of ye air may be hereafter allowed for when the proportion of ye variation
by these causes shall be known.[3]

A day or two before I left London I dined wth Mr Halley & had much
discourse wth him about ye Moon. I told him of ye *Parallactick Equation*[4]
amounting to about 8′ *or* 9′ *or at most* 10′ & of *another equation wch is greatest in ye*
Octants of ye Moons Apoge & might there amount to about 6′ or 7′, thô I had
not yet computed any thing about it. He replied that he beleived there might
be also an equation depending upon ye Moons Nodes.[5] To wch I answered
that there was such an equation, but so little as to be almost inconsiderable.
But what kind of equation this was I did not tell him & I beleive he does not
yet know it because it is too little to be easily found out by observations or by
any other way then ye Theory of gravity. He told me some years ago his
correction of ye Moons excentricity & repeated it when I was with him last at
London & this made me free in communicating my things wth him. *By your*
Observations I find it to be a very good correction. I recconed it a secret wch he had
entrusted me wth & therefore never spake of it till now. Upon my saying that
I hoped to mend ye Moons Theory by some Observations you had communicated to me & that those Observations made ye Parallactick Equation in ye
Quadratures between 8′ & 10′, he was desirous to view them but *I told him he*

must not take it ill if I refused him that because I stood engaged to communicate them to no body wthout your consent. I am very glad that there is like to be a new correspondence between [you] & hope it will end in friendship.

The Parallactick Equation depends not upon the common center of gravity of ye Earth & Moon but upon another center whose distance from ye center of ye earth is as ye square of ye diameter of the Moons Orbit, & therefore makes that equation proportional to that diameter. But this equation is less then I took it to be when I saw you last. Tis so involved wth other equations that I cannot determin its just quantity[6] till I have your Observations in other positions of the Moons Apoge.

In that new synopsis of Observations you are drawing up, *pray insert ye distance of ye Moon from the Sun wth the Variation.*[7] For I must correct ye Variation,[8] wch I cannot well do wthout your numbers. In the second of those two synopses you communicated to me I was fain to compute it, but that was not so well as to have those very numbers by wch you computed the Moons place. *Pray insert also ye columns wch relate to ye Moons Latitude*[9] because the Theory of her Latitude needs some amendment.

I am

Your very humble Servant

Is. Newton

Trin. Coll. Octob. 24
    1694
For Mr John Flamsteed
at the Observatory in
Greenwich neare
London.[10]

NOTES

(1) C. 361, no. 49. Flamsteed has inserted the date 'Octob 24. 1694' at the head of the letter. The postmark is OC/26. In this letter the words and sentences in italics were underlined by Flamsteed.

(2) Newton was particularly anxious to have a table of refractions near the horizon, for it enabled him to determine the different refractive powers of the atmosphere at low altitudes. Flamsteed believed that 'ye Refractions are [not] so irregular near ye horizon as they are commonly esteemed'. See Letter 474.

(3) Baily (p. 137 note) has observed: 'It would have been fortunate had Flamsteed attended to this important hint given by Newton, to observe the thermometer and the barometer. Not that it would have made much difference in the places of the stars in the *British Catalogue*; because the major part of the observations, from which they were deduced, were made prior to the period here mentioned.'

(4) See Letter 470, note (9), and vol. II, p. 466, note (9). In a letter dated 9 July 1695 Newton declared that its value scarcely exceeded two or three or, at most, four minutes. Bürg (*Mécan. Cél.* Tom. III, p. 282) gives it as 2′ 2″.38. This equation is omitted in the second edition of the *Principia* and Biot suggests reasons for the omission (Edleston, p. lxvii, n. 119).

(5) The Moon's path changes in such a way that the nodes move slowly backwards (east to west) along the ecliptic, performing a complete revolution in 6794·4 days, or about 18½ years. By considering the retrograde motion of the Moon's nodes Newton was led to an explanation of the cause of the precession of the equinoxes. See Letter 472.

(6) In the margin, at each line of this paragraph, excepting the last, quotation marks have been added, presumably by Flamsteed.

(7) A marginal note by Flamsteed states: 'these were not in ye first Synopses when I shewed ym to him but were added after'. See Memoranda 468. The new synopsis, the third, was sent on 29 October 1694. See note (10) below.

(8) See *Principia*, Book III, Prop. 29: *Invenire Variationem Lunæ* (To find the variation of the Moon). See Glossary, and Letter 474, note (3).

(9) A marginal note by Flamsteed says: 'these were in ye first'.

(10) Below the address, on the cover, is a note in Flamsteed's hand: '1st about refractions. October 24 1694. Answered ye 29th lunæ & sent him a Synopsis of 50 lunar observations with ye Calc ye Synopsis of ye χ of ye ☉ & ♀ refractions & a Copy of Mr Caswells letter about Magnetisme & a letter of my owne about ye ☽ & refr.'
For Caswell on magnetism see Letters 476, 480 and 481.

476   FLAMSTEED TO NEWTON
25 OCTOBER 1694
From the original in the University Library, Cambridge.[1]
In further reply to Letter 473; for answer see Letter 478

The Observatory October 25. 1694

Sr

I am heartily glad you find your Theories so well confirmed by my observations. there is one[2] of them Feb: 27. 1691, wherein ye ☽s place is miscomputed; it ought to be 3s.07°.56′.08″ & so the error −9′.38″ which in your copy is above 17′. twas an hasty fault of my servant   pray correct it

He is transcribeing another Synopsis of ye Moones observed places & elemts of ye Calculations for you, which had beene in your hands last week but that I found it necessary to repeate some of the calculated places which cost me 3 or 4 days paines & time.

If you find any of ye observations not agree wth your Theorys please to informe me freely   I will repeate them; for tho I use all care & diligence imaginable yet I find that sometimes small faults are committed & indeed the Elements being so many tis allmost impossible to perceave them or avoyd them without a repetition of ye worke.

On Monday morneing [29 October] next God Willing I shall cause my Servant to deliver ye Synopsis of lunar Observations to the Cambridg carrier Mr Martin for you, & with it another Synopsis of all the refractions observed,[3] from which I derived the little empeiricall table of them I sent you in my last.

Last post I receaved from Mr Caswell an account of some experiments he made concerneing Magnetisme a copy of it shall beare them company.

Yesterday at London I had a great deal of talke wth Mr Halley about ye $\mathbb{)}$s Motion. he affirmed the Meane Motion to have been swifter in ye time of Albatani[4] then at present and that the cause of it was by reason that the bulke of all the planets continually encreased.[5] I gave him the heareing and at last told him that this Notion was yours; he answered in truth you helpt him wth yt.[6]

Hee affirmes further That the Moones Apoge Moves swifter in Winter yn in Summer, & that the greatest equations[7] of it are biggest when ye Sun is Perige. That they are as big as Copernicus makes ym, yt is 13°.9′. this smells too of your Theorys. I remember yt you affirme all ye æquations biggest when ye Earth is nearest ye $\odot$.

PP. pag 428.

I should be glad to hear yt you had found in what proportion ye æquations of ye Apoge & ye Excentricitys altere, & what are their greatest differences in ye last, the quantity of ye first, how ye Variation alters & yt you would please to impart yt to me.

That so hereafter I may calculate on sure grounds & compare not an apparently erroneous but a true Theory with my observations. whereby its faults may be corrected.

Mr Paget[8] I hear is ill of a feavor: I am heartily sorry for it. If he should die I know no person fit to succeed him but Mr Caswell who wants his talents of drawing & writeing Neatly. in others is much his superior. but I hope he may recover tho he has buried three of his pupills of this distemper all health is wished you by Sr

Your affectionate freind & Servant

J: FLAMSTEED

To
Mr Isaack Newton.
Fellow of Trinity Colledge
    in Cambridge these
        prt
        Cambridg.

NOTES

(1) Add. 3979, fo. 20.

(2) It occurs in the second synopsis. See Memoranda 468, note (4), p. 8.

(3) The third synopsis. It was sent on 29 October 1694. See Letter 477.

(4) Al Battani (Albategnius), an Arab prince who died A.D. 929; so called from his birth-place, Battan, in Syria. He was an astronomer and he calculated anew the Ptolemaic tables which were long used by the Arabs as superior to any others. He was greatly esteemed by Halley who had lately examined a Latin translation by Plato of Tivoli, *Albatenij Observationes Astronomicæ* (Nuremburg, 1537). He detected more than thirty considerable faults in a few pages by calculating tables from Al Battani's principles. See *Phil. Trans.* **17** (1691–3), 913. (*Emendationes ac Notæ in vetustas Albatênii Observationes Astronomicas, cum restitutione Tabularum Lunisolarium ejusdem Authoris. Per Edm. Halley. S.R.S.*)

(5) 'Halley said that Mr Newton had lately told him that there was reason to conclude that the bulk of the Earth did grow and increase in magnitude by the perpetuall accession of new particles attracted out of the Ether by its gravitating power, And he supposed and pro-posed to the Society, That this encrese of the Moles of the Earth would occasion an accelera-tion of the Moons motion, she being at this time attracted by a stronger Vis Centripeta than in remote Ages' (Journal Book of the Royal Society of London, 31 October 1694).

(6) Newton had visited London to discuss a new edition of the *Principia* (see Letter 472). Before leaving, he dined with Halley and 'had much discourse wth him about ye Moon'. See Letter 475.

(7) A reference to evection (see Glossary; and Godfray, *A Treatise on Astronomy*, 1886, pp. 92–3).

(8) Edward Paget (1656–1703); appointed mathematical master at Christ's Hospital in 1672 (see vol. II, p. 374, note (1); vol. III, p. 366, note (2) and p. 368, note (2). He was suc-ceeded by Samuel Newton, no relation to Isaac. See Letter 499, note (5), p. 104.

477   FLAMSTEED TO NEWTON
29 OCTOBER 1694
From the original in the University Library, Cambridge.[1]
In reply to Letter 475; for answer see Letter 478

The Observatory Octob: 29. 1694

Sr

I have yours of ye 24th instant & here[2] send you the sunopses of lunar observations & refractions I promised you. by compareing the refractions of Feb: with those of Apr. 26 you will find your thoughts confirmed. I have more by me deduced from ye ☉s heights taken in his riseing & setting in June which I shall send you ere long.

In the lunar Synopsis I thinke there is all Inserted you desire. the result of your thoughts on them will be Welcome to me.

I have by me 100 places of the Moone[3] calculated exactly by me & my servant to as many times of observations taken with ye sextant in ye yeares 1676. 77. 78. 79. 80. the first of those happen 2 revolutions of ye Apoge agone. When I get a little leasure I shall deduce her places from ye observations yt happen in ye Syzugies & Quadratures which will shew whether ye errors be the same in ye same Annuall Argument & distance from ye Sun, or not, much better then those that have been deduced by places of the fixed stars that were uncertain or faulty:

Observations of ye New or full moones when ye Apogee was in □ of ye Sun will help us to ye lowest of ye greatest æquation & of ye Quartile moones (when ye Apoge was in ☌ or ☍ to him) to the highest of them, as also to the quantity of the greatest parallactick æquation[4] & therefore I shall not fayle to mind these Cheifly.

I find, on reviewing your Theory you make the meane Variation 35′ minutes onely which I use in my tables 38′,[5] but this I see is variable. If you let me know what you determine of it at your leasure you will oblige me.

After all nothing certeine can be determined by observations of ye ☽ without the true places of ye Fixed Stars first stated, & therefore whilest you are determineing the inæqualitys of the Moone I shall proceed wth them haveing layd a sound & firme foundation for yt Worke In ye little tract[6] I shewed you wherein I corrected the Motions of ye Earth & determined the places of ye Aphelion & ye quantity of ye greatest æquations.

Mr Hanway[7] of your Colledge is now wth me & applys himselfe to ye Study of Mathematicks wth good successe. he with my servant examined ye Copies. So I hope tho they be not so cleane as they ought they are nevertheless Just: he presents you wth his humble service. this in hast is from

Your affectionate freind & Servant

J: FLAMSTEED

For Mr Isaack Newton

these

Synopsis Refractionum ab observatis Veneris a Sole distantiis deductarum cum veris distantiis puncti Solis observati a vertice ad quamlibet deductionem[8]

a Johanne Flamsteedio Grenovici

Altitudo Veneris ad occasum ⊙is	35° 37'	39°.14'	37°.17'	36°.46'	36°.12'	35° 23'
Puncti observat: distantia a vertice vera	Feb die 23	Apri 11	21	23	25	26
61 18	0' 30"	' "	0' 30"	' "	' "	' "
62 23			0 30			
64 39	1 00					
64 54	1 20					
76 41					1 40	
77 06					2 00	
78 58					2 32	
79 00			3 13			
79 20			3 14			
79 49	4 00					
80 08	4 15					
80 27			3 40			
80 32						3 10
80 48						3 10
81 02	4 20					
81 07						
81 41			4 10			
81 54	5 20					
82 21			5 06			
82 48	5 45					
83 09		5 50				
83 20		6 15				
83 36					5 50	
83 37	6 10					
83 46					6 00	
84 16					6 40	
84 33	7 40					
84 44						6 35
84 47			7 17			
84 48					7 20	
84 56						7 00
85 02					7 30	
85 17					7 40	
85 20						7 35
85 32	9 15					
85 37					8 20	
85 50			9 15			
85 58			9 30			
86 01					9 05	

distantia a vertice vera	Febr 23	Apr 11	21	23	25	26
87 54	' "	' "	' "	' "	' "	13' 40"
87 57			14 53			
87 59				14 30		
88 04					15 10	
88 06	16 12					
88 07				15 16		
88 11						14 35
88 24	17 20					
88 26			16 42			
88 27						16 20
88 30				16 50		
88 31					16 25	
88 42			17 40	17 21		17 15
88 53	19 30		18 45			18 30
88 58					18 50	
89 00				19 25		
89 06						19 10
89 09					19 40	
89 12			21 10			
89 13		20 45				
89 14				20 35		
89 16						20 10
89 20					21 15	
89 24			21 45			
89 24				21 40		
89 25						21 06
89 33			23 20			
89 34				23 00		
89 36	23 20					
89 37					23 30	22 25
89 41			24 15			
89 44						23::
89 46					24 15	
89 48			25 05			
89 51		26 10				
89 53				25 25		24 25
89 54					25 30	
89 57			26 25			
89 59						25 25

86 10	10 30						90 03	27 10				26 00
86 14						9 00	90 04			26 45	27 05	
86 26			,		9 35		90 07		27 10			
86 29		10 30					90 10					27 00
86 31						9 30	90 11				28 35	
86 39		10 50					90 12		28 30			
86 42	11 45						90 15	29 20				28 00
86 45			10 30				90 17			29 10		
87 04				11 20			90 19					29 24
87 08	13 06						90 21			29 50		
87 10		12 15		12 05			90 25	30 40	30 00			30 00
87 19			11 22				90 30		31 25			
87 25			12 55				90 33			32 15		
87 38	14 27						90 34	33 40	31 45			
87 39					12 35		90 38					
87 40				13 40								
87 41			13 18									
87 44					12 55							

NOTES

(1) Letter: Add. 3979, fo. 21. Table: Add. 3967, fo. 15.

(2) Two documents were enclosed, one being the third synopsis of 56 lunar observations (see Memoranda 468, note (2)), the other the table of refractions reproduced here. Newton returned the former to Flamsteed on 18 November 1694 (see Letter 480, 1st para.), after having taken a copy of it. See also Flamsteed's addition to Newton's letter of 1 November 1694.

(3) In a note (undated) in the University Library, Cambridge (see Letter 473, note (6)), Flamsteed refers to 400 places of the Moon calculated from 'Observations taken with ye Sextant betwixt 1675 & 1689'. Some of these were sent with his letter of 15 March 1694/5 (Letter 497).

(4) See Letter 470, note (9), p. 15.

(5) See Letter 474, note (3), p. 30. In the *Principia*, Book III, Prop. 29, p. 443 (*Invenire Variationem Lunæ*), Newton quotes Halley's figures: *Halleius autem recentissimè deprehendit esse 38′ in Octantibus versus oppositionem Solis, & 32′ in Octantibus Solem versus. Unde mediocris ejus magnitudo erit 35′: quæ cum magnitudine à nobis inventa 35′.9″ probe congruit.* (Halley very recently discovered it (the variation) to be 38′ in the octants facing the Sun's opposition and 32′ in the octants facing the Sun, whence the mean magnitude will be 35′, a result which well agrees with that found by us (35′.9″).)

(6) See Letter 474.

(7) John Hanway, or Hanney (born 1671), son of Sir William Hanway; educated Westminster School and Trinity College, Cambridge, where he was admitted pensioner 28 June 1690. Scholar, 1691, B.A., 1693/4. See also Letter 493.

(8) The title, when translated, reads: 'Synopsis of Refractions deduced from the observed distances of Venus from the Sun with the true distances of the observed position of the Sun from the vertex to any deduced.' The table was drawn up by a copyist but it bears a few later entries. It was returned three weeks later with Letter 480.

Newton's figures at the top of the table are transcribed from Flamsteed's letter of 2 March 1694/5. The table is mentioned in Letter 474.

478   NEWTON TO FLAMSTEED
1 NOVEMBER 1694
From the original in Corpus Christi College Library, Oxford.[1]
In reply to Letter 477; for answer see Letter 479

No: 1: 1694

Sr

A day or two after I wrote to you I received your letter with an emendation of ye Observation Feb. 27, 1691.[2] You say ye Moons place at that time ought to be 3s. 7gr.56'.8" & so ye error −9'.38". I suppose it should be 3s.17gr.56'.8" & so ye error −10'.33". For her observed place in ye synopsis[3] you gave me is ♋ 17°.45'.35" unless you have corrected it by the new places of ye fixed stars.[4]

There are some other faulty observations, particularly those of Feb. 21. March. 12. Apr. 7. May 22. July 1. July 30. Sept. 6 1690. But whether ye faults lye in the calculated places[5] or in the observed ones or in ye places of the fixed stars I cannot tell.

Mr Caswells magnetical observations[6] you need not send me, for I have no occasion for any thing of that kind. Neither need you send me your larger synopsis of ye refractions.[7] That short one wch I have already is sufficient for me. I desire only such Observations as tend to perfecting the Theory of the Planets in order to a second edition of my book & would not give you ye trouble of superfluous communications. The greatest equation of physical parts[8] I told you was by my calculation *13'*, and now by your Observations I find it is about *12'* or *13'. 1.*[9] *The Variation in Spring & Autumn is about 36' or 35½',*[10] *in winter tis greater & and in summer less by two or three minutes: And in ye Moons Apoge tis greater by two or three minutes then in her Perige. 2. The excentricity & equation of ye Moons Orbit is sensibly greater in winter then in summer & seems to be sometimes as great as Mr Halley makes it,* but ye law of its increase I am not yet master of, nor can be till I have seen ye course of the Moon as well when her apoge is in ye summer signes as in ye winter ones. For those Observations you gave me at London contein only her course when her Apoge is in the winter ones. *The equation wch depends upon ye Moons nodes is too little to be sensible by your*

Observations till they are corrected by the new places of the fixt stars. I only see in general by my Theory that there is such an equation, & by your Observations that the Theory and the heavens agree so far as I have been able to compare them hitherto. In my two letters I quite forgot to explain to you the menstrual parallax of ye Sun.[11]

Let S be ye Sun, T be ye earth, L ye Moon in ye first quarter & C the common center of gravity of the Moon & Earth. This common center of gravity whilst ye Moon and Earth move about it moves regularly in ye orbis magnus: so that when you have computed the place of the earth you are to place the point C in yt place & set ye earth T forwarder by adding ye angle CST to ye computed place. But if the Moon be in the last quarter you are to substract that angle. The quantity of this angle I do not yet know certainly.

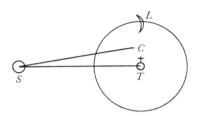

Tis not so great as I thought when I was in London. If you assume it to be 16″ or 20″ & find that by such an assumption ye greatest errors of ye suns place are diminished you may retain yt quantity, till it shall be determined more exactly. I am

Sr

Your faithfull friend

& humble Servant

Is. NEWTON

Cambridge
Novem. 1.
1694.
For Mr John Flamsteed at the
Observatory in Greenwich
London.

[Note written on the letter by Flamsteed] No[v] 25 ♃[12] I wrote to Mr Newton yt I would send him the Synopsis of Refractions. I sent them on ☽ 29. with ye 3d Synopsis of lunar observations he writes me this answer No[v]: 1.♃ when he had receaved them. if ye Carrier performed his duty. q[ue]ry why he says nothing of the receipt.

NOTES

(1) C. 361, no. 51. The date has been inserted by Flamsteed. See below, note 12.

(2) See the first paragraph of Letter 476.

(3) The second synopsis, which Newton returned, and which was acknowledged on 7 September 1694 by Flamsteed, who declared that he was then at work 'on ordering ye fixed stars'. See also Letter 495.

(4) In the margin Flamsteed has here inserted the figures: 17.45.35[;] 17.56.08[;], 10.33.

(5) Newton wrote 'placed'.

(6) See Letter 476.

(7) Flamsteed had already sent these communications on 29 October. See the opening paragraph of Letter 477.

(8) The equation of the physical parts is that portion of the equation of time which is due to the ellipticity of the orbit; in the case of the Earth it is distinguished from a second equation which is due to the obliquity of the ecliptic. Horrocks called them the Keplerian and Tychonian equations respectively.

(9) The reference numbers 1 and 2 are inserted by Flamsteed who has also underlined the passages and figures in italics. In the margin he has added the words: 'greatest physicall parts 13.00.' Lower down, also in the margin at the appropriate places, he has added the words: 'Variation', 'Excentricity': 'Mr Halleys', and a few lines lower: 'New æquations near ye ℧s [nodes] to small to be perceaved by ye old places of ye fixed * *.'

(10) See Letter 477, note (5), p. 41.

(11) This is now called the lunar equation of the Sun. It is equal to

$$\frac{\text{Mass of Moon}}{\text{Mass of Earth}} \cdot \frac{\text{distance of Moon from Earth}}{\text{distance of Sun from Earth}} \cdot \sin \text{(difference of long. of Moon and Sun)}.$$

The coefficient is given 8″·83 in Bürg *Mécan. Cél.*, Tom. III, p. 108. Newton in this letter states that he has not yet ascertained its magnitude but that it may be assumed to be 16″ or 20″ until it be determined more exactly (Edleston, p. lxvii, note 120).

(12) Nov. 25 was Sunday, not Thursday as stated by Flamsteed. He appears to have written Nov. 25 for 25 October which was Thursday. This is confirmed by the postmark which is 2 November. '☽ 29' is Monday the 29 i.e. of October and 'No[v]: 1. ♃' is Thursday, 1 November, because in Letter 479 Flamsteed refers to 'yours of ye 1st instant'.

479   FLAMSTEED TO NEWTON

3 NOVEMBER 1694

From the original in the University Library, Cambridge.[1]
In reply to Letter 478; for answer see Letter 480

The Observatory ♄ *No*: 3 1694

Sr.

I had yours of ye 1st instant last night. On Monday before I sent you some papers which I promised you in one dated ye 24th instant.[2] I hope you have them tho you mention not their safe arriveall.

I have been ill of a cold this 3 days & therefore not beeing fit for work that require much intention I have this morneing turned over my book of observations & Calculations to enquire what might be reason of ye fault you suspect on ye days you mention.[3]

for ye observation of Feb 27. 1691 I gave you in my letter the ☽s place by my repeated Calculation from my Tables the error was committed through hast, you correct it right

The true places of ye Moon May 22. July 1. & 30 were gott by compareing ye true times of her transits over ye meridian with the Suns, who Was much nearer ye Vertex in his transits yn ye ☽ in hers no proper fixed stars were noted on those days & therefore you may esteeme ye observed places faulty in these & lay them aside as dubious

But if there be any error in ye others tis in the place calculated from ye tables for ye observed is strongly confirmed by ye transits of More stars yn one the same night I employed a servant to repeate ye calculations at yt time yt was more carefull of his own business yn mine & often made too much hast in that work & would tell me he knew I would repeate ym all when ye fixed stars were replaced. I cannot esteeme them but very little erroneous & therefore shall cause them to be recalculated if you persist in your opinion yt they are faulty.

I suppose you have not compared above one halfe ye three Synopses[4] with your corrections when you wrote to me I shall desire you to let me know what others you find suspicious that I may correct them altogeather & send you ye emendations at once.

When you read Mr Caswells letter[5] about Magnetisme I persuade my selfe You Will find something in it not common. he is a person of great Ingenuity an Excellent Geometrician & Wonderfully modest. & all yt know him except one person who has injured him higly, give him the same charecter. I sent it to you onely to have your opinion of it which I will expect from you at your leasure

I sent you the Synopsis[6] of ye differences of Refraction of ye Sun & ♀ collected from their observed distances yt you might see your owne suggestions concerneing them confirmed you will find ye Refractions of Feb: 23 exceed those of Aprill every where a minute or More in the same distance from ye vertex, & therefore I thought these communications would not have been superfluous.

I perceave you are as yet onely tryeing how my observations will consent with your Emendations & yt you have not as yet limited them to your Mind therefore urge you no further for them. when you have determined what corrections or additions are to be made to that theory which it was my Good fortune to Meet with & usher into ye light,[7] I doubt not but you will impart them to me as freely as I did the observations Whereby you limit or confirme them to you.

The parallactick Equation of ye Sun is so small[8] it will scarce be sensible by observation a single vibration of ye pendulum is equall to it & it will be impossible With our grosse sences to determine the Suns place to yt exactnesse wth ye best & largest instruments yt can be built, as I thinke may be easily proved. I am

Your most affectione freind & servant

JOHN FLAMSTEED

To
    Mr Isaack Newton
    at his Chamber in
    Trinity Colledge
    Cambridg.
        these present

NOTES

(1) Add. 3979, fo. 22. In the first line a full stop followed by a capital has been inserted to make clear what Flamsteed meant. 'Monday before' means 'Monday last' (29 October). See Letter 478, note (12), p. 44.

(2) Letter 476. Flamsteed dated it 25 October 1694.

(3) Letter 478.

(4) See Memoranda 468. Newton had not received the package containing the third synopsis and table of refraction. See opening paragraph of Letter 480.

(5) See Letter 476.

(6) See the table, pp. 40–1, sent with Letter 477.

(7) The Horroxian theory (see Letter 474, note (5), p. 31).

(8) The value of the parallactic equation, according to H. Spencer Jones (1944), is $8'' \cdot 790$.

480   NEWTON TO FLAMSTEED
17 NOVEMBER 1694
From the original in Corpus Christi College Library, Oxford.[1]
In reply to Letter 479; for answer see Letter 481

No: 17 1694

Sr

The Carrier came wthout your parcel[2] but had it sent after him & I received it about three hours after I had sent away my last letter to you. I like Mr Caswels experimts[3] well. They deserve to be made publick. I have taken copies of your other papers[4] & designed to return them the last week but that I could not get my copies collated soon enough. They shall be sent to morrow, together wth a Table of refractions[5] wch I have computed by applying a certain Theorem to your Observations.[6] For being at a stand about ye Moons Theory I set my self to compute this Table. The first column expresses ye refraction in mid winter in time of a gentle frost, & agrees almost wth your observations of Feb. 23. The third column expresses the refraction in ye usual heat of July, & agrees almost wth your observations of April 26. The middle column expresses ye refraction in a middle degree of heat & agrees wth your observations of April 21, 23 & 25, most nearly. The proportion of the first to the third I determined by ye difference of the rarefaction of the air in winter & summer wch I found some time ago by certain experimts to be as 8 to 9 or thereabouts. You may communicate this Table to Mr Halley if you think fit.

I beleive there may be more faulty Observations in your Synopsis then I have yet discovered & I suspect that of Sept. 30 1690, thô I cannot well judge of it because there are no other Observations near it to compare it with. So also yt of Feb. 6 1691 seems faulty. Pray see in your book if these Observations be not dubious. For as for ye places calculated from ye Tables I will give you no trouble about them. My servant has lately learnt Arithmetick, & if I go on with this business of the Moon he shall learn Astronomical calculations & examin them & I will send you his corrections.

I beleive you have a wrong notion of my method in determining the Moons motions.[7] For I have not been about making such corrections as you seem to suppose, but about getting a generall notion of all the equations on wch her motions depend & considering how afterwards I shall go to work wth least labour & most exactness to determin them. For the vulgar way of approaching by degrees is bungling & tedious. The method wch I propose to my self is first to get a general notion of the equations to be determined & then by accurate observations to determin them. If I can compass the first part of my designe I do not doubt but to compass the second & that made me write to you that I hoped to determin her Theory to ye exactness of two or three minutes. But I am not yet master of ye first work nor can be till I have seen something of ye Moons motions when her Apoge is in ye summer signes. And to go about ye 2d work till I am master of ye first would be injudicious, there being a complication of small equations wch can never be determined till one sees the way of distinguishing them & attributing to each their proper phænomena. *Sr if you can have but a little patience wth me till I have satisfied my self about these things & made the Theory fit to be communicated wthout danger of error I do intend that you shall be the first man to whom I will communicate it.*[8]

And because I would give you as little trouble as may be if you please to communicate to me the Right Ascentions & apparent meridional altitudes of ye Moon as you have found them in your observations without allowing for the refraction & parallax I will take care of all the rest, & *return you synopses of her Longitudes & Latitudes* &c.[9] But I desire her right Ascensions by ye correct places of the fixt stars. For otherwise your Observations will not reach to distinguish & determine those small equations wch remain to be found out, & I would not have the work to do over a second time. This may give you a little trouble at present but it will save you ten times the trouble wch you must otherwise undergo here after & that perhaps without bringing the Moons Theory to half that perfection wch I think I have a prospect of. If you please to do me this favour, then I desire that you would send the right Ascensions & Meridional Altitudes of the Moon in your Observations of the last six months. You may do it in three columns under these titles

Tempus apparens Grenovici.	Lunæ Ascentio recta observata.	Lunæ Altitudo meridiana apparens.

And for the trouble you are at in this business, besides the pains you will save of calculating (& that upon an erroneous hypothesis as I must do) the Observations you communicate to me, & the *satisfaction you will have to see the Theory you have ushered into the world brought (as hope) to competent perfection & received by Astronomers; I do intend to gratify you to your satisfaction,* tho at present I return you only thanks, as I do heartily for what you have already communicated. I am

Your affectionate & humble Servant

IS. NEWTON.

Cambridge Novem. 17*th*
     1694.
I sent your *papers*[10] back by ye
Carrier yesterday & this Letter
should have been sent by ye
Post before.
For Mr John Flamsteed
at the Observatory in
    Greenwich   [Added by Flamsteed]
    neare London.   No: 17. about refractions
                        ye Table

TABULA REFRACTIONUM[11]

Altitudo apparens gr. '	Refractio æstiva ' "	Refractio verna et autumnalis ' "	Refractio hyberna ' "	Alt. appar gr.	Refractio verna et autumnalis ' "	Alt. appar gr.	Refractio verna et autumnalis ' "
0. 00	31. 30	33. 20	35. 10	31	1. 28	61	0. 29
0. 30	26. 06	27. 45	29. 24	32	1. 24	62	0. 28
1. 00	21. 50	23. 12	24. 34	33	1. 21	63	0. 27
1. 30	18. 51	20. 2	21. 13	34	1. 18	64	0. 26
2. 00	16. 27	17. 29	18. 31	35	1. 15	65	0. 24
2. 30	14. 31	15. 23	16. 15				
				36	1. 13	66	0. 23
3. 00	12. 52	13. 40	14. 28	37	1. 10	67	0. 22
3. 30	11. 32	12. 15	12. 58	38	1. 8	68	0. 21
4. 00	10. 25	11. 4	11. 43	39	1. 5	69	0. 20
4. 30	9. 29	10. 5	10. 41	40	1. 3	70	0. 19
5. 00	8. 40	9. 13	9 46				
				41	1. 1	71	0. 18
6. 00	7. 24	7. 52	8 20	42	0. 59	72	0. 17
7. 00	6. 27	6. 51	7 15	43	0. 57	73	0. 16
8. 00	5. 42	6. 3	6 24	44	0. 55	74	0. 15
9. 00	5. 5	5. 24	5 43	45	0. 53	75	0. 14
10. 00	4. 36	4. 53	5 10				
				46	0. 51	76	0. 13
11. 00	4. 11	4. 27	4. 43	47	0. 49	77	0. 12
12. 00	3. 51	4. 5	4. 19	48	0. 48	78	0. 11
13. 00	3. 33	3. 46	3. 59	49	0. 46	79	0. 10
14. 00	3. 18	3. 30	3. 42	50	0. 44	80	0. 9
15. 00	3. 4	3. 16	3. 28				
				51	0. 43	81	0. 8
16. 00	2. 52	3. 3	3. 14	52	0. 41	82	0. 7
17. 00	2. 42	2. 52	3. 2	53	0. 40	83	0. 6
18. 00	2. 33	2. 42	2. 51	54	0. 38	84	0. 5
19. 00	2. 24	2. 33	2. 42	55	0. 37	85	0. 5
20. 00	2. 17	2. 24	2. 33				
				56	0. 35	86	0. 4
21. 00	2. 9	2. 17	2. 25	57	0. 34	87	0. 3
22. 00	2. 2	2. 10	2. 18	58	0. 33	88	0. 2
23. 00	1. 57	2. 4	2. 11	59	0. 32	89	0. 1
24. 00	1. 51	1. 58	2. 5	60	0. 30	90	0. 0
25. 00	1. 46	1. 53	2. 00				
26. 00	1. 42	1. 48	1 54				
27. 00	1. 37	1. 43	1 49				
28. 00	1. 33	1. 39	1 45				
29. 00	1. 30	1. 35	1 40				
30. 00	1. 26	1. 31	1 36				

NOTES

(1) C. 361, no. 53. The date has been added by Flamsteed.

(2) See Letter 477 and the Table of Refractions which accompanied it.

(3) See Letter 476.

(4) The third synopsis, and the Table of Refractions. See Letter 477, note (2), p. 41.

(5) See Table following this letter.

(6) See Letters 483 and 486.

(7) Marginal note by Flamsteed: 'I had & he of me & still has.'

(8) Underlined by Flamsteed who has added in the margin: 'as much as he pleases I have waited 5 years for ym.'

This passage is in reply to a remark which Flamsteed had made, in a preceding letter, that Dr Halley had asserted 'that Mr Newton had done the theory of the Moon', whereupon, says Flamsteed, 'I wrote to him for the performance of his engagement, not taking any notice he had forgot it'. On this answer of Newton, Flamsteed makes the following comment: 'Satisfied herewith that Mr Halley's talk was only boast I troubled him no more about it: though I found he had forgot his first engagement, as he had done his *intention* (for so he termed it) since' (R.G.O., vol. 35, fo. 152).

(9) The passages in italics were underlined by Flamsteed.

(10) At this point Flamsteed has noted: '*papers*, ye Synopses of the])s places calculated & compared togeather with ye elements of ye Calculations Ye 2 synopses conteining above 100 places of ye moon observed compared with my old tables.' For a list of the observations sent by Flamsteed see Memorandum 516.

(11) This is a copy of the table of refractions mentioned in the letter. It was found among some of Flamsteed's papers in the Library of Corpus Christi College, Oxford. The original is in the University Library, Cambridge (Add. 3967, fos. 18 and 19), but it is very difficult to read on account of damage at the edges of the paper. At the foot of the original, Newton has written: *Construitur hæc Tabula a gradu primo ad ultimum quærendo an...er proportionem sinuum* 7000 *ad* $7000 - 11 = 6989$, *ac diminuendo inventos in ratione data.* (This table is constructed from the first degree to the last by seeking...the proportion of the sines 7000 to 6989, and diminishing those found in the given ratio.) The gap is due to the damage referred to above.

481   FLAMSTEED TO NEWTON
27 NOVEMBER 1694
From the original in the University Library, Cambridge.[1]
In reply to Letter 480; for answer see Letter 482

The Observatory No: 27. 1694

Sr.

I receaved yours of ye 17th in due time but the papers you sent by ye carrier with ye table of refractions came not to my hand till Sunday last in ye morneing by a freind that lives in London whom I wrote to to enquire for them & who tells mee he had been wth & sent to ye Carrier 4 or 5 times before he could have them. I have been very ill of a cold ever since I wrote to you last. & have had great paines of my head nor am I yet free of them but I hope to get to London this week on ye election day[2] not haveing been their this moneth before I shall then acquaint Mr Halley that I have a new table of refractions from you that answers my observations. but I hold it adviseable for you Not to let it goe abroad as yet for they seeme bigger ascending towards the vertex then you make them which causes me to thinke you make ye height of ye Aire less yn it ought to be Cassini in his table of refractions supposes it about $2\frac{6}{10}$ miles the Barascope (supposeing it every where of equall density) about 5 miles: tis knowne very well in what proportion its density decreases as it removes from ye earth. but admitting wth you a double refraction & the height about 5 miles I conceave the refractions will come bigger upwards I shall trie in a day or two & you shall know ye result.

I have some more observations in my hands of refractions yt have something very remarkable in them yt may be imparted to you when I have better leasure. & more perfect health

I have not time to give you an account of ye lunar Observation of Sept 30. 1690.[3] but shall erelong. onely I find I determined her place from one star onely, whereas severall were observed ye same night[:] till I write to you againe let it be marked dubious: the same I say for yt of Feb: 6. 1691.

Included I give you a Catalougue[4] of all the dayes when the moone has been observed from ye Conclusion of ye last or 3d sheet of the Synopsis till now: the hour is added to ye day that you may see Whether ye ☽ were near ye octants quadratures or opposition. Chuse what times you please out of them. I shall send you the observation adjusted as you desire to your times.

Mr Halley[5] I am told is for printing what he has to say concerneing ye Moon & tis thought we shall have it in some transaction.

You needed not to have returned ye Synopsis of ye Moon or observed refractions. those papers I tould you were onely transcripts. & not expected back. I am Sr

Your affectionate freind & humble Servant

JOHN FLAMSTEED

I desire you to let me know whether Mr Hally did not about 5 or 6 yeares agone present you with a Geometricall peice of Viviani's[6] in quarto. J.F:

To Mr Isaack Newton
   Fellow of Trinity
      Colledge in Cambridge
         these present.
            Cambridge

NOTES

(1) Add. 3979, fo. 23.

(2) 30 November, the date upon which the President of the Royal Society was elected.

(3) See Letter 480, 2nd para.

(4) This has not been found. But possibly it is U.L.C. Add. 3967, fo. 5.

(5) In 1721 Halley communicated to the Royal Society a refraction table which he declared to have been drawn up by Newton 'such as I long since received it from its Great Author; it having never yet, that I know of, been made publick'. (*Phil. Trans.* **31** (1721), 169.) Halley does not state how it was constructed, whether purely by observation, or by calculation.

(6) This may well be *De Locis Solidis Secunda Divinatio Geometrica, in quinque Libros...Autore Vincentio Vivianio.* According to David Gregory (R. S. Greg. MS. fo. 75) 'He wrote other *Five Books de Locis Solidis*'. See also *Phil. Trans.* **24** (1704–5), 1607.

482   NEWTON TO FLAMSTEED

4 DECEMBER 1694

From the original in Corpus Christi College Library, Oxford.
In reply to Letter 481; for answer see Letters 483, 484

Dbr 4 1694

Sr

The Table of refractions[1] I sent you I do not designe to publish. Tis not so accurate as it may be made & I beleive ye refractions above ye altitude of 15gr are something too litle, but if you go to examin it by the Hypothesis of refraction being made at ye top of the Atmosphere you are upon a wrong bottom. For this Table was computed upon a better foundation. However there being a certain circumstance omitted in computing it, I intend to examin it wth allowance for that circumstance When I have set it right I will send you a new copy[2] of it. Perhaps for determining the difference of ye refractions in winter & summer, it would not be amiss to observe the refractions of a fixed star in the altitude of three or four degrees[3] sometime this winter in frosty weather.

I thank you for complying wth my request of sending me the observed right ascentions & meridian altitudes of ye Moon & for ye cataloge[4] you have given
* me of your observations. *If* [5] *you please to send me those of August September &*
* *December 1692, & those of January, March, April & October 1693 & all those of the*
* *year 1694 except the three first, that is ye Observation of Jan 25 & all those that follow you will oblige me, Also in the year 1693 add ye Observations of Sept. 30 & Novemb. 2.*

I am glad your cold is going off. I hope you are pretty well recovered of it before this time. Pray this next Moon make all ye observations you can & begin your observations when ye Moon is in the first Octant if you can. For ye position of the Apoge in ye Sun's Opposition in midd winter is a case of great moment & will not return in many years. The Observations in ye full & both ye Quadratures are of greatest moment but all ye rest are useful & my method does best where ye Observations are continual. A little diligence in making frequent Observations this month & another month or two hereafter, will signify more towards setting right ye Moons Theory then ye scattered observations of many years. I am in hast[6]

Your very humble Servant

Is. NEWTON.

Cambridg Dec 4th
    1694:
For Mr John Flamsteed at
the Observatory in
Greenwich neare
London

NOTES

(1) C. 361, no. 63. See Table sent with Letter 480.

(2) See Letter 496 and the accompanying table.

(3) In a note, in Newton's hand, found in the University Library, Cambridge (Add. 3965 (17), fo. 645), are the words: 'In ye altitude of 3 gr ye refractions of ye starrs by Mr Flamsteeds & Mr Halleys observations are 14′ 0″ & a table of refractions made on a horizontal plane according to this proportion answered all observations.'

(4) See Letter 481, note (4), p. 52.

(5) The rest of this paragraph is underlined, probably by Flamsteed, who has also inserted the asterisks.

(6) Flamsteed added, after the word 'hast': '***I was ill now of ye headake not being able to calculate I sent him ye observations yt he might compute ye ☽s places from ym himselfe. My worke of ye fixed stars was interrupted also by my distemper.' He also added the date at the head of the letter.

483   FLAMSTEED TO NEWTON
6 DECEMBER 1694

From a draft in the British Museum.[1]
In reply to Letter 482; for answer see Letter 485

The Observatory December 6. 1694

Sr.

I am glad I did not impart your table of Refractions[2] to any body since I find you have better considered & thinke of altering it. Since you were not pleased to impart the foundations on which you calculated it to me[3] I have been seekeing of them and at last found a way of Answering them admitting 2 Spheres of vapors one ye usuall height about $2\frac{1}{2}$ miles the other much less with two horizontall refractions & with little labor have answered those under 5 degrees within halfe a minute those above much nearer.

By the Way I have examaned the tables of Refractions of Kepler Cassini Picart Boucher & La Hire.[4] the four first I find built all on the same foundation which supposes the refractions made in an equable Sphere of vapors about 2 miles & an halfe high. some more, others a little less. but la Hires, the last, is not built on the same principle, for his refractions from ye horizon upwards are all too big for his horizontall, & more at a distance yn neare it. If you have not his table I will send you a Copy of it in my next.[5]

Considering the uncerteinty of these refractions I contrived to find out the Inæqualitys of ye earths motion without any consideration either of them or the latitude of the Observatory but now I come to settle the distances of the fixed stars from ye visible Pole I must determine them and therefore should be glad to know the foundations of ye table which if you please to impart I shall as a suitable returne afford you what other Observations I have made of them which are no less considerable then those I have already imparted.[6]

I know very Well the æquations of the Moons motion are the highest this moneth & the next that they can be againe this 9 yeares & had therefore determined to let slip no opportunitys of observeing her. My indisposition has not hindred me but the foggs & clouds have kept her from my view since the first quadrature of the last moneth till now. the Clouds seeme to break, & if it proves frost I promise my selfe fair Weather & frequent opportunitys of determineing her place in the Meridian. which you need not doubt but will be imparted to you. But I must entreat you to be patient & beare wth me for a little time for I must visit my Cure[7] at Xtmas & prepare for my Jorney to it, which will employ me some days so yt I cannot give you the places of ye moon you desire till after ye holidays. But then you shall have them if God spare me life & health and wthout any consideration or recompense but such communications as usually are made betwixt persons conversant in ye same sorts of studies. I admire ye P.S. of your last letter[8] yt mentions another sort of recompense but I considered that you might be possest with ye Charecter which a malitious false freind[9] has spread of me, and so resolved then to take no notice of but in my next, when I was lesse moved to assure you that I never received any reward for any thing I imparted to any Ingenious person and allways scorned the thought of it. I am a freind I confess to frugality, but not for ye spareing of mony but to avoyd ostentation useless disturbance, and especially as much as I may for ye saveing of my time which is very precious with me by reason of my frequent indispositions & avocations by company &

54

visitors to which this place subjects me. I have all ways had monys, more perhaps then I desired, at my command; I bless God for it for I never took any thing of any for communicateing of my skill or paines, except of those who forced themselves upon me to devour my time & could not make me any other recompense other ways then by their pay. pray therefore lay by any prejudiciall thoughts of me, which may have crept into you by malitious suggestions & assure your selfe that without the prospect or thought of any other reward then like communications you may & shall ever freely command the paines of

      Sr

        Your affectionate & sincere freind & servant

                JOHN FLAMSTEED.

P.S. I designe to bestowe a little paines againe on ye correction of the satellits motions & should therefore be very much obliged to you if you would mind a request I made to you in one of my first letters after you went hence[10] what the physicall partes are in each of ye satellits

In my last I desired to be informed whether you had not been presented with a latin geometricall tract of Vivianis in qto:[11] You gave no Answer if you please to afford me one I shall make no ill use of it & it will much oblige me

You requested of me the places of ♄ observed these three last yeares & their differences from ye Rudolphin tables.[12] I have calculated them dureing my sickness & unakt againe present you wth them. These you may adde to the first large table of ♄s computed & observed places compared. I thinke I gave you the Observation of May ye 5th 1691 some time since. If I did not[,] acquaint me. & it shall be sent you in my next by   yours   J F

    To
Mr Isaack Newton
    Fellow of Trinity
      Colledge in Cambridge
        these present
Cambridge

NOTES

(1) This (Add. MSS. 4292, fo. 119) is probably only a draft of the letter which follows (10 December), for it bears no postmark. It is unlikely that it was ever sent. There is a copy in the hand of an amanuensis (Add. MSS. 4292, fo. 96), but it is badly damaged.

(2) Sent on 17 November 1694. See Letter 480.

(3) From now onwards there is a steady deterioration in the friendly relations which had hitherto existed between Newton and Flamsteed. However, in order to mitigate Flamsteed's annoyance, Newton sent him a fortnight later (Letter 485) the theorem on which he had computed his table, together with the reason why he had not complied earlier with Flamsteed's request. In the final paragraph of the same letter he repudiates the suggestion that he had ever had 'a mean opinion' of Flamsteed.

(4) See Letter 474, notes (12), (13), (14) and (16).

(5) There is, in Flamsteed's hand, a small table of refractions which include the figures of Bucher and La Hire (U.L.C. Add. 3967, fo. 6) but it was not sent until several months later.

(6) There are some investigations and tables by Flamsteed relative to refraction in the Royal Greenwich Observatory (vol. 33, fo. 24).

(7) At Burstow, in Surrey.

(8) See the final sentence of Newton's letter to Flamsteed (Letter 480).

(9) Halley. From now onwards Flamsteed never missed an opportunity of expressing his profound distrust of Halley.

(10) A reference to Newton's visit on 1 September 1694 (Memoranda 468).

(11) See Letter 481, note (6), p. 52. Newton wisely ignored this request. Flamsteed however referred to it again on 31 December (Letter 486) when he made it the occasion of another bitter attack on Halley. The book had been entrusted to Halley for transmission to Flamsteed, who, not having received it, suspected that it had not been given to Newton and he gave reasons for his concern (Letter 486). Newton later admitted having seen the book (Letter 489), but excused himself for not having answered Flamsteed's inquiries 'because I feared it might tend to widen the breach between you & Mr Halley wch I had rather reconcile if it were in my power'

(12) The new planetary tables (*Tabulæ Rudolphinæ*) upon which Kepler had worked for many years, were so called in honour of the Emperor Rudolph II, who after Kepler's appointment as Tycho's successor in the Observatory at Uraniborg (1602), continued to be his patron. They were published in 1627 at Ulm. See J. L. E. Dreyer, *Planetary Systems* (1906), p. 404. The table asked for is reproduced in Letter 484.

484   FLAMSTEED TO NEWTON
10 DECEMBER 1694[1]

From the original in the University Library, Cambridge.[1]
In reply to Letter 482; for answer see Letter 485

The Observatory Dec 10. 1694

Sr

I am glad I did not impart your Table of Refractions as you allowed me since I find by yours of ye 4th Instant yt you intend to review & correct it. Your concealeing the foundations on which you computed it from me has caused me to bestow some time & paines on yt subject. I have examined all the tables I have by me of Refractions to see on what fundamentalls they were built. Tychoes are onely empeiricall as are allso those of Wing[2] & Street[3] that are onely copied from him & cease when ye appearance is 20 degrees high. Kepler is ye first that Thought of a Theory, and all since have been beholden to him for his ground worke tho they doe not acknowledge it. he supposes the cœlestiall refractions made in a sphere of Vapors about $2\frac{1}{2}$ miles high. but to make them vanish or[4] become insensible in ye midway betwixt ye horizon & vertex he involves ye secondes of angles of Inclination into the Angles themselves. Cassini[5] has the same height of ye Air for his Summer refractions. but a bigger for his æquinoctiall & Winter which seemes absurd. Picart[6] & one Mr Boucher[7] our Countriman make ye height of their aggregate refractive air something less. but Monsr La Hires table[8] seemes built on other fundamentalls for his refractions from the horizon upwards are all too big for his horizontall admitting ye usuall Theory. If you have it not I shall therefore send it you.

Supposeing two different spheres of vapors one the same height with Keplers & Cassinis in summer, the other very near our earth, & two horizontall refractions the one $\frac{1}{4}$ or $\frac{1}{5}$ of the other I can answer all your table very nearly. that is those next ye horizon within halfe a minute those higher much nearer.

I have a many more observations of refractions by me that suggest something more then was to be learnt from those I imparted to you tho they bee not altogeather so accurate When ever you please to impart to me the limitations of that theory on which you computed yours as a suitable return I shall impart these to you

I have foreseen that the lunar observations of the last this & ye next moneth would be of great importance because that now all the æquations are the highest they can be till her Apoge returns to ye place it is now in againe & that will not be till this time 9 yeares.[9] & therefore I began to observe her as early as the heavens would allow me the last moneth, but since the 8th day of it the heavens have been so continually covered wth clouds that I could very seldome see her & never in the meridian. I shall omit no opportunity of observeing her till Christmas eve which will be after ye opposition but then I must visit my cure 20 miles of & it will cost me some little paines to prepare for my Jorney. So yt I shall not have time to determine the Moones Right Ascentions just[10] as you desire till after my returne beare with me till after the holidays patiently & they shall be the first work I will fall upon.

But I am displeased wth you not a little for offering to gratifie me for my paines either you know me not so well as I hoped you did or you have suffered your selfe to be possest with that Charecter which ye malice & envy of a person from whom I have deserved much better things has endeavord to fix on me & which I have disguised because I knew he used me no other ways then he has done the best men of ye Ancients nay our Savior his Apostles & the Scriptures permit me to give you a truer Charecter of my selfe & which you shall allways find me answer. I dare boldly say I was never tempted with Covetousness God allwayes blest me wth more monys then I know well how to dispose of & those that know me even those who calumniate me know how free I have been of it on good occasions. besides I have allways made my Instrumts at my owne no small Charge my Arch haveing cost me above 120 *lib*, & mainteind my own assistant to bring my worke the sooner to perfection & avoyd the trouble of long attendance to sollicit allowances which I might have got had I not valued my time more then my Monys. My freinds are allwayes heartily welcome to me whenever they please to oblige me with their good Company but I am I profess it a freind to frugality not for the avoyding of expence so much as the preserveing my often endangered health, to avoyd ostentation & pride & for spareing my time which considering my small assistance you must needs be sensible is precious wth me. I have taught more people gratis then ever I took monys for I sell not my paines or skill to any but those who can make me no other recompense for the time they devour but by their pay. All the return I can allow or ever expected from such persons wth whom I corresponded is onely to have ye result of their Studies imparted as freely as I afford them the effect of mine or my paines. I have told you my disposition plainely & if hereafter you offer me any other then this just reward. I shall thinke as meanely of you as I feare you have been persuaded to thinke of mee by false & malitious suggestions.

My distemper has been very sharp. my heareing on one side is very much prejudiced by it I have used few remedies. finding those I employed in-effectuall. I hope that time through Gods blessing will recover it. tho I have still paines of my head but not so great as to hinder me from pursueing my business if God sends us fair weather of which now ye ☿ in ye baroscope begins to fall I have some hopes.

You requested of me when last here to give you the errors of the Rudolphine Numbers from ye heavens in the planet ♄. I had not then calculated his places from ye observations but I have since dureing my sickness & compared them with ye tables. the result I here give you & expect your thoughts of them at your leasure I have found ye collation of them in a figure which I could not meet wth when you were here

TABLE

Ann. Mens.	die h ′	♄ Anom Med s ° ′ ″	♄s Logm. Dist. Curtatæ	☉ s ° ′ ″	♄ Loca per Tab s ° ′ ″	♄ Loca per obser s	χæ ′ ″
1691 Maij	5.12.10	10.26.30.01	5998421	♉ 25.09.07	♏ 27.23.38	♏ 27.17.44	−5.54
1692 Maij	16.12.10	11.09.07.51	6000645	♊ 6.26.09	♐ 8.49.17	♐ 8.43.30	−5.47
	19.11.57½	11.09.13.53	6000658	♊ 9.17.04	♐ 8.35.34	♐ 8.30.15	−5.19
1693 Junij	2.11.48	11.21.53.42	6001887	♊ 22.26.43	♐ 19.46.47	♐ 19.41.35	−5.02
1694 Junij	15.11.43	0.04.32.31	6002109	♋ 4.34.51	♑ 1.00.09	♑ 0.56.10	−4.09

I designe ere long a review & correction of my Tables of ♃s Satellits it will take up but little time you will much oblige me if you please to remember a request I made to you in one of my first letters after you left London last & let me know how much you state the acceleration of ye meane motion (or ye greatest quantity of the Physicall parts(11) as I call them in ye Moone) of each satellit whilest ♃ moves from his Aphelon to ye Quartile of it.(12) I told you in that letter yt Monsr Cassinis new Satellit Tables were printed & yt I had them. they will be reprinted I am told ere long here in the transactions. I hope I may represent the Eclipses & motions not onely of the first but 3d & 4th Satellits much better then ye French have done & I despaire not of ye 2d.

I desired you in my last to let me know if you had not been presented some yeares agone with a geometricall tract of Vivianis in 4to Latin. you have given me no Answer pray be free wth me & let me have one it will much oblige Sr

Your sincerely affectionate freind &

humble Servant

JOHN FLAMSTEED

Bullialdus(13) is lately dead at Paris. After ye holidays I intend God willing to set those to ye Restitution of ye fixed Stars, & in order to it first to review my restitution of ye Earths motion. I hope to have your thoughts about refrac-

tions before for yn they must be applyed in determineing the stars true
distances from ye visible Pole. twill be a weighty worke but I hope not long.
J F:

To Mr Isaack Newton
   Fellow of Trinity Collidge
     in Cambridge there
       these present
   Cambridge

NOTES

(1) This letter (Add. 3979, fo. 24) is very similar to Letter 483, with which it should be
compared.

(2) Vincent Wing (1619–68) wrote *Astronomia Britannica* (London, 1652 and 1669) and
Harmonicon Cœleste (London, 1651). In the former he explained the sexagesimal notation of
integers. In the latter he introduced the colon (:) to denote ratio. Newton, in the *De Analysi*
and in the *De Quadratura*, followed Wing although Wallis and Barrow both showed a preference for Oughtred's dot notation (·). See vol. II, p. 395, note (21).

(3) See Letter 474, note (7), p. 32.

(4) The phrase 'or become insensible' is marked by Flamsteed for insertion from the
margin.

(5) See Letter 474, note (12), p. 33.

(6) See Letter 474, note (16), p. 33.

(7) See Letter 474, note (14), p. 33.

(8) See Letter 474, note (13), p. 33.

(9) Kepler knew this. It is the regular advance of 3° per lunar month. See Introductory
Note on Lunar Theory.

(10) The words, which are superimposed upon one another, are difficult to decipher.

(11) See Letter 474, the penultimate paragraph.

(12) See Newton's reply, Letter 485 (20 December 1694).

(13) Bullialdus (Ismael Boulliau, 1605–94), French astronomer; wrote *Astronomia Philolaica*
(Paris, 1645), perhaps the most important work on planetary systems between that of Kepler
and the *Principia*. Newton declared that Bullialdus, like Borelli, had arrived at the laws of
gravitation between two planets in which the force was proportional inversely to the square of
their distance apart (see vol. II, p. 440, note (12)). 'At Paris Bulialdus hath a booke in the
Presse against Dr Wallis his *Arithmetica Infinitorum*' (Collins to James Gregory, 29 September
1670. *G.M.V.* p. 106).

485   NEWTON TO FLAMSTEED

20 DECEMBER 1694

From the original in Corpus Christi College Library, Oxford.[1]
In reply to Letter 484; for answer see Letter 486

Cambridge Decemb 20th 1694

Sr

The foundation of the Table of refractions I concealed not as a secret but omitted through ye hast I was in when I wrote my last letter. But since you desire it I will now set it down.

Let *AKL* represent the globe of the earth, & suppose this globe is covered wth an Atmosphere of Air whose density decreases uniformly from ye earth upwards to the top wch is here represented by the circle *MON*.[2] And let *SO* be a ray of light falling on ye top of this Atmosphere at *O* & in its passage from thence through ye Atmosphere to the spectator at *A*, continually refracted & bent in ye curve line *OBA*. From any point of this curve line *B* to ye center of the earth draw the right line *BC* cutting the surface of the earth in *D* & take *CF*

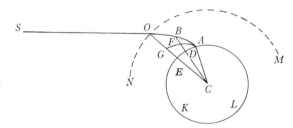

a mean proportional between *CB* & *CD* & let *AFG* be ye Locus of the point *F*, that is the curve line in wch ye point *F* will be allways found: & if this curve line *AFG* cut the right line *OC* in *G*; the whole refraction of ye ray in passing from *O* to *A* will be proportional to the area *AFGC* & the refractions in passing through any part of that line *OB* or *BA* will be proportional to the corresponding part of the area *GFCG* or *FACF*. This Theorem is Geometrically demonstrable[3] but the demonstration is too intricate to be set down in a Letter.

Now as my Table of refractions computed from this Theoreme agrees much better with your Observations then the vulgar ones, so I beleive you will allow ye Theorem it self to be a better foundation then the vulgar ones of a single refraction on ye top of an uniform Atmosphere.

What you desired me about ye equations of ye mean motions of Jupiters Satellites I did not understand in your first letter, & thank you for putting me in mind of it again in your last, because it tends not only to perfect ye Theory of Jupiters Satellites, but also to confirm ye Theory of Gravity.[4] The Rule for determining those Equations is this.

Let *A* be the time of ye earths revolution (wch is a yeare, or $365\frac{1}{4}$ days) *B* the time of Jupiters revolution (wch is 12 years, or more exactly 4320 days &

61

14½ hours) C the time of the Moons revolution (wch is 27 days 7 hours 43 minutes) D the time of ye revolution of a satellite, P the greatest equation of the Earths Orbe (wch is 1°.56′.20″) & Q ye greatest equation of Jupiters Orbe (wch is about 5°.32′.0″:) & the greatest equation of the mean motion of a Satellite shall be to the greatest equation of the mean motion of our Moon as $A \times D$ quad $\times Q$ cub to $B \times C$ quad $\times P$ cub. When if the greatest equation of our Moons mean motion be assumed 12′.0″, the greatest equation of the mean motion of ye outmost Satellite (whose periodical time is 16 days 18 hours 5 minutes) will be 8′.52″. And this being found its easy to find the equations of the mean motions of ye rest of the Satellites. For they are as ye squares of the times of their revolutions. Whence ye greatest equation of ye second satellite will be found 1′.37″. That of the third 24″ & that of ye 4th or innermost 5″.

These are minutes & seconds of degrees in ye Orbits of ye satellites: wch being converted into time give the greatest equations of ye times of the Eclipses of the satellits to be added & subducted when ♃ is in the Quartile of his Orbe: That is, to be added when ♃ is [in] the Quartile ascending from his Perihelium to his Aphelium & subducted when he [is] in ye opposite quartile descending from his Aphelium to his Perihelium. And this greatest equation in the outmost Satellite will be found 9′ 54″ of time, wch must be very sensible. But in the next satellite it will be only 46″: In the next between 5″ & 6″ & in the innermost not much more then half a second. These computations I have done but once, knowing that you'l examin them.

* *I intend to determin ye Orb of* ♄ *within a few days & then Ile' send you ye result.*[5] And before you return to ye work of the fixt stars I hope to have the Table of Refractions ready. *But pray let me have your Lunar Observations as soon as you can*[6] that I may be about ye Moon whilst you are about other things.

What you say about my having a mean opinion of you is a great mistake. I have defended you when there has been occasion but never gave way to any insinuations against you. And what I wrote to you proceeded only from hence that you seemed to suspect me of an ungrateful reservedness, wch made me begin to be uneasy. But if you please to let all this pass & concur with me in promoting Astronomy I'le concur wth you being

Your faithful friend

to serve you

Is. NEWTON.

For Mr John Flamsteed at the
Observatory in Greenwich
neare London

NOTES

(1) C. 361, no 65. Flamsteed had repeatedly asked for the foundations on which the table of refractions, sent by Newton on 17 November 1694, was built.

(2) Newton wrote *MCN*.

(3) The determination of the areas is based on the lemma: *Invenire lineam curvam generis Parabolici, quæ per data quotcunque puncta transibit.* See Letter 487, note (4), p. 69.

(4) See Letter 484. The problem raised by Flamsteed, and here solved by Newton, confirms the theory of gravity by showing that the same principle operates for a satellite of Jupiter as for the satellite of the Earth, namely, the Moon, when the adjustment is made for the eccentricities of the orbits of the Earth and of Jupiter. But it is not apparent how he arrives at the result which expresses the ratio of the greatest equation of the mean motion of a satellite to the greatest equation of the mean motion of our Moon. Using Newton's figures this ratio is 0·7375, so that if the greatest equation of the Moon's mean motion be assumed to be 12′.0″ that of the outermost satellite will be 8′·850, which agrees closely with Newton's 8′.52″.

At this period, only four of the satellites of Jupiter (*the Medicean Planets*) had been discovered. The 'outmost satellite' is Callisto. Its period is 16 d. 16 h. 32 m. The periods of the other three satellites, from the innermost, are 1 d. 18 h. 28 m., 3 d. 13 h. 14 m., 7 d. 3 h. 43 m. If Newton's formula is correct, the equations of the mean motions of these would clearly be 'as ye squares of the times of their revolutions'.

(5) This passage was underlined, the words 'Ile' send you ye result' being doubly so, probably by Flamsteed, who has added in the margin the words: 't'was never sent. q[uer]y if done'.

(6) Newton's impatience was due to the fact that in order to test his theory he had need of the most accurate observations, and Greenwich alone could supply these.

486   FLAMSTEED TO NEWTON
31 DECEMBER 1694
From the original in the University Library, Cambridge.[1]
In reply to Letter 485; for answer see Letter 487

The Observatory Dec ☽ 31. 1694

Sr

Your determination of ye greatest quantity of ye physicall parts in each of ♃s satellits is ye more acceptable to me in that it shews me I need not feare any effect of ye Gravitation of ♄ & ♃ on each other. in changing ye Revolutions of the Satellits of ye last since ye great alteration of ♃s distance from ye Sun has so little. I give you thanks for cleareing this point to me

I esteeme what you say about cœlestiall refractions to be the onely proper ground for a table of them. tis altogeather rationall & agrees with ye nature of our Aire which is no equable medium as it has hitherto been proposed[2] but such as you assume it Since your Theoreme & ye proof of it is too prolix for a letter I shall rest content till you have made it shorter or can find leasure for your servant to transcribe it for me. but I would gladly know how you compute ye quantity of those Areas yt measure these refractions to every degree of elevation.[3]

I see yt in elevated places they will be less yn in those yt stand lower on plaines & greater when ye ☿ in ye Torricellian tube is high then when tis low. I grant allso yt ye aire next our earth is warmer yn that of ye upper regions which may be of a nearly constant coolenesse that it [is] replenished with great variety of vapors breathd & exhaled from our earth. & on ye faith of your experiments that the Winter to ye Summer expansions of it may be as 8 to 9. & the refractions reciprocally proportionall as you make ym in your table.

I shall now discharge my promise & afford you some further observed refractions yt were taken by ye help of my pendulum clocks & a quadrant of onely 40 inches radius but accurately divided by my selfe. Tho I esteeme not these so exact as those that were deduced from ye distances of ye ⊙ & ♀ measured wth ye sextant of 7 foot Radius yet I thinke them preferable to any the french have made for I [am] satisfied their instruments were not so well accommodated for this purpose as mine & much beyond Tychoes with plaine sights.

I have a great Number of the like observations by me: but only these have been examined. by which

1[.] you see yt on ye 12 of June when at 89°.40′ distance from ye vertex ye refraction was in ye morning 32′; in ye evening at ye same distance it was but 26′

2[.] That ye morneing refractions generally exceed ye evening by reason of the greater quantity of vapors breathd from ye earth. in ye night & digested by ye heat of ye day. this remarke was printed in *Trans* but you were not told whence it was derived as also

3[.] And yet nevertheless when ye ⊙ is 3 deg high both morneing & evening

1678. Junij	die ⊙s limb Infr a vert ° ′	12. mane Ref. ′	17. mane Ref ′	21. mane Ref ′	12 vesp Ref ′	20 vesp Ref ′	26 vesp Ref ′
	89.50	34½	36				
	40	32	32		26		30½
	30	29	28	27⅓	25⅓		27
	20	28	27				26
	10	26	25	23	22⅔		24½
	89.00	23½		20⅓			23
	88.50	23		17½			
	40	20					
	30	19				20⅔	
	20						
	10	18					
	88.00	18	:: 17½	16½			16¼
	87	14⅓	:: 12⅔		14	14½	14
	86	11½	9		10⅓	11½	10⅔
	85	9½	8		9	9½	8⅔
	84	8	6½		6	8	7
	83	6⅔	5⅔		5⅓	7	6½
	82				5	6	5¼
	81	6	5¼		:: 4¼		5
	80	5			3½		
	79	4¼			3¾		4

the refraction is nearest 14 minits

64

4. It seemes very probable to me that the horizontall evening refraction was more yn 30 minutes & that when ever it is more as it was yt morneing and on ye 17th day it was occasioned by vapors breathd & exhaled in ye night from ye earth. yt lay very low & consequently had little effect when ye ☉ was 3 degrees high both morneing & afternoon. & therefore ye pure horizontall refraction ought to be stated no bigger but at 3 degrees high about 14 minutes which ye contracted distances of ye ☉ from ♀ also allow very near

Perhaps you will fear this may make the refraction in the upper altitudes too little because I told you formerly they ought as I then thought to be encreased let it be as it will. So wee have ye truth of what your Theory & Theorems give. Some other way may be found to salve these small differences [which] I thought might have been made out by greater refractions nearer ye Vertex. perhaps this ordering of them as I propose may help as much as Is requisite

From ye 16 day of this moneth to ye 27 we have had continuall Misty & Cloudy weather & not much better before so yt I could only observe the ☽ on ye 11th 13 & 16 dayes. & since on ye 28. ♄ & this morneing from ye 2 first I have derived her places allready. I shall give you as much of her as you require

Dec. 11 at 4h.31'.10" T.A. ye Right ascention of her Center was 338°.38'.50"
                            its apparent distance from ye visible Pole. 94 .18 .00
                            this last correct onely by ye refractions

13 at 6h.03'.06" TA ye R.A. of ye ☽s center againe   3 .52 .10
                            dist from ye pole correct onely by refraction 82 .37 .50"[4]

The others are not yet examined: the clouds not haveing permitted me to examine how many beates ye pendulum made in a sydereall day. before I went into ye Country I have found since. & shall determine what you desire in a few days

You may set your servant to work with these data. I feare he will not prove either quick or expert without a great deale of excercise & therefore when ever you desire it I shall give you the places of ye moon calculated hence by my selfe either from ye observation or Tables as you intimate.

Dec 11. & ye 13 ye equation of ye Orbit was affirmative & in the latter near ye greatest it could be but your parallactick æquation being Negative diminished it. The cœlestiall places of ye ☽ are forwarder yn my Tables by about $3\frac{1}{2}$ minutes in both so ye greatest winter æquations exceed those of my Tables by *$3\frac{1}{2}$ min + your parallactick æquation*[5] which is more yn I expected

I must beg your pardon for haveing urged you twice about Vivianis book. I shall tell you ye occasion & give you no further trouble. Mr Rook[6] being in Italy 6 yeares agone receaved one of them directed to me by the authors owne hand which he sent to EH.[7] with other things. who I am told presented it to you & himselfe denies not that he sent it you. Now I am not concerned for the loss of this book at all, if you had one from him pray keep it [(]either as his gift or mine) but because I have great reason to suspect a book of much greater valew directed to me has been disposed of for advantage by a freind & acquaintance of his this last summer. & if the first had been brought to light the latter might have been made evident. but I desire to concerne you no further with it. & therefore shall move you no more nor expect any answer on this particular. being ever desirous to make my freinds as easy as I can I am Sr

Your most affectionate freind &

humble Servant

JOHN FLAMSTEED

I am goeing to spend two or 3 days in London. wthin a week after my return you may expect to hear from me again. J: F:

To
Mr Isaack Newton:
Fellow of Trinity
Colledge in Cambridge
these there present
Cambridge.

NOTES

(1) Add. 3979, fo. 25.

(2) By G. D. Cassini. See note on Atmospheric Refraction, p. 96.

(3) See Letter 487, note (4), p. 69.

(4) Newton has inserted the figures '83.18.40' in the margin.

(5) The excess here stated ($3\frac{1}{2}$ min. etc.) is underlined by Flamsteed.

(6) Possibly a son of Lawrence Rooke (1622–62), astronomer and geometer, who became Gresham Professor of Geometry in 1652 (see J. F. Scott, *The Mathematical Work of John Wallis* (1938), p. 210). The person referred to in the above letter is described by Newton as 'an Oxford Gentleman, a student in Mathematicks' (Letter 489). Hooke mentions him twice in the *Diary*: '29 November 1672; With Rook at Tower Hill, signed his bill for eight pounds 10sh' (p. 15); '15 April 1680, At Jonathans, Hally and Rook' (p. 443).

(7) Halley. See Letter 483, note (9), p. 56.

487   NEWTON TO FLAMSTEED

15 JANUARY 1694/5

From the original in Corpus Christi College Library, Oxford.[1]
In reply to Letter 486; for answer see Letter 488

Jan 19 1695/4

Sr

I presume you are by this time returned to Greenwich. You need not fear that ♄ *can sensibly disturbe the motions of ye satellites*[2] of ♃. The Theoreme of refractions I sent you[3] has this fault that it makes ye refracting power of ye Atmosphere as great at ye top as at ye bottom. This has put me upon thinking on a new Theoreme & I think I have found one but intend to consider it a little further. The areas in that Theoreme I sent you are to be determined by the 5t Lemma of my Third Book of *Principia Math.*[4] But ye calculation is intricate.

I thank you for your Observations about ye morning & evening refractions. The reason why the former are greater in summer then the latter I take to be nothing but ye different heat & coldness of ye air. For ye Air cools all night & is coldest at sunrise & heats all day & is hottest about 12 hours after sunrise. The cold condenses ye Air & makes its refraction greatest at sunrise & ye heat rarefies it & thereby diminishes its refractive power in the evening.[5]

I thank you for ye two Observations you sent me & since you have calculated ye Moons places in these & ye other three Observations of ye last Month, you will oblige me by a synopsis of ye calculations. But for ye rest of your Observations, I desire you would leave ye calculations to me & only send me your naked Observations. For otherwise I cannot correct ye errors wch sometimes happen in ye calculated places, nor can I go over ye calculations again as perhaps I may do when I have carried the Theory a little further & know better how to allow for ye refraction & parallax. For you make ye horizontal parallax too great by a minute or above, but how to rectify it I do not yet know exactly enough. I thank you for your offering to be at ye pains of these calculations, but I will give you no other trouble about them then to send me ye Moons right ascentions & meridian altitudes according to your observations without any allowance for ye refraction or parallax.

In trying to compute ye mean motion of ye Moon from ye tempus apparens in some of your observations. I find that ye mean motion gathered by my computations differs sometime from that in your Synopses 5″ or 6″ or above. Which makes me suspect that in determining ye tempus apparens, your servant followed some tables wch are not sufficiently exact: such as are those wch

Tompion[6] uses for his watches. For those err sometimes 6″ or 8″. For avoyding this inconvenience I desire you would instead of the Tempus apparens, use sidereal hours counted from ye appulse of the Equinoctial point to ye meridian, or rather from that of the Dogg starr or of some other notable star. If you use canicular hours counted from ye appulse of the Dogg-starr to ye meridian you may note them thus [1695 Jan 7. 5*hc*. 44′.15″.][7] putting hc for hora caniculari. By this means ye equation of time will be wholy avoyded wch is troublesome to calculate & makes ye work liable to errors. For where ye equation of time must be considered it must be twice, computed; first to get ye tempus apparens & then to get ye tempus verum, wch is to go a great way about. For in ye way wch I propose, ye time in canicular hours is ye tempus verum & its found wthout any other labour then by seeking ye right ascention of ye Dogg-star from ye meridian & turning it into time. But that there may be no mistake in ye day to wch ye canicular hours belong I would count them from ye appulse of that starr to ye meridian next after the midnight which precedes ye day, that is, wch comes between ye midnight wch begins & ye midnight wch ends ye vulgar day. So yt for instance, the canicular day wch begins at any time between midnight & midnight on Jan 7 shall be called Jan 7: & so of all other days.

I should say something to you about your book:[8] but ye Post is going & I must reserve it to ye next

Your most sincere friend &

humble Servant

Is. NEWTON.

Jan 15 1694/5:

[Notes, written on the letter, by Flamsteed]

Rec ♃s 17 at dinner.

1. I grant the heat & cold of ye Aire is the cheife cause of ye change of its refractive power but not all for other days were as hot as that yet had less refractions

The Sun rose then over ye Thames & adjacent Marshes. set over a drie hill on ye west end of London. it was a misty morneing and a great fogg over ye Meadows

I should be glad to see this business of refractions finished. it will be of use to me

I have not time to send the Synopsis now may doe it hereafter but would gladly see what places you have derived from ye given AR [?Right Ascension] first. Shall give mine after

The Semdiameters of ye Luminarys in my doctrine of ye Sphere are just being setled by my owne observations excepting what error may be caused in ye ☽s by ye fault or defects of ye Theory

But the horizontall parallaxes were setled by ye observations of ye Lunar Eclipse of Octob 18. 1678 in whose duration the French & I agree very nearly. The[y] ought to be setled by Eclipses observed when the Sun is in ye Meane distance. This I guess a plaine Reason why in all ye Moneths from Feb: to october my parallaxes may be too big.[9] but from October to Feb. perhaps they will be found too little some small matter.

Tompions a true table of æquations but made for a particular year perhaps fitts not ye present. Those in Parkers Almanacks[10] true supposeing ye Old Solar æquations just. but now I have changed them & translated the Aphelion they may be 6. 8. or 10″ erroneous. being made for 4 years may serve for an Age with a small table I have made to correct them.

The old way of Numbring days not to be left. I have a Clock for yt purpose.

NOTES

(1) C. 361, no. 69. Newton inserted the date, 'Jan. 15 1694/5', at the foot of the letter. At the head of the letter Flamsteed has inserted 'Jan 19 1695/4' and in a note above Newton's signature are the words, in Flamsteed's hand: 'Answer'd Jan 19 1695 ♄. but no observations imparted.' This is followed by his notes which are reproduced above. Flamsteed appears to have been unsure of his dates for his reply (Letter 488) is dated 'Jan 18. 1695/4'.

(2) See the first paragraph of Letter 486, and Letter 495. This passage in italics was underlined, presumably by Flamsteed.

(3) See the Table of Refractions sent with Letter 480.

(4) This famous lemma, *Invenire lineam curvam generis Parabolici, quæ per data quotcunque puncta transibit* (To find a curved line of the parabolic sort which shall pass through a number of given points), follows Proposition 40, Book III; *Cometas in Sectionibus conicis umbilicos in centro Solis habentibus moveri, & radiis ad Solem ductis areas temporibus proportionales describere* (The comets move in conic sections having their foci in the centre of the Sun, and by radii drawn to the Sun describe areas proportional to the time). In the Corollary Newton goes on to state (translated): 'Hence the areas of all curves can be ascertained approximately. For if several points are found of any curve whose area is sought, and a parabola is understood to be drawn through the same points, the area of this parabola shall be approximately the same as the area of that curve to be squared. But the parabola can always be squared geometrically by well-known methods.'

A complete solution of the problem is given in the lemma without, however, any indication of the method by which the result was obtained. This solution is recognized as a fundamental proposition of the calculus of finite differences. For an exhaustive treatment of Newton's methods, see *Isaac Newton, 1642–1727* (edited by W. G. Greenstreet, 1927), 'Newton and Interpolation', by Duncan C. Fraser. See also vol. II, p. 161.

(5) It was generally accepted at this time that refracting power ($\mu - 1$, μ being the refractive index) was proportional to the density. Jean Picard had already connected diurnal and seasonal refraction with atmospheric temperature. See Letter 496, Note on Atmospheric Refraction, p. 96.

(6) Thomas Tompion (1638–1713), a renowned maker of clocks and mechanical instruments. Sir Jonas Moore commissioned him to make two clocks for Flamsteed's use in the new Observatory in 1675. In his search for still greater accuracy, Tompion designed the clocks with approximately 14-foot pendulums, beating two seconds, with periods of a year. These two clocks were installed in the Octagon Room of the Old Royal Observatory. The originals are, one in the British Museum and the other in the possession of the Earl of Leicester. See H. Alan Lloyd, *Some Outstanding Clocks* (1958), pp. 76, 78. Tompion also made a sextant, fitted, on the advice of Hooke, with telescopic sights for Flamsteed's use in the Observatory. See also *Some Account of the Worshipful Company of Clockmakers of the City of London* (1881), p. 174, by S. E. Atkins and W. H. Overall.

(7) The square brackets are Newton's.

(8) Probably Viviani's book. See Letters 481, 484, 486, 489, and 493.

(9) A note, in Newton's hand, in the University Library, Cambridge (Add. 3965, 17, fo. 645), reads: 'The meane distance of ye Moon wch Mr Flamsteed makes $59\frac{1}{4}$ semidiameters of ye Earth must be 60 or $6\frac{1}{2}$ [*sic*] semidiameters & ye horizontal parallax at that distance must be decreased from 58′ 00″ to 57′.00″ or 56′.45″ or thereabouts as I find by his observed latitude of ye Moon.'

(10) John Parker (1651–1743), astrologer and almanac maker; published Ephemerides which were considered the best of their day. Parker, like many other almanac makers, included mathematics in his publications, e.g. Flamsteed's Equation of Time, tide tables, lunar tables, etc. See E. G. R. Taylor, *Mathematical Practitioners*, p. 265.

488   FLAMSTEED TO NEWTON

18 JANUARY 1694/5

From the original in the University Library, Cambridge.[1]
In reply to Letter 487; for answer see Letter 489

The Observatory Jan 18. 1695/4

Sr

I hope you will bring ye business of refractions to a speedy and happy conclusion. & that your New Theorem will be more serviceable to you then ye old. a good Table of them with the grounds of it will be exceeding acceptable because very usefull to me I forbear employing my thoughts on that subject till I see what success you have not doubting but that your paines will spare me mine in this particular for which I have supplied you plentifully with my best Materialls.

I grant that the change of heat & cold in ye aire next our Earth may be in great Measure ye cause of those alterations in the morneing and evening refractions I have found by my experimts, but Not all. Sr. Jonas Moore[2] has often told me that whilest he lived in the fens he has often seene the beasts feeding in the meads raised to his sight very much by ye vapors or fog that lay betwixt him & them when they were neare him so as their bodys appeared raised by ye vapors, their backs above ym. I have heard ye late king Charles & our Sea Captaines talking togeather about the Sea Air, & relateing how standing upon Dover beach at highe water they saw the Steeples of Calis very plaine. but whilest they stood as the Water sank these objects sank & at last disappeared. you have a like story in the *Transactions* concerneing the Cattle seen on ye Isle of Doggs from ye Coast of Greenwichside at high water but not when the tide is sunke. tho you are not told from whom it proceeded. you may assure yourselfe tis true all these Instances hint that vapors filling our Aire encrease its refractive power.[3] & I took particular notice that the 11th of June 1678 in ye Morneing when I found the refractions biggest a great mist lay both on ye River & adjacent Marshes over which the Sun rose, & I doe not remember yt it was an hotter day then any others when ye morneing refractions were less ye evening bigger.

I have not time to copy my calculations at present I shall send you them some time hereafter; in the meane time if you send me ye places deduced from ye Right Ascentions & Polar distances of ye Moon I gave you. I shall returne you mine.

I doe not wonder yt you find ye æquation of Days a little different from what I have [sent] you. ye æquations of ye Earths orbit turned into time make one parte of it, & since I have encreased these 80″ it will Make them 5 or 6 seconds of time bigger yn formerly I used; the translation of ye Aphelion 35′ minutes forwarder may cause in some places a change of 4 or 5″ more: & therefore you are too quick on Mr Tompions table[4] which I gave him & is one of those 4. I used till now yt I have changed the elements of it. for these as ye Ephemerides of the Suns places return to be the same very nearly after 4 yeares & a small reduction which I have made a table of will make my Quadrienniall æquations of time[5] serve an Age. If you used the table of æquations that was not proper for ye yeare it might cause a greater error in the Meane Motions then yt you mention, which is but a trifle. The Tables in Parkers allmanack[6] are those I have hitherto used.

You advise [me] to use ye Sydereall hours in the room of those usuall & Vulgar. You may remember I have a New Contrived Motion for yt purpose in my Quadrant house that Numbers the degrees minutes & sixths of ye equator, provided this two yeares for yt purpose. but to vary from ye Common way of reckning onely to shun ye account of ye æquation of days is needless when it is so easily found and allowed for. I intend to try how it will succeed when ye weather grows a little warmer yt I may be sure ye frost shall not by thickning the oyle spoyle ye goeing of my Movement but I shall Count not from ye Dog Star but ye Vernall equinox because it will be most convenient on severall accounts yt will occur to you on ye first thought. but still I shall not omit ye old usuall Way of Numbring the Solar hours with it.

You say I make the Moons horizontal parallaxes too big. I beleive it the Suns & her Semidiameters were setled not as Mr Horrox[7] gave them but made agreeable to my owne observations so yt there can be no error in them save what is caused by ye faults of ye lunar Eclipse of October ye 18. 1678. wherein the French agreed admirably with me in determineing ye Duration.[8] The Moone passed near ye Center of ye Shadow but it was a Winter Moone & Shee not two signes from her Perige. hence you will easily guess ye cause & remidie of this fault but I must tell you further that I find my Tables agree nearly to the Durations of lunar totall Eclipses tho in the small ones both of her & the Sunn I have found very remarkeable errors of which I may say more to you hereafter & therefore you will be cautious how you remove her much further of then I have placed her except your Theorys draw her back to answer the durations of these eclipses

I am now falleing upon my great worke[9] as fast as I can but I shall not for it omit any thing that requires to be observed as usually. it will cost me I feare more time then I suspected to prepare Necessary tables for shortning the Worke of Calculation. but when they are made I shall shun above two thirds of my usuall labor. I want nothing but your table of refractions which I can stay for till you have finished pardon a hasty Conclusion for the post will be gone If I defer to add more yn that I am allways

Your affectionate Freind & humble Servant

JOHN FLAMSTEED

To
Mr Isaack Newton
Fellow of Trinity
Colledge in Cambridge
    these present
  Cambridg.

NOTES

(1) Add. 3979, fo. 27.

(2) Sir Jonas Moore (1617–79), mathematician and civil engineer. On Flamsteed's appointment as Astronomer Royal (March 1675), Moore presented him with an iron sextant of 7-foot radius, two great clocks, and a telescope object glass. See vol. II, p. 427, note (15); also MacPike, *Hevelius, Flamsteed, Halley*, pp. 21–2.

(3) See Letter 487, note (5), p. 70.

(4) See Letter 487, note (6), p. 70.

(5) The non-uniformity of apparent solar time causes a varying difference between the apparent and mean time. The equation of time is a correction which must be applied to mean time to give apparent (i.e. sun-dial) time. See Glossary.

(6) See Letter 487, note (10), p. 70.

(7) See *Opera Posthuma* (1673) of Horrocks, chapter v, especially p. 104.

(8) For a detailed account of Cassini's observations of the eclipse see *Phil. Trans.* **12** (1677/8), 1015: 'Monsieur Cassini's Observation of the Lunar Eclips on the 29 Octob. 1678.' See also Letter 501.

(9) Flamsteed did not live to see the publication of his great work, the *Historia Cœlestis Britannica*. It was published in 1725, having been edited by Joseph Crosthwait and Abraham Sharp, five years after Flamsteed's death. A copy unauthorized by Flamsteed appeared in 1712 under the editorship of Halley.

489   NEWTON TO FLAMSTEED
26 JANUARY 1694/5
From the original in Corpus Christi College Library, Oxford.[1]
In reply to Letter 488; for answer see Letters 490 and 493

Jan 26 94/5

Sr

That wch I would have said in my last about Viviani's book[2] was only this, That about 3 or 4 months before Dr Gregory was made Astronomy Professor in Oxford,[3] an Oxford Gentleman a student in Mathematicks (I think his name was Rook)[4] called on me in his way from London & shewed me a new book published either by Viviani or some other Italian but I think by Viviani. He offered to leave it with me to peruse. Whereupon I turned over the leaves & then returned it to him again & he took it away with him, I think to Oxford & I saw it no more. I forbare to answer your first enquiries about it because I feared it might tend to widen the breach between you & Mr Halley wch I had rather reconcile if it were in my power. And now I hope that what I have told you will not be made use of to that purpose least it should also do me an injury. For your offering to present me wth Viviani's book I thank you as much as if it had been left with me.

The equation of time I derived not from your new Theory of the Sun but from the old one. For having by me two of Mr Tompions Tables of ye Equation of time: I examined them by your Tables printed in Sr Jonas Moors works[5] & found some difference. And at the same time I tried whether the equation of time you used in computing the mean motion of ye Moon in your Synopses would agree wth ye rest & found the like differences. Yet these differences were of no great moment being seldome above 4″ or 5″. I have got your Table in Parkers Almanach for this year.

Whether ye sidereal hours be counted from the equinoxial point or from ye Dog star, is of no great moment. The computation is readier from ye equinoxial point, but the star is a point better defined & immoveable, & its proper to refer the motions & positions of things moveable to ye positions of things immoveable.

The diminution of the horizontal parallax by about a minute I seemed to collect from your observations in your first two synopses. For those in your third I have not yet considered. But if a minute be too much, what think you of half a minute? Would that, or the parallax in your printed Theory agree best with total eclipses of Sun & Moon?

I agree with you that the dense vapors wch always stagnate upon the surface of the Sea & and often upon Fenny places cause a strong refraction. And its probable that those wch rise to a greater height may increase the refraction of the horizontal Sun. But can you tell whether the refraction of the Sea-vapours or Fen-vapours be greater in hot weather or in cold? at morning or at night?

To make a new Table of refractions has taken up almost all my time ever since the holy-days. And I have hitherto lost my pains in fruitless calculations by reason of the difficulty of the work. For considering that such a Table is the foundation of Astronomy & very necessary for your great work & that you have taken so great pains in providing materials for it: I was desirous to complete it that I might have something to present you with for the pains that you have taken for me about your Observations. Yet I have not wholy lost my labour, for I have found a new Theoreme wch makes ye calculation very easy, & wch I must content my self with if I can think of nothing better. At present I am a little indisposed but hope in a few days to be well enough again to finish this business. I hope you have your health perfectly.

I think Venus was about 36 degrees high at sunset Feb 22 & Apr. 11 & about 32 or 30 degrees high at sunset on the following days of April. If I had her altitude on every day at sun set more exactly, I could make the Table of refractions more accurate.

The places of the Moon from your two Observations I have not yet computed. For I thought it superfluous to do what you had done to my hands & desired a copy of your computations only to save my self that labour. But since I perceive you have a mind to see whether we can compute exactly, if you please to send me the Latitude of Greenwich I'le send you what you desire. The sun's greatest declination I think you make 23 degr. 29'. I had rather make it 23 degr. 29'. 12".[6] The difference is inconsiderable.

I told you in Autumn that it would be necessary for me to have about one half of the Observations in your Synopses set right by the correct places of the fixt starrs. If you please to do it at your leisure I'le send you a catalogue of the Observations. *And because to perfect the Theory of the Moons Parallax,* besides the subduction of some seconds, *there is requisite* an equation wch sometimes amounts almost *to a minute & wch I know exactly*: I'le make a Table of it *& send it to you*: And then you'l be perfect in that part of the Moons Theory wch consists in computing her longitude & latitude from Observations & wch is the foundation of all the rest.

One thing I did not consider. The Observations being yours perhaps you had rather have them perfectly your own in all respects by determining the Moons longitude & Latitude from them all your self. If so (for that's what you have a very just right unto) I will stay your time. And when I have got a little further in the Theory & satisfied my self about something *I am yet in the dark in, I'le make a new Table of the Moons excentricities & equations of her Apoge for finding her mean Anomaly & send you a copy of it,*[7] *to be used instead of that printed in Sr Jonas Moor's works pag 94,* provided *you will keep it to your self till I have perfected the Moons Theory* because it will need correction. Chuse you therefore whether you will compute the Moon's places from the observations or leave that work to me.

Three or four Observations in the end of this Moon & as many opposite to them in the beginning of the next would be very significant. I am

Your most affectionate humble Servant

Is. NEWTON.

Cambridge
Jan. 26. 1694/5.

[Added by Flamsteed]

Rec Jan. 29 & answered ye same day[8] in hast but more fully Feb 7.[9] & then sent him The 4 tables of ye æquations of days & ye lunar observations of June &c 92 & december & Jan last.

1692 Maij 16. 17. 19. Junij 13. 15. 16 : 1694 Δer 28. 30. 31 ☽ obsr communicavi in responso ut et 1695 Jan 8. 9. 11. 12. 13. 14. 18 cum differentiis a Meis tabulis

	h ′ ″	Arg A	Anom med		° ′ ″	° ′ ″	
1692 Maij 16	8.59.11	2.19.17.00	6.28.59.12	≃ 23 10 02	199.39.20	103.27.47	

	A R	D a p		′ ″	° ′ ″	′ ″
	199.39.20	103.22.50	≃ 23.10.58	+ 0.56	4.41.40	−2 50[10]

For Mr John Flamsteed
  at the Observatory in
      Greenwich
  neare London

NOTES

(1) C. 361, no. 73. The date was inserted by Flamsteed, who has underlined the passages in italics. There is a draft copy in the British Museum (Add. MSS. 4292).

(2) See the final paragraph of Flamsteed's letter of 31 December 1694, p. 66. Other references to this work are to be found in Letters 481, 484, 487, 493.

(3) David Gregory was appointed Savilian Professor of Astronomy on 6 February 1692, largely through the instrumentality of Newton.

(4) See Letter 486, note (6), p. 66.

(5) See Letter 488, note (2), p. 73.

(6) The following note, in Newton's hand, is to be found in the University Library, Cambridge (Add. 3965, 17, fo. 645): 'Flamsteed makes ye Suns greatest declination 23°.29′. Latitude of London 51°.29′.'

(7) This was sent on 23 April 1695 (Letter 500).

(8) Letter 490.

(9) Letter 493.

(10) See Letter 493 for the complete table.

490   FLAMSTEED TO NEWTON
29 JANUARY 1694/5
From the original in the University Library, Cambridge.[1]
In reply to Letter 489; for answer see Letter 494

The Observatory Jan: ☽ 29 1694/5

Sr.

I am very well satisfied in what you tell me about Vivianis book & you may conclude what you are to think of Mr Halley from this that he told me before a club of the Society[2] that you had it I find you understand him not so well as I doe. I have had some years experience of him & very fresh instances of his ingenuity with which I shall not trouble you tis enough that I suffer by him I would not that my freinds should & therefore shall say no more but that there needs nothing but that he shew himselfe an honest man to make him & me perfect freinds & that if he were candid there is no body liveing in whose acquaintance I could take more pleasure: but his conversation is such yt no modest man can beare it no good man but will shun it

I shall take care to have my 4 yeares tables of the æquations transcribed[3] and sent you with my first leasure.

As also my Calculations of ye Moones places to the observations of December last & this present January Wherein by reason of our foggs & cloudy weather I could only take her on ye Meridian of ye 8. 9. 11. 12. 13. 14 & 18th days the places are all computed. & by them I perceave the Latitudes follow the laws of ye Longitudes and are biggest in ye Winter quarter as are the other equations.

I intend to give you with these about 9 places of the Moone[4] observed & computed from my Tables in ye Summer of the yeare 1692. which are very proper to compare wth them. they shall all be ranged into one Sunopsis & sent you a week hence

for at present both my selfe and servant are just set to calculate a large table of Longitudes & latitudes Answering to every degree of Right ascension & distance from the Pole I have thought of a very compendious way of doeing it not by ye resolution of oblique Angled but onely a set of right-angled Sphericall triangles. & when this is done to sixty degrees from ye æquator. It will serve to give me the Longitudes & latitudes not of ye ☽ & planets onely but of 9 tenths of all the stars I have observed by Inspection with the variation of their Right ascensions & declinations in any space of time proposed. I shall acquaint you with my fundamentalls for this table hereafter. at present I cannot thinke of leaveing the Worke we are set into this morneing (haveing prepared for it last Week) till we have made some progress in it I hope to finish 15 degrees of it on each side ye Ecliptick within a Weeke as soon as this is done ye Numbers for ye Synopsis shall be sent you with what more I promise you here. & pray trouble not your selfe to send me your computed places for my observations. I stand not on punctilios wth my freinds I love to deale freely wth them. & as candidly as I desire they should doe wth me.

When I have made these new tables it will be very easy for me to examine my calculations of the Moons observed places over againe & therefore I desire you to send me a note of all that you suspect in ye Synopsis, & I shall soone give you a good account of them

In computeing the moones places from my owne observations I have hitherto used the places of the fixed stars according to a Catalogue I made of those I employd to ye begining of the yeare 1686.[5] which I thinke was not much amisse I beleive was no where two minutes erroneous. When this Table is finished I shall quickly have a new which shall be imparted to you as soone as setled. in the meane time as oft as I observe ye Moon I make use of those places for determineing hers with necessary allowances for the time lapsed since. when the stars are replaced this fault will be easily corrected by ye help of my New Tables.

The paines you have taken in setling the Theory & formeing a table of refractions I think as well imployd as I doe my own in provideing ye Materialls to trie it by. & therefore I shall be very glad to have not onely ye Table it selfe but the ground of it & I thinke I have satisfied you by my managemt in one particular of this that I am not hasty to impart what you send me. I assure you I have onely one freind Mr Caswell[6] to whom besides yourselfe I can communicate any thinge freely. & without your allowance I shall not even to him I know our first thoughts are often corrected & amended by further meditation. Seldome any thing is perfect in its first production. & therefore I am not hasty to expose my own first thoughts much less my freinds

The Latitude of this Observatory allowing the refractions of ye table you sent me is 51°.29′.00″ but if you make the Refractions bigger at yt height you diminish it so much as you encrease them. if you lessen them you make ye Latitude more

By your 66th proposition of your *Prin.* & Corollarys[7] the obliquity of the Ecliptick is greatest at ye Equinoxes least at the Solstices. but ye differences you say are Inconsiderable. pray let me know whether this of 23°.29′.12″ which you propose be a meane betwixt ye least & greatest or one of them.[8]

Your table for correcting my horizontall parallaxes will be very acceptable to me. when I have it with my owne New tables of Longitudes & latitudes to ye Right Ascentions &c. all the observed places will be easily corrected &

restored. but it will be convenient to forbeare doeing it till such time as I have got the places of the fixed stars I have emp[l]oyed anew rectified. you are sensible this will require time to accomplish I hope not so much as I at first thought it would tis my great concern to get it done. & when tis done you may assure your selfe you shall have them with what more I have by me not yet⁽⁹⁾ examined as I get leasure.

Tho the heavens favor me not now in the Morneings yet they did in ye end of ye last moneth as you will find when I send you ye Synopsis of this & the last moneths observations which I design you by thursday sevennights post if not before: I adde no more least ye⁽¹⁰⁾ post should be gone. I am ever your sincere freind

JOHN FLAMSTEED

I am sorry for your Indisposity & pray God send you your health againe. I can I blesse him for I endure ye cold of ye nights as well as formerly but I often take cold when [I] sit long to my Numbers within doores & have scarce recovered⁽¹¹⁾ of one yet yt I got in my first days work last week when I was very intent I feare you suffer in ye saim manner with my selfe. tis as much requisite to defend our selves against the Injurys of ye Air in our studys as on a Jorney our bodys being every day less able to bear ye injurys of cold as our days increase. J F:

I have wrote in great hast pardon a length that you reape little profit when I write next I will take care be shorter. & give you what may make amends for the defects of this

        To
Mr Isaack Newton Fellow
   of Trinity Colledge
      in Cambridg. there
            these
                  present.
   Cambridge

NOTES

(1) Add. 3979, fo. 28.

(2) See Hooke, *Diary*, p. 205: '1675/6 Saturday, January 1st. We now began our New Philosophicall Clubb. And Resolvd upon Ingaging ourselves not to speak of anything that was revealed *sub sigillo* nor to declare we had such a meeting at all.' There are several references in the *Diary*, as well as in Evelyn and Pepys, to a club which suggest that a few Fellows met and dined together, but there is no evidence that it survived long. In a footnote in the *Diary* (p. 205) it is stated that this was the first attempt to form a club within the Royal Society. Prominent

among its members were Sir Christopher Wren and Doctor William Holder, the distinguished divine, and an early fellow of the Royal Society. The present Royal Society Club dates from 1743. See Sir A. Geikie, *Annals of the Royal Society Club* (1917).

(3) See Newton's request in Letter 482.

(4) Sent on 7 February 1694/5 (Letter 493).

(5) See Newton's letter to Flamsteed, 10 August 1691 (vol. III, note (3) p. 164), and Flamsteed's reply, 24 February 1691/2 (Letter 386, vol. III).

(6) For Flamsteed's opinion of Caswell see Letters 476, 479 and 497.

(7) See Manuscript 467, note (8), p. 6.

(8) Note by Newton, in the University Library, Cambridge (Add. 3965, 17, fo. 645): 'The obliquity of ye ecliptic must be encreased from $23°.29'.00''$ to $23°.29'.10''$ & ye Moons greatest declination decreased from $5°.0'.0''$ & $5°.17'.46''$ (in my Theory) to $\begin{matrix} 5°.0'.11'' \\ 4°.59.45 \end{matrix}$ & $\begin{matrix} 5°.18'.0'' \\ 5.17.30 \end{matrix}$ or $5°.0'.20''$ & $5°.18'.6''$'
The obliquity of the ecliptic (1950) is $23°\ 26'\ 45''$.

(9) The word 'yet' is repeated.

(10) Flamsteed wrote 'be' for 'ye'.

(11) The word is abbreviated to four letters. The sense suggests an abbreviation for 'recovered'.

491   MEMORANDA BY FLAMSTEED
2 FEBRUARY 1694/5
From the original in the Royal Greenwich Observatory[1]

Next to Mr I.N.

All I impart to him is under his hand that he shall not communicate it to any without my hand till I print it
I study not for present applause. Mr Ns approbation is more to me then the cry of all the Ignorant in ye World.
That one hint about Refractions & the different expansion of ye aire near our earth in Winter & Summer[2] was enough for me to build a Correct table on yet shall be glad to have his further thoughts as freely as he has mine
Memd To Mr I.Nn. Feb 2 1695
yt I omitted severall things in my last
Syrius a fit Star for being seen in ye day unfit by reason of the parallax of ye orbe. Syrius nearer us yn ye rest becaus his light [is] briskest
This very sensible in my observations of a remarkable star none of ye biggest
The light of ♃ faint neare ye Sun brisk in ye opposition.

I shall mind my business of ye fixed Stars. & give him an account of my progress whilest he is employd on ye Moon & shall be very well pleased wth an account of his success

That I shall not impart any thing I receave from him wthout his leave. & expect the same kindness from him.

about E.H.[3] That he is very much mistaken in him. that I never found any thing so considerable in him as his Craft & forehead. his art of filching from other people & makeing his workes their owne as I could give instances. but that I am resolved to have nothing to doe wth him for peace sake

That I beleive he told me Mr Newton had Viviani on purpose to worke a division betwixt us

That I forbore not his company till I found that a part of his Character was thrown upon me, & that I had rather be without his acquaintance to purchase it wth ye loss of an honest reputation

   Answer to Mr Bossley of Feb. 25. 1695/6 & dated March 3. 96

1. That I esteeme all my paines in makeing ye observations of ye planet ♄ well before. since he has found by them how to Answere all the Inæqualitys of yt planets motion so exactly by their help & tho we doe not at present see the reason why this Anomaly that helps to represent its motion make its returne in 72 yeares yet I doubt not but since it is certeine that a restitution is made in neare yt time I doubt not but that we may hereafter find it or those that succeede us will be assisted by it to discover ye true cause which yet I apprehend to have some dependance on ye returnes of those planets to each other since halfe a revolution is made in 30 yeares in which time ♃ makes 3 Revolutions & ♄ about 3 signes more yn one or halfe ye Variation of ye Anomaly.

2 I am heartily glad & rejoyce for ye good success of his often repeated paines

NOTES

(1) These Memoranda were found in the Royal Greenwich Observatory (vol. 62 B). The first appears to be a draft of an answer to Newton's letter of 26 January. At the foot of that letter Flamsteed had written: 'Ans same day and on Feb. 7.' The second memorandum, dated 2 February, is probably a draft of the letter which Flamsteed sent to Newton on 7 February, since the penultimate paragraph of the draft bears a very close resemblance to the opening paragraph of the letter where Flamsteed complains that Halley's object was 'to create a difference betwixt you & me'.

The third memorandum 'Answer to Mr Bossley of Feb. 25, 1695/6 & dated March 3. 96' refers to a letter which has not been found. A number of letters which passed between Flamsteed and Bossley are in the Royal Greenwich Observatory. They deal mainly with matters relating to Flamsteed's estate, or with the computation of stellar positions.

William Bossley, apothecary of Bakewell, acted as intermediary between Flamsteed, who had been appointed Astronomer Royal in March 1674/5, and Luke Leigh (1658–1707), whom

Flamsteed employed occasionally from 1695 to assist in calculating star positions which were eventually published in the *Historia Cœlestis* in 1725. Although it is more than likely that Bossley assisted in this undertaking, no mention is made of him in the *Prolegomena* (*Historia*, III, pp. 152–3) where there is a generous reference to Luke Leigh, Abraham Sharp, and James Hodgson. On the other hand, in a letter to Bossley, dated 8 March 1704/5 (Baily, p. 236) Flamsteed wrote: 'I hope now, in good time, to have an opportunity to acknowledge your pains in correcting the motions of Saturn and Jupiter...it will be a pleasure to me to let the world know how much, and without any view of your own advantage, you have obliged, Sir, Your real friend and servant, John Flamsteed.'

(2) See Letter 487.

(3) The reference is to Halley. See the first paragraph of Letter 489.

492   A MEMORANDUM BY DAVID GREGORY
3 FEBRUARY 1694/5
From the original in the Library of the Royal Society of London[1]

Newtonus Condidit tabulas Refractionis Syderum pro quibusdam observationibus sibi exhibitis a Flamstedio, quæ tamen tabulæ minime congruunt cum aliis ab eodem factis, unde aliud nunc comminiscitur systema ad syderum refractionem explicandam.

3 febr. 1694/5

Translation

Newton built up tables of the refraction of the stars in accordance with some observations shown to him by Flamsteed, which tables however show but little agreement with others made by the latter: whence he is now devising another system for explaining the refraction of the stars.

NOTE
(1) This note follows Gregory's memorandum of a talk with Halley on 19 January 1694/5. The sheet, R.S. Greg. MS. fo. 96, is mentioned in Gregory's Index (U.L.E. Greg. quarto A) as '40. Notanda Phys: et Math: Janr 1694/5'. It contains the following:

Problema D. Halleij
19 *Janr.* 1694/5

Si Globus datus in datâ recta, data cum velocitate ab alio Globo dato oblique percutiatur, Quæruntur gyrationis motus in Globis utrisque a Collisione producti? (If a given sphere moving in a given straight line, with a given velocity, is struck obliquely by another given sphere, the rotary motions of each sphere produced by the impact are to be sought.)

493   FLAMSTEED TO NEWTON
7 FEBRUARY 1694/5

From the original in the University Library, Cambridge.[1]
In further reply to Letters 487 and 489; for answer see Letter 494

The Observatory   Feb: 7   1695

Sr.

I left some particulars of your last not fully answerd the defects of that I shall supply in this letter. What you acquaint me concerneing Vivianis booke[2] might have been told at first. Mr Rook intimated as much to me long since. but I knew not how to credit it, E H haveing answered me positively that You had it, when I asked him of it before a club[3] of our Society. Judge you what reason he had to say it, or for what purpose he designed this Answer; if it was to banter me that was not ingenious; if to create a difference betwixt you & me he is no fit man for me to use as a freind. You know him not [.] I doe and Many of his unsincere & disingenious practises. particularly the best part of that stock of lunar observations where of he boasts are mine & severall things besides he is very unwilling to owne, & hence it comes that he represents me as one so very eager to have acknowledgmts made, as if I required more then was due. I can be content with less, & as for what you have receaved of me I am not sollicitous what you say of them being well assured you will say nothing but what is just & fit & for his aspersions I doe not valew them. I have done with him, who has almost ruined himselfe by his indiscreet behaviour. & you shall hear no more of him from me till wee meet when I shall tell you his history which is too foule & large for a letter.[4]

I send you included a Copy of my four tables of the æquations of Naturall days for four yeares with an additionall Table for makeing them serve for other yeares past or to come they are fitted to the Solar tables in my Doctrine of the Sphere[5] when I make new as I must ere long for my New limitations I shall send you them. they will not be above 10 or 12 seconds different from these.

Tis true the æquation of dayes may be avoyded by accounting of time in sydereall hours. but this is a change that none but those who are versed in Astronomy would easily understand & therefore not to be made because it departs too far from ye sence of the common people & reproaches them with ignorance which would cause them againe to deride us. but further I doe not thinke that hereby wee should avoyd all appearance of error for I find the parallax of the orbe is sensible even in ye pole star,[6] tho it be but of ye 3d light, Whence it will be argued that it is much more so in Syrius, the briskness of whose light & the bigness of his body shew that he is much nearer to us then the rest of the largest stars in our hemisphere.[7] yet I argue his nearness to us more from the briskness of his light then the bigness of his body, for I thinke the difference in the strength of ♃s light when he is nearest the earth & when remotest from it, is much more then in proportion to ye difference of his distances.

The parallax of the Orbe in the pole star is about hafe a minit.[6] I cannot at present find the first copys of my Synopses of refractions observed but as I remember the heights of ♀ which I had marked in them were nearly as you say I shall cause them to be computed or calculate them my selfe & send them you in my next.

I give you on the following pages the places of the Moone I promised you. I thought it needless to adde more of my calculations because yt would have added to ye bulke of my letter & have been of little use to you since you now have new tables[8] of æquations of ye Apoge & Excentricitys which I presume you will try them by & compare ym wth I would gladly have your table for correcting ye parallaxes of ye ☽ for the reason you mention
As also yt of refractions when you have finished it with ye New Tables of ye Excentricitys & æquations of ye ☽s Apoge
You may assure your selfe I shall not communicate them without your leave first obteined. & I desire ye same kindness from you that you would not permit my observations or letters to goe out of your owne hands & in case of Mortality take care they may be safe returned to me
The day after I received your Last Mr Hanway[9] brought me News from London yt you were dead but I shewed him your letter which proved the contrary. he had it from Sr C. Wren to whom he wrote immediately to satisfye him of the falsehood of yt report I bless God for your life & pray for your perfect health

I have made a good progresse in my Worke haveing calculated ye æquations of ye right Ascention from 0 to 10 deg of latitude allready. So yt this first decade of my table will determine ye places in Longitude & latitude of all the planets & such stars as lie not above 10 degrees from ye Ecliptick from the given right ascension & distance from ye pole by Inspection & a little trouble of proportioning it will now be an inconsiderable labor to find ye ☽s places from ye observation & therefore when I have your tables for correcting ye parallaxes. I shall give you the moones places & if you please you may repeate them for certeinty.
I am ever Sr

Your very affectionate & sincere freind

JOHN FLAMSTEED

Annus Men die	Tempus Appar d h '	☽ A Rect ° ' "	☽ dist a P ° ' "	Longitud s ° ' "	Latitudo ° ' "	diff: a Tab Flam Longit ' "	Latit ' "	
1692 Maij 16	8.59.11	199.39.20	103.22.50	♎ 23.10.58	4.41.40 A	+0.56	−2.50	
17	9.52.05	213.56.50	108.55.50	♏ 8.01.45	4.59.22 A	+2.13	−2.16	
19	11.46.15	244.36.00	116.04.40	♐ 7.15.47	4.34.50 A	+2.07	−1.24	
Junij 13	7.41.09	208.56.00	107.17.40	♏ 2.58.28	5.04.58 A	+2.06	−2.20	
15	9.30.42	238.28.30	115.15.15	♐ 1.40.05	4.49.25 A	+8.18	−2.18	
16	10.28.37	254.00.00	116.56.20	♐ 15.44.06	4.14.45 A	+7.50	−2.04	
1694 Dec. 28	17.30.36	192.18.40	100.34.55	♎ 15.26.18	4.52.15 A	−4.31	−2.24	
30	19.12.57	220.08.30	109.14.30	♏ 14.31.38	3.25.25 A	−3.40	−1.19	
31	20.10.47	235.42.40	112.05.05	♏ 28.29.53	2.16.48 A	−6.02	−2.09	:: ob diei lucem
1695 Jan. 9	3.44.49	358.43.00	84.53.50	♈ 0.51.43	5.11.25 B	+2.11	+1.46	
11	5.18.09	24.12.40	75.13.30	♈ 27.49.40	4.21.13 B	+7.13	+1.01	
12	6.05.04	37.02.40	71.31.00	♉ 10.40.02	3.37.03 B	+8.02	+1.21	
13	6.52.55	50.04.50	68.47.20	♉ 23.12.33	2.41.23 B	+8.31	−0.17	
14	7.41.26	63.18.00	67.04.30	♊ 5.32.35	1·41·03 B	+8.13	+0.08	
18	10.54.32	115.51.00	71.17.00	♋ 24.25.25	2.35.53 B	+7.17	+0.42	
omissa interponatur 1695 Jan. 8	2.57.07	345.42.00	90.31.50	♓ 16.38.05	5.09.34 B	−0.03	+0.44	

differentiæ ostendunt quantum lunæ longitudines et latitudines observatæ supputatas e meis tabulis excedunt vel ab iis deficiunt

J.F:

Altitudo Poli Grenovici 51°.29'.[10]

To
Mr Isaack Newton
   Fellow of Trinity Colledge
     in Cambridge these
       there.
          present.

NOTES

(1) Add 3979, fo. 29.

(2) See, *inter alia*, the first paragraph of Letter 489.

(3) See Letter 490, note (2), p. 79.

(4) Flamsteed's hostility towards Halley, which by now had amounted almost to a passion, may well have contributed to the deterioration in the Newton–Flamsteed relations.

(5) The reference is to the tables for The Equation of the Apparent Time which appear in Flamsteed's *Doctrine of the Sphere* (1680), pp. 78, etc. See Letter 488, note (5), p. 73.

(6) The parallax is far too small for Flamsteed to have detected with the instruments at his disposal. The discrepancy which he supposed to be due to parallax was actually due to aberration, as Bradley discovered in 1726. See also Wallis's letters dated 9 and 25 May 1695 and 13 August 1698, for table of parallaxes of Polaris and other stars.

(7) See Flamsteed's Memoranda 491, p. 80.

(8) See Letter 489. Newton sent these tables on 23 April 1695 (see p. 107).

(9) See Letter 477, note (7), p. 41. For other rumours of Newton's death see also Letter 501.

(10) This note is in Newton's hand.

494   NEWTON TO FLAMSTEED
16 FEBRUARY 1694/5
From the original in Corpus Christi College Library, Oxford.[1]
In reply to Letters 490 and 493; for answer see Letter 495

Cambridge Feb. 16. 1694/5.

Sr

I received your two last letters wth your Tables of ye Equation of time & your Observations of Decemb. & January, for all wch I thank you. I have been ever since I wrote to you last, upon making a new Table of refractions & have not yet finished it.[2] Tis a very intricate & laborious piece of work. Yet something I have done towards it. For supposing ye Atmosphere to be of such a constitution as is described in the 22th Proposition[3] of my second book (wch certainly is the truth:) I have found that if the horizontal refraction be 34′ the refraction in ye apparent altitude of 3gr. will be 13′. 3″, & if the refraction in ye apparent altitude of 3gr. be 14′ the horizontal refraction will be something more then 37′. So that instead of increasing the horizontal refraction by vapors we must find some other cause to decrease it. And I *cannot think of any other cause besides ye rarefaction of ye lower region of ye Atmosphere by heat.*

And indeed the rarefaction & condensation of ye Air by heat & cold seems to have a much greater hand in ye phænomena of refractions, then we are yet aware of.[4] For even the[5] very refractions wch you have ascribed to ye Sea-vapours & Fen-vapours, seem to me upon second thoughts to arise from ye condensation of ye Air by cold. For in travelling we find it always colder upon ye water then upon the land & that very considerably & therefore ye water doth cool ye Air to ye height of some fathoms above it & by cooling condenses it & increases it's refractive power. This therefore is certainly one cause of those refractions & I take it to be a sufficient cause. But as for vapours we have now no one experimt that I know of to prove that they encrease ye refraction of ye Air, unless perhaps where they cool it. And were ye Air upon ye sea overloaded wth vapours it would scarce be so transparent as to let Callais wth its buildings & church-steeples be seen through it cross the Channel.

I am still labouring at a new Table of refractions & as soon as that's done I intend to make the Table I promised you for ye Moons Parallax (for this will be quickly done) & after that as soon as I can get time from some other occasions wch begin to press me I will make the new Table of ye Moons excentricities & ye equations of her Apoge This last Table I shall make more for your use in determining ye Moons longitude & Latitude then for my own. For when I enter upon ye work of determining the Moons motions I shall stick to no Tables but alter the Equations dayly as I shall see occasion till I have made them exact.

As for your Observations you know I cannot communicate them to any body & much less publish them wthout your consent. But if I should perfect ye Moon's Theory &
* you should think fit to give me leave to publish your Observations with it:[6] you may rest assured that I should make a faithfull & honourable acknowledgmt of their Authour with a just character of their exactness above any others yet extant. In ye former edition of my Book you may remember yt you
* communicated some things to me & I hope ye acknowledgmts[6] I made of your communications were to your satisfaction: & you may be assured I shall not be less just to you for ye future. For all ye world knows yt I make no observations my self & therefore I must of necessity acknowledge their Author: *And If I do not make a handsome acknowledgment, they will reccon me an*
* *ungratefull clown.*[6] And for my part I am of opinion that for your Observations to come abroad thus wth a *Theory wch you ushered into ye world* & wch by their means has been made exact would be much more[7] for their advantage & your reputation then to keep them private till you dye or publish them wthout such a Theory to recommend them. For such a Theory will be a demonstration of their exactness & make you readily acknowledged ye Exactest Observer that has hitherto appeared in ye world. But if you publish them wthout such a Theory to recommend them, they will only be thrown into ye heap of ye Observations of former Astronomers till somebody shall arise that by perfecting ye Theory of ye Moon shall discover your Observations to be exacter then the rest. But when that shall be God knows: I fear not in your life time if I should dye before tis done. For I find this Theory so very intricate & the Theory of Gravity so necessary to it, that I am satisfied it will never be perfected but by somebody who understands ye Theory of gravity as well or better then I do. But whether you will let me publish them or not

may be considered hereafter. *I only assure you at present that without your consent I will neither publish them nor communicate them to any body whilst you live, nor after your death without an honourable acknowledgment of their Author.*

When I have finished the Table of Refractions I will endeavour to make you understand ye grounds of it as far as I can. But ye demonstrations being very intricate I have not yet set them down in writing. I am very glad you have got so far on your great work as to be able to rectify the places of ye fixt stars wthin 10 degrees of ye Ecliptick on both sides & by them to set right your Observations of the Moon. I shall make what hast I can to furnish you wth what I am about, being

Sr

Your most affectionate Friend to serve you

Is. NEWTON

Pray till April be ended make what Observations you can in ye last Quarter of Each Moon opposite to those you make in first Quarters.

For Mr John Flamsteed at
the Observatory in
Greenwich neare
London

[Addition by Flamsteed]

Answered March 2. 1695
& given him the heights of ♀ at Sun set

Feb 23.	35.	10
Aprill 11...39.	14	
21.	37.	17
23.	36.	46
25.	36.	12
26.	35.	33

togeather wth ye Refractions observed at Cape Sete by Monsr Picart.[8]

Vindication of my selfe for not imparting my observations. & an account of my Northerne Correspondent JF.

NOTES

(1) C. 367, no. 77. Sentences in italics were underlined, probably by Flamsteed.

(2) The table was sent on 15 March 1694/5 (Letter 496).

(3) *Principia*, Book II, Prop. 22. 'Sit Fluidi cujusdam densitas compressioni proportionalis;

& partes ejus a gravitate quadratis distantiarum suarum a centro reciproce proportionali deorsum trahantur: dico quod si distantiæ sumantur in progressione Musica, densitates Fluidi in his distantiis erunt in progressione Geometrica.' (Let the density of any fluid be proportional to the compression, and its parts be attracted downwards by a gravitation inversely proportional to the squares of the distances from the centre: I say that if the distances be taken in harmonic progression, the densities of the fluid at these distances will be in geometrical progression).

(4) See Letters 487 and 488.

(5) Newton wrote 'to'.

(6) At these points asterisks have been inserted in the margin, probably by Flamsteed.

(7) This word, which is indistinctly written, seems to be what Newton intended.

(8) See Letter 474, note (16), p. 33.

495   FLAMSTEED TO NEWTON
2 MARCH 1694/5
From the original in the University Library, Cambridge.[1]
In reply to Letter 494; for answer see Letter 496

The Observatory March 2d: 1695/4

Sr

I am glad to find yt my Quadrenniall Tables of æquations wth ye lunar observations came safe to your hands The cloudy weather has not permitted me to observe her above once this forthnight & I have not had leasure as yet to examine the obse[r]vation or deduce her place from it my time haveing been wholly employd in carrying on my table of Long. & Latit. to 65 degr of declination. wherein I have made some progress. but not so great as I expected to have done in this time by reason my servant about 10 days agone ran away from me without any provocation in hopes as I suppose of prefermt at sea for his skill. I have not yet heard of him. but have hopes he may returne because he is bound to me by Indenture. if he should not tis a great loss to me tho he be of a capricious humor.

I have never read your 2d book. but lookeing on your 22th prop.[2] it seemes to me a proper foundation for a Table of refractions but I cannot thinke how you should ground a calculation upon it Without a great deale more consideration then I have leasure for under my present Circumstances. I assent to what you say about ye refractions being altered by the change of the heat or cold of ye Air.[3] & that vapors or exhalations alter them only by cooling & condenseing it Not onely ye aire on ye Water is cooler yn yt on ye

89

land. but that on ye hills is generally cooler yn in ye valleys & I beleive that ye eveining refractions of June. 11 might become less yn ye morneings by ye raies being in parte restored or bent the contrary way to what they were in ye cooler upper Region of the Aire. I am desirous to see your New table of refractions when finished. it will be of great use to me. for without these the observed Meridionall distances of the stars from ye Vertex are faulty & their places uncerteine. I give you at the foot of this letter Mons Picarts observed refractions at Cape Sete transcribed from ye *Voyages Astronomique*[s].[4] they were made with an Instrument of something a less radius yn my Quadrant & will confirme mine if you make ye refractions in 3 degrees height to be 13 min or a little less then observations with ye Sextant will allow it. & you ought rather to trust to them then those made with ye Quadrant.

I hope you are sensible now I had good reason to delay the Publishing of any of my observations, & to be very carefull to whom I communicated them.[5] Since they will be of no use without a correct Catalogue of the fixed stars and that could not be made till I had a Competent Number of Observations to ground it on. which I had not till Now. And tho In the Meane time I have made a Rectification of about 150. it was onely to trie ye observed places of the Moon & planets, to set them near ye truth, & see how much the Theorys have erred. God has blest my endeavors beyond my expectation. You have (by the helpe of those I have imparted to you) found that the Moones Motions answer to ye laws of gravity & those corrections of her Theory which expunge the greatest part of the Error yt my observations shewed in the Tables, I pray for your life that this parte may be perfected by you whilest I am goeing on with ye fixed stars. in whose places if I make any alterations I am confident it will be very easily allowed for I hope in a few months to have made a small Catalogue that shall be a ground worke for ye great one & shall conteine principally those stars from which I have observed ye Moone or wherewith I have compared her. & this as soon as finished I shall send you. the rest will require no small time to Compleat them.

Had I imparted my observations to those who made so much noyse about them & calld for them so earnestly I had then afforded them what they desired an opportunity of finding fault with & censureing them for yt the Meridionall Zenith distances taken with my former slender Arch were not so exact as they ought to be. the distances measured in the heavens with the Sextant were of little use & troublesome to Manage. My new arch is sub-stantiall curiously wrought & exactly divided & tho the wall on which it is fastened sinkes a little yet it is so gradually and so little that tis scarce to be valued especially since tis easily found at all times & allowed for. Now I am in this good posture I hear no more Noyse about my observations which I

shall continue as God spares me health whilest I am workeing on ye Catalogue
till I have finished; what I derive from them I shall very freely impart to you
as your occasions call for my help. as for acknowledgmts assure your selfe I
am not so covetous of them as I have been represented[.] I know those to
whom I impart my paines if they be just, as you are allways, will make such as
are really requisite. more I desire not & I am sure you will not blame me for
witholding [*sic*] my paines from those Who endeavor to render me esteemed
an idle & lazy person, by concealeing what they had from me; or worse, by
misrepresenting it or detracting from it

To make me some recompense for the Want of my servant the good providence of Heaven has afforded me the helpe of a Couple of very Ingenuous
men who liveing at great distance hence have set themselves to make new
tables of ye Planets motions & sollicited me for some observations of ♄.
I sent them onely my owne but at the same time put them of all their false
principles & Theorys[:] they have returned me the places of ♄ calculated
from New Tables wherin they Now make ♄s aphelion move forward 3°.20'
his Node 1°.20' & abate ye mean motion about 14' in 100 yeares. & hereby
they represent his observed places always within 2 minutes but most commonly much nearer. from ye year 1676 to 1694. I shall send them next post
the Hevelian & Tychonick. corrected by my owne places of ye fixed stars.
with directions to compare them with old observations & when this is done I
shall encorage them to goe on to Jupiter if they find the same agreemt it will
persuade me that ye effects of the Intermutuall gravitation of ♄ & ♃ are
much more regular then they were thought[6] however it proves I doubt not
but we shall have Numbers from them much more proper for discovering the
effects of the planets gravitations yn we expected

I owe you an account of ye ground of my My New Tables of Longit &c but
that not being of any great use to you I defer till I have more leasure.

		d.	°	'
The heights of ♀ which I pro- mised y[ou] for ye Synopsis of Refractions were at Sun Set as here	1681 Feb	23.	35.	37
	Apr	11.	39.	14
		21.	37.	17
		23.	36.	46
		25.	36.	12
		26.	35.	23

I beg your pardon for not haveing given you them sooner I hope you will have
leasure to make ye little table for correcting ye ☽s parallaxes with ye New
equations of ye Apoge & Excentricitys in a short time if not send me your
Law & ye limitations I shall doe it for you[7]. I looke on your first determinations

to be used onely for tryall. they will be nearer the truth tho perhaps they may
not be perfect at first I have altered my Solar Numbers now 5 times the lunar
3ce & every time as I thought for the best yet the first may perhaps still admit
of some very small alterations what the last want you know well. he that will
represent the heavens must not be ashamed to change his Numbers as oft as he
finds he can doe it for ye better. Fare well I am ever Sr

Your affectionate freind & servant

JOHN FLAMSTEED

		Ref		Ref
		′ ″		
under ye horizon 10′:		34.00		
				° ′ ″
in ye horiz. upper l.		..37.05	17.	2.28
lower l.		36.50	18	2.00
		′ ″	21	1.43
height	2. 0	17.42	22	1.39
	3. 0	12.56		
At Cape Sete.	4. 0 ...	10.40		The table goes No farther &
deduced from	5. 0	9.00		seemes to be too fine for
observt by	6. ...	7.25		their Instruments & the Way
Monsr Picart	7. ...	5.50		of observeing tis I fear a
	8.	5.36		little assisted but I have not
	9.	4.56		time to examine I think I can
	10.	4.38		make it out that with their 3
	11.	4.10		foot Instruments so as they
	12.	3.40		were fixed. they Could not,
	13. ...	3.20		did not take an height nearer
	14. ...	3.10		then to a single minute.
	15.	2.50		
	16.	2.30		

To
Mr Isaack Newton.
  Fellow of Trinity
    Colledge in Cambridge
      there these
          present
  Cambridge

NOTES

(1) Add. 3979, fo. 30.

(2) See Letter 494, note (3), p. 88.

(3) See Letters 487, 488, and 494, particularly the last.

(4) Flamsteed had received the *Voyages Astronomiques* towards the end of 1694. See Letter 474, note (15), p. 33.

(5) See Flamsteed's 'vindication' written at the foot of Letter 494, p. 88.

(6) See the first paragraph of Letter 486.

(7) See Letter 494, p. 88. Flamsteed had made the original tables of the equations of the apogee and eccentricities (*Æquationes Apogæi, & Excentricitates Orbitæ Lunaris* which is printed in *Horroccii Opera Posthuma*, p. 480) from such a 'Law & ye limitations'.

If the table here reproduced be compared with that which Flamsteed sent on 16 February (Letter 494) it will be noted that the observations given for 23 February and 26 April do not agree.

496   NEWTON TO FLAMSTEED
15 MARCH 1694/5
From the original in Corpus Christi College Library, Oxford.[1]
In reply to Letter 495; for answer see Letter 497

Sr

The last week about three or four days before I received your letter I wrote to your Treasurer Mr Hawes[2] about a successor to Mr Paget[3] & proposed three persons Sr Collins[4] of this University Mr Caswel of Oxford & Mr Newton[5] late of Yarmouth. Sr Collins has Mathematicks enough, but is young & unexperienced. If they chuse him it will be requisite that the Governours oblige Mr Paget (if they can) to inspect ye school next winter & teach him to designe & draw, & then if he take hold of that advantage to improve himself & continue as industrious as they of his College tell me he has hitherto been, I beleive he will prove a good Mr. But because he is young I added Mr Caswel & because I knew not whether he would accept of ye place I named also Mr Newton. I remember Mr Caswel's character pretty well since his competition wth Dr Gregory[6] & am satisfied that he's a man of very good morals & great industry, & so well skilled in teaching Mathematicks, that could he have drawn Prospects I would have recommended none but him till he had refused to accept of ye place. However I gave him a recommendation to ye following purpose, That he is sober indoustrious & well skilled in ye Mathematicks & will make a good Master if he will accept of ye place & that

93

Dr Wallis[7] & you are able to give a fuller account of his abilities. I would have sent you the words of ye recommendation I gave Mr Caswel but yt I have lost ye copy. However you may see it in Mr Hawes's hands. Mr Newton[8] I am a stranger to but had an opportunity about two years ago of knowing his abilities. To ye best of my remembrance he wants Algebra. In other respects he has mathematicks enough having taught Navigation for some years at Yarmouth.

I have now finished the Table of Refractions, & send you enclosed a copy of it.[9] In a regular sky when in ye altitude of 3 degrees ye refraction is 13′.20″, you may rely upon it yt ye Table is exact to a second minute for all altitudes above 10 degr. & that in ye altitudes between 3 & 10 degrees ye greatest error cannot be above 2 or 3 seconds. If the refraction in ye altitude of 3 degrees be greater or less than 13′.20″, it must be increased or diminished in ye same proportion in all altitudes above 3 degrees. Within a few days I will send you ye other Tables I promised.
I am

Your affectionate Friend

to serve you

Is. NEWTON

Cambridge Mar. 15
1694/5.
For Mr John Flamsteed
at the Observatory in
Greenwich neare
London.

NOTES

(1) C. 561, no. 81. With this letter was enclosed Newton's Second Table of Refractions. Newton's autograph copy of this table is in the University Library, Cambridge (Add. 3967, fo. 19), where there is also an incomplete copy by Flamsteed (Add. 3969, fo. 6). There is also a copy by David Gregory in the Library of the Royal Society, 'Tabula Refractionis Newtoniana' (R.S. Greg. MSS. fo. 23). At the foot of the Gregory copy are two tables: 'Hevelii Tabula Refractionum Solarium', and 'Hevelii Tabula Refractionum Stellarum'. The table should be compared with that which Newton sent with his letter of 17 November (Letter 480).

This is the table, *Tabula Refractionum Siderum ad Altitudines Apparentes*, afterwards published by Halley in the *Philosophical Transactions*: 'Some Remarks on the Allowances to be made in Astronomical Observations for the Refraction of the Air. By Dr. Edm. Halley, R.S.S. Astronomer Royal. With an Accurate Table of Refractions' (*Phil. Trans.* **31** (1721), 169).

(2) Nathaniel Hawes was Treasurer of Christ's Hospital from 1683 to 1699. As a result of severe criticism by Pepys on the way in which the mathematical school was conducted he resigned. See vol. III, p. 366, notes (1) and (2); and Letter 499, note (5), p. 104.

94

Altitudo apparens gr. ′	Refractio [min. sec.]	Altit. appar. gr	Refractio	Altit. appar. gr	Refractio
0. 0	33. 45	21	2. 18	56	0. 36
0. 15	30. 24	22	2. 11	57	0. 35
0. 30	27. 35	23	2. 5	58	0. 34
0. 45	25. 11	24	1. 59	59	0. 32
1. 0	23. 7	25	1. 54	60	0. 31
1. 15	21. 20	26	1. 49	61	0. 30
1. 30	19. 46	27	1. 44	62	0. 28
1. 45	18. 22	28	1. 40	63	0. 27
2. 0	17. 8	29	1. 36	64	0. 26
2. 30	15. 2	30	1. 32	65	0. 25
3. 0	13. 20	31	1. 28	66	0. 24
3. 30	11. 57	32	1. 25	67	0. 23
4. 0	10. 48	33	1. 22	68	0. 22
4. 30	9. 50	34	1. 19	69	0. 21
5. 0	9. 2	35	1. 16	70	0. 20
5. 30	8. 21	36	1. 13	71	0. 19
6. 0	7. 45	37	1. 11	72	0. 18
6. 30	7. 14	38	1. 8	73	0. 17
7. 0	6. 47	39	1. 6	74	0. 16
7. 30	6. 22	40	1. 4	75	0. 15
8. 0	6. 0	41	1. 2	76	0. 14
8. 30	5. 40	42	1. 0	77	0. 13
9. 0	5. 22	43	0. 58	78	0. 12
9. 30	5. 6	44	0. 56	79	0. 11
10. 0	4. 52	45	0. 54	80	0. 10
11. 0	4. 27	46	0. 52	81	0. 9
12. 0	4. 5	47	0. 50	82	0. 8
13. 0	3. 47	48	0. 48	83	0. 7
14. 0	3. 31	49	0. 47	84	0. 6
15. 0	3. 17	50	0. 45	85	0. 5
16. 0	3. 4	51	0. 44	86	0. 4
17. 0	2. 53	52	0. 42	87	0. 3
18. 0	2. 43	53	0. 40	88	0. 2
19. 0	2. 34	54	0. 39	89	0. 1
20. 0	2. 26	55	0. 38	90	0. 0

(3) Edward Paget, Fellow of Trinity College, and mathematical master at Christ's Hospital. See vol. III, p. 366, note (2), p. 368, note (2); and Letter 499, note (5), p. 104.

(4) William Collins (Sr = Scholar). Scholar of St Catharine's College, Cambridge; B.A. 1693/4; M.A. 1698; ordained priest, 1700. Possibly a son of John Collins (1625–83). See vol. II *passim*.

(5) Samuel Newton kept a mathematical school at Wapping (see E. G. R. Taylor, *Mathematical Practitioners*, pp. 118 and 290). He was not related to Isaac although he was one of the three persons proposed by him for the mathematical post at Christ's Hospital rendered vacant by the resignation of Paget. He succeeded to the post, and he continued in it until 1709, despite the fact that he does not appear to have been satisfactory. In a letter to Sharp (24 March 1708/9) Flamsteed wrote: 'Mr Newton ye Math Master at Xts Hospitall has

resigned. yt is, turned out for insufficiency & James Hodgson succeeds him. & has been in that Schole ever since Xtmas' (from a letter in the possession of F. S. E. Bardsley-Powell, Esq.). For an account of the election see Flamsteed's letter to Newton (Letter 499), note (5), p. 104.

(6) Caswell was an unsuccessful candidate for the Savilian Chair of Astronomy (1692) when David Gregory was appointed, largely through the instrumentality of Newton. He was eventually appointed to the Chair in 1709, shortly after the death of Gregory. See also vol. III, p. 168, note (15).

(7) Caswell was evidently well known to Wallis, for he had been teaching mathematics at Oxford since 1677. With his *Algebra* (1685), Wallis published *A Brief (but full) Account of the Doctrine of Trigonometry, both Plain and Spherical by John Caswell, M.A.*

(8) On 14 June 1695 (Letter 514, p. 132), after the appointment had been made, Newton wrote to Hawes: 'As for Mr Newton I never took him for a deep Mathematician, but recommended him as one who had Mathematicks enough for your business, wth such other qualifications as fitted him for a Master in respect of temper and conduct as well as learning...I was almost a stranger to him when I recommended him'. Despite extensive search no letter of Newton's specifically recommending his namesake has been found.

(9) The copy which accompanied this letter, should be compared with the *Tabula Refractionum* (480) which Newton sent to Flamsteed on 17 November 1694.

Note on Atmospheric Refraction

The atmosphere which surrounds our Earth may be regarded as consisting of a series of infinitely thin concentric layers, the density of which increases continuously as we approach the surface of the Earth. On account of this increasing density, a ray of light from a star does not travel in a straight line, but follows a curve on passing to an observer on the Earth's surface, and the apparent place of a star, as seen by the observer depends upon the direction of the ray as it enters the eye. Hence the effect of atmospheric refraction is to cause the star to appear at a greater elevation above the horizon than it would if the Earth possessed no atmosphere. As astronomical instruments became more precise, a knowledge of the magnitude of the correction which must be applied to the observed altitude of a star in order to determine the true altitude became of increasing importance. From the sixteenth century, this correction was effected by means of refraction tables. These served to show, for any apparent altitude, the appropriate correction for refraction.

Towards the end of the sixteenth century, Tycho Brahe (born 1546) constructed the earliest empirical table. He believed that refraction was not operative right up to the zenith, but faded out at some arbitrary elevation. In 1662 G. D. Cassini constructed a table on the assumption that the Earth was enveloped in an atmospheric layer of limited height and uniform density. His results agreed well with observations up to a zenith distance of nearly 80°, despite the artificiality of his hypothesis. Further progress awaited the discovery of a relation between the variation of the atmospheric density with height under isothermal conditions; even then, the precise connection between density and refractive power (refractive index -1) was the subject of conflicting hypotheses. Nevertheless it became generally accepted that the refractive power of the air was proportional to its density.

Newton had made several attempts to complete a satisfactory table of refraction, but without any great success.

In order to investigate the path taken by a ray through our atmosphere, he had not only to establish a relation between the density of the air and its refractive power $(\mu - 1)$, he had also to determine the law according to which the density of the air diminished with distance from the surface of the Earth. He rightly assumed the refractive power of the air to be proportional to its density. To establish a relation between density and altitude was more complicated since it involved a knowledge of the physical constitution of the atmosphere. Newton began by assuming the atmosphere above the surface of the Earth to consist of an infinite number of concentric spherical strata, increasing in density from the uppermost layer to the surface of the Earth, so that the problem was reduced to the investigation of the effect produced by an infinite number of media, all of which possessed different degrees of refractive power. He first supposed the density to diminish by equal amounts with equal increments of altitude: 'suppose this globe [the Earth] is covered wth an Atmosphere of Air whose density decreases uniformly from ye earth upwards to the top' (Letter 485, p. 61). Upon this hypothesis he compiled the three tables which accompanied his letter (480) of 17 November 1694 (p. 49). As Flamsteed pointed out, these did not accord with the results of observation, and Newton himself soon became aware of the fact that the law of density upon which they rested was faulty, inasmuch as it supposed 'ye refracting power of ye Atmosphere as great at ye top as at ye bottom' (Letter 487, p. 67). Accordingly he abandoned it and proceeded to the consideration of another hypothesis based upon a conception of the constitution of the atmosphere described in the 22nd proposition of the second book of the *Principia*, namely, the densities of the air being proportional to the pressures, and its parts being attracted downwards by a gravitational force inversely proportional to the square of the distance from the centre, then, if the distances from the centre be in harmonic progression, the densities at these distances will be in geometrical progression. This was more in accord with the variation of the density of the atmosphere, nevertheless the results based upon this assumption were no closer to those obtained by observation, since no account was taken of the temperature at different distances from the Earth, and this clearly affects the density. Consequently he found that for low altitudes where the atmospheric temperature was higher, the calculated refractions were too great: 'For supposing ye Atmosphere to be of such a constitution as is described in the 22th Proposition of my second book (wch certainly is the truth:) I have found that if the horizontal refraction be 34' the refraction in ye apparent altitude of 3gr. will be 13'.3", & if the refraction in ye apparent altitude of 3gr. be 14' the horizontal refraction will be something more then 37'. So that instead of increasing the horizontal refraction by vapors we must find some other cause to decrease it. And I *cannot think of any other cause besides ye rarefaction of ye lower region of ye Atmosphere by heat*' (Letter 494). Having now calculated a table of refractions upon this hypothesis he transmitted a copy to Flamsteed with the above letter. See R. Grant, *History of Physical Astronomy* (1852), pp. 325–7, where the question of the identity of this table with that communicated by Halley to the Royal Society in 1721 (*Phil. Trans.* **31**, 169) is exhaustively discussed. See also W. M. Smart, *Text-Book of Spherical Astronomy* (1949), p. 62, for an investigation of the general formula for refraction.

497   FLAMSTEED TO NEWTON

21 MARCH 1694/5

From the original in the University Library, Cambridge.[1]
In reply to Letter 496

The Observatory. March 21. 1695

Sr.

I have yours with the Table of refractions included which I find differs but little from yt you imparted to me formerly.[2] tis a signe to me of its goodness that both agree so well togeather: & therefore I shall be glad to know your Manner of calculateing them when you are at leasure In the meane time I give you thanks for both of them & shall retaine them to my selfe till you allow me to impart them to others

I have not calculated the Moons places from my last observations of her by reason I have been employed in continueing the Tables I told you of in my last & have layd a foundation for carrying them on to 65 degrees of Latitude. the previous calculations are finished so yt now I have nothing to doe but to make ye Longitudes from the prosthaphereses[3] at every 5 degrees of Latitude & interpole the rest. which I doubt not but to effect in a moneth or two.

For I am upon agreemt for a new servant[4] that is something acquainted with these things already[:] soone after he comes to me if he prove what I expect I shall send you the result of my last lunar observations & perhaps some more if the Weather favor me

You have much obliged me by the Charecter you have given Mr Tresurer Haws of Mr Caswell[5] who I assure you deserves it I am not so earnest to have him into ye vacant place because he is my acquaintance but because I know he is able to discharge yt duty[6] fully & will conscientiously mind his duty being a person no less religious & honest then modest & skillfull. you doe ye greatest kindness imaginable in recommending him to ye Hospitall. I have not seen your Namesake[7] but I know ye methods of the ordinary teachers of Navigation & that any of them would ruin our schole. You will permit me therefore to oppose him till I am well satisfied of his Abilityes & fitness which if ye [i.e., they] are more yn Mr Caswells I shall freely depart from my acquaintance to serve yours & promote him.

I am goeing to my Cure[8] for 6 or 7 days & send this that you may not thinke me forgetfull of you or unthankfull for the obligation you have layd on Sr

Your affectionate freind & humble servant

JOHN FLAMSTEED

[Along the margin] Mr Hally has got himselfe recommended by Mr Pepys,[9] who understands little of ye business, but his ill moralls & abuseing religion has been objected so yt he injures himselfe onely by putting in I beleive he perceives it & will desist. J:F:

To
Mr Isaack Newton
Fellow of Trinity Colledge
  in Cambridg there
        these
          present
Cambridge.

NOTES

(1) Add. 3979, fo. 31.

(2) On 17 November 1694 (Letter 480).

(3) *Prosthaphæresis*: this word, which was frequently misspelt both by Newton and Flamsteed, is formed from πρόσθεσις, addition, and ἀφαίρεσις, subtraction. It originally signified the equation of the centre, and in this sense it was used by Tycho Brahe. According to *O.E.D.*, it is the correction necessary to find the 'true', i.e. actual apparent, place of a planet, etc., from its mean place. It is equal to the angle subtended at a planet by the eccentricity of its orbit. In the figure S is the Sun, P the position of a planet in its orbit APB whose centre is C,

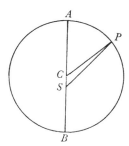

the angle CPS is the *prosthaphæresis*. See Henry Gellibrand, *A Discourse Mathematical on the Variation of the Magneticall Needle* (1635), p. 5: 'And moreover its absolutely necessary for that Seaman who sailes by his Compasse, continually to search the variation, that so by the *Prosthaphareticall* application thereof, the true point of the compasse (which is his principall Mercurialist) may be rectified'. *Mercurialist*, formerly one under the influence of the planet Mercury, later guide or director.

(4) Possibly James Hodgson, who served Flamsteed as assistant from 1696 to 1702. Luke Leigh also assisted Flamsteed about this time.

(5) See Letter 496; also Letter 499, note (5), p. 104.

(6) The word 'business' was first written (it is an insertion between the lines) and then it was altered to 'duty'.

(7) Samuel Newton. See Letter 496, note (5), p. 95.

(8) At Burstow, in Surrey.

(9) Pepys was one of the governors of Christ's Hospital. See vol. II, p. 374, note (5) and Letter 499, note (5), p. 104. There is no evidence that Halley was a candidate for the post.

498   WALLIS TO NEWTON
10 APRIL 1695
From the original in the Library of the Royal Society of London[1]

Oxford, Apr. 10. 1695.

Sir,

I was in hopes of seeing you in Oxford last summer; which made me neglect sending you (by the Carrier) two Cuts which belonged to the Volume[2] you had before. They were not wrought off at ye Rolling-Press when you had the rest; but are easy to be inserted in their proper places. I send them now, with the other Volume; which I desire you to accept.

I understand (from Mr Caswell) you have finished a Treatise about Light, Refraction, & Colours;[3] which I should be glad to see abroad. 'Tis pitty it was not out long since.[4] If it be in English[5] (as I hear it is) let it, however, come out as it is; & let those who desire to read it, learn English. I wish you would also print the two large Letters of June & August 1676.[6] I had intimation[7] from Holland, as desired there by your friends, that somewhat of that kind were done; because your Notions (of *Fluxions*) pass there with great applause, by the name of *Leibnitz's Calculus Differentialis*. I had this intimation when all but (part of) the Preface to this Volume was Printed-off; so that I could onely insert (while the Press stayd) that short intimation thereof which you there find. You are not so kind to your Reputation (& that of the Nation) as you might be, when you let things of worth ly by you so long, till others carry away the Reputation that is due to you. I have endeavoured to do you justice in that point; and am now sorry that I did not print those two letters *verbatim*.

I understand you are now about adjusting the Moons Motions; and, amongst the rest, take notice of that of the *Common Center of Gravity* of the Earth & Moon as a conjunct body: (a notion which, I think, was first started by me, in my Discourse of the Flux and Reflux of the Sea.)[8] And it must

needs be of a like consideration in that of Jupiter with his Satellites, & of
Saturn with his. (And I wonder we have not yet heard of any about Mars.)
But Saturn & Jupiter being so far off, the Effects thereof are less observable
by us than that of the Moon. My advise upon the whole, is, that you would
not be too slow in publishing what you do. I am

Sr

Your very humble servant,

JOHN WALLIS.[9]

For Mr Isaac Newton
Fellow of Trinity College,
& Professor of Mathematicks
in Cambridge.
    With a Book.

NOTES

(1) The original of this letter is in the Library of the Royal Society of London (W. 2. 48). It
is printed in Edleston (pp. 300–1) and part of it also appears in Raphson, *History of Fluxions*
(1715), p. 120.

(2) Volume II of the *Opera*, which appeared two years before 'the other Volume', i.e.
volume I. See the 'Intimation' below, note (7).

(3) *Opticks: or, a Treatise of the Reflexions, Refractions, Inflexions and Colours of Light.* It appeared
in 1704, together with two short tracts in the same volume, in Latin, entitled *Enumeratio
Linearum Tertii Ordinis* and *Tractatus de Quadratura Curvarum* (An Enumeration of Lines of the
Third Order, and a Treatise of the Quadrature of Curves).

(4) Wallis missed no opportunity of advocating the immediate publication of treatises by
his fellow-countrymen. To Boyle he wrote (25 April 1666): 'But there be two reasons by which
you have prevailed with me, at last to do something, *First*, because it is the common Fate of
the *English*, that out of a modesty, they forbear to publish their Discoveries till prosecuted to
some good degree of certainty and perfection; Yet are not so wary, but that they discourse of
them freely enough to one another, and even to Strangers upon occasion; whereby others, who
are more hasty and venturous, coming to hear of the notion, presently publish something of it,
and would be reputed thereupon, to be the first Inventors thereof; though even that little,
which they can then say of it, be perhaps much lesse, and more imperfect, than what the true
Authors could have publish'd long before, and what they had really made knowne (publikely
enough, though not in print) to many others' (Library of the Royal Society of London, W. 1,
18); and in the following year he wrote to Oldenburg: 'Onely I could wish that those of our
nation were a little more forward, than I find them generally to bee, (especially the most considerable) in timely publishing their owne discoverys, and not let strangers reap the glory of
what those amongst ourselves are the Authors' (Letter Book of the Royal Society of London,
21 March 1666/7). In a further attempt to overcome Newton's reluctance to appear in print
he went so far as to write to Waller, the Secretary of the Royal Society, requesting him to

persuade Newton to 'do right to himself, & to our Nation'. See his letter to Waller (Letter 502), p. 114, and Waller's reply, Letter 509, p. 127.

(5) Latin was declining sharply and the practice of publishing works in the vernacular was now becoming increasingly common. Notable examples are Galileo's *Discorsi e Dimostrazioni Matematiche* (1638) and Descartes's *Discours de la Méthode* (1637), as well as the Works of Boyle and Hooke. As an international language English was increasing in popularity. See Letter 514.

(6) The letters are printed in vol. III of Wallis's *Opera* (*Epistolarum Collectio*) with the titles: (i) *Epistola Prior Cl. Viri D.* Isaaci Newton, *Matheseos Professoris in Celeberrima Academia Cantabrigiensi; ad D.* Henricum Oldenburg, *Regalis Societatis Londini Secretarium;* 13 *Junii* 1676, *cum Illustrissimo Viro D.* Godefredo Guilielmo Leibnitio (*eo mediante*) *communicanda. Literis* Oldenburgii (26 *Junii*) *ad* Leibnitium *missa* (pp. 622–9); (ii) *Epistola D.* Newtoni *posterior, ad D.* Oldenburgium, *Octob.* 24 a 1676 *cum D.* Leibnitio *communicanda* (pp. 634–45). The letter was dated 24 October 1676, not August. See vol. II, Letters 165 and 188.

(7) The intimation referred to was published by Wallis in the Preface (*Opera*, vol. I, 1695): 'Quæ in Secundo Volumine habentur, in Præfatione eidem præfixa dicitur. Ubi (inter alia) habetur *Newtoni* Methodus, de *Fluxionibus* (ut ille loquitur,) consimilis naturæ cum *Leibnitii* (ut hic loquitur) *Calculo Differentiali*, (quod, qui utramque methodum contulerit, satis animadvertat, utut sub loquendi formulis diversis,) quam ego descripsi (*Algebræ Cap.* 91. &c. præsertim *Cap.* 95.) ex binis *Newtoni* literis (aut earum alteris) *Junii* 13. & *Augusti* 24. 1676, ad *Oldenburgium* datis, cum *Leibnitio* tum communicandis (iisdem fere verbis, saltem leviter mutatis, quæ in illis literis habentur,) ubi methodum hanc *Leibnitio* exponit tum ante *decem* annos, nedum plures, ab ipso excogitatam. Quod moneo, nequis causetur, de hoc *Calculo Differentiali* nihil à nobis dictum esse.' (What is contained in the second volume is stated in the preface to the same, where (among other material) is included Newton's Method of Fluxions (to use *his* term) which, to one who has compared the two methods, is of a similar nature to Leibniz's Differential Calculus (to use *his* term); this is something quite apparent, allowing for the different forms of expression. Newton's Method of Fluxions I described in Ch. 91 etc., and especially in Ch. 95 of my *Algebra*, basing myself on two letters of Newton (or on one of them) dated 13 June and 24 August 1676, which were sent to Oldenburg to be communicated then and there to Leibniz (in words almost the same, or only slightly altered, as those contained in the said letters) where he explains this method to Leibniz, as having been at that date excogitated by him, ten years before, if not more. I utter this warning, lest anyone should allege that I have left his *Calculus Differentialis* unmentioned.) See also Wallis's letter to Newton (Letter 503).

(8) 'Mr OLDENBURG read part of a letter of Dr WALLIS to him, dated at Oxford, 18 August 1666, desiring that certain observations about the tides might be made by the Society's order, to see how matter of fact would agree or disagree with his hypothesis.' The letter is printed in *Phil. Trans.* 1 (1665/6), 297 with the title: 'Some Inquiries and Directions concerning Tides, proposed by Dr Wallis for the proving or disproving of his lately publish't Discourse concerning them.'

Wallis's treatise *De Æstu Maris Hypothesis Nova* was published in 1668. It is remarkable for the sagacious assumption that the Earth and Moon might be regarded as a single body whose mass was concentrated at their common centre of gravity.

(9) Newton replied to Wallis's request on 21 April (see opening paragraph of Letter 503). Unfortunately this letter has not been found.

499   FLAMSTEED TO NEWTON
20 APRIL 1695
From the original in the University Library, Cambridge.[1]
For answer see Letter 500

The Observatory April 20. ♄. 1695

Sr

I have yours[2] & should be very glad to have your New tables of ye Moons horizontall parallaxes the æquations of the Apoge & excentricitys.[3] that I may hereafter give you ye moons places deduced from my observations such as shall need no emendation. I had thoughts to have falne upon makeing some my selfe but forbeare in hopes to have yours for since you have begun this work & gone on so far your selfe I thinke none so fit to finish it. I Would have the correction of ye lunar Theorys be all your owne & I wish you may have as good success with it as you have met with in ye business of Refraction in limiting of which I concur with you

I wish you would give me leave to impart your Table to some few freinds of mine who will be obliged by it.

My Country men have made a good progress wth ♄. but compareing their Numbers with Mine & Tychoes observations & fitting them so as to Answer both very nearly they find them too swift for most of ye Hevelian by 8 or 9 min. I am putting them on ♃. when they have finished this planet we shall better be able to judge what the effect of these two planets gravitations on each other is[4]

The election at Xts Hospitall[5] was over on Tuesday the 2d Instant a fortnight before ye date of your letter[6] and Mr Newton chosen. One of ye Governors & ye Tresurer both told me that too much learning Made their Masters proud. that the Youths were proud & troublesome to ye seamen. with their *quillities*[7] so he Cald them. & that Mr Newton was a person cut out purposely for them. Mr Caswells Certificate from Sr Ch: Wren. Drs Wallis Bernard[8] & Gregory I could never get read but was told I was prejudiced. So I desisted & Mr Caswell returnd to Oxford 3 days before ye election. he is pleased that he is clear of them. So I beleive is Mr Paget who has perfectly answered the Charecter my freind gave me & I you of him I am satisfied in that I have done my duty. Mr Newton yt he has the place & now I have done wth them. but I would advise you as a freind be very cautious how you concerne your selfe any further wth them. for their is a party yt governes amongst them yt is neither of your mind nor mine Mr Newton is of theirs. & for yt very reason was pitcht upon.

103

I have got but halfe a score observations of the Moone this 3 moneths. Sr Ch: Wren has helpt me to a servant of whom I have very good hopes. if Calculation work be tiresome to you send me onely your limitations yt labor shall now be saved you by Sr.

Your affectionate sincere freind & servant.

JOHN FLAMSTEED.

To Mr Isaack Newton
   fellow of Trinity
      Colledge in Cambridge
        there these present
          Cambridge.

NOTES

(1) Add. 3979, fo. 32.

(2) Sent on, or about, 16 April 1695; see note (6) below.

(3) See Letter 494. The tables were sent on 23 April 1695 with Letter 500.

(4) See the opening paragraph of Flamsteed's letter to Newton, 31 December 1694 (Letter 486).

(5) Briefly, the history of the election is as follows. Paget had been appointed mathematical master at Christ's Hospital in 1682 (see Letter 258, vol. II, pp. 373–5), against the vigorous opposition of Pepys, who was one of the governors. On 20 February 1694/5 he wrote resigning his post. The reason he gave was ill-health, but according to Flamsteed there were other reasons which made him unfitted for the post (*Ebrietati deinde post annos* 7 *nimium addictus immemor officij, pueros neglexit*) (then after seven years, having become too much addicted to drunkenness, unmindful of his duty he neglected the boys).

On 13 March 1694/5 the Committee of Schools were informed that Mr Samuel Newton (see Letter 496, note (5), p. 95) wished to apply for the post. 'Mr Treasurer [Hawes] informed the Committee that he hath lately received a letter from Mr Isaac Newton the Mathematical Professor at Cambridge, wherein he gave the said Samuel Newton a good Character, which letter was now read.' This letter has not been found.

On 18 March 1694/5 'Mr Flamsteed proposed to the Committee that Mr John Caswell is a very fit person to succeed Mr Paget, as Mathematical Master, having the recommendations of divers very eminent and learned men'.

Five candidates attended the meeting of the Committee on 2 April following. They were Arnold, Ward, Samuel Newton, Atkinson and Caswell. Newton alone appears to have accepted the conditions imposed by the Committee, and so the Committee proposed his appointment to the Council. Newton was thereupon appointed, and he continued in the office till 1709. But he was never considered satisfactory. The Governors had doubts about his fitness, and within six months of the appointment wrote to Isaac Newton, who had proposed his namesake (see Letter 496), to which Newton replied that he really knew very little about the man (Letter 514).

Pepys, who was violently antagonistic to Hawes, seems to have been prejudiced against Samuel Newton, possibly because the latter had been introduced by Hawes, and he wrote to the Committee on 29 November 1695 alleging that Newton was not working to the syllabus laid down. Newton thereupon sent in his resignation, but was persuaded to withdraw it on 15 January 1695/6. Pepys wrote to the Committee insinuating that there was a lot of gossip about the way in which the Mathematical School was managed; it is not possible to discover to what extent these charges were justified. Trouble broke out again with Trinity House in 1708; as a result Samuel Newton sent in a letter of resignation on 8 December of that year. He later asked to withdraw the letter, which he said he had sent 'rashly and inadvertently', but the Committee would not allow him to do so, and his services at the School terminated early in 1709. He was succeeded by James Hodgson, F.R.S., who was the husband of Flamsteed's niece, and who assisted in the publication of the great posthumous record of Flamsteed's observations. Hodgson continued in office until 1755. See, however, Pepys's letter to Newton (Letter 508).

(6) This helps to fix the date of Newton's letter, to which this is a reply. It is most unfortunate that Newton's letter is nowhere to be found.

(7) Quillity = a quillet or quiddity. The word is now obsolete.

(8) Doctor Edward Bernard (1638–96), astronomer, linguist; educated at St John's College, Oxford. He was a pupil of Wallis. He succeeded Wren as Savilian Professor of Astronomy (1673) and was succeeded by David Gregory in 1692. He made important contributions to our knowledge of the mathematics of classical times. See vol. I, p. 235, note (2).

500   NEWTON TO FLAMSTEED

23 APRIL 1695

From the original in Corpus Christi College Library, Oxford.[1]
In reply to Letter 499; for answer see Letter 501

Cambr.
Apr. 23. 1695.

Sr

I now send you ye Tables I promised. They are accurate enough for computing ye Moons Parallax & thence her longitude & Latitude from Observation. The little Table of the Equation of ye Moons Parallax is founded on ye 28th Prop. of ye 3d Book of my Principles[2] where I shew that the Moons Orb (wthout regard to her excentricity) is Oval & that her distance in the Quadratures is greater then her distance in ye Octants in ye proportion of 70 to 69.[3] In ye Table of her horizontal parallaxes, I make her horiz. parallax[4] in the syzygies less then you make it in your printed Tables by about half a minute, & in ye Quadratures I make it less then you do by about $1\frac{1}{3}'$. Were the French mensuration of ye earth to be confided in as exact,[5] these parallaxes ought to be still less: but I am unwilling to diminish them any further as yet. In computing the Moons mean Anomaly for finding her Parallax, add 12 minutes to the mean motion of her Apogee.

When *I set my self*[6] wholy to calculations (as I did for a time last Autumn & again since Christmas in making the Table of Refractions) *I can endure them & go through them well enough. But when I am about other things, (as at present) I can neither fix to them wth patience nor do them wthout errors.* Which makes me let the Moons Theory alone at present wth a designe to set to it again & go through it at once when I have your materials. I reccon it will prove a work of about three or four months & when I have done it once I would have done with it forever. In Autumn when I was tracing the Moons motions by your Observations, I found that where they were continued, two or three or four of them would agree with one another to half a minute or less: & then would follow two or three others wch would again agree with one another but disagree from ye former 2 or 3 minutes, & whether to follow the former or latter I knew not, & so could not come to ye conclusions I would have made by reason of their disagreemt, but wrote to have your Observations set right by your new places of ye fixt stars. And I am glad your work is now so far onward.

Upon Mr Paget's resignation[7] I understood that a great interest had been made among ye Governors (by ye Seamen as I presume) for a Tarpolian Master,[8] which would have ruined the School. To stop that Gap I recommended three persons,[9] & I beleive ye Tarpolian interest struck in more readily wth Newton then they would have done wth any University man. Concerning the Table of Refractions I will write to you in my next

Yours
Is. NEWTON

For Mr John Flamsteed at the
Observatory in
         Greenwich
neare
         London.

The Equations of the Moons Apoge & the Excentricities of her Orbit in such parts as the radius is 1000000.[10]

Annual Argument	Add the Equations of the Apoge						Annual Argument
	Sign 0 / 6 (° ′ ″)	Excentr. (parts)	Sign 1 / 7 (° ′ ″)	Excentr (parts)	Sign 2 / 8 (° ′ ″)	Excentri (parts)	
0	0 0 0	66850	9 22.50	61855	11.32.17	50406	30
1	0.20.54	66845	9.36.57	61537	11.22.59	50022	29
2	0.41.46	66827	9.50.31	61211	11.12.37	49645	28
3	1. 2.38	66798	10. 3.40	60878	11. 1.10	49274	27
4	1.23.27	66757	10.16.14	60438	10.48.39	48908	26
5	1.44.12	66705	10.28.17	60192	10.35. 2	48551	25
6	2. 4.54	66638	10.39.47	59838	10.20.21	48201	24
7	2.25.31	66562	10.50.41	59479	10. 4.36	47859	23
8	2.46. 0	66475	11. 0.58	59113	9.47.47	47527	22
9	3. 6.24	66375	11.10.40	58742	9.29.55	47204	21
10	3 26.41	66265	11.19.42	58366	9.10.59	46891	20
11	3.46.50	66146	11.28. 5	57986	8.50.58	45266	19
12	4. 6.48	66012	11 35.46	57600	8.29.57	45040	18
13	4.26.37	65870	11.42.44	57211	8. 7.57	44829	17
14	4.46.15	65716	11.48.58	56819	7.44.58	44633	16
15	5. 5.41	65549	11.54.27	56422	7.21. 1	44452	15
16	5.24.55	65373	11.59.11	56023	6.56. 8	44287	14
17	5.43.53	65185	12. 3. 6	55622	6.30.23	44138	13
18	6. 2.38	64988	12. 6.12	55218	6. 3.49	44824	12
19	6.21. 9	64779	12. 8.28	54814	5.36.28	44628	11
20	6.39.22	64562	12. 9.53	54408	5. 8.22	44447	10
21	6.57.20	64343	12.10.25	54001	4.39.34	44283	9
22	7.14.55	64094	12.10. 1	53595	4.10. 8	44134	8
23	7.32.14	63847	12. 8.43	53190	3.40.10	44003	7
24	7.49.11	63590	12. 6.28	52784	3. 9.41	43888	6
25	8. 5.47	63323	12. 3.16	52381	2.38.45	43789	5
26	8.22. 0	63046	11.59. 6	51980	2. 7.27	43709	4
27	8.37 51	62761	11.53.58	51581	1.35.51	43647	3
28	8.53.17	62467	11.47.45	51185	1. 4. 2	43602	2
29	9. 8.17	62165	11.42.14	50794	0.32. 3	43575	1
30	9.22.56	61855	11.32.17	50406	0. 0. 0	43566	0
	Sign 5 / 11		Sign 4 / 10		Sign 3 / 9		
Substract the Equations of the Apoge							

calculo proprio
correcti sequentes
46891 numeri
46588   JF
46298 Maij 4 ♄
46019   1695
45753
45500
45260
45034
44823⅓

2

57·02 = 15′.44″

107

Moon's mean Anomaly		Moon's Horizontal Parallax.	
		Excentricity.	
		4356	6685
s.	deg.	′ ″	′ ″
0.	00	55. 05	53. 54
	06	55. 05	53. 54
	12	55. 08	53. 57
	18	55. 13	54. 02
	24	55. 18	54. 09
1.	00	55. 24	54. 18
	06	55. 31	54. 29
	12	55. 39	54. 42
	18	55. 49	54. 55
	24	56. 00	55. 11
2.	00	56. 11	55. 28
	06	56. 25	55. 47
	12	56. 39	56. 07
	18	56. 54	56. 29
	24	57. 08	56. 51
3.	00	57. 23	57. 14
	06	57. 39	57. 39
	12	57. 55	58. 04
	18	58. 11	58. 29
	24	58. 26	58. 53
4.	00	58. 41	59. 16
	06	58. 56	59. 40
	12	59. 11	60. 02
	18	59. 24	60. 23
	24	59. 34	60. 41
5.	00	59. 43	60. 57
	06	59. 51	61. 11
	12	59. 58	61. 22
	18	60. 02	61. 30
	24	60. 05	61. 35
6.	00	60. 06	61. 37

Moon's distance from Conjunction or Opposition	An Equation to be subducted from the Moon's horizontal Parallax.	
	Horizontal Parallax	
	least 53′ 54″	greatest 61′ 37″
	Subduct	Subduct
degr.	″	″
00	0	0
10	8	9
20	16	18
30	23	26
40	30	34
50	35	40
60	40	46
70	44	50
80	46	52
90	46	53

These tables are grounded on the supposition that the mean distance of the Moon in the Octants is $60\frac{1}{5}$ semidiameters of the Earth, & by consequence her horizontal Parallax in that mean distance 57′.5″.39‴. And that her mean distance in the syzygies is less in the proportion of 69 to $69\frac{1}{2}$, & in the Quadratures greater in the proportion of 70 to $69\frac{1}{2}$. And that her mean distance in the syzygies is to her greatest & least distance in the syzygies as 1000000 to 1066850 & 933150.

NOTES

(1) C. 361, no. 83. The original of the table *The Equations of the Moons Apoge & the Excentricities of her Orbit* is in the University Library, Cambridge (Add. 3969, fo. 18). A copy, now in Corpus Christi College Library, Oxford (MS. C. 361, no. 87), was sent to Flamsteed. The other tables (*Moon's Horizontal Parallax* and *An Equation to be subducted from the Moon's Horizontal Parallax*) are in Newton's hand; the original is also in Corpus Christi College Library, Oxford (MS. C. 361, no. 85).

(2) *Invenire diametros Orbis in quo Luna absque excentricitate moveri deberet* (To find the diameters of the orbit in which, there being no eccentricity, the Moon should move).

(3) *Est igitur distantia Lunæ à Terra in Syzygiis ad ipsius distantiam in Quadraturis (seposita scilicet excentricitatis consideratione) ut 68$\frac{11}{12}$ ad 69$\frac{1}{12}$, vel numeris rotundis ut 69 ad 70.* (Therefore the Moon's distance from the Earth in the Syzygies is to her distance in the quadratures (not allowing for eccentricity) as 68$\frac{11}{12}$ to 69$\frac{1}{12}$, or, in round numbers, as 69 to 70) (*Principia*, p. 441).

(4) Newton here substituted the words 'horiz. parallax' for 'mean distance' which he wrote first.

(5) In France Jacques Cassini was misled by somewhat inaccurate geodetic measurements, taken over a comparatively short meridian, to the belief that the Earth was elongated at the poles. Later geodetic measurements made in Lapland by Maupertuis and Clairaut showed a flattening of the Earth at the poles as required by Newton, who on theoretical grounds found the diameter of the Earth at the equator to be to its diameter at the poles as 230 to 229. See Cajori, *Principia*, p. 664, Todhunter, *History of the Mathematical Theories of Attraction and the Figure of the Earth* (1873), Chapter 1, and F. Rosenberger, *Isaac Newton und seine Physikalischen Principien* (1895).

(6) Flamsteed underlined 'I set my self' and the remainder of the sentence following the parenthesis. In the margin he has added the words: 'ye same wth me'.

(7) See Letter 499, note (5), p. 104.

(8) The Tarpaulian captains had spent their lives at sea; by contrast a few army commanders obtained naval commands through influence. See Weld, *History of the Royal Society* (1848), I, 80: 'Every Tarpaulian shall then with ease Saile any ship to the Antipodes.' Later the word was used as a nickname for a mariner or sailor, especially a common sailor, from his canvas hat which had a covering of tar to make it waterproof.

(9) The three persons proposed by Newton were Collins, Caswell and Samuel Newton. See opening paragraphs of Letter 496.

(10) In this table Flamsteed has crossed out the entries in the third column of eccentricities from 45266 to 44138 and has substituted the figures in the margin, as shown, and by the side he has written: 'calculo proprio correcti sequentes numeri JF Maij 4 ♄ 1695' (the following numbers are corrected by my own calculation). He has also written in the margin '2' opposite the entry 44003, and at the foot of the table '57.02 = 15'.44"'. The error in the uncorrected table is apparent on noting the abrupt changes among the first differences at the sequence of entries indicated. For a similar reason the entry at the foot of the first column has been changed from 9° 22' 50" to 9° 22' 56", the 50" of the copy agreeing with the entry in the original autograph.

In Letter 506, Flamsteed sent Newton his corrections, saying, 'The excentricitys in your table from 2s. 10d. to 2s. 17° for 7 degrees are either false-computed or mistranscribed by your servant.'

By the side of the table (U.L.C. Add. 3969) referred to in note (1) above, is a table of Flamsteed's which includes a column of Newton's figures deduced from the tables which Newton sent to Flamsteed on 15 March 1694/5.

501   FLAMSTEED TO NEWTON
27 APRIL 1695

From the original in the University Library, Cambridge.[1]
In reply to Letter 500; for answer see Letter 505

The Observatory Aprill 27. 1695

Sr

I receaved yours of ye 23 very seasonably on Wednesday last [24 April] for being at London ye day following It served me to assure your freinds that you were in health they haveing heard that you were dead againe.[2] I thanke you for ye Included table of ye Equations of ye Apoge & Excentricitys. these I suppose are the Meane. but when ye earth is not in its meane distance from ye Sun you apply a Correction. I would gladly know whether you still find yt all ye æquations of ye lunar Systeme are as ye Cubes of ye Suns apparent diameters or whether you correct this proportion, & how.

Your Table of horizontall parallaxes you think may serve for ye calculateing of her places from my observations. I have made an experiment this morneing that will not allow it & I shall impart it to you as I use to doe other things, freely.

October ye 19. 1678. being then but newly recovered from a great fit of sickness I observed an Eclipse of the Moone yt was the most proper yt could be for determineing ye diameter of ye shadow & Consequently her horizontall parallax. My eyes were then a little weake & therefore I could not take ye times of ye transits of ye spots into ye shadow so well as I desired. but I had Mr Hally then with me & compareing my observations with his & ye French that were published soon after in ye *Phil. Transactions* Numb 141[3] I found I had no such cause to suspect my observations as I feared. by my tables in ye *Doctrine of ye Sphere*.[4]

There are some little incongruitys in these but much less yn usually are seen in ye observations of ye same eclipse by different & distant persons; but all agree in this that the duration was longer yn ye tables allow by above a minute, the Mora by 4 minutes at least

Now should I apply your parallax it would make both the duration & mora each at least 2 minutes lesse yn it is so yt it is evident the parallax of your tables is now 2 minutes too litle & needs not to be diminished but must be so much encreast at this time

I have examined the times of Mareotis & Meotis[7] falling into ye shadow & Emerging from it by ye french observations & abateing ye Moons Diameter. they give ye same Moras & semidiameter of ye shadow with ye other appearances so that the time of ye Mora seemes indubitable at ye time of ye ♂. hor. 8h.47′.45″

Octob 19 observed	At Greenwich		At London
1678	J.F.	E.H.	Mr Heynes
	h ′ ″	h ′ ″	h ′ ″
The begining	6.33.36	6.33.39	6.33.12
Immersion	7.31.34	7.31.44	7.31.09
Emersion	9.12.46	9.12.21	9.10.10
End	10.11.06	10.10.30	10.10.17
hence ye Mora.(5)	1.41.12	1.40.37	1.39.01
Duration	3.37.30	3.36.31	3.37.05

	Mrs Cass. & Pic.	At Paris Roemer	ye Jesuits(6)
	h ′ ″	h ′ ″	h ′ ″
ye begining	6.43.30	6.43.40	6.43.54
Immers.	7.40.41	7.41.00	7.41.41
Emersi	9.21.30	9.21.30	9.21.05
End	10.20.00	10.20.10	10.20.42
Mora	1.40.49	1.40.30	1.39.24
Duration.	3.36.30	3.36.30	3.36.48

	s ° ′ ″
the ☉s Meane An.	4.01.37.24
his place	♍ 6.48.01
the ☊	♍ 6.48.00
ye Arg. Annu	2.02.08.11
ye ☽s Meane An.	5.15.28.26
hence her horiz. Pallax	59.57
☉ semid. 16′.14″ ye shadow	43.48
	h
semidur. of ye Eclipse	3.33.50
of ye Mora	1.35.10
But by your tables	
The Moons parall is	59.23

So ye semiduration &⎫
semimora ought to be⎬
each one Minute lesse⎭

Tho we might erre in determineing ye begining of the Eclipse by makeing it about halfe a minute too late the End as much too soone by reason of ye diluteness of the true shadow neare the limbe & indeed I much suspect wee all Commonly doe.

Allow ye Mora to have been onely 1h.39′.00″ the horizontall parallax must be encreast a minit, i.e. it will be 1½ min more yn your Correction makes it

This you say proceeds from ye Earths being in the Perihelion semicircle of her Orbe. She was now neare 60 degrees from ye Perihelion whereon ye error must consequently be more. tis already as big as ye whole correction you apply even in ye Quadratures but ye contrary way & ye ☽s place deduced from my observations sometimes may & will be so much faulty in Latitude still

Except you please to helpe me further with a small table yt may give ye variation to every 6th or 10th degree of ye Earth's Anomaly.

1682 Feb: 11. I observed another Lunar Eclipse at Greenwich which was also observed by the same persons at Paris[8] & herein also the duration & Mora observed exceeding the Calculated times by my tables intimate that the parallax was not too big but too little. The Earth againe not 60 deg. from ye Perihelion. but I have not time to repeate ye old calculations againe so say no more of it till I have it onely you may see ye observations in ye *Phil. Transactions* Nom 145.[9] & 146.[10] 1687.

I have a Solar Eclipse observed in May yt will require a parallax less yet yn ye correction gives by reason the Earth was then nearer ye Aphelion.

After Whitsontide I intend God willing to get Close to my business of the fixed stars. & as soone as I get those yt be neare ye Ecliptick restored I shall give you an account of them that you may goe on with your Correction of ye Moones Motions In the mean time I am prepareing my servant for yt business.

As for the Xts hospitall business[11] I have done with it I am sorry I must tell you that wee have light of such a master as you feared I had onely one halfe hours conversation wth him wherein I found he understood not latin so well as to construe an ordinary Mathematicall Author nor the reason of ye Ortho-graphicall projections of ye Sphere.[12] I tried him not in Algebra because you had acquainted me yt 2 yeares agone he knew nothing of[.] he resolves sphericall triangles The Tarpaulin[13] way by ye Lord Napiers Axiom's[14] but the Treasurer dropt his secret why he was chosen before he was aware for he complaind seriously to me that he found by Mr Paget *that too much learneing Made their Masters proud of it.* if the want of it will render them obsequious he is sure now of one that is *cut out for his purpose.* The old Gentleman[15] is a stiffe formall Churchman. & a freind of those who dissent from ye present establishmt. the New Master is a great man yt way. these are the true reasons why no other person could be lookt upon but he. The Schole will decay under his as much I feare as it did formerly under Dr Woods[16] Managemt. Young Sr Collins may live to restore it whom therefore you may doe well to encorage to mind these studies, I doubt not but he will be well versed in Algebra, yt was his fathers talent. the Astronomy will be most usefull in the Schole. Our teachers in town understand little of it. pray advise him to study the Theorys of the planets & to make himselfe expert in calculation. Tho I never saw him yet for his fathers sake my good freind & his owne good report he shall find me allways ready to serve him. I am Sr

Your affectionat sincere freind & servant

JOHN FLAMSTEED[17]

To
Mr Isaack Newton
  Fellow of Trinity
  Colledge in Cambridge
     these prt
    Cambridge

NOTES

(1) Add. 3979, fo. 33.

(2) Newton's health at this time was evidently causing grave concern and there had been rumours of his death. See Flamsteed's letter of 7 February 1694/5 (Letter 493). It is also referred to in Wallis's letter to Newton (Letter 503).

(3) 'Monsieur Cassini's Observations of the Lunar Eclips on the 29 Octob. 1678' (*Phil. Trans.* **12** (1677/8), 1015).

(4) *Doctrine of the Sphere*, Part II, Section VIII, p. 69.

(5) Mora (Latin, *mora* = delay), a period of time, in this case the interval between immersion and emersion.

(6) Jean Charles Gallet (1637–1713), a noted astronomer. See vol. II, p. 339, note (2).

(7) Palus Mareotis and Palus Mæotis were terrestrial lakes whose names occur in Pliny. They are respectively Lake Maryût, just west of the Nile delta, and the Sea of Azov. Hevelius named several lunar features after terrestrial ones, and in his *Selenographia* (1647) he mentions these two frequently. Both are near the limb of the Moon, and Hevelius was interested in their appearance in connection with his study of libration. Hevelius's names did not come into general use and Riccioli, in his *Almagestum Novum* (1651), introduced the nomenclature much of which still survives: Palus Mareotis became Mare Grimaldi, and Palus Mæotis became Mare Crisium.

(8) Cassini, Picard, La Hire, Fonteney.

(9) 'Observations of the Eclipse of the Moon Feb 11/21 p.m. at the Royal Observatory at Greenwich...by Mr. John Flamsteed, Math. Reg.' (*Phil. Trans.* **13** (1683), 89–92).

(10) 'Observations of the Eclipse of the Moon Feb 11/21, 1682 by divers learned Astronomers at divers places' (*Phil. Trans.* **13** (1687), 113).

(11) See Letter 499, note (5), p. 104.

(12) Pepys, a governor of the School, wrote: 'I find him [Flamsteed] to insist upon the expediency of having a man of learning for the Master there' (E. G. R. Taylor, *Mathematical Practitioners*, p. 118).

(13) See Letter 500, note (8), p. 109.

(14) Napier made considerable improvements in spherical trigonometry, particularly his general rule by which he resolves all cases of right-angled spherical triangles very simply.

(15) Probably Hawes. See Letter 496, note (2), p. 94.

(16) Robert Wood (1622–85), mathematician. He was a pupil of William Oughtred (1574/5–1660), whose *Clavis Mathematicæ* (1631) he translated into English (*The Key to the Mathematics*, 1647). He was elected F.R.S. in 1681, and in the same year he was appointed mathematical master at Christ's Hospital on the death of Perkins, but he only held the post for a year. He was a strong supporter of the Commonwealth, and on the Restoration (1660) he was deprived of his fellowship at Lincoln College, Oxford, to which he had been appointed by order of the Parliamentary Commissioners in 1650. He retired to Ireland and became Chancellor of two dioceses, one of which was of Meath (Anthony à Wood, *Athenæ Oxoniensis* (1691–2), vol. 2, p. 590). He wrote *The Times Mended* (1681). See *D.N.B.*

(17) Below Flamsteed's signature Newton has added: 'The parallax wch Mr Flamsteed puts 59′.57″ should be 60′.49″.'

502   WALLIS TO WALLER
30 APRIL 1695
From the original in the Library of the Royal Society of London.[1]
For reply see Letter 509

Oxford, Apr. 30. 1695.

Sir,

Yours of Apr. 23. I have well received; & in pursuance thereof I write the two inclosed; which you may please to do me the favour to send forward. I sent last week (by Mathews our Oxford carrier) the other volume (now finished) for the Royal Society, to accompany that I sent before. It was directed to Mr Hunt[2] at Gresham College for the Royal Society; & a letter with it, desiring him to present it to them when they meet, which I hope was done. I now send you a short Abstract of it, that, if you think fit to mention it

in the *Transactions* amongst your Account of Books, it may save you the labour
of collecting it. I hear Mr Newton of Cambridge hath by him a Treatise of
Light, Colours, & Reflexions,[3] finished & fairly transcribed some while
since. I wish he were called upon to print it without further delay. Perhaps
Mr Halley[4] may prevail with him so to do. I hear it is in English; but that
need not hinder the printing of it as it is. I wish also that he would print two
letters of his,[5] directed to Mr Oldenburg, (to be communicated to M.
Leibnitz) dated June 13 & Aug. 24.[6] 1676. of which I give some Abstracts in
my Algebra Cap. 91. &c. But he would do right to himself, & to our Nation,
to print them at large. For what he there delivers, passeth now with great
applause, under the name of Leibnitz's *Calculus Differentialis*.[7] If you think
fit to insert them in the *Transactions* (for they deserve it) they are too long for
one, but you may insert them, by parts, in two or three. I am sorry I did not
insert them *verbatim* in my Algebra;[8] but that is now too late. I have copies of
them, which I can send you; but I had rather you have them from himself, for
fear of errors or mistakes in my Transcripts. But, rather than fail, I will send
you transcripts from mine. I have heretofore suggested to you in a former
letter (& I know not whether it be proper to repeat it now,) whether the Society
will please to excuse my contributions, (& instead thereof to accept my
readyness to serve them,) as do some other members who do not live amongst
them.[9] For though I shall be allways ready to serve them the best I can;
'tis very seldome I have the opportunity or advantage of waiting on you, to
enjoy the content or benefit of your meetings; & then it shal be at their
choice, whether to continue or not continue my name in their list. But I
submit myself to them. And am

Sr

Your very humble servant

JOHN WALLIS.

These
For Richard Waller Esqr[10]
Secretary to the Royal Society,
in Crosby Court over against
 Gresham College,
        London.

NOTES

(1) W. 2.49. The letter was read to the Society on 8 May 1695 and entered in the Letter
Book on the 12th.

(2) Harry Hunt (died 1713). He became assistant to Hooke and to the Royal Society as a
boy, and served the Society, which was then housed in Gresham College, in the capacity of

operator, housekeeper and draughtsman. See Hooke, *Diary*, p. 498. He was possibly related to Richard Hunt (died 1690), Gresham Professor of Rhetoric.

(3) See Wallis's letter to Newton, Letter 498, note (3), p. 101.

(4) It was due to Halley's zeal that the *Principia* came to be published. See Letters 285, 287 and 289, vol. II.

(5) The 'Epistola Prior' and the 'Epistola Posterior' (Letters 165 and 188, vol. II).

(6) The date of the 'Epistola Posterior' was 24 October 1676.

(7) See Letters 498 and 503.

(8) The *Algebra* was published in 1685. The letters were printed in vol. III of the *Opera*. See Letter 498, note (6), p. 102.

(9) Wallis was now approaching his 80th year, and although Pepys, in a letter to Dr Charlett, Master of University College, Oxford, wrote: 'This good old gentleman is now as fresh and vigorous for any new undertaking (of any sort) as if he had never put pen to paper' (Tanner, *Correspondence of Samuel Pepys*, I, 171–2), there is no doubt he was feeling the weight of his years. Moreover he was living at Oxford and the journey to London would be no easy one for a man of his age. For Waller's reply, see Letter 509.

(10) Richard Waller (died 1714) was elected F.R.S. in 1691 and was Secretary to the Society from 1687 to 1709, and again from 1710 to 1714. He was the author of several physiological papers in the *Philosophical Transactions*. See also *Record of the Royal Society of London*, p. 141.

503   WALLIS TO NEWTON
30 APRIL 1695
From the original in the University Library, Cambridge.[1]
In reply to a letter from Newton, dated 21 April 1695

Oxford Apr. 30. 1695

Sir

I thank you for your letter of Apr. 21. by Mr Conon. But I can by no means admit your excuse for not publishing your Treatise of Light & Colours.[2] You say, you dare not *yet* publish it. And why *not yet*? Or, if not now, when then? You adde, least it create you *some trouble*. What trouble *now*, more then at another time? Pray consider, how many years this hath lyen upon your hands allready: And, while it lyes upon your hands, it will stil be some trouble. (for I know your thoughts must needs be still running upon it.) But, when published, that trouble will be over. You think, perhaps, it may occasion some Letters (of exceptions) to you, which you shal be obliged to Answer. What if so? 'twill be at your choise whether to Answer them or not. The Treatise will answer for itself. But, are you troubled with no letters for not publishing it? For, I suppose, your other friends call upon you for it, as well as I; & are as

little satisfyed with the delay. Mean while, you loose the Reputation of it, and
we the Benefit. So that you are neither just to yourself, nor kind to the publike.
And perhaps some other may get some scraps of ye notion, & publish it as his
own; & then 'twil be His, not yours; though he may perhaps never attain to ye
tenth part of what you be allready master of. Consider, that 'tis now about
Thirty years since you were master of those notions about *Fluxions* and *Infinite
Series*;[3] but you have never published ought of it to this day, (which is worse
than *nonumque prematur in annum*.)[4] 'Tis true, I have endeavoured to do you
right in that point.[5] But if I had published the same or like notions, without
naming you; & the world possessed of anothers *Calculus differentialis*,[6] instead
of your *fluxions*: How should this, or the next Age, know of your share therein?
And even what I have sayd, is but playing an After-game for you; to recover
(precariously *ex postliminio*)[7] what you had let slip in its due time. And, even
yet, I see you make no great hast to publish those Letters, which are to be my
Vouchers for what I say of it. And even those Letters at first, were rather
extorted from you, than purely voluntary. You may say, perhaps, the last piece
of this concerning Colours[8] is not quite finished. It may be so: (and perhaps
never will.) But pray let us have what is. And, while that is printing, you may
(if ever) perfect the rest. But if, during the delay, you chance to die,[9] or those
papers chance to take fire (as some others have done,)[10] 'tis all lost, both as to
you, & as to the publike. It hath been an old complaint, that an Englishman
never knows when a thing is well. (But will still be over-doing, & thereby
looseth or spoils many times what was well before.) I own that Modesty is a
Vertue; but too much Diffidence (especially as the world now goes) is a Fault.
And if men will never publish ought till it be so perfect as that nothing more
can be added to it: themselves & the publike will both be loosers. I hope, Sir,
you will forgive me this Freedome (while I speak the sense of others as well as
my own,) or else I know not how we thus forgive these delays. I could say
a great deal more: But, if you think I have sayd too much allready, pray forgive this kindness of

Your real friend & humble servant

JOHN WALLIS

Dr Gregory gives you his service.[11]

NOTES

(1) Add. 3977, fo. 15. It is unfortunate that Newton's letter, sent in reply to Wallis's letter
of 10 April 1695, has not been found. But its contents may well be surmised from Wallis's
reply.

(2) This letter provides further evidence of Newton's reluctance to appear in print.
Newton had made notable advances in his study of the composite nature of white light when

he was at Woolsthorpe during the plague years. In his first published paper, some years later, he wrote: 'In the beginning of the year 1666 (at which time I applyed my self to the grinding of Optick glasses of other figures than *Spherical*,) I procured me a Triangular glass-Prisme, to try therewith the celebrated *Phænomena* of *Colours*.' With this he established the fact that 'light consists of rays differently refrangible'. The criticism of Hooke and of Linus (vol. II, Letter 220) had irritated Newton and it is believed that after the disputes of 1675 in regard to Hooke's claim to have discovered the inverse square law (Letter 285, vol. II, p. 431) he had resolved to publish no more whilst Hooke lived. In any case, it was in 1704, the year after Hooke died, that his great work *Opticks* appeared (see Letter 498, note (3), p. 101). In the Preface (*Advertisement*) Newton declared: 'To avoid being engaged in Disputes about these Matters, I have hitherto delayed the printing, and should still have delayed it, had not the Importunity of Friends prevailed upon me.' See Preface to *Opticks*, 1 April 1704 (No. 672).

(3) In 1716 Newton wrote: 'I invented the method of series and fluxions in the year 1665, improved them in the year 1666, and I still have in my custody several mathematical papers written in the years 1664, 1665, 1666, some of which happen to be dated.' (E. N. da C. Andrade, *Sir Isaac Newton*, 1961, pp. 49–50.)

(4) 'Let it be kept back until the ninth year.' The quotation is from Horace, *Ars Poetica*, line 388. Horace is cautioning the young poet against rushing into publication.

(5) Wallis printed the letters on fluxions in the third volume of his *Opera* (1699), pp. 622 and 634 (see Letter 498, note (6), p. 102). A further short account was added in the preface to the first volume as it was going to press in 1695 (*ibid.* note (7)).

(6) See Letter 498.

(7) A right to recover one's old rank and privileges.
The effect of the capture of a Roman as a prisoner of war was that all his civil rights were in abeyance. If he regained his freedom, either by escape or ransom, his rights could be restored by the doctrine of *postliminium*. Wallis means that after thirty years he had recovered for Newton the credit of being the original author of certain theories, details of which Newton had neglected to publish, and so had temporarily lost by default.

(8) *Opticks: or a Treatise on the Reflexions, Refractions, Inflexions and Colours of Light.* See note (2) above and Letter 498, note (3), p. 101.

(9) Here lies the true cause of Wallis's anxiety. Newton's health had already given his friends cause for grave alarm and there was a widespread fear that either his mental health or his death should leave the work uncompleted. There had already been rumours of his death, which were happily unfounded. See Flamsteed's letters of 7 February 1694/5 (Letter 493) and 27 April 1695 (Letter 501).

(10) See vol. III, pp. 369–70, note (2); also article by E. N. da C. Andrade (*Proc. Phys. Soc.*, vol. 55, p. 137, f.n.): 'The rumour of a fire, once having been started, grew so mightily that Sturmius, the mathematician of Altdorf, reported that Newton's house and all his goods had been burnt.'
Abraham de la Pryme, in his *Diary* (1870), p. xxiv, discusses the matter at length and states that it was Newton's book of colours and light, and several other valuable writings that were utterly consumed.

Wallis has an allusion to the fire, in his *Algebra* (1685), p. 347: 'But I here only give some *Specimen*, of what we hope Mr *Newton* will himself publish in due time. And it was, I hear, near ready for the Press in the year 1671. But most of those Papers have since (by a mischance) been unhappily burned'.

(11) David Gregory could hardly fail to pass on to Wallis, his colleague at Oxford, something of what he had gathered from Newton on his visits to him in May 1694 and subsequently. See vol. III, p. 380.

504   MORLAND[1] TO NEWTON
2 MAY 1695
From the original in Corpus Christi College Library, Oxford[2]

Hammersmith 2 May 95

Sr

I know not how to make an Apology for this boldness. It being so many years since I had the honor to wait on you at your House to see your Ingenious Astronomical Contrivances. Besides that Sr. Rob. Murray,[3] Sr Jonas Moor,[4] & all our acquaintance are long since dead. And I have been quite blind for near four years last past.

The favour I now beg, is that you will please to let me know, in 2 lines, if you havent in your House a Well, so deep, that at the bottom thereat, a man standing, may easily discern the stars at noon day.

Or if you have not such a well. What must be ye Depth of such a Well, at the bottom of wch ye stars may bee best seen in the Day time And whether [you] have ever seen that Artificial Well in ye Observatory at Paris.[5] Hereby you will extremely oblige

Your most humble & faith: Servt

S MORELAND

I am forced to write by ye
help of a Frame

NOTES
(1) Sir Samuel Morland (1625–95), a man with an outstanding talent for scientific invention. See vol. I, p. 8, note (19).

(2) C. 361, no. 139.

(3) Sir Robert Moray (d. 1673), statesman and scientist. He was knighted by Charles I, whose son, Charles II, he followed into exile. He was one of the group out of whose meetings the Royal Society took its rise; he was elected President in March 1660/1 and thereafter, monthly, till 1662. He had great influence with Charles II, whose interest he secured for the benefit of the newly formed society.

(4) See Letter 488, note (2), p. 73.

(5) The Paris Observatory was established in 1671 under the direction of G. D. Cassini. Germain Brice (*Description nouvelle...de Paris*, 1684, ii, 99–103) describes the building, including the well: 'Cet espace que l'on nomme ordinairement *le Puits*'. He says it was constructed so that the stars could be seen in daylight, but so far no one had seen them.

505   NEWTON TO FLAMSTEED
4 MAY 1695
From the original in Corpus Christi College Library, Oxford.
In continuation of Letter 500; for answer see Letter 506

Apr. 25. 1695.[1]

Sr

The table[2] of ye Equations of the Apoge & Excentricities serves for all ye year, winter & summer as well as spring & autumn without any correction. For ye equations of the Parallax which arise from ye Earth's being in its Aphelium or Perihelium can never amount to above three or four seconds in excess or defect & therefore I consider them not.

The actions of ye Sun for varying ye Lunar motions I reccon to be as ye cubes of the Suns apparent diameter & the menstrual Equations wch arise from thence to be nearly in ye same proportions: but the annual ones[3] & those of longer periods arise from a mixture of impressions in summer & winter spring & autumn compounded together & observe such laws as I cannot yet determin. Nor have I been considering this point since I wrote to you last about it.

The Table of horizontal Parallaxes was made by such limits as I gathered in Autumn from your two first synopses of Observations. I do not pretend to be accurate in it. But what you object from Lunar Eclypses overthrows it not because these & ye Solar ones disagree. You think to reconcile them by supposing yt ye parallax is greater in the Suns Perige less in his Apoge: whereas ye contrary is true. The Sun in his Perige *draws ye Moon off from ye earth*[4] & thereby diminishes her parallax in winter & on ye contrary encreases it in summer thô not sensibly. The reason therefore why the Lunar Eclipses make the Parallax greater then ye solar ones do, is to be enquired. One reason you hint, namely yt ye diluteness of ye shadow neare ye limb makes it seem broader then it is. Another may be that all ye Suns light wch passes through ye Atmosphere within 20 or 24 miles of the earth is scattered by ye refraction of ye Atmosphere & goes not to ye edge of the shadow. A third may be some mistakes in your calculation. For you make ye ☽s mean Anom. 5s. 15°. 28′. 26″ & thence her horiz. Parallax 59′. 57″: you should have said

60′. 49″.[5] For 59′ 57″ is ye Parallax agreeing to the mean Anom. 4s. 15°. 28′. See therefore if there be not some such mistake in your calculations.

But yet if my Table satisfy you not, you may use your printed one, & only apply to it that little menstrual equation wch I sent you.

As for ye late election[6] it belongs not to me to enquire what made ye governours so much against Mr Caswel; but now Mr Newton is in, ye best way is to make the best of it. I am

Your humble servant

Is. NEWTON.

I suspect yt Mr Caswell put in to late & yt the Governours were afraid least that should come to pass wch you tell me did come to pass; I mean that Mr Caswel having another more easy way of living should upon any occasion be glad to be clear of them.

For Mr John Flamsteed
at the Observatory in
        Greenwich
neare   London

NOTES

(1) C. 361, no. 91. At the head of the letter Flamsteed has noted: 'Rec May. 6 ☽æ I suppose it is misdated & ought to have been dated May 4. ♄.' This is confirmed by the postmark, MA/6. There is a draft of this letter in the British Museum (Add. MSS. 4292).

(2) See Letter 500.

(3) Newton first wrote 'to be as ye cubes' but crossed it out.

(4) The words in italics were underlined by Flamsteed.

(5) See Tables sent with Letter 500.

(6) See Letter 497, and Letter 499, note (5), p. 104.

506 FLAMSTEED TO NEWTON

6 MAY 1695

From the original in the University Library, Cambridge.[1]
In reply to Letter 505; for answer see Letter 515

The Observatory May 6. 1695

Sr.

Your last has fully satisfied me in what I desired & I am pleased to find by it that ye New table of the æquations of ye Apoge & Excentricities will serve for the whole year. I shall hereafter make use of it onely in all my calculations of her places from my observations.

I have been ill againe & much troubled with paines in my head from ye time I wrote last to you till this day which is ye reason I have not calculated the moones places from my observations of the last 3 moneths. My servant has them under his hand. he is as yet unexpert. as soon as he has finished if God send me health I shall repeate perfect & send them you.

The excentricytys in your table from 2s. 10d. to 2s. 17° for 7 degrees, are either false-computed or mistranscribed by your servant[2] I have calculated them anew & give you them in the annexed little table the æquations standing against them need no correction being all true within a second or two. the rest I have not examined because their differences seeme proportionall & give me no reason to suspect any Error.

°	Sign. 2 8	
11	46588	19
12	46298	18
13	46019	17
14	45753	16
15	45500	15
16	45260	14
17	45034	13
	Sig 3 9	

When I argued against your diminution of ye ☽s parallaxes in my last I did it onely for truths sake & with no other designe but to find out. Some such encrease of ye parallaxes in the Winter full moones & a like decrement in ye Summer New Moones as my observations of 2 or 3 remarkeable Eclipses seemed to require & methinkes your assertion that when the Sun is [in] perige he draws the moone towards him favors my Conjecture. for then he must draw ye moone most from ye Earth in the conjunction, towards it in the opposition, whereby ye parallaxes of ye winter full moons will be encreased & of the summer New moons diminished, but whether this will be so much as my observations require I have not time to examine. being wholly intent on the Fixed stars whose restitution will take up all my time so yt the Moon is wholly at present in your hands & if I give you my thoughts of her at any time tis freely to assist you what I can. not to derogate from your paines. as I fear you suspect when I consider ye Stile of your Answer.

You complaine in your last save one to me that when I gave you the observations of many days togeather you could represent 3 or 4 togeather within a minute & then ye next 3 or 4 would differ 3 or four minutes, & you thinke the cause to be in the erroneous places of ye fixed stars. I have compared the differrences of ye Right Ascentions of my old tables with those of my new & find them seldom differ above one minute so yt I fear the cause is not to be found in their places. & I give you notice of it that you may seek for it else where.

I assent to your reasons why the Earths Shadow appears larger then really it is. both of them have been long in my mind. and seeme confirmed if you compare the time of Incidence with ye observed Mora in ye Eclipse of 1678.

But there is no mistake in my limitation of the ☽s parallax. If you have copied right from my letter I gave you the æquated place of the Apoge 5s. 15°. 28′. 26″ instead of ye Meane Anomaly yt was by my old tables 7s.17°.03′.54″ but by your correction 7s.16°.26′.53″ to which my parallax is 59′.57″ but yours onely 59′.25″; but a very small fault I find I committed by forgetting to adde ye Suns horizontall parallax (10″) to ye Moones which would have encreased the Mora about 40″ of time & made it agree so much better with ye observations.

Mr Caswell did not put into too late at Xs hospitall[3] nor were the Governors averse at all to him but onely the starcht Treasurer[4] all that I write to you concer[n]ing yt affaire is really & punctually true. nor is Mr Caswell in such Circumstances as you Imagin & Mr Pagit may perhaps have informed you he gets a Narrow livelyhood by his paines at Oxford. is not much aforehand with the World[5] but content & I think pleased with his poverty I did not move so heartily for him I assure you by reason of our freindship or acquaintance but because I knew him to be the ablest & fittest to discharge yt duty of all the Mathematicians I was acquainted with. Interest never sways me in thinges of this nature nor affection. but the benefit of the place which I have solemly swore to endeavor to promote when I was called to be one of ye Governors,[6] did. I had not sayd one worde more of this business but that your postscript forct it from me. Let it be as it will it ought not to make any difference neither shall it betwixt you and Sr

Your humble Servant

JOHN FLAMSTEED

To
Mr Isaack Newton
Fellow of Trinity
Colledge in Cambridge
     there these
            present
Cambridge.

NOTES

(1) Add. 3979, fo. 34.

(2) See 'The Equations of the Moon's Apoge & Excentricities' which accompanied Letter 500, and note (10) of the same letter.

(3) See Newton's postscript to Letter 505.

(4) Hawes. See Letter 496, note (2), p. 94.

(5) The word 'world' is repeated.

(6) A comma has been inserted after the word 'Governors', and a full stop after the word 'did'.

507　WALLIS TO FLAMSTEED

9 MAY 1695

From the original in Corpus Christi College Library, Oxford[1]

Oxford May 9. 1695.

Sir,

I am told by Mr Caswel (which I am very glad to hear,) that you seem to have observed a discernable Parallax (as to the Earths Annual Orb) in the Polar Star.[2] which is a noble Observation if you make it out. And therefore I ernestly desire you to pursue that business with great care. I have suggested somewhat to that purpose in my Latin Algebra;[3] & somewhat more fully in some letters to Mr Molineux;[4] which (if I do not mis-remember) are printed in the *Transactions.* I think we may reasonably presume, the[5] Fixed stars are not all at an equal distance from us, but some of them at a distance vastly greater than others. So that some (which are nearer) may have a discernable Parallax, though the Remoter have not. And those may be *reasonably thought nearest*[6] (though we are not sure of it) which *look biggest & brightest.* And if we fail in some one of them, we may try some others, & more than one or two. Again, stars though equally near, are not all equally fit for this purpose; but those fittest which are nearest to the Pole of the Zodiac (For those in, or near, the Zodiac, will have none or little Parallax, though that of those which are farther from it may be discernable.) Now, in both respects, the Polar Star is very proper; as being of the second Magnitude[7] (*though not very bright*) and not far from the Pole of the Zodiac. And therefore I am well pleased that you pitch upon that. There is yet another, in the Shoulder of the Little Bear, which is nearer to the Pole of the Zodiac, and (I think) a brighter star. I would advise you therefore carefully to observe that allso. (And, if you think fit, some others.) And set down carefully the times of their Observations, & ye meridional Altitudes observed. *And they are, I hope, such stars as by your Telescopic Sights you may discern in the day-time.* (which will be a great advantage.) The greatest Meridional Altitude (above the Pole) should be at the Summer-Solstice, and the least at the Winter-Solstice; & therefore most carefully to be observed about those times: But, in the intermediate times also, to see at what rate they vary & change their Altitudes. But I would advise you also, not to

124

talk too freely of it (unless to trusty friends) till you have a considerable number of observations to justify it; and, when you have, to print them; least others perhaps supplant you.[8] And I would not have you loose the Reputation of it. Except this freedome of

Sr   Your friend & humble servant

JOHN WALLIS.

These
For Mr Flamstead, at
the Observatory in
    Greenwich.

[Added by Flamsteed]
Answerd May 13 1695.

NOTES

(1) C. 361, no. 128. This, and a number of letters which follow, are included since they closely affect the relations between Newton and Flamsteed.

(2) Flamsteed's reply was sent in a letter to Wallis, who published it in his *Opera* (III, 701): *Epistola D. Johannis Flamsteed (Mathematici Regii Grenovici) ad D. Wallisium Dec. 20. 1698. De Parallaxi Orbis Annui Telluris Observata*, where he says: 'Indeque manifesto ostendi, quod Parallaxis Orbis Annui Telluris sit in ea Stella notabilis; quodque apparens Diameter Circuli, quem circa Polum describit ea stella, major sit in Mensibus *Septemb.* & *Maio* quam in Mense *Decemb.* uno minuto circiter....Petebam ab eo ut hasce Tibi ostenderet literas; simulque significaret (quo tibi plenius satisfieret) quod quasi eadem Parallaxis confirmata fuerit ex Observationibus septem Annorum continue consequentium' (*ibid.* p. 703). (Whence I have clearly shown that the parallax of the annual orbit of the Earth is remarkable in that star [Polaris] and that the apparent diameter of the circle which that star describes about the Pole is greater in the months of September and May than in the month of December by about a minute....I asked of him that he would show this letter to you and at the same time that he would signify (in order to give you the more complete satisfaction) that he had established this same parallax from observations over seven years continuously.)

Wallis's knowledge of astronomy, it should be noted, was not very profound.

(3) *De Algebra Tractatus, Historicus et Practicus*. This was published in the second volume of the *Opera* (1693). The work had already appeared in English in 1685.

(4) William Molyneux (1658–98), scientist; friend of Locke and Flamsteed. He lived in Dublin and was a member of the Irish Parliament. He was Surveyor General of the King's Buildings, and he held many other public offices. He was elected F.R.S. in 1685, and he has several papers in the *Philosophical Transactions*. His *Dioptrica Nova* was published in 1692. See vol. II, p. 427, note (13), and *D.N.B.*

(5) The word 'the' is repeated.

(6) Words and phrases in italics were underlined, probably by Flamsteed.

(7) Between the lines Flamsteed has inserted, after the word 'Magnitude', the words 'tho not so large yet its light is strong'.

(8) See Wallis's letter to Newton (Letter 503).

508   PEPYS TO NEWTON
13 MAY 1695
From the original in the British Museum[1]

York-Buildgs. May 13. 95.

Hond. Sr.

The Bearer Mr. Saml. Newton[2] intending You a Visitt, gives mee an opportunity of kissing Your hand & enquiring after Your health. It leads mee naturally too to ye telling You (& that wth a great deal of Satisfaction) that as farr as my short Conversation wth him can warrant it, I have mighty hopes of seeing the Royll Foundation recover through ye Industry, Practice, & Sobriety of this Gentn, what it has for sevll years been loosing from ye opposite Defects of its late Maister[3] Whose Following of ye Camp has (I fear) rendred him as little fitt, as hee now seems inclined, to any Service but that & the Libertys of it. But I say again; as far as the redeeming what is lost lyes within ye Power of a Successor, I doe in every respect hope it from Mr Newton and that hee will therein fully make good ye Character You were lately pleased to give him[4] to ye Hospitall. Where his Vertues shall want no Encouragemt or assistance within my power of rendring them. To which give mee leave to add in reference to yourselfe, my most sincere Acknowledgmts of your many Favours, & the assurance of my being most respectfully

Your true Honorer & most

faythfll & Obednt Servt

S Pepys.

Mr. Newton.

NOTES

(1) Add. MSS. 20732, fo. 114.

(2) See Letter 496, note (5), p. 95. Pepys became a governor of Christ's Hospital in 1676 and held the office of vice-president from 1699 till his death in 1703. This accounts for his interest in the appointment. See vol. II, p. 374, note (2).

(3) Edward Paget, who resigned his post as mathematical master in February 1694/5 and was succeeded by Samuel Newton. See Letter 499, note (5), p. 104. Shortly after the appointment Pepys wrote to his cousin, Major Thomas Aungier, a member of the 'Committee of Schooles', 'Mr. Newton (the Person chosen) was indeed recommended to mee, and I have no reason of doubting his being a learned & virtuous man' (*Correspondence of Samuel Pepys relating to Christ's Hospital* 1694–5. British Museum, Add. MSS. 20732, fo. 99, under date 14 April 1695). Yet

according to Taylor (*Mathematical Practitioners*, p. 118), Pepys considered Newton's methods quite out of date.

Two years after the appointment, complaints were made about the poor standard reached by the boys at Trinity House. See Flamsteed's letter to Sharp (Letter 496), note (5), p. 95.

(4) Newton wrote (15 March 1694/5, p. 94): 'Mr Newton I am a stranger to', and on 14 June 1695, in his letter to Hawes, he declared: 'As for Mr Newton I never took him for a deep Mathematician, but recommended him as one who had Mathematicks enough for your business wth such other qualifications as fitted him for a Master in respect of temper and conduct as well as learning'. This is clearly the substance of Newton's report in the missing letter to Hawes (Letter 496, note (8), p. 96); the candidate knows no algebra but he has mathematics enough to teach navigation and, according to the above letter Newton believed him to be a man of good character, and it is on the strength of this, not his skill in mathematics, that Newton recommended him.

In the Pepys Correspondence mentioned above, there is a letter dated 27 February 1694/5, over a signature which is quite illegible, which states: 'And I am well assur'd Doctor Newton (if there be time enough to obtain it) will concur in the character I gave him [S. Newton] in the request I make in his behalf.'

509   WALLER TO WALLIS

15 MAY 1695

From a copy in the Library of the Royal Society of London.[1]
In reply to Letter 502; for answer see Letter 512

Reverend Sir,

Yours of ye[2] past with your Letters to Snr Viviani and Sturmius[3] we recieved wch were sent as you desired by ye Poste. your Book likewise sent by ye Carryer came safe to us, and thankes are ordered you by ye Society for it, the acct therof you were pleased to send me, shall be inserted in ye *Transact*. As to what you mention concerning your payments, to ye Society, I am ordered to let you know that out of Respect to your worth, and in consideration of your being a Mathem: Professor, they will not expect it from you, tho' it may not be so convenient to make a publick order for it, but keep it private least others should expect the same favour. Mr Halley has promised to write to Mr Newton concerning those Letters you mention, I hope they may be procured from him, & thank you for your intimation therof. I am

Sr Your most humble servt R.W.

NOTES

(1) 'A Letter from Mr. R. Waller to Dr. Wallis May 15: 1695.' (Letter Book of the Royal Society of London, XI (2), p. 43.)

(2) Waller has left a gap here.

(3) Johann Christoff Sturm (1635–1703), a celebrated mathematician and philosopher. He became Professor of Mathematics and Natural Philosophy at the University of Altdorf (now defunct). He was the author of several useful works, including *Mathesis Juvenilis* (1699–1701). The work upon which his reputation mainly rests is his *Collegium Experimentale sive Curiosum, in quo primaria hujus Seculi Inventa & Experimenta Physico-Mathematica, A. 1672, quibusdam Naturæ Scrutatoribus spectanda exhibuit & ad causas suas naturales demonstrativâ methodo reduxit Johannes Christophorus Sturmius, Mathem. ac Phys. in Academ, Altdorfina.* (*Phil. Trans.* **10** (1675), 509.) He died at Altdorf. See vol. III, p. 168, note (14), also Hutton, *Mathematical Dictionary*, II, 465.

510   WALLIS TO FLAMSTEED

25 MAY 1695

From the original in Corpus Christi College Library, Oxford.[1]
In continuation of Letter 507

Marston near Brackly, Northamptonshire.
May 25. 1695.

Sir,

I am well pleased to understand by yours of May 14 (which came to my hands yesterday where I am at present) that you are so well stored with Observations for ye Earths Annual Parallax.[2] And therefore desire you will suffer them to be made publike forthwith: (for the Reasons mentioned in my last, which I need not repeat.) The brisk light of the Star mentioned, though small, may argue it to be nearer perhaps than some which seem bigger. And its situation is very convenient. Yet I would not have you neglect the Shoulder of the Lesser Bear. Perhaps, by darkening your Room, you may gain a sight of them in the day-time. I am not displeased to find you distinguish between the Threed's *Touching* and *Covering* a star; which seems to argue they have some *apparent Magnitude*, contrary to what some good Observers would persuade us. Your diligence to perfect your Catalog of Fixed Stars, I approve likewise; & your Reasons for it are good. Yet I would not have you delay that of the Parallax. For, the Observations being allready made; it will require no great time to digest them. And if you think that little time may not be spared for it: be pleased to furnish us with materials, & either Mr Caswell or I will help to digest them for you. And if you be excluded elsewhere, we will take care to have them printed at Oxford. You are to be careful also (in so nice a point) that ye motion in longitude (as it is wont to be called) of ye Fixed Stars do not impose upon you; and that some very small variation of ye *Measuring Thread* may not deceive you. I could suggest some other things in order to this Parallax, which I may hereafter do at leisure; But I would not distract you from what is before you. I am

Sr

Your Friend to serve you

JOHN WALLIS.

These
For Mr John Flamstead
   at the Observatory in
     Greenwich.

NOTES

(1) C. 361, no. 129.

(2) For the various references to this see the Index.

511 WALLIS TO NEWTON
30 MAY 1695
From the original in the University Library, Cambridge.[1]
In continuation of Letter 503; for answer see Letter 519

May 30. 1695.

Sir

I have taken the pains to transcribe a fair copy of your two letters, which I wish were printed. I send it you with this, because I suspect there may be some little mistakes either in ye Calculation or Transcribing in some places, which therefore I desire you will please carefully to peruse, & correct to your own mind, & then (if you please) remit to me. I would have subjoined them (with your good leave) to the second volume of my *Opera Math*:[2] if I had thought of it a little sooner, before that had been sent abroad; but 'tis now, I think, too late. If any of your Book-sellers will undertake the Printing of it; I think Oxford the most convenient place for it; Because here we have most of the Cuts allready, & furniture fit for it; & our Compositors are acquainted with this kind of troublesome work; which to others unacquainted with it will seem strange. And Mr Caswell or I will see to the correcting of the Press. But I find that these letters do refer to two Letters of Leibnitz,[3] which I have never seen. If you have copies of them by you, it would be proper to print those with these.

But while I suggest this, I would not have you neglect or defer printing your Treatise of Light & Colours,[4] even though your third part be not quite finished. And you will, after, have time enough to adjust the Moons Motions. I gave you my Reasons against great delay in my last, which I need not repeat. I have taken the liberty in this transcript, for *Collinsius*, to substitute all along *Collinius*; because it is a softer sound, & (I think) more proper. For *Collin, Robert, Richard, Roger, Henry, William*, &c being originally Proper Nounes, I take *s*, in *Collins, Roberts, Richards*, &c, to be the Formative of a Patronymick, (and to signify the same as *Collinson, Robertson*, &c) for which, in Latine, I would supply a Latine Formative *ius* (as in *Martius* from *Mars, Martis*, & many others) and would therefore change to say *Collinius, Robertius, Richardius, Rogerius, Henricius, Guilielmius*, &c. rather than *Collinsius* &c; And so *Hobbius, Hugenius*, rather than *Hobbesius, Huginsius*. But if you like the other better; I am content. You may please to let me know, what time it was that you first lighted on these notions of Infinite Series:[5] I guessed formerly (being not near you to ask) that it was about ye year 1663; If it were sooner, you may please to rectify me therein. Mean while, I am

Sr

Your very humble servant

JOHN WALLIS.

For Mr Isaac Newton Fellow
of Trinity Colledge & Professor of
Mathematics in
      Cambridge

NOTES

(1) Add. 3977, fo. 44.

(2) Extracts only of the letters were printed in the volume referred to.

(3) Letters 158 and 172, vol. II.

(4) See Letters 498 and 502.

(5) See Letter 503, note (3), p. 118.

5I2   WALLIS TO WALLER
31 MAY 1695
From the original in the Library of the Royal Society of London.[1]
In reply to Letter 509

May 31. 1695.

Sir

I have been absent from home for some time. I do not hear of any thing there received from you, since I sent you the two Letters for Sturmius[2] & Viviani. I have, since, one from Sturmius, which signifies that he had, some weeks before, received the Book I sent him. He sends me word of a Rumor amongst them concerning Mr Newton, as if his House and Busks[3] & all his Goods were Burnt,[4] & himself so disturbed in mind thereupon, as to be reduced to very ill circumstances. Which being all false, I thought fit presently to rectify that groundless mistake...

Sir,

Your very humble servant,

JOHN WALLIS

NOTES

(1) W. 2. 50.

(2) See Letter 509, note (3), p. 128.

(3) The word looks like *busks*: it can only mean books. Pryme, *Diary*, p. 23, states that several valuable writings were utterly consumed.

(4) There were other occasions when fire among Newton's papers and books was reported. See Letter 503, note (10), p. 118.

5I3   CHRISLOE[1] TO NEWTON
6 JUNE 1695
From the original in the University Library, Cambridge[2]

London ye 6 June 95

Sr

I have sent you mr Tutt receatt of 108*lib* witch I hope you will recuife safe from him. thatt is youre mostt humble searvantt to Command

RICH CHRISLOE

For Isaac Newton Esq att
Trinnity Collidge in
Cambridge

NOTES

(1) 'It was a misfortune that Mr Chrichloe, apothecary there, dy'd but a little while before I fixed my abode at that place [1726, Grantham]. He, with whom my brother was apprentice, was 84 years old, Sir Isaac's schoolfellow, a very sensible man and a scholar; who could have largely furnished me with informations. Sir Isaac had a very great esteem for him, and always inquired after his health, when he knew I had been at Grantham, and desired me to carry his service to him, when he knew I went that way, saying he was the chief acquaintance he had left in that place, except Mrs Vincent' (*Memoirs of Sir Isaac Newton's Life by William Stukeley*, M.D., F.R.S. 1752, London, 1936, p. 23).

(2) Add. 3961.2, fo. 16v.

514 NEWTON TO HAWES[1]
14 JUNE 1695
From the original in the University Library, Cambridge

Cambridge June 14. 1695.

Sr,

I should have writ to you by Mr Newton[2] but that I stay'd to consider further of ye scheme of Mathematical learning before it be established.[3] For the last Article seemed too indefinite to be subscribed, and in the forme it is there set downe, has noe books written of it, & therefore I have changed it into the last Article of the scheme I now send you enclosed in this Letter. For this last Article conteins as much of the other, as has been hitherto reduced to a certain science and something more, and is definite, soe that the Master may know what he subscribes, and the Governors what the Master is obliged to by his subscription. It has alsoe books written upon every parte of it to make it more fit for the school. As for Mr Newton I never took him for a deep Mathematician, but recommended him as one who had Mathematicks enough for your business, wth such other qualifications as fitted him for a Master in respect of temper and conduct as well as learning. But I reckon two yeares too short a time for this scheme of learning, and Dr Gregory who taught Mathematicks in Scotland wth very good success, and was here last weeke, tells me that by the time he spent in teaching he should reckon three yeares little enough for this scheme. Mr Newton may try if he can compass it sufficiently in two yeares but if that be found too little, perhaps the wisdome of the Governors may soe order things as to allow him halfe a yeare more from the other schooles. For were it not for some Mathematicall bookes in Latine, I should think that language of soe little use to a Seaman as not to deserve four or five yeares of the childrens time, while Mathematicks are allowed but two;

I thank you for your concerne and pains in behalfe of Mr. Newton, and am very glad to understand that he behaves himselfe so well. For tho' I was almost a stranger to him when I recommended him, yet since he was elected, I reckon myselfe concerned that he should answer my recommendation. The ill will you may have got by your acting for him I perceive is but of little extent and cannot hurt you. Mr Caswel's friends at Oxford blame his freind[4] neere London, and some of them think the place would not have suited with his humour, soe that I am satisfyed you made the best choice.

Sr,

Your most humble &

most obedient Servant,

Is. NEWTON.

NOTES

(1) Nathaniel Hawes, Treasurer of Christ's Hospital. See Letter 496, note (2), p. 94.

(2) Samuel Newton, successor to Paget at the Mathematical School, Christ's Hospital. See Letter 499, note (5), p. 104.

(3) For 'A New Scheme of Learning proposed for the Mathematical Boys at Christ's Hospital', see vol. III, p. 365, and Edleston, p. 292.

The new scheme, with Newton's modifications, was sent to Wallis and David Gregory at Oxford, who gave their 'opinion and advice' regarding it in a joint paper dated 13 June 1694, and advised the addition of a course in mechanics and hydrostatics. On 25 June it was agreed to adopt the new scheme, and it was ordered that 'humble & hearty thanks be returned to Mr Newton, Drs Wallis & Gregory for their extraordinary pains & kindness in this affair' (Edleston, p. 296).

(4) Flamsteed, who had strongly recommended Caswell. See Letter 497.

515   NEWTON TO FLAMSTEED

29 JUNE 1695

From the original in Corpus Christi College Library, Oxford.[1]
In reply to Letter 506; for answer see Letter 517

Sr

I received your solary Tables & thank you for them. But these & almost all your communications will be useless to me unless you can propose *some practicable way or other of supplying me wth Observations.* For as your health[2] & other business will not permit you to calculate the Moons places from your

133

Observations, so it never was my inclination to put you upon such a task, knowing that *the tediousness of such a designe will make me as weary wth expectation as you with drudgery.* I want not your calculations but your Observations only. For besides my self & my servant, Sr Collins[3] (whom I can employ for a little money, wch I value not) tells me that he can calculate an Eclips & work truly. I will therefore once more[4] propose it to you to send me your naked Observations of the Moons right ascentions & meridional altitudes & leave it to me to get her places calculated from them. If you like this proposal, then pray send me first your Observations[5] for the year 1692[6] & *I will get them calculated & send you a copy of the calculated places.* But if you like it not, then I desire you would propose some other practicable method of supplying me wth Observations, or else let me know plainly that I must be content to lose all the time & pains I have hitherto taken about the Moons Theory & about the Table of refractions.[7]

I am glad you betake your self to riding for your health rather then to physick. Tis certainly the best & safest remedy for an ill habit of body arising from bad blood in most cases, & therefore you may do well to continue it. I am

> Your humble Servant
>
> Is. NEWTON

Cambridge June 29
     1695.

For Mr John Flamsteed
at the Observatory in
Greenwich neare
    London.

[Added by Flamsteed]

Answerd July ye 2d wth an offer of ye Lunar observations made with ye Sextant from 1679 to 90. Wrote ye same day to Mr Bossley[8] & sent him my new Solar Tables.

NOTES

(1) C. 361, no. 93. Words and phrases in italics were underlined, probably by Flamsteed.

(2) See Letter 506. Flamsteed had never been of robust constitution and after his appointment as Astronomer Royal he was afflicted with severe headaches as well as other disorders.

(3) William Collins, whom Newton had proposed for the post of mathematics master at Christ's Hospital. See Letter 496, note (4), p. 95.

(4) Newton had already repeatedly requested Flamsteed's 'naked observations'. See Letters 480, 496.

(5) See Flamsteed's Memorandum, which follows this letter.

(6) In his reply (Letter 517) Flamsteed offered Newton not only the mural observations of 1692 but the sextant observations from 1679 to 1690.

(7) Newton has been criticized for the show of impatience displayed in this letter, and perhaps not without reason. But it must be remembered that great issues were at stake. Newton's gravitational theory was slow in acceptance on the Continent, largely on account of the pictorial nature of the Vortex Theory of Descartes, and only its agreement with the best observations available could establish Newton's theory. Flamsteed alone could supply these observations, and this fact may help to explain Newton's impatience. See *Remarks on an Article in Number CIX of the Quarterly Review by the Rev. William Whewell, M.A.* (Cambridge, 1836), pp. 5–6.

(8) See Flamsteed's Memoranda (491), note (1), p. 81.

5ı6   A MEMORANDUM BY FLAMSTEED

29 JUNE 1695

From the original in Corpus Christi College Library, Oxford[1]

Lunar observations & ye $\mathbb{D}$s Calculated places imparted to him

In ye 1   Synopsis 52 from No: 16. 1689 to June 11 (90) Inc[lusive]

2d	50	to	Ap 27	9	91
3d	55		Ap 16		92

6 in a letter Feb 7. 94/5 to to June 16.   92

toto. 163

26

189

1694	Sept.	15	Feb	3	$\mathbb{D}$ obsr 1695
		21			
		22	(1695 .	14	
		27		16	
	Oc	1		25	
		16			
		18	Martij	8	
		21		9	
		22		13	
		30		14	
		31		16	
	No	17	April.	6	
		18		9	

Let ye world Judg whether Mr Newton had any cause to complaine of want of obser-vations when all these were imparted to him I was ill of ye headake all this summer which ended in a fit of the stone yet forbore not as I was able to serve him without reward or prospect of any. I contend it

					1st Synopsis . .	52
	Dec	11	17		2d	50
		13	18		3	55
		16	20		in a letter	6
		28	:: 23		more as above	26
		30 Maij	13		more pag 81. & 83	
		31	15		*libr Calc.*	12
1695	Jan	8	24		201 Calculated	
		9	26		Uncalculated ta[ken w]ith	
		11	27		ye Arch	30
		12			taken with ye Sextant—	
		13 Junij	6		1677 Januarij	16
		14	11		Feb	8
		18	12		Martij	27

toto. 26 calculated

& imparted to Mr Newton

	29
	30
14 Aprilis	28
15 Maij	30
16 Junij	10
17 .	27
18	28
25 Julij . .	4
30 not calculated:	6

besides appulses & Eclipses 243
publisht.
places of ♄
♃
refractions calculated

(1) This Memorandum was written on the back of Letter 515.

517   FLAMSTEED TO NEWTON
2 JULY 1695
From the original in the University Library, Cambridge.[1]
In reply to Letter 515; for answer see Letter 520

The Observatory July 2. 1695

Sr.

That the large Number of observations I have already imparted to you may not be useless to you as you complaine[,] they will to others[:] before you tell me of it I have thought of an easy & practicable way of supplying you with such more as shall I hope fully content you. I have a large Number of ye Moons distances from fixed stars taken with ye sextant betwixt ye yeares 1676 & 90 by me; they are very accurate but more troublesome to manage then those I use because they require the Calculation of parallaxes. You have a servant & Sr Collins[2] to calculate for you whose paines & time will cost you little. I shall impart to you what you please of these; you may cause them to examine the same observations both togeather, & by their agreemt you will judge of their care & exactness. & as I get others & leasure wth health to examine them I shall impart them to you:

But whereas you desire my *Naked observations of the* ☽*s Right ascentions* &c. I must mind you that the Right ascentions are not given as ye Meridionall distances but deduced from ye observed differences of ye times of ye transits of ye ☽s limb & stars whose Right ascentions are known by Calculation. & that I am minding to calculate about 30 I observed since Feb. last[3] that I may see how much nearer your new tables of æquations of the Apoge & Excentricitys[4] will agree with the heavens then my old & what is ye Quantity of ye greatest parallactick equation in the Quadratures.

I will not say that your communicated table of Excentricitys &c will be useless to me except you adde further some limitations for this. but I must tell you that you will save me some needless labor by informeing me what the greatest parallactick æquation is in ye Apoge & Perige Quadratures. Whether ye table of Excentricitys you have imparted is sufficient to find the Moons place in her orbit at ye Syzigies. or whether you apply some other necessary æquations & of what Nature if you doe. I have saved you a great labor by my Calculations in ye Synopsis tis but equitable & just you should save me some in this & let me know what you derive from my observations as I impart them. Your determinations will be safe in my hands you may alter them as you please or revoke them. And they will be still no less yours in my hands yn my observations are still my owne in yours

I desired your consent to impart your table of refractions[5] to a freind or two of mine yt are usefull to me. about 2 moneths agone you returne me no Answer I had unwarily given Mr Caswell a Copy of yt you first sent me. but on your Intimation of your desire not to have it pass abroad I acquainted him with it tis as safe as in your owne deske of ye Second no one has yet any copy from me nor shall have with out your leave. If you write any thing on that subject I have some other necessary Tables relateing to those observations to impart to you.

If you desire ye lunar observations made with the sextant I shall afford you a copy of The Table of Fixed Stars yt is necessaryly to be used in deduceing her place from the observations. & a Nonagesimary table[6] for callculateing ye parallactick Angle.

You see how willing I am to accomodate you with what is necessary for cleareing the Motion of ye Moon & how small a returne I desire: that is onely to know what æquations you use at present in ye ☽ & what limitations you give them not that I have any desire or designe to meddle with the Restitution of her Motions my selfe: but onely to satisfie my owne Curiosity & not to be ignorant of the use you have made of what you imparted to me. as I told you before

Onely I must desire you to acquaint Mr Bently[7] (whom I know not) but who I am told complaines that your The 2d Edition of your *Principia* will come out wthout the ☽ because I do'nt impart my observations to you. that I *shall furnish you to your Satisfaction in yt particular.* had I heard of it from your selfe I had told you the contents of this letter some days since: & assured you the fault should not be layd to my Charge.

Sweateing moderately has given me some small releife for My headake but I have it this afternoon whilest I write to you I pray God keep you in health & am. Sr

                                                      Your humble Servant

                                                      JOHN FLAMSTEED

What one freind may Justly
expect from another you
shall ever command from
        Yours J:F:

    To
Mr Isaack Newton
  Fellow of Trinity
     Colledge in Cambridge
          these
             present
  Cambridge

NOTES

(1) Add. 3979, fo. 35. See preceding Memorandum by Flamsteed (516), p. 135.

(2) See Letter 496, note (4), p. 95; Letter 515, note (3), p. 134.

(3) It would appear from this that Flamsteed had also sent with the above Memorandum the thirty observations he had made from 3 February to 25 June of the current year. But Newton makes no mention of having received them, merely saying: 'When you have computed your 30 observations, you'l know no more of it [the parallactic equation] then at present' (Letter 520).

(4) Sent on 23 April 1695, Letter 500.

(5) See Letter 499.

(6) Sent with Letter 524.

(7) Richard Bentley (1662–1742), scholar and critic. Appointed Chaplain to Edward Stillingfleet, Bishop of Worcester, 1690. Keeper of the Royal Library in St James's Palace, 1694; F.R.S., 1695; Chaplain in Ordinary to King William, 1695; Master of Trinity, 1700–42. Was ordered to be deprived of his Mastership for having committed a number of encroachments on the privileges of the fellows. See Monk, *Life of Bentley* (1830), *D.N.B.* and vol. III, p. 156, note (1).

518 WALLIS TO NEWTON

3 JULY 1695

From the original in the University Library, Cambridge.[1]
In continuation of Letter 511; for answer see Letter 519

Oxford July 3. 1695.

Sir,

About a month or five weeks since I sent you a letter, & with it a Transcript of your two letters which I wished might be printed. And because I suspect there were some mistakes (particularly in some of those Examples wch you give of what you call your Theorema Primum in your second letter) I desired you would please to consider & correct them to your own mind. And I would desire allso that you would please to explain those words which (in two places) you have conceled by transposing the letters.[2] I hoped that by the first or second return of the Carrier you would have favoured me with a return of those papers so corrected. If you do not think fit to proceed to print them: I would yet desire you would favour me with them so corrected that I might at le[a]st leave them reposited in the Savilian Library amongst other Manuscript Papers; which will be no dishonour to you, but confirm to you the reputation of your having discovered these notions so long ago.[3] And if in this or ought else I may be in a capacity of serving you I shal readyly do it; & do still continue to importune you to be so just to your self, & kind to the publick as to let those things come abroad which you have in so great readyness as I hear you have. And what is not yet ready, may come afterward in due time. I am

Sr

Your very humble

servant

JOHN WALLIS

For Mr Isaac Newton,
Fellow of Trinity College
in Cambridge: & Professor
of Mathematics in that
University

NOTES

(1) Add. 3977, fo. 17.

(2) See vol. II, p. 159, note (72). The anagram when transposed and translated reads: 'One method consists in extracting a fluent quantity from an equation at the same time involving its fluxion; but another by assuming a series for any unknown quantity whatever, from which the rest could conveniently be derived, and in collecting homologous terms of the resulting equation in order to elicit the terms of the assumed series.'

(3) See Letter 498, note (4), p. 101.

519   NEWTON TO WALLIS

? JULY 1695

From a draft in the University Library, Cambridge.[1]

In reply to Letters 511 and 518

Sr

I am very much obliged to you for your pains in transcribing my two Letters of 1676[2] & much more for your kind concern of right being done me by publishing them. I have perused your transcripts of them & examined ye calculations & corrected some few places wch were amiss. The chief was in pag. 13 lin 29 where $\theta + 1 = r$ was written for $\dfrac{\theta + 1}{\eta} = r$. Which mistake made ye examples in ye next page seem faulty, tho they were not so. In ye end of ye

20th page & beginning of ye next, it may be convenient to print ye words after this manner

Possum utique cum sectionibus conicis Geometrice comparare curvas omnes (numero infinities infinitas) quarum ordinatim applicatæ sunt[3]

$$\frac{dz^{\eta-1}}{e+fz^{\eta}+gz^{2\eta}} \quad \text{vel} \quad \frac{dz^{2\eta-1}}{e+fz^{\eta}+gz^{2\eta}} \quad \&\text{c}$$

$$\frac{dz^{\frac{1}{2}\eta-1}}{e+fz^{\eta}+gz^{2\eta}} \quad \text{vel} \quad \frac{dz^{\frac{3}{2}\eta-1}}{e+fz^{\eta}+gz^{2\eta}} \quad \&\text{c}$$

$$\frac{d}{z}\sqrt{e+fz^{\eta}+gz^{2\eta}} \quad \text{vel} \quad dz^{\eta-1}\sqrt{e+fz^{\eta}+gz^{2\eta}} \quad \&\text{c}$$

$$\frac{dz^{\eta-1}}{\sqrt{e+fz^{\eta}+gz^{2\eta}}} \quad \text{vel} \quad \frac{dz^{2\eta-1}}{\sqrt{e+fz^{\eta}+gz^{2\eta}}} \quad \&\text{c}$$

The explications of the two sentences wch were concealed in letters set out of order pag 13 & pag 24 will be best set in ye margin.[4] And in pag 13 over against the words [quam solertissimus Slusius ante annos duos tresve tecum communicavit, de qua tu (suggerente Collinio) rescripsisti eandem mihi etiam innotuisse[5]] may be set in the margin this Note. Hoc intellexit Newtonus ex Epistola Collinij, die 18 Junij 1673, ad ipsum data, cujus hæc sunt verba,[6] *As to Slusius method of Tangents it was by him well understood when he published his book de Mesolabio but he did not then divulge it because he would not prevent his friend Riccio[7] who afterwards declining mathematical studies desired Slusius to divulge it, who, not obteining leisure to write of it at large promised to send it to Mr Oldenburgh to publish it in ye Transactions. Before it arrived I writ to you to understand what you knew of it & having received your Answer I imparted it to Mr Oldenburgh to send to Slusius to let him know that it was understood in England tho perchance not so long or so soon as himself had attained it.[8]*

As[9] to the time of my finding the method of converging series, the exactest account I can give of it is this, That in the year 1664 between Michaelmass & Christmass I borrowed & read your works & found ye intercalation of your series that winter. For in the notes I then took out of your *Arithmetica Infinitorum* I find this intercalation set out for squaring ye circle & ye method of reducing quant[it]ies into converging series by division & extraction of roots & thereby of squaring all curves. And then (that is in ye beginning of the year 1666) I retired from the University into Lincolnshire to avoyd the plague.[10]

NOTES

(1) Add. 3977, fo. 3. This is a rough draft in Newton's hand. The date is conjectural, but it clearly relates to the queries raised by Wallis in Letters 511 and 518. This is part of a table which appears with slight modifications of punctuation in Wallis, *Opera*, iii, p. 639.

(2) See Letter 511, note (2), p. 130.

(3) 'I am indeed able to compare geometrically with the conic sections all the curves, infinitely infinite in number, whose ordinates are....' See vol. II, pp. 119 and 138.

(4) See vol. II, p. 153, note (25) and p. 159, note (72), also Wallis, *Opera*, iii, 639.

(5) 'Which the most ingenious Sluse communicated to you, two or three years before, concerning which you, at the suggestion of Collins, replied that the same was already known to me.' The square brackets are Newton's.

(6) 'Newton understood this from Collins's letter given to him on 18 June 1673, of which these are the words.'

(7) Michael Angelo Ricci (1619–82). See vol. I, p. 49, note (3), and *G.M.V.* p. 210.

(8) The passages in italics are underlined, no doubt by Newton.

(9) A number of the words in this paragraph are surmised. They occur at the edge of the manuscript, which here has decomposed. The line preceding 'quanties into converging series' is almost completely missing. The restored line appears to indicate what Newton meant. See note (1) above and the 'Epistola Posterior', Letter 188, vol. II.

(10) Newton's notebook in the Fitzwilliam Museum in Cambridge, as well as the buttery books in Trinity College (see Edleston, p. xliii, note (8)), make it quite clear that Newton left Cambridge for Lincolnshire in June 1665 (and his college was closed on 8 August following). He returned on 20 March 1665/6, left again in June of that year, and returned on 22 April 1667. See also U.L.C. (Add. 4000.140):

'July 4th 1699. By consulting an accompt of my expenses at Cambridge in the years 1663 & 1664 I find that in ye year 1664 a little before Christmas I being then senior Sophister, I bought Schooten's Miscellanies & Cartes's Geometry (having read this Geometry & Oughtreds Clavis above half a year before) & borrowed Wallis's works and by consequence made these Annotations out of Schooten & Wallis in winter between the years 1664 & 1665. At wch time I found the method of Infinite series And in summer 1665 being forced from Cambridge by the Plague I computed ye area of ye Hyperbola at Boothby in Lincolnshire to two & fifty figures by the same method

Is Newton'

After a lapse of thirty years, Newton appears to have become a little unsure of his dates. See W. G. Bell, *The Great Plague* (1924), p. 323.

Wallis's transcript of the 'Epistola Posterior' exists in the Portsmouth Papers (3977.1): in it Newton has corrected in his own hand '$\theta+1=r$' into '$\frac{\theta+1}{\eta}=r$'.

520   NEWTON TO FLAMSTEED
9 JULY 1695

From the original in Corpus Christi College Library, Oxford.[1]
In reply to Letter 517; for answer see Letters 521 and 522

Sr

After I had helped you where you stuck[0] in ye three great works, *that of the Theory of Jupiters Satellites*[1], that of your *Catalogue of the fixt stars*[2] & that of calculating the *Moons places from Observations*[3], & in all these things freely communicated to you what was perfect in it's kinds (so far as I could make it) & *of more value then many Observations*[4] & what (in one of them) cost me above two months hard labour wch I should never have undertaken but upon your account, & wch I told you I undertook that I might have something to return you for the Observations you then gave me hopes of, & yet when I had done saw no *prospect of obteining*[5] them or of getting your *Synopses rectified*[6], I despaired of compassing ye Moons Theory, & had thoughts of giving it over as a thing impracticable & occasionally told a friend so who then made me a visit. But now you offer me those Observations wch you made before ye year 1690 I thankfully accept of your offer & will get as many of them computed as are sufficient for my purpose. As to ye greatest Parallactick Equation I know no more of it then when I wrote to you last about it. Tis but a small equation scarce exceeding 2 or 3 three or at most 4 minutes & so involved wth other equations that when you have computed your *30*[7][2] observations you'l know no more of it[3] then at present. I have no thoughts of writing about refractions. The Table of Refractions I would not have yet communicated.[4] The Observations I shall chiefly want are those when ye Moons Apoge is within 12 degrees of the seventh degree of ♈, ♋, ♎, ♑, both in antecedentia & consequentia, & those of the years 1687 & 1689 when the Apoge is wthin 12 degrees of ye 22th degree of ♌ & ♍. I am

Your most humble Servant

Is. NEWTON.[5]

Cambridge Jul. 9:
1695.

For Mr John Flamsteed
at the Observatory in
   Greenwich neare
      London

143

[At the top of the letter Flamsteed has written] I was ill all this Summer & could not furnish him as I had done formerly he mistook my illness for design & wrote this hasty artificiall unkind arrogant letter.[5]

[At the foot of the letter, above the date (Cambridge Jul. 9: 1695) Flamsteed has written] Answered ♄ [Saturday]. July 13 & sent him the lunar observations from Jan. to July 1677 marked N. in ye Margent of ye book.

[Below the date Flamsteed has written]

0   I know not yt I stuck any where: all my 3 workes goe on wthout him
1   An answer to my Query whether the physicall partes were sensible in ye Satellits motions or not he answered me not
2. A table of refractions for correcting their Merid. Zenith distances.
3   The table of ye æquations of ye Apoge & Excentricitys wth that of horizontall parallaxes & their Correction
4   without my observations he had never found them out.
5. 7. my sickness has hindred:
6   I have desired him to shew what he thinks faulty he has not yet

NOTES

(1) C. 361, no. 95. The numbers 0, 1, 2, 3, 4, 5, 6, 7 in this letter refer to the unbracketed numbers which Flamsteed inserted between the lines at the places indicated in the above letter. Flamsteed has also underlined the words and passages of the letter which are printed in italics.

(2) See the second paragraph of Flamsteed's letter to Newton (Letter 517) and note (3) of the same.

(3) That is, the parallactic equation.

(4) Flamsteed acknowledged Newton's table of refractions on 21 March 1694/5, and he added that he would retain them until Newton gave him permission to impart them to others.

(5) Brewster (II, 176) notes that there is nothing in the seven letters which Flamsteed wrote to Newton between February and 2 July to justify the tone of this letter. Flamsteed had expressed, only a week before, his willingness 'to accomodate you with what is necessary for cleareing the Motion of ye Moon & how small a returne I desire'.

521   FLAMSTEED TO NEWTON

13 JULY 1695

From the original in the University Library, Cambridge.[1]
In reply to Letter 520; for answer see Letter 523

The Observatory July 13. 1695

Sr

The paines of my head continue tho I bless God for it they are much abated on takeing ye Waters I am goeing to ride out a few miles so have only leasure to tell you that the Included observations were made in a yeare when as now the Moones apoge was in the same signe wth the Suns: The Tables of the places of ye fixed Stars & Nonages[i]mary table shall be sent you this day seven-night or sooner[2] If I can procure them copied. A report is industriously spread in towne that I have refused to impart any more observations to you.[3] I heare that he who spreads it intends you a visit ere long. I hope you will take notice of his disingenuity in this particular since tis onely my violent distempers & your owne silence that were the cause of mine I shall Answer yours God willing more fully next weeke. I am Sr

Yours to serve you

JOHN FLAMSTEED

To Mr Isaack Newton
Fellow of Trinity
Collidge in
Cambridge
These present

OBSERVATIONES LUNÆ. LUNÆ DISTANTIÆ À FIXIS

1677 Jan		Temp app h ′ ″	♂ Januarij 16die 1677	° ′ ″
☽ a ☉ 9s. 14°	♂ 16	16.31.28	☽ᵃᵉ Limbus seq & prox a Spica ♏s	27:25:03
		35.24	. . . . rep.	27.26.39
		40.02	. . . . iter.	27.28.30
		53.17	☽ Limb: prox ab Ophiuchi Yed	22.50.37
		55.45	. . . . rep.	22.49.43
		17.00.25	. . . . iter	22.49.12
		7.35	☽ᵃᵉ Limb. prox a Spica ♏s . . . . . iter	27.41.12
		16.55	. . . . rep.	27.44.03
2s. 17°	☽ 29	6.42.43	☽ limb. prox ab Alamech Androm	25.34.45
		45.18	. . . . accur rep	25.34.45
			Alamech Androm et ad ρ borealis	26.25.10
			. . . . Australis	26.43.54
		7.09.33	☽ limb: prec. rem. a Cornu ♉′ boreo	36.30.58
		13.03	. . . . rep	36.29.53
		8.20.26	Australis duarum ad ρ ♈tis erat in recta per Cuspides ducta	
3s. 11°	☿ 31	7.12.40	☽ᵃᵉ limb. præcedens a Polluce	43.18.27
		17.20	. . . . rep	43.17.07
		19.25	. . . . iter	43.15.10
		31.48	☽ᵃᵉ limbus boreus a genu Persei. ε	19.30.03
		37.57	. . . . rep	19.30.53
		45.30	☽ᵃᵉ limb. rem: & præce. a Polluce	43.07.09
		48.42	. . . . rep	43.06.00
		50.38	. . . . iter	43.05.24
		7.56.28	Pollux a genu. Persei. ε	47.51.18

Feb:	15.54.35	☽ᵉ limb sequens a Spica ♍	30.37.51
♃. 8 die	16.01.38	. . . . rep	30.34.39
	11.28	☽ᵉ limb: remot: a Cauda Leonis	28.24.42
	14.10	. . . . rep	28.24.42
	22.30	☽ᵉ limb. seq & prox a Spica ♍s	30.24.58
	27.50	. . . . rep	30.22.47
1s. 23° Martij	8.17.50	☽ᵉ limb: rem a sequente humero Orionis	20.56.13
♂ 27 die	20.08	. . . . rep	20.55.28
	27.43	☽ᵉ limb. Infr. rem: a Cornu ♉i boreo	10.31.28
	31.46	. . . . rep	10.30.10
	39.20	☽ᵉ limb præc: rem. ab humero Orionis	20.48.27
2s 17 Martij	8.15.43	☽ᵉ limb. prox a Capella	30.13.31
♃ 29⁽⁴⁾	18.35	. . . . rep	30.14.30
	25.48	☽ᵉ limb. rem & Aust a Polluce	18.00.03
	28.40	. . . . rep	17.59.26
	29.35	. . . . iter	17.59.02
	34.42	☽ᵉ limb. prox a Capella	30.18.57
	36.45	. . . . rep	30.19.27
	38.25	. . . . iter	30.17.24
	43	*Sequens humerus Orionis a cornu Tauri boreo	22.16.57
	53	Capella a Polluce	34.18.03
		. . . . rep	34.17.58
2s. 29° ♀ 30	8.35.18	☽ᵉ limb: prox a Calce Castoris	15.09.49
	40.15	. . . . rep	15.11.46
	46.05	☽ᵉ limb: rem a Polluce	11.14.54
	48.25	. . . . rep	11.14.52
	55.30	☽ᵉ limb. prox a Calce Castoris	15.17.54
	57.31	. . . . rep	15.18.46
Aprilis	8.27.06	☽ᵉ limb. prox præc. sup. a Regulo	11.37.25
♂ 3 die	31.42	. . . . rep	11.39.39
	37.47	☽ᵉ limb. præc. . . a Cauda ♌s	21.42.43
	39.23	. . . . rep	21.12.25
		tunc Nubes usque	
	9.32.24	☽ᵉ limb. prox a Regulo. . . . . . . iter	12.01.52
	34.35	. . . . rep	12.02.37
5s. 25° ♀ 6	8.34.42	☽ᵉ limb. præc. a Spica ♍s	4.21.12
	37.23	. . . . rep	4.20.24
	44.44	☽ limb. Supior a Cingulo ♍sδ	15.51.49
	48.45	. . . . rep	15.50.24
	53.47	☽ limb: præced. rem. & Spica	4.14.33
	55.52	. . . . rep	4.13.50
		Spica ♍s a Cingulo ♍δ	16.17.28
⁽⁵⁾	9.52.22	☽ᵉ altæ 25 gr Diam: 6480=32′: 19½″	
		. . . . rep 6469=32. 16	
♄ 28 die	9.54.50	☽ᵉ limb. præced. prox a Polluce	21.09.31
	57.45	. . . . rep	21.10.44
	10.04.09	☽ᵉ limb. rem. a juba ♌s γ	23.34.00
	07.10	. . . . rep	23.33.00
	18.02	☽ᵉ limb. præc. a Polluce	21.19.15
	20.11	. . . . rep	21.20.06

* pro Observation-
ibus 27æ Noctis

146

3s 23° Maij ♃ 30	8.34.37 55.28 58.21 10.01.36 7.17 10.31 16.16 19.07	☽ altæ 27½ Diam 6305 = 31.27 ☽ limb. rem. a Spica ♍s . . . . rep . . . . iter ☽ limb. prox a Cauda ♌s . . . . rep ☽ limb. rem. a Spica ♍ s . . . . rep	 14.40.34 14.39.26 14.37.54 25.05.30 25.06.22 14.31.24 14.30.18
+6s 20° Junij ☿ 6 die	12.18.15 21.50 39.45 42.47 48.28 50.11 56.20 57.22 13.12.06 14.32	☽ limb seq. rem: & Ophiuci quæ ad genu η . . . . rep . . . . iterum . . . . iterum ☽ in *Nonagesimo* ☽ᵃᵉ limb Supior ab Aquila . . . . rep ☽ᵃᵉ Diameter . . . . . .6740 = 33′. 37″ rep. altæ ea 18 gr 6737 = 33 . 36	35.21.31 35.23.16 35.31.39 35.32.54 35.35.30 35.36.21 27.31.01 27.30.42
8s 10° ☉ 10⁽⁶⁾	14:52:34 58.25 10.01.10 12.40 14.15 26.04 27.28	☽ limbus infer vel Austrinus a Markab Pegasi in Nonagesimo Markab a κ Pisciam [Piscium] ☽ limb. boreus & Markab ☽ limb seq a lucida ♈tis . . . . rep κ, ✕ium vix. ♃s diam: a limbo. lucida limbum tetegit ad montem Alabastrinum in rectatum Creta & Porphyritide cujus puncti distantia a Cuspide boreo 2980 = 14′. 52″ ☽ altæ 33½ Diameter 6441 = 32′.08″	15.27.10 14.58.23 14.54.36 44.46.57 44.46.17
+3s. 5° ☿ 27	9.19.04 29.32 31.50 37.12 39.56 46.28 56.30	☽ limb. præc. rem: ab. Antare. . . . . rep . . . . inclius [melius] iter ☽ limb: infer. ab Arcturo . . . . . super. ab Arcturo ☽ limb. rem. ab Antare ☽ altæ 6° Diam: 6165 = 30.45	51.20.12 51.15.48 51.14.37 36.24.18 35.52.46 51.07.34
♃ 28	9.18.41 49.55 52.09 53.42 10.00.54 3.36 8.14 11.46	☽ altæ 13° Diam: 6347 = 31.40 ☽ limb: rem: ab Antare . . . . rep . . . . iter: ☽ limb: bor. ab Arcturo . . . . rep ☽ limb: rem: ab Antare . . . . rep	 37.52.46 37.51.58 37.51.10 35.34.12 35.24.12 37.44.12 37.42.41

[In Newton's hand, on the reverse of the last sheet]

Apogæum ☽ᵃ in ♋ 7½ gr	Apogæum [☽]ᵃ in ♑ 7½ gr	Apogæum ☽ᵃ in ♈ 7½ gr	Apogæum in ♎ 7½ gr	Argumentum annuum 1677.	Novilunum medium anno 1677
1642 Jan 3	1646 June 6			Mar 9d. 270°	Jan. 22d. 22h.
1650 Nov 8	1655 Apr. 12			Jun 22. 00°.	Feb. 21. 11
1659 Sept 14	1664 Feb 15			Oct 5. 90°.	Mar. 23. 0.
1668 Jul 20	1672 Dec 21			1678	Apr. 21. 12½
1677 May 24	1681 Oct 26			Jan 21. 180.	May 21. 1¼
1686 March 30	1690 Aug 31				Jun 19. 14
1695 Feb 2	1699 Jul 6				Jul 19. 2¾
					Aug. 17. 15½
					Sep. 16. 4⅕
					Oct 15. 17
					Nov. 14. 5⅔
					Dec. 13. 18⅖
					Jan 12. 7

Notes on the Table

♂ Januarij 16 die 1677 = Tuesday, 16 January 1677.

☽ a ☉   = The angular distance of the Moon from the Sun.

	h ′ ″	
16 January	16.31.28	The Moon's westerly and nearer edge from Spica (α Virginis)
	53.17	The Moon's nearer edge from Ophiuchi Yed (δ Ophiuchi)
29	6.42.43	The Moon's nearer edge from Alamech Andromedæ (γ Andromedæ)
	7.09.33	The Moon's easterly edge remote from the northern horn of Taurus
31	7.12.40	The Moon's easterly edge from Pollux (β Geminorum)
	31.48	The Moon's northerly edge from the knee of Perseus ε
	7.56.28	Pollux from the knee of Perseus ε
8 February	16.11.28	The Moon's distant edge from the Lion's tail
27 March	8.17.50	The Moon's edge remote from the westerly shoulder of Orion (Betelgeuse α Orionis)
	27.43	The Moon's lower edge remote from the northern horn of Taurus
	39.20	The Moon's easterly edge remote from the shoulder of Orion
29	8.15.43	The Moon's edge next to Capella (α Aurigæ)
	25.48	The Moon's southerly edge from α Pollux (β Geminorum)
	43	The westerly shoulder of Orion from the northern horn of Taurus
	53	Capella from Pollux
30	8.35.18	The Moon's edge nearest to the heel of Castor (α Geminorum)
3 April	8.27.06	The Moon's upper easterly edge nearest to Regulus (α Leonis)
	37.47	The Moon's easterly edge from the Lion's tail
6	8.44.44	The Moon's upper edge from the Virgin's girdle
	55.52	Spica (α Virginis) from the Virgin's girdle
28	10.04.09	The Moon's distant edge from the Lion's mane
30 May	10. 7.17	The Moon's near edge from the Lion's tail

6 June	12.56.20	The Moon's upper edge from Aquila (κ Piscium)
10	14.52.34	The Moon's lower, or southerly edge from Markab Pegasi (α Pegasi)
	58.25	Markab from κ Piscium (Aquila)
	10.01.10	The Moon's northerly edge from Markab
	12.40	The Moon's westerly edge from the bright star of Aries (Alcyone)
27	9.19.04	The Moon's distant easterly edge from Antares (α Scorpii)
	37.12	The Moon's lower edge from Arcturus (α Boötis)

NOTES

(1) The observations referred to are given in the enclosed table (U.L.C. Add. 3969, fo. 11). The heading is taken from Flamsteed's *Historia Cœlestis Britannica*, vol. I, p. 212, where the complete table, covering the period 1676 to 1689, appears. The entries on the left of the margin (distances of Moon from Sun) are in Newton's hand.

In his letter of 2 July 1695 (Letter 517), Flamsteed wrote: 'I have a large Number of ye Moons distances from fixed stars taken with ye sextant betwixt ye yeares 1676 & 90 by me.' The accompanying observations cover the period 16 January to 28 June 1677. On 18 July (Letter 522) he wrote: 'I shall send you another sheet of them.' This was sent with his letter dated 6 August (Letter 527). This covers the period 2 October 1685 to 29 May 1686. The two tables are briefly described in the printed *Catalogue of the Portsmouth Papers* at Section XII (Astronomy), p. 9: U.L.C. Add. 3969, fo. 11, 12v. 'Lunar distances in 1677 and in 1685 by Flamsteed'.

(2) These were sent on 23 July 1695 (see Letter 524).

(3) See opening paragraph of Letter 523 and of Letter 525.

(4) Opposite the asterisk in the entry for 29 March, the words 'pro Observationibus 27ᵃᵉ Noctis' have been inserted in the margin.

(5) In the entry for 6 April, Flamsteed in his *Historia* (p. 216) wrote:

9.52.22   Lunæ altæ 25 gr. Di-   6480 part per Mic. = 32' 19½"
           ameter

           ——————————————— rep. 6469   ,,   ,,   ,,   = 32 16

(6) Opposite the entry for 10 June, the following has been inserted in the margin:

h   '   "
14.   42.   42

☽ lim: prox
& seq. a
lucida ♈
h   '   "
45.   00.   36
h   '   "
14.   44.   57
              rep
44.   59.   07

522   FLAMSTEED TO NEWTON
18 JULY 1695
From the original in the University Library, Cambridge.[1]
In further reply to Letter 520; for answer see Letter 523

The Observatory July 18. 1695.

Sr

I have sent you the Catalogue of the fixed stars I promised you. & now haveing this you may Calculate ye places of the Moon from ye observations I sent you your selves. My Indisp[os]ition has hindred the transcription of ye Nonagesimary Table.[2] but you shall have it as soone as I can get it Copied for you.

When you have gathered the moones places from the observations I sent you last let me know it I shall send you another sheet of them. I have calculated none from any of the observations made since January last the paines of my head not permitting me to sit to that worke tho they seldome have hindred me from observeing her when ye heavens were cleare.

I have just cause to complaine of the stile & expressions of your last letter, they are not freindly but that you may know me not to be of yt quarrelsome humor I am represented by ye Clerk of ye Society.[3] I shall wave all save this expression *that what you communicated to me was of more valew yn many observations.*[4] I grant it. as ye Wier is of more worth then ye gould from which twas drawne: I gathered the gould melted refined & presented it to you sometimes unasked[5] I hope you valew not my paines ye less because they became yours so easily. I allow you to valew your own as high as you please & require no other reward for what assistance I sometimes afford you, but that I may now & then see some of your workemanship. & if that be not ready when I desire it or if you thinke it not fit to favor me wth it I can easily be contented nor doe I take it amiss yt you often take no notice of some small particulars wherein I have desired to know what you have determined since I know very Well that in thinges of their Nature. it is difficult to determine & wee often change what at first wee thought would need no alteration or correction. I have altered my solar Numbers 5 times & would not be ashamed to change againe If I saw reason for it; If you answer me that you have not determined whether any other then the usuall æquations are to be used in ye Syzigies, if you are not resolved how the))s meane motion is to be corrected you may say it. I shall urge you no further & nevertheless when ever you let me know that it lies in my power to serve you I shall doe it freely but you will not complaine of me to others without Cause & thereby added to ye afflictions I suffer from my obstinate distempers & ye Calumnies of disingenuous & impudent people if you have any valew for Your freind

& humble Servant

JOHN FLAMSTEED

To
Mr Isaack Newton fellow
  of Trinity Colledge
    in Cambridge.
      these present

NOTES

(1) Add. 3979, fo. 37. See Letters 517 and 521, and the Table of Letter 521.

(2) The table was sent on 23 July 1695 (Letter 524).

(3) Halley. He was appointed Clerk, or Assistant Secretary, to the Royal Society on 27 January 1685/6, and he continued in that office till 1696. See Birch, IV, 450 note, for Halley's letter to William Molyneux dated 27 March 1686 which relates to his appointment.

(4) This passage was underlined by Flamsteed. In view of his oft-repeated offer to supply Newton with all the observations he required, Newton's remark seems needlessly uncivil. But see the opening paragraph of the following letter.

(5) See Brewster, II, 178, f.n.: "'Machin told me' says Conduitt 'that Flamsteed said "Sir Isaac worked with the ore he had dug" to which Sir Isaac replied "if he dug the ore, I made the gold ring."'"

523   NEWTON TO FLAMSTEED
20 JULY 1695
From the original in Corpus Christi College Library, Oxford.[1]
In reply to Letters 521 and 522; for answer see Letter 524

Sr

The report you mention was much against my mind & I have writ to put a stop to it. I thank you for your communications of ye Table of fixt stars & your Lunar Observations. So soon as I have got some business off my hands I intend to get such of them calculated as I have need of & send you ye places. The Moons mean motion is not much amiss & may be retained as you printed it till I can determin it more exactly. I beleive there is an equation requisite in ye syzygies but I am not yet master of it. Such niceties I have not yet determined, & you must have patience wth me till I can compass them, otherwise I must desist, as your impatience had once made me resolve to do. The Horroxian Theory by the Table of excentricities & equations of ye Apoge wch I sent you,[2] never errs above 10 or 12 minutes, & so is twice as exact as your

printed Tables wch erre sometimes 20 or 21 minutes; but I would not advise
you to spend your time calculating by it till I have compassed the small
equations, wch I cannot do till I have observations for a sufficient number of
cases. Such expostulations or expressions in your last & some other letters as
tend to a difference I pass by. Pray take care of your health. Dr Battely[3]
(chapplain to Archbishop Sancroft) was much troubled wth violent headachs
& found it a certain cure to bind his head strait wth a garter till ye crown of
his head was nummed. For thereby his head was cooled by retar[d]ing the
circulation of the blood. Tis an easy remedy if your pain be of the same kind.
I am

                                        Your humble servant

                                                Is. NEWTON

Cambridge Jul. 20th
        1695.

For Mr John Flamsteed at the
    Observatory in Greenwich
        neare
            London.

                        NOTES

(1) C. 361, no. 101.

(2) Sent with Letter 500.

(3) John Batteley (1647–1708), D.D., chaplain to William Sancroft, who became Archbishop of Canterbury in 1678. See *D.N.B.*

        524   FLAMSTEED TO NEWTON
                23 JULY 1695
    From the original in the University Library, Cambridge.[1]
        In reply to Letter 523; for answer see Letter 525

                        The Observatory. July 23 1695
Sr.

I send you here the Nonagesimary Table.[2] the originall is calculated to
seconds, but so much preciseness being of no use I omitted them in this copy.
it will be of great use to you in Calculateing the parallaxes of longitude &
latitude for turneing the Moons visible places into ye true. I would not have
you want any thing that lies in my power to save you ye labor of Calculation.
& therefore have presented you with it before you demand it.

                        152

Yours of ye 20th Instant I have receaved this morneing[:] it setts all right betwixt us.[3] I have as great a Stock of patience & as good an one as I have of observations & tis allwayes drawn out on every occasion to serve my friends. My indisposition hindred me from serveing you as I desired[:] you mistooke ye reason of my silence, & I hope [you] will have ye patience on My account that You demand of me on yours. By the waters & excercise my Malady is something abated. but it returnes every day by 10 or 12 a clock & holds me till 4 or 5. It begins sooner if I set to write in the morneing which makes me thinke some small doses of ye Jesuits powder[4] will be the most proper remedy[.] I have an aversion to it but I feare must use it least it turn to an Ague in ye Autumn that may be difficult to be removed. Were my case the same wth Dr Battelys I should gladly trie that experiment. however I thanke you kindly for your account of his remedy. it shews your freindly concern for my wellfare.

I have rode near 40 miles since yesterday morneing & find such good effect of it that I resolve to trie it tho not in such long Jorneys every day this week. the next I am goeing to my parsonage:[5] but I shall take care to have you furnisht wth another sheet of observations before. If you had rather have any other then the remaines of ye year 1677 let me know it I shall fit you according to your desires: tho I beleive they Answer your demands & when I consider your letter I find that very few will be omitted. for allwayes the moones Apoge will be neare ye Suzigies Quadratures or Octants of the Suns. & therefore except you direct other ways shall copy them for you as they lie in my bookes.

I am glad to find by your last save one[6] that the p[ar]allactick equation will be diminished to 2. 3 or 4 minutes, since hence it seemes to me to follow that the small parallax of ye Sun may be reteined which my observations of Eclipses will scarce allow to be encreased especially since you have diminished the Moones parallaxes. By frequent trialls and alterations of his contrivances to answer ye appearances of the heavens Kepler found out the true theory of ye planetary motions[:] you need not be ashamed to own yt you follow his example. when ye Inæqualityes are found you will more easily find the reason of them yn he could to whom but little of ye Doctrine of gravity was known. I wish you all good success & health which is now the more precious because seldome enjoyed by Sr Your Servant.

JOHN FLAMSTEED

    To
Mr Isaack Newton
fellow of Trinity Colledge
  in Cambridge these
     there
           present.
   Cambridge.

A Nonagesimary Table for ye Observatory.   Latitude 51°.29′

AR mc	Nonages	Altit	AR mc	Nonages	Altit	AR mc	Nonages	Altit	AR mc	Nonages	Altit	AR mc	Nonag:	Altit	AR mc	Nonages	Altit
0	♈.26.35	44.08.	60	♊.8.53	59.49	120	♋.21.07	59.49	180	♍ 3.25	44.08	240	♏. 0.27	21.10	300	♒ 29.33	21.10
1	27.19	44.29.	61	9.35	59.57.	1	21.49	59.40.	1	4.09	43.47.	1	1.54	20.49.	1	♓ 0.59	21.30.
2	28.03	44.50.	62	10.17	60.06.	2	22.31	59.30.	2	4.53	43.25.	2	3.23	20.29.	2	2.22	21.51.
3	28.47	45.12.	63	10.59	60.13.	3	23.13	59.22.	3	5.38	43.03.	3	4.54	20.09.	3	3.43	22.13.
4	29.30	45.32.	64	11.41	60.21.	4	23.55	59.12.	4	6.23	42.41.	4	6.28	19.50.	4	5.04	22.34.
5	♉.00.13.	45.53.	65	12.24	60.29	125	24.37	59.02¼	185	7.07	42.19.	245	8.03	19.31	305	6.22	22.56.
6	00.57	46.14	66	13.06	60.36.	6	25.19	58.52.	6	7.51	41.57.	6	9.41	19.12.	6	7.39	23.18.
7	1.40	46.34.	67	13.48	60.42.	7	26.01	58.42.	7	8.36	41.35.	7	11.21	18.54.	7	8.55	23.40.
8	2.23	46.54.	68	14.30	60.49.	8	26.43	58.31.	8	9.22	41.13.	8	13.03	18.36.	8	10.09	24.03.
9	3.06	47.15	69	15.12	60.55.	9	27.25	58.20.	9	10.08	40.50.	9	14.48	18.19.	9	11.21	24.25.
10	3.49	47.35.	70	15.54	61.01	130	28.07	58.09.	190	10.54	40.28.	250	16.35	18.02.	310	12.32	24.48.
11	4.32	47.54.	71	16.37	61.07.	1	28.39	57.58.	1	11.40	40.05.	1	18.24	17.46.	1	13.42	25.11.
12	5.14.	48.14	72	17.19	61.12.	2	29.31	57.46.	2	12.26	39.43	2	20.16	17.31.	2	14.51	25.34.
13	5.57	48.33.	73	18.00	61.17.	3	♌ 0.13	57.34.	3	13.12	39.20.	3	22.11	17.16.	3	15.58	25.57.
14	6 40	48.53	74	18.43	61.22.	4	0.55	57.22.	4	13.59	38.57.	4	24.08	17.01.	4	17.05	26.20.
15	7.22.	49.12	75	19.25	61.27	135	1.37	57.10	195	14.46	38.34	255	26.07.	16.46.	315	18.10	26.44.
16	8.05	49.31.	76	20.08	61.31.	6	2.19	56.57.	6	15.33	38.11	6	♐ 28.09.	16.35.	6	19.14	27.07.
17	8.47	49.49.	77	20.50	61.35.	7	3.00.	56.44.	7	16.21	37.47.	7	0.14	16.22.	7	20.18	27.31.
18	9.30	50.08	78	21.32	61.39.	8	3.42	56.31.	8	17.09	37.24.	8	2.21	16.11.	8	21.20	27.54.
19	10.12	50.26.	79	22.15	61.42.	9	4.24	56.18.	9	17.57	37.01.	9	4.30	16.00.	9	22.21	28.18.
20	10.54	50.45.	80	22.57	61.45.	140	5.06	56.04.	200	18.46	36.37.	260	6.41	15.50.	320	23.22	28.42.
21	11.36.	51.02.	81	23.39	61.48.	1	5.38	55.50.	1	19.35	36.14	1	8.54	15.41.	1	24.21	29.05.
22	12.19	51.20.	82	24.21.	61.50.	2	6.30	55.36.	2	20.24	35.50.	2	11.10	15.33.	2	25.20	29.29.
23	13.01	51.38.	83	25.04	61.52.	3	7.12	55.22.	3	21.14	35.27.	3	13.27	15.26.	3	26.18	29.53.
24	13.43	51.55.	84	25.46	61.54.	4	7.54	55.07.	4	22.04	35.03.	4	15.46	15.20.	4	27.16	30.17.
25	14.25	52.12	85	26.28	61.56	145	8.35	54.53	205	22.54	34.39	265	18.06	15.14	325	28.12	30.41

154

Tables of Houses (six cusp panels). Footnote markers at head of first panel: a 27.36, b 2.24, b 2.54.

Panel 1 — Nos. 326–360

No.	♑	Ascen.	2	3 ♈
6	15.10	20.27	31.05	29.08
7	15.06	22.49	31.29	0.03
8	15.04	25.12	31.53	0.58
9	15.02	27ª.86	32.16	1.52
330	15.02	0.00	32.40	2.46
1	15.02	2.54	33.04	3.39
2	15.04	4.48	33.28	4.31
3	15.06	7.11	33.52	5.23
4	15.10	9.33	34.16	6.15
335	15.14	11.54	34.39	7.06
6	15.20	14.15	35.03	7.56
7	15.26	16.33	35.27	8.46
8	15.33	18.51	35.50	9.36
9	15.41	21.06	36.14	10.25
340	15.50	23.19	36.37	11.14
1	16.00	25.30	37.01	12.03
2	16.11	27.39	37.24	12.51
3	16.23	29.46	37.47	13.39
4	16.35	1.51	38.11	14.27
345	16.48	3.53	38.34	15.14
6	17.01	5.52	38.57	16.01
7	17.16	7.49	39.20	16.48
8	17.31	9.44	39.43	17.34
9	17.46	11.36	40.05	18.20
350	18.02	13.25	40.28	19.06
1	18.19	15.12	40.50	19.52
2	18.36	16.57	41.13	20.38
3	18.54	18.39	41.35	21.24
4	19.12	20.19	41.57	22.09
355	19.31	21.57	42.19	22.53
6	19.50	23.32	42.41	23.38
7	20.09	25.06	43.03	24.22
8	20.29	26.37	43.25	25.07
9	20.49	28.06	43.47	25.51
360 ♈	21.10	29.33	44.08	26.35

Panel 2 — Nos. 266–300

No.	♑	Ascen. §§
6	15.10	2.27
7	15.06	2.
8	.04	2.
9	15.02	2.
270	15.02	0.
1	15.02	2.54
2	15.04	4.48
3	15.06	7.11
4	15.10	9.33
275	15.14	11.54
6	15.20	14.15
7	15.26	16.33
8	15.33	18.51
9	15.41	21.06
280	15.50	23.19
1	16.00	25.30
2	16.11	27.39
3	16.23	29.46
4	16.35	1.51
285	16.48	3.53
6	17.01	5.52
7	17.16	7.49
8	17.31	9.44
9	17.46	11.36
290	18.02	13.25
1	18.19	15.12
2	18.36	16.57
3	18.54	18.39
4	19.12	20.19
295	19.31	21.57
6	19.50	23.32
7	20.09	25.06
8	20.29	26.37
9	20.49	28.06
300	21.10	29.33

Panel 3 — Nos. 206–240 (♏)

No.	col 1	col 2
6	34.16	23.45
7	33.52	24.37
8	33.28	25.30
9	33.04	26.21
210	32.40	27.14
1	32.16	28.08
2	31.52	29.02
3	31.29	29.56
4	31.04	0.52
215	30.41	1.48
6	30.17	2.44
7	29.53	3.42
8	29.29	4.40
9	29.05	5.39
220	28.42	6.38
1	28.18	7.39
2	27.54	8.40
3	27.31	9.43
4	27.07	10.46
225	26.44	11.50
6	26.20	12.55
7	25.57	14.02
8	25.34	15.09
9	25.11	16.18
230	24.48	17.28
1	24.25	18.39
2	24.03	19.51
3	23.40	21.05
4	23.18	22.21
235	22.56	23.38
6	22.34	24.56
7	22.13	26.16
8	21.51	27.38
9	21.30	29.01
240 ♏	21.10	0.27

Panel 4 — Nos. 146–180 (♍)

No.	col 1	col 2
6	54.38	9.17
7	54.22	9.59
8	54.07	10.41
9	53.51	11.23
150	53.35	12.05
1	53.19	12.47
2	53.03	13.29
3	52.46	14.11
4	52.29	14.53
155	52.12	15.35
6	51.55	16.17
7	51.38	16.59
8	51.20	17.41
9	51.02	18.24
160	50.44	19.06
1	50.26	19.48
2	50.08	20.30
3	49.49	21.13
4	49.31	21.55
165	49.12	22.38
6	48.53	23.20
7	48.33	24.03
8	48.14	24.45
9	47.54	25.28
170	47.34¾	26.11
1	47.15	26.54
2	46.54½	27.37
3	46.34	28.20
4	46.14	29.03
175	45.53	29.47
6	45.32	0.30
7	45.12	1.14
8	44.51	1.57
9	44.29	2.41
180 ♍	44.08	3.25

Panel 5 — Nos. 86–120 (♒)

No.	col 1	col 2
86	61.57.	27.12
87	61.58.	27.53
88	61.59	28.35
89	62.00	29.18
90	62.00	♒: 0.00
91	62.00	0.42
92	61.59	1.25
93	61.58	2.07
94	61.57.	2.48
95	61.56	3.32
96	61.54.	4.14
97	61.52.	4.56
98	61.50	5.39
99	61.48	6.21
100	61.45	7.03
1	61.42	7.45
2	61.39	8.28
3	61.35	9.10
4	61.31.	9.52
105	61.27	10.34
6	61.22	11.17
7	61.17.	11.59
8	61.12.	12.41
9	61.07	13.23.
110	61.01.	14.06
1	60.55.	14.48
2	60.49	15.30
3	60.42.	16.12
4	60.36	16.54
115	60.29	17.36.
6	60.21.	18.19
7	60.13¾	19.01
8	60.06	19.43
9	59.57.	20.25
120	59.49	21.07

Panel 6 — Nos. 26–60 (Π)

No.	col 1	col 2
26	52.29.	15.00
27	52.46.	15.49
28	53.03	16.31
29	53.19	17.13
30	53.35	17.55
31	53.51	18.37
32	54.07	19.19
33	54.22.	20.01
34	54.38	20.43
35	54.53	21.25
36	55.07.	22.06
37	55.22	22.48
38	55.36	23.30
39	55.50.	24.12
40	56.04	24.54
41	56.18	25.26
42	56.31	26.18
43	56.44	27.00
44	56.57	27.41.
45 Π	57.10	28.23.
46	57.22.	29.05.
47	57.34.	29.47
48	57.46	00.29
49	57.58	1.11
50	58.09.	1.53
51	58.20.	2.35
52	58.31.	3.17
53	58.42	3.59
54	58.52	4.41
55	59.02.	5.23
56	59.12.	6.05
57	59.22.	6.47
58	59.31.	7.29
59	59.40.	8.11
60	59.49	8.53

155

NOTES

(1) Add. 3979, fo. 38.

(2) The Table was enclosed with this letter. In breaking the seal Newton has destroyed a fragment of the table, but has replaced the missing entries by the side. Against the entry 27.86 he has placed 27.36, and two lines lower, against 2.54, he has placed 2.24. The nonagesimal point is that point of the ecliptic which is highest above the horizon at any given time, being 90° above the point at which the ecliptic intersects the horizon; also called the mid-heaven. See *O.E.D.*

(3) In spite of Flamsteed's conciliatory letter, the hope of a lasting peace between the two men was not realized.

(4) 'Jesuits powder', a febrifuge introduced into Europe about 1633 from Loxa in Ecuador, was prepared from the bark of the 'fever tree' (cinchona): it was used so extensively by the Jesuits that it came to be known as 'Jesuits' bark', or 'Jesuits' powder'. It has often been confused with the inferior febrifuge obtained from the Peruvian balsam (*quina-quina*), the name quinine being applied to cinchona. See A. W. Haggis, *Bull. Hist. Med.* **10** (1941), 417–59 and 568–92.

(5) This was at Burstow, near Reigate in Surrey, about twenty miles from Greenwich.

(6) See Letter 520.

525   NEWTON TO FLAMSTEED

27 JULY 1695

From the original in Corpus Christi College Library, Oxford.[1]
In reply to Letter 524; for answer see Letter 526

Sr

The other day I had an excuse sent me[2] for what was said at London about your not communicating, & that the report should proceed no further. I am glad all misunderstandings are composed. I thank you for your nonagesimal table. I designed to make such a Table & it saves me ye labour. You may continue your observations if you please till Octob 10th 1677.[3] But I had rather you would send me those from Aug 24th 1685 to July 5th 1686, when ye Aphelium was in the same position as in ye year 1677. For when I see all your Observations together in this position of ye Aphelium I can tell better what to select for this case. The transcribing of these things gives your servant trouble & for encouraging him I shall order Will Martin the Cambridge Carrier (who lodges every week, from 9 in ye Morning on Saturday till 3 in ye afternoon on Munday, at the Bull in Bishopsgate street) to pay him [two guineas if you please to let him call for them, or to pay it to his or your order in London if you please to let me know where].[4] I shall not have time to go through all your observations, but will send you the times for wch I would have them when I have done wth these for this position of ye Aphelium. I am

Your thankfull humble

servant

Is. NEWTON

Cambridge. Jul. 27.
      1695

    For Mr John Flamsteed at
    the Observatory in Greenwich
    neare London.

NOTES

(1) C. 361, no. 103.

(2) According to Brewster (II, 180) the 'excuse' came from Bentley or Halley.

(3) In the table (521) the entries terminated at 28 June 1677.

(4) The words within editorial brackets are crossed out in the manuscript after the word 'guineas' had been altered to 'shillings', all apparently by Flamsteed. The words after 'for them' are conjectural, the rest of the sentence having been very effectively crossed out.

In the copy of this letter in the British Museum, the words 'two shillings' appear with the following note: 'Mr Flamsteed altered it so for the word *guineas*, which is in the original, as is evident from the erasure.' See Brewster, II, 180 note and Edleston, p. lxviii note 125.

526   FLAMSTEED TO NEWTON

4 AUGUST 1695

From the original in the University Library, Cambridge.[1]
In reply to Letter 525

The Observatory August 4. ♄ 1695

Sr.

    That ye want of a letter from me may not cause any misapprehensions as it did lately betwixt us I send this to acquaint you I have had employment for my servant this week yt has hindred him from copying the Observations[2] I desired for you. I intend he shall transcribe them the begining of the next that the[y] may be sent you by Thursdays post.

    I take it very kindly that you acquainted me with your intent to gratifie him for his paines before you did it but I must entreat you to forbeare. He is payd all ready.⁽³⁾ a superfluity of monys I find is allways pernicious to my Servants it makes them run into company & wast their time Idly or worse. I take care he wants nothing[:] if you send him verball acknowledgmts of his paines & commendations for his care & fidelity in copying it will be a reward for him & encouragemt ye best you can give him & further I can not allow.

I am goeing into ye Country to my Cure on Wednesday next but if you have occasion to write to me You may direct your letters as formerly I shall order them to be sent to me & you shall have Answers as sure tho not perhaps so speedily as from this place.

My distemper continues but I blesse God my paines tho they are not gone of are something abated. I am Sr

Your affectionat freind & Servant

JOHN FLAMSTEED

Pray say nothing to any body of your proposall:

To
Mr Isaack Newton.
Fellow of Trinity Colledge
in Cambridge   these
      present   there.

NOTES

(1) Add. 3979, fo. 39.

(2) That is, to October 1677. See note (3) of previous letter.

(3) Up to 1696 the assistants at the Observatory were paid (£26 per annum) by the Board of Ordnance.

527   FLAMSTEED TO NEWTON
6 AUGUST 1695
From the original in the University Library, Cambridge.⁽¹⁾
In continuation of Letter 526; for answer see Letter 530

The Observatory August 6. 1695

Sr

Yesterday & this day I blesse God for it I have only had some small grudge-ings of my headache so yt I hope now In a little time to be cleare and able to follow my studys as formerly. I am goeing into the Country this afternoon

158

where I intend to travel often for the Confirmation of this health. & at my returne to fall close to that business which has been interrupted by ye most uncomfortable kind of distemper yt I ever had. for dureing ye Stone and In a consumption I had the satisfaction of enjoyeing the pleasure of my thoughts but this would not permit me that. The fixed stars will be my whole worke[2] when I returne if God Continues my health. & after that I shall thinke of Correcting the Lunar Synopsis whereby it will be evident that in 6 Yeare last past I have done more towards the Restitution of Astronomy then has been done in some ages before: I have sat here now 19 yeares but for the 14 first I have not had such Instruments as were necessary for my worke[3] and had been without them till this day if I had not made them at my owne Charge. I write this purposely to you because I know a Sparke is with you that complaines much I have lived here 20 yeares & printed nothing. I doe not intend to print a St. Helena Catalogue[4] & for that reason I defer the printing of any thing thus long that when I doe print it may be perfect as by the Grace of God it shall. I have sent you a sheet of observations included[5] & wish you health to employ them which has long been wanted by Sr

Your Freind to serve you

JOHN FLAMSTEED

Mr Newton:

To
Mr Isaack Newton
fellow of Trinity
Colledge in Cambridge
these present
Cambridge

♀ Octobris 2. 1685

		h ′ ″		° ′ ″
☽ a ☉ 5s. 29°.22′	1685 ♀ Octob 2	10.21.58	☽æ limb: prox: a lucida Pleiad:	30.42.29
		24.15	. . . . rep:	30.41.30
		28.25	☽æ limb: rem. a Cap. Androm	34.32.54
		30.07	. . . . rep	34.33.03
		33.44	☽æ limb. prox. a lucida Pleiad.	30.38.06
		35.45	. . . . rep	30.37.20
	♀ 30	9.19.52	☽æ limb. prox. ab Aldebare	31.17.53
		23.14	. . . . rep	31.17.00
		29.53	☽æ limb. rem a lucida ♈tis	14.07.12
		32.00	. . . . rep	14.06.51
			☽ in Nonag. hora 9h. 35′ Vere	
		35.38	☽æ limb. prox. a Pallilicio	31.11.58
		38.10	. . . . rep	31.11.17
	♄ 31	9.23.49	☽æ limb. rem. a Pallilicio	19.08.05
		26.55	. . . . rep	19.06.38
		34.00	☽ limb. prox. a Mandib. Ceti.	11.01.28
		37.08	. . . . rep	11.02.37
		42.28	☽æ limb. rem. a Pallilicio	19.00.45
		44.29	. . . . rep	19.00.00
		46.42	Pallilicium a Mandib. Ceti	26.07.03
	Novem	17.14.32	☽ limb. rem. a Calce Castoris	31.06.25
	♀. 6	17.00	. . . . rep	31.07.20
		24.40	☽æ limb. prox. a Clara Seq. in Capite Hydræ [ζ]	16.15.00
	☽ in Non	18.15	. . . . rep	16.14.03
		32.55	☽ limb. rem. a Calce Castor	31.12.58
		34.53	. . . . rep	31.13.51
		40.03	Calx Castoris a Clara Seq in Capite Hydræ [ζ]	40.18.58
	☽ 23	5.36.55	☽ limb. rem. a lucida ♈tis	44.51.40
		38.40	. . . . rep	44.50.44
		44.26	☽ limb. prox. ab ore Pegasi	36.29.48
		46.11	. . . . rep	36.30.07
		49.41	☽ limb. rem. a lucid ♈tis	44.46.10
		52.03	. . . . rep	44.45.32
	☿ 25	6.25.19	☽ limb. rem. a lucida Pleiad	39.29.39
		28.05	. . . . rep	39.28.55
		33.33	☽ limb. prox. a Cap. Andro	31.42.51
		36.13	. . . . rep	31.42.15
		40.01	☽ limb. rem. a lucida Pleiad	39.23.51
		43.00	. . . . rep	39.23.00
		48.00	Lucid. Pleiad. a Cap. Androm	48.48.42
	Decem. ☉s 6	17.30.26	☽æ limb rem a Polluce	46.26.40
		32.33	. . . . rep	46.27.20
		38.22	☽ limb rem a lucida in Lumb: [♌s]	9.38.14
		41.34	. . . . rep	9.38.25
		46.55	☽æ limb rem a Polluce	46.32.58
		49.09	. . . . rep	46.33.43
		51.54	Pollux a lucida in lumbis ♌s	47.44.58

			h ′ ″			° ′ ″
[1685] [Decem.]	☽	7	17.50.17	☽ limb. rem. ab η ♌s		24.51.16
			53.08	. . . . rep		24.50.02
			57.46	☽ᵃᵉ limb. rem. a Cauda ♌s		7.56.01
			18.00.25	. . . . rep		7.56.07
			5.24	☽ limb. rem. ab η ♌s		24.57.10
			8.11	. . . . rep		24.58.33
			12.44	Cauda Leonis ab ejusdem η		24.36.33
1686 Januarij	♃	21	6.37.11	☽ limb. rem. a Calce Castoris		43.45.39
			40.30	. . . . rep		43.44.30
			48.04	☽ᵃᵉ limb. rem. ab Algol Med		26.04.22
			50.49	. . . . rep		26.04.13
			55.32	☽ᵃᵉ limb. rem. a Calce Castoris		43.38.36
			7.01.54	. . . . rep		43.36.39
			4.38	. . . . iterum		43.35.40
			11.31	Calx Castoris ab Algol Medusæ		44.38.13
	☽	25	8.58.04	☽ limb. prox a Pallicio		31.18.12
			9.00.25	. . . . rep		31.19.03
			5.52	☽ limb. rem a Procyone		23.01.39
			8.18	. . . . rep		23.01.07
			12.40	☽ limb. prox a Pallicio . . . iter		31.23.30
			14.58	. . . . iter		31.24.13
			21.46	Pallilicium a Procyone		46.19.52
Februarij	♂	16	6.29.10	☽ limb. rem. a Pallilicio		35.30.51
			31.50	. . . . rep		35.29.44
			38.10	☽ limb. prox a Cingulo Andro		29.56.08
			41.00	. . . . rep		29.56.03
			45.10	☽ limb. rem. a Pallilicio		35.23.55
			48.00	. . . . rep		35.22.42
	♄	20	8.48.23	☽ limb. rem. a Calce Castoris		10.58.15
			50.44	. . . . rep		10.58.07
			57.05	☽ limb. prox a Cornu ♉ i. Bor.		5.27.29
			59.53	. . . . rep		5.27.40
			9.04.56	☽ limb. rem. a Calce Castoris		10.52.48
			7.07	. . . . rep		10.52.05
			11.34	Cornu ♉ i Bor. a Calce Castoris		14.09.14
	☉	21	6.34.13	☽ limb. rem. a Polluce		18.36.06
			36.39	. . . . rep		18.35.45
			42.07	☽ limb. rem. a Calce Castoris		8.42.58
			44.18	. . . . rep		8.42.50
			49.28	☽ limb. rem. a Polluce . . . iter		18.31.10
			51.35	. . . . rep		18.30.45
	♂	23	8.45.25	☽ limb. prox. a Calce Castoris		25.18.30
			47.35	. . . . rep		25.19.22
			52.12	☽ limb. prox. a Castore		12.17.44
			53.57	. . . . rep		12.18.22
			57.28	☽ limb. prox. a Calce . . . iterum		25.23.00
			59.51	. . . . rep		25.23.40

			h ′ ″		° ′ ″
	♃	25	8.26.50	☽ limb. prox a Calce Castoris	49.52.32
			29.44	. . . . rep	49.53.36
			34.00	☽ limb. rem. a Regulo	6.09.05
			36.04	. . . . rep	6.08.28
			40.43	☽ limb. prox. a Calce Castoris	49.58.05
			42.52	. . . . rep	49.58.57
[1686] [Februarij]	♄	27	9.42.33	☽ limb. rem. a Regulo	22.14.02
			45.11	. . . . rep	22.15.03
			51.51	☽ limb. prox a Cauda ♌s	7.29.12
			53.56	. . . . rep	7.27.45
			58.55	☽ limb. rem. a Regulo	22.22.00
			10.01.38	. . . . rep	22.23.03
Martij	☽	2	15.01.12	☽ limb rem. a Vindem.	27.41.21
			3.25	. . . . rep	27.42.13
			8.21	☽ limb. prox a Lucida in⎫ Collo Serpentis ⎭	28.19.13
			10.36	. . . . rep	28.18.55
			14.59	☽ limb rem. a Vindemiatrix	27.46.51
			16.46	. . . . rep	27.47.36
			21.38	Lucida in Collo Serpent a Vindem	40.14.21
		18	7.46.53	☽ Limb. rem. a Calce Castoris	27.49.20
			51.17	. . . . rep	27.47.40
			56.59	☽ Limb. rem. a Pallilicio	4.59.25
			59.00	. . . . rep	4.59.10
			8.03.18	☽ Limb. rem. a Calce Castoris	27.42.10
			8.00	. . . . rep	27.40.05
Aprilis	☽	19	8.43.43	☽ᵃ Limb. rem. a Regulo	26.20.12
			45.49	. . . . rep	26.19.12
			51.32	☽ᵃ Limb. prox. a Procyone	20.42.44
			54.23	. . . . rep	20.42.57
			59.34	☽ᵃ Limb. rem. a Regulo . . iter	26.13.33
			9.01.46	. . . . rep	26.12.36
Maij	♂	. 18	9.46.37	☽ Limb. rem. a Regulo	6.40.40
			49.16	. . . . rep	6.39.46
			56.41	☽ Limb. rem. a Lucida in Collo ♌s	6.53.16
			58.59	. . . . rep	6.52.25
			10.02.36	☽ Limb. rem. a Regulo . . . iter	6.34.10
			4.20	. . . . rep	6.33.30
	♃	. 29	9.09.12	☽ Limb. rem. a Spica ♍s	35.34.14
			11.38	. . . . rep	35.33.18
			18.05	☽ᵃ Limb. rem. a Cauda ♌s	8.23.16
			20.36	. . . . rep	8.23.00
			25.48	☽ Limb. rem. a Spica ♍s. . iter	35.27.06
			27.51	. . . . rep	35.26.14
			32.11	Spica a Cauda ♌s	35.02.07

[In Newton's hand on the last sheet]

Novilunia media 1685		Argumentum annuum 1685	
Aug.	19d.4h	Aug.	26d.90gr
Sept.	17.17	Dec.	5.180
Oct.	17. 5⅗		
Nov.	15.18⅓	1686	
Dec	15. 7	Mar.	13.270
		June.	29. 0
1686		Octob.	12. 90
Jan	13.19⅘	Apogæum ☽ᵐ Mar 30	
Feb.	12. 8½	1686 in ♋ 7½gr.	
Mar	13.21¼		
Apr.	12.10		
May.	11.22¾		
Jun.	11.11½		
Jul.	10. 0⅕		

Notes on the Table

♀ Octobris 2 1685 = Friday, 2 October 1685.
☽ a ☉ = Moon's angular distance from the Sun.

1685	h ′ ″	
2 October	10.21.58	The Moon's nearer edge from the bright star in the Pleiades (Alcyone).
	28.25	The Moon's further edge from the head of Andromeda.
30	9.19.52	The Moon's nearer edge from Aldebaran (α Tauri).
	29.53	The Moon's further edge from the bright star of Aries.
	35.38	The Moon's nearer limb from Pallilicium (Hyades).
31	9.34.00	The Moon's nearer edge from the jaw of the Whale.
	46.42	Pallilicium from the jaw of the Whale.
6 November	17.14.32	The Moon's further edge from the heel of Castor.
	24.40	The Moon's nearer edge from the bright westerly star in Hydra's head.
23	5.44.26	The Moon's nearer edge from the mouth of Pegasus.
25	6.48.00	The bright star of the Pleiades from the head of Andromeda.
6 December	17.38.22	The Moon's further edge from the bright star in the loin [of the Lion].
	51.54	Pollux from the bright star in the loin of the Lion.
7	17.50.17	The Moon's further edge from η Leonis (Algol Medusæ).
	57.46	The Moon's further edge from the Lion's tail.
1686		
21 Jan.	6.37.11	The Moon's further edge from the heel of Castor.
	48.04	The Moon's further edge from Algol Medusæ (η Leonis).
	7.11.31	Heel of Castor from Algol Medusæ.
25	8.58.04	The Moon's nearer edge from Pallilicium.
	5.52	The Moon's further edge from Procyon (α Canis Minoris).
	21.46	Pallilicium from Procyon.
16 Feb.	6.29.10	The Moon's further edge from Pallilicium.
	38.10	The Moon's nearer edge from Andromeda's girdle.

20	8.57.05	The Moon's nearer edge from northern horn of the Bull.
	9.11.34	Northern horn of the Bull from the heel of Castor.
21	6.34.13	The Moon's further edge from Pollux.
25	8.34.00	The Moon's further edge from Regulus (α Leonis).
2 Mar.	15.01.12	The Moon's further edge from Vindemiatrix (ϵ Virginis).
	8.21	The Moon's nearer edge from bright star in the throat of the Serpent.
	21.38	Bright star in the throat of the Serpent from Vindemiatrix.
29	9.32.11	Spica (α Virginis) from the Lion's tail.

NOTES

(1) Add. 3979, fo. 40.

(2) When Flamsteed took up his duties at Greenwich (see note (3) below) he found that the whole of the views then current and the tables of the heavenly bodies then in use were faulty. He thereupon undertook the task of revising them. 'My chief design', he wrote to Dr Seth Ward on 31 January 1679/80, 'is to rectify the places of the fixed stars.' For the many references to Flamsteed's anxiety to accomplish this work, see the Index to the present volume.

(3) Flamsteed entered upon his duties at Greenwich in July 1676 having been appointed 'King's Observator' on 4 March 1674/5. On taking up his appointment he found that he was provided only with a sextant and two clocks. He apparently took other instruments with him from Derby where he had settled in a small living after his ordination. Sir Jonas Moore provided him with an iron sextant of seven feet radius, two great clocks, a telescope object glass of 52 feet focal length, and some books (MacPike, *Hevelius, Flamsteed, Halley*, p. 22). All other instruments and tubes, he declared, were provided at his own expense (Baily, p. 60). He himself constructed a mural quadrant of 50 inches radius at his own cost which he set up and divided in 1683 (MacPike, *op. cit.* p. 23).

(4) Halley sailed for St Helena in November 1676, arriving there the following February. On account of the adverse weather conditions he was not able to make nearly as many observations as he had wished; in fact, to secure the positions of 300 stars not entered in Tycho's Catalogue was as much as he could do in twelve months. In a letter from St Helena to Sir Jonas Moore, dated 28 November 1677, he wrote: 'Such hath been my ill fortune, that the horizon of this island is almost always covered with a cloud, which sometimes for some weeks together hath hid the stars from us, and when it is clear is of so small continuance that we cannot take any number of observations at once' (Library of the National Maritime Museum). The catalogue referred to is entitled: *Catalogus Stellarum Australium sive Supplementum Catalogi Tychonici exhibens longitudines & latitudines Stellarum fixarum, authore Edmundo Halleio e Col. Reg. Oxon. 1679.* This gives the celestial longitude and latitude for 341 stars, grouped by constellations, compared with the observed distances.

(5) The table which follows the letter. The double folio sheet (U.L.C. Add. 3969, fos. 12–13v), like the table communicated earlier (with Letter 521), was made with the sextant. The initial marginal entries, showing the angular distances of the Moon from the Sun, are by Newton. So, too, are the short columns of lunar dates at the end. The complete table appears in Flamsteed's *Historiæ Cælestis, Libri Duo* (1712), p. 330 *et seq.*

528 HALLEY TO NEWTON
7 SEPTEMBER 1695
From the original in the University Library, Cambridge[1]

London Sept. 7 1695

Ever Honoured Sr

Since I left you[2] I have been desirous to make some triall how I could obtain the position of the Orb of the Comet[3] of 1683, and after having gotten some little direction from a course Construction, I took the pains to examine and verifie it by an accurate Calculus, wherin I have exceeded my expectation, finding that a parabolick orb limited according to your Theory will most exactly answer all the Observations Mr Flamsteed and my self formerly made of that Comett, even within the compass of one minute. If that of 1664 be but well observed,[4] I doubt not of the like success therin, but I fear, for want of Telescope sights[5] Hevelius could not sufficiently distinguish the Nucleus therof. It is no great trouble and if you think it requisite I will by the same method compute exactly some or all of Mr Flamsteeds observations[6] of that of 1680/1, which you say, you did *per operationes partim Graphicas,*[7] being desirous as far as you will permitt it, to ease you of as much of the drudging part of your work as I can, that you may be the better at leisure to prosecute your noble endeavours. I begg your pardon that I have not yett returned your Quadratures of Curves, having not yet transcribed them, but no one has seen them, nor shall, but by your directions; and in a few days I will send you them. I should be glad of the favour of a line or two from you, to receive your commands in any thing wherin I may be able to shew my self Sr

Your most faithfull Servant

EDM. HALLEY[8]

To his Honoured friend
Mr Isaac Newton
in Trinity College
in Cambridge

NOTES

(1) Add. 3996, fo. 115.

(2) Halley was with Newton early in August (see the opening paragraph of Letter 530). We have no information as to what was discussed at their meeting, but an interesting and important correspondence upon the existence of elliptical orbits for certain comets follows.

According to the Keplerian view current at the time, a comet travelled in a straight line,

165

and so a comet observed in November 1680 to be passing near the Earth towards the neighbourhood of the Sun, and one observed two months later to be passing near the Earth but
away from the Sun were thought to be two different comets moving in nearly parallel but
opposite directions. Flamsteed rejected the traditional view and believed them to be one and
the same comet. After some consideration Newton concurred with Flamsteed (see Letter 281,
vol. II, p. 419). But it still remained to be decided whether the path was parabolic or elliptical,
for there would be little difference in the neighbourhood of the Sun (and therefore of the
Earth) between a parabola and a very elongated ellipse. Newton, basing his results on three of
Flamsteed's observations (21 December, 5 January, 25 January; see Manuscript 529), worked
out a parabolic path for the comet as it moved about the Sun as a focus. Halley however
inclined to the view that the path was elliptical. 'I find certain indication of an Elliptick Orb
in that Comet [i.e. that of 1680/2] and am satisfied that it will be very difficult to hitt it exactly
by a Parabolick' (Letter 532).

(3) Hevelius communicated the details of his observations of this comet, made from 30 July
to 4 September 1683, to the Royal Society. ('Joh. Hevelii Historiola Cometæ Anni 1683',
Phil. Trans. **13** (1683), 416–24.)

(4) The comet seen from 2 December 1664 to 17 March 1664/5 is described by Auzout,
Phil. Trans. **1** (1665), 3, 18, 19, 107; by English astronomers, pp. 108 and 150; and by G. D.
Cassini, p. 18. Newton records having seen it on eleven nights: first 'a sleighty Observation'
on 10 December, then a detailed observation on the 17th at 4.30 a.m., and then 'On fryday
before midnight Decemb 23 1664 I observed a Comet whose rays were round her, yet her
tayle extended it selfe a little towards ye east parallel to ye Ecliptick. The * itselfe was not
seen onely it looked like a little cloude The altitude of Sirius at ye time of observation was
16 d. The comet was then entering into ye whales mouth at ye nether jaw being distant from
Aldebaran 23d 21' & as much from Rigel. Therefore ye longitude of it was 48° 4'. its latitude
22°, 3', 44". At about 9 hr 23' at night' (U.L.C. Add. 3996, fo. 115).

(5) Hevelius relied upon plain sights although Hooke maintained (*Phil. Trans.* **15** (1685),
1164/5), 'it were not possible with these *Sights* (be the Instruments never so large or accurate,)
to make Observations nearer then to Two or Three whole Minutes: But himself could, with
Telescopick Sights, (by an Instrument but of a Span breadth,) make observations, Thirty, Forty,
Fifty, yea Sixty times more accurate than could be done the other way, with the most Vast
Instruments.' See E. N. da C. Andrade, 'Robert Hooke' (Wilkins Lecture), *Proc. Roy. Soc.*,
1950, A, 201, p. 453.

(6) See the table of Flamsteed's observations in vol. II, Letter 252, pp. 348/9, and *Principia*,
III, 490. Halley fulfilled his promise; see Letter 532.

(7) See Letter 529, note (3), p. 167.

(8) Though no answer to this letter has been found, it is known from Halley's next letter
(Letter 532) that Newton had replied in the meantime. We infer that he accepted Halley's
offer to compute again Flamsteed's observations and that he mentioned his intention of
revising his own computations. See Manuscript 529.

529   A MANUSCRIPT BY NEWTON
13 SEPTEMBER 1695
From the original in the University Library, Cambridge[1]

Sept. 13. 1695. Constructio orbis Cometæ qui annis 1680 & 1681 apparuit.[2]

Tempus verum		Long. ☉is	Dist ☉is a ⊖	Long. Cometæ	Lat Com[et]æ.
1680	Dec. 21d. 6h. 36.′ 59″	9. 11. 6.44	983112	10. 5. 7.38	21.45.30
1681	Jan 5. 6. 1. 38	9. 26.22.17	983966	0. 8.49.10	26.15.26
	Jan 25. 7. 58. 42	10. 16.45.35	986795	1. 9.55.48	17.56.54

Jam ad orbem Cometæ determinandum selegi ex Observationibus hactenus descriptis tres quas Flamsteedus habuit Dec 21, Jan 5 & Jan 25.[3]

NOTES

(1) Add. 3965, fo. 170.

(2) In the table here reproduced, the numbers 9, 10, 0, 1 in the column of longitude are used in place of the signs ♑, ♒, ♈, ♉, respectively. As Halley pointed out (see Letter 532, p. 171) the figures 1s.9°.55′.48″ of longitude for 25 January should read 1s.9°.36′.0″ to agree with those of Flamsteed. He also pointed out that there was an error in Newton's own observation of 25 January. See Table in note (3) below.

(3) This Manuscript indicates some of Newton's work on the computation of the path of the great comet which appeared in November 1680, and which was observed until March following. Newton had begun to study the motion of this comet some ten years earlier; so much is clear from his letter of 19 September 1685 (Letter 281, vol. II, p. 419). 'I have not yet computed ye orbit of a comet, but am now going about it: & taking that of 1680 into fresh consideration, it seems very probable that those of November & December were ye same comet.... My calculation of ye orbit will depend only on three observations & if I can get three at convenient distances exact to a minute or less I hope ye orbit will answer exactly enough not only to ye observations of December January February & March but also to those of November before ye Comet was conjoyned with ye Sun'. Much of the work on the revised computation of the path of the comet was due to Halley, as is clear from the opening paragraph of Letter 532.

Newton was led to consider whether a comet's motion could not be explained, like that of a planet, in terms of a gravitational force towards the Sun. There are four steps in his exposition:

(i) He first established the fact that the comets move in conic sections having their foci in the centre of the Sun, and by radii drawn from the comet to the Sun, describe areas proportional to the times. This is Prop. 40, Book III: *Cometas in Sectionibus conicis umbilicos in centro Solis habentibus moveri, & radiis ad Solem ductis areas temporibus proportionales describere.*

(ii) The cometary orbits approximate so closely to parabolas, that parabolas could be used in place of them without sensible error. This is Corollary 2 of Prop. 40: *Orbes autem erunt Parabolis adeo finitimi, ut eorum vice Parabolæ absque erroribus sensibilibus adhiberi possunt.*

(iii) If the orbit of the comet is assumed to be parabolic, it may be determined from three observations of it. This is Prop. 41: *Cometæ in Parabola moventis Trajectoriam ex datis tribus observationibus determinare.*

(iv) Having now determined the orbit in this way, it remained to show, by considering other observations, that the comet did truly move along it: *Tandem ut constaret an Cometa in Orbe sic invento verè moveretur, collegi per operationes partim Arithmeticas partim Graphicas, loca Cometæ in hoc Orbe ad observationum quarundam tempora: uti in Tabula sequente videre licet* (p. 494).

Newton had before him the observations of the comet from 20 December to 3 February 1681 which had been recorded by Flamsteed. Flamsteed had sent these observations to Newton on 7 March 1680/1 (see Letter 252, vol. II, p. 354). These observations, with slight modifications, appear in the *Principia* (p. 490) and are reproduced.

		Tem. appar	Temp. verum.	Long. Solis		Long. Cometæ		Lat. Cometæ
1680 *December*	12	4.46	4.46.00	♑	1.53.2	♑	6.33. 0	8.26. 0
	21	6.32½	6.36.59		11. 8.10	♒	5. 7.38	21.45.30
	24	6.12	6.17.52		14.10.49		18.49.10	25.23.24
	26	5.14	5.20.44		16.10.38		28.24. 6	27.00.57
	29	7.55	8.03. 2		19.20.56	♓	13.11.45	28.10.05
	30	8. 2	8.10.26		20.22.20		17.37. 5	28.11.12
1681 *January*	5	5.51	6. 1.38		26.23.19	♈	8.49.10	26.15.26
	9	6.49	7. 0.53	♒	0.29.54		18.43.18	24.12.42
	10	5.54	6. 6.10		1.28.34		20.40.57	23.44.00
	13	6.56	7. 8.55		4.34. 6		25.59.34	22.17.36
	25	7.44	7.58.42		16.45.58	♉	9.55.48	17.56.54
	30	8.07	8.21.53		21.50. 9	♉	13.19.56	16.40.57
February	2	6.20	6.34.51		24.47. 4		15.13.48	16.02.02
	5	6.50	7. 4.41		27.49.51		16.59.52	15.27.23

To these are added some observations by Newton.

		Temp. appar.	Cometæ Longit.	Com. Lat.
Febru.	25	8h.30′	♉ 26.19′.22″	12.46⅞
	27	8 .15	27. 4 .28	12.36
Mart.	1	11 . 0	27.53 . 8	12.24³⁄₇
	2	8 . 0	28.12 .29	12.19½
	5	11 .30	29.20 .51	12. 2⅔
	9	8 .30	♊ 0.43 . 2	11.44⅗

Following the table, Newton in the first edition (p. 491), pays tribute to Flamsteed's skill as an observer (In his observationibus *Flamstedius* eâ usus est diligentiâ); this tribute, however, is omitted from subsequent editions.

From the above observations, Newton selected three, namely, those for 21 December, 5 January and 25 January, and from these he worked out a parabolic path for the comet as it moved about the Sun as focus. Having accomplished this, he investigated its places, partly by arithmetical operations, partly by scale and compasses in this orbit (*Principia*, p. 494; Cajori, *Principia*, p. 512) to the times of some of the observations. The results for 12 December, 29 December, 5 February and 5 March are here shown:

	Distance from Sun	Longitude computed	Latitude computed	Longitude Observed	Latitude Observed	Difference Longitude	Difference latitude
Dec 12	2792	♑ 6.32	$8.18\frac{1}{2}$	♑ 6.33	8.26	−2	$-7\frac{1}{2}$
29	8403	♓ $13.13\frac{2}{3}$	28.0	♓ $13.11\frac{3}{4}$	$28.10\frac{1}{12}$	+2	$-10\frac{1}{2}$
Feb. 5	16669	♉ 17.0	$15.29\frac{2}{5}$	$16.59\frac{7}{8}$	$15.27\frac{2}{5}$	0	$+2\frac{1}{5}$
Mar. 5	21737	♉ $29.9\frac{3}{4}$	12.4	♉ $29.20\frac{6}{7}$	$12.2\frac{2}{3}$	−1	$+1\frac{1}{3}$

The table was later modified by Halley. In the second edition (p. 458) Newton observes: *Postea vero Halleius noster Orbitam, per calculum Arithmeticum, accuratius determinavit quam per descriptiones linearum fieri licuit.* (But afterwards our countryman Halley determined the orbit more accurately by an arithmetical computation than could be done by the description of lines.) This acknowledgment also appears in the third edition (Cajori, *Principia*, p. 512, where the corrected figures are shown).

In the second edition (p. 462) are two tables, the one showing the observed places of the comet, the other showing the corresponding places in the parabola as described above. If these tables are compared it will be seen that they agree very closely. Newton observed (pp. 496–7) that such slight differences as occur could easily arise either from errors of observation, or from errors in defining the orb by scale and compasses, and therefore he concluded that the observations of this comet from the beginning to the end agree so perfectly with the motion of the comet in the orbit just now described, as the motions of the planets do with the theories from which they are calculated, and by this agreement plainly evinces that it was one and the same comet that appeared all the time, and also that the orbit of that comet is here rightly defined.

530 NEWTON TO FLAMSTEED

14 SEPTEMBER 1695

From the original in Corpus Christi College Library, Oxford.(1)
In reply to Letter 527; for answer see Letter 543

Sr

When I received your last Mr Halley was with me about a designe of determining the Orbs of some Comets for me. He has since determined ye Orb of ye Comet of 1683 by my Theory & finds by an exact calculus that it answers all your Observations & his own to a minute.(2) I am newly returned from a journey I lately took into Lincolnshire & am going another journey so that I have not yet got any time to think of ye theory of the Moon nor shall have leisure for it this month or above: wch I thought fitt to give you notice of that you may not wonder at my silence.(3) I hope you get ground of your distemper & that I shall ere long heare that you are well recovered. I am

Your humble Servant

Cambridge Is. NEWTON.
Sept 14. 1695.

[Added below the date, in Flamsteed's hand]
Answerd Sep 17.

[Added on the reverse, still in Flamsteed's hand]

It commends & Confirmes your Theorys of the Comets that they will represent exact observations of them within a Minit. Mr Halley has set a freind of mine to desire some observations of the Comet of ye yeare 1682 from me. If I am not mistaken I imparted them to him as well as those of 83. Whatever he may say to you to the contrary his behaviour towards me has been the most impudently & Ingratefully base. I know him & you doe not therefore am resolved to have no further concerne wth him but if you want any of that comet I shall give you them & leave to employ them as you please.

My distemper abates[:] the paines of my head are not greate, but I am rarely free from them when I am travelling. I am setting on that worke that was interrupted by them in the spring. my excercise will devour no small parte of my time & therefore I shall desire my freinds to excuse me if I answer not their letters so fully nor readily as formerly[:] however when you want more of my lunar observations I shall cause them to be transcribd, & it will be no trouble to

NOTES

(1) C. 361, no. 105.

(2) See Letter 528.

(3) With this letter, Newton closes the correspondence. Four months later (Letter 543) Flamsteed asked the reason of Newton's silence, and reminded him of his offer to supply further observations of the Moon. Newton appears to have ignored this request.

531 FLAMSTEED TO NEWTON
19 SEPTEMBER 1695
From the original in the University Library, Cambridge[1]

The Observatory Sept. 19. 1695

Sr:

I returned hither 10 days agone my distemper I bless God being something abated but not wholly removed by travelling and excercise: I have still every day some paines of my head most in ye afternoones but not so great as to hinder me from resumeing the Worke interrupted by them last spring in which I hope still to make some good progress by ye next.

The impossibility of answering accurate observations of Comets by any Theorys that are not built on ye laws of gravity shews them all false[:] the near agreement of yours with them demonstrates its truth & confirmes the Theory of Gravity at ye same time

I shall be forced to lose no small part of my time in travelling to confirme my health but I hope nevertheless to have some small share of it left wherewth to serve my freinds & to supply you with what is wanting to finish your theorys of ye Lunar motions which I heare you doubt not now but to render very nearly agreeable to ye heavens. I pray God give you health to finish it & good success & am ever Sr

Your humble Servant

JOHN FLAMSTEED

To

Mr Isaack Newton
fellow of Trinity
Colledge in Cambridg
    these present
      Cambridge

NOTE

(1) Add. 3979, fo. 41.

532 HALLEY TO NEWTON

28 SEPTEMBER 1695

From the original in the University Library, Cambridge.[1]
In continuation of Letter 528

London Sept 28° 1695

Honoured Sr

I have been hard at work to serve you, and having done the Comet of 1683,[2] which I can represent most exactly; and that of 1664,[3] (wherin I find Hevelius has not been able to observe with the exactness requisite,) as near as I conceive it possible; I fell to consider that of 1680/1 which you have described in your book,[4] and looking over your Catalogue of the observed places, I find in that of the 25th of January 1681,[5] that there is a mistake of 20 minutes, in the Latitude that day, of 56 minutes for 36, and so I have it in a lre Mr Flamsteed sent me when I was at Paris. I thought fitt to advertise this, because you wrote me you designed to undertake to correct what you had

171

formerly determined about the Orb therof; and that day is one of those you have taken to define the orb by.[5] I find certain indication of an Elliptick Orb in that Comet and am satisfied that it will be very difficult to hitt it exactly by a Parabolick.[6] When I have computed all the Observations, I shall send you what I have done. If you have not gotten Vlacq's great Table of Sines and Tangents to 10 Seconds,[7] I believe I can procure it for you, and shall be glad to serve you therin, well knowing how much it will ease you in the Computations you are at present engaged in; If you still want the book pray lett me know. I must entreat you to procure for me of Mr Flamsteed what he has observed of the Comett of 1682 particularly in the month of September, for I am more and more confirmed that we have seen that Comett now three times, since ye Yeare 1531,[8] he will not deny it you, though I know he will me.

I am

Sr Your most obedient Servant

EDM. HALLEY

For his Honoured friend
  Mr Isaac Newton
    in Trinity Coll:
      Cambridge
        These

NOTES

(1) Add. 3982, fo. 2.

(2) See Letter 528, note (2), p. 165.

(3) Newton recorded his own observations of the comet of 1664 during December and January 1664/5. See vol. II, p. 367, note (9). The comet of 1664/5 was considered in the second edition of the *Principia*, pp. 471–8.

(4) The reference is to the first edition of the *Principia*, p. 490 *et seq*.

(5) See Newton's Manuscript 529, note (3), p. 167. Halley wrote *Longitude* for *Latitude*.

(6) Because it was returning at intervals of approximately 75 years. See note (8) below.

(7) Vlacq published two works, (i) *Adriani Vlacq: Tabulæ Sinuum, Tangentium & Secantium et Logarithmorum Sinuum, Tangentium et Numerorum ab 1 ad 100000*, (ii) *Trigonometria Artificialis sive Magnus Canon Logarithmicus*. It is probably the latter work to which Halley is referring, for in Letter 533 he says: 'I have gotten for you *Canon Magnus Triangulorum*...it cost me eight Shill.' Although Halley wrote (Letter 534): 'I hope you rec[eive]d the Vlacq's Tables which I sent you on Munday sennight' (i.e., 7 October), this might well be the *Trigonometria Artificialis* since this work also contains the logarithms of sines, etc.

172

Vlacq (1600–67) was born at Gouda, in Holland. He lived in London for some years as a bookseller and publisher; later he settled in Paris. He completed the *Arithmetica Logarithmica* (1624) of Henry Briggs by calculating the logarithms of the numbers 20,000 to 90,000 to ten places of decimals. In the above work (*Tabulæ Sinuum*) Vlacq adopted the system in which the logarithm of unity is 0, and that of 10 is unity.

(8) 'Halley produced the Elements of the Calculation of the Motion of the two Comets that appear in the Years 1607 and 1682, which are in all respects alike, as to the place of their Nodes and Perihelia, their Inclinations to the plain of the Ecliptick and their distances from the Sun; whence he concluded it was highly probable, not to say demonstrative, that these were but one and the same Comet, having a Period of about 75 years; and that it moves in an Elliptick Orb about the Sun, being when at its greatest distance about 35 times as far off as the Sun from the Earth' (Journal Book of the Royal Society of London, 3 June 1696). Halley correctly predicted that it would return in 1759.

533   HALLEY TO NEWTON
7 OCTOBER 1695
From the original in the University Library, Cambridge[1]

Honoured Sr

In answer to yours of the 1st October,[2] I give you many thanks for your Communication of the Observations of the Comet of 1682 which next after that of 1664 I will examine, and leave it to your consideration, if it were not the same with that of 1607,[3] and when your more important bu[s]iness is over, I must entreat you to consider how far a Comets motion may be disturbed by the Centers of Saturn and Jupiter, particularly in its ascent from the Sun, and what difference they may cause in the time of the Revolution of a Comett in its so very Elliptick Orb.

I have gotten for you Vlacq's *Canon magnus Triangulorum* and will send it you the beginning of next week, it costs me eight Shill. and I am very glad I can accommodate you therwith.

As to the Comet of 1680/1 I was only desirous to trie the method I used in that of 1683, in this also, taking your limitation for an Hypothesis and I found I could not stirr the Nodes or Inclination; that the Angle between the Aphelion and descending Node was 9°.20′ and the Latus rectum of the parabola .0243 of such parts as the mean distance of the Sun is 1,0000, hence the Aphelion or Axis of the parabola is directed into ♊ 27°.22½′ with 8°.11′ North Latitude: and in this is the principall fault of your first determinations: The time of the perihelion I see no cause to alter but that it was December 8°.0h 4′ p.m.[4]

From these data by an exact Calculus I derived the following Table to the moments of Mr Flamsteeds Observations.[5]

		Dist Comet☉	Long Comp.	Lat. Comp.	Long Obs.	Lat Obs.	Error Long.	Error Lat.
							′ ″	′ ″
1680. Decemb.	12	28028	♑ 6.29.25	8.26. 0	♑ 6.33.00	8.26.	−3.35	+0.00
	21	61076	♒ 5. 6.30	21.43.20	♒ 5. 7.38	21.45.½	−1.8	−2.10
	24	70008	♒ 18.48.20	25.22.40	18.49.10	25.23.⅔	−0.50	−0.48
	26	75576	28.22.45	27. 1.36	28.24. 6	27. 0.57	−1.21	+0.39
	29	84021	♓ 13.12.50	28.10.10	♓ 13.11.45	28.10. 5	+0.55	+0.5
	30	86661	17.40. 5	28.11.20	17.37. 5	28.11.12	+3.0	+0.8
81. Januar	5	101440	♈ 8.49.49	26.15.15	♈ 8.49.10	26.15.26	+0.39	−0.11
	9	110959	18.44.36	24.12.54	18.43.18	24.12.42	+1.18	+0.12
	10	113162	20.41. 0	23.44.10	20.40.57	23.44. 0	+0.3	+0.10
	13	120000	26. 0.21	22.17.30	25.59.34	22.17.36	+0.47	−0.6
	25	145370	♉ 9.33.40	17.57.55	♉ 9.35.48	17.56.54	−2.8	+1.0
	30	155303	13.17.41	16.42.07	13.19.36	16.40.57	−1.55	+1.10
Feb.	2	160951	15.11.11	16. 4.15	15.13.48	16. 2. 2	−2.37	+2.13
	5	166686	16.58.25	15.29.13	16.59.52	15.27.23	−1.27	+1.50
	25	202570	26.15.46	12.48. 0	26.19.22	12.46.⅞	−3.36	+1.⅛
Mart.	5	216205	♉ 29.18.35	12. 5.40	29.20.51	12. 2.40	−2.16	+3.0

Thus you see how near your Theory agrees with the observed Motion and where the errours are greatest the Observation may justly be suspected; for in the first, the Comet was just setting within the uncertainties of Refraction, and its place collected from the Suns by Mr Flamsteeds tables which he now has altered as much as amounts to two Minutes about the Winter Tropick, and I belive that that causes so much of the errour in Longitude on December 12°, and on December 30th I am apt to suspect that either the observed place was 2′ more or 17.39 of ♓,[6] or else that the time was 10′ later. And in Mr Flamsteeds lre to me Jan. 6 1680/1, he writes that at 8h. 30′ he observed the distance of the Comet from Andromeda's head 19°.55′.40″, and from Markab pegasi 8°.52′, whence he deduces the Longitude 17°.51¼ ♓ with Lat 28°.14′, tho afterwards a copy he sent me agrees with Yours, However it be, it is impossible for my numbers to err 2 minutes in one diurnall motion, when I agree so well both before and after.

As to the signs of the Orb's being Elliptick, I think that the bending therof towards the Node, towards the End, is an argument of a greater Curvity compared with the other errours, for the Latitude was greater than computed at first and lesser afterwards, which seems to require that the perihelion was somewhat earlier. but that cannot be by reason of the errours in Longitude on December 29.30 and Jan 5 which would be encreased therby. So that I conclude the Angle at the Sun to have been greater than by our Calculation, in an orb of the same Latus Rectum, and consequently Elliptick. perhaps Your sagacity may discover how to adjust the matter so as to remove the greatest part of those errours which upon severall attempts I found to hard for me.

I compared this Comets motion in the other parts of its orb, whilest it descended and could wish that Mr Flamsteed had seen it at least once for all the rest are very course observers. the computed places are these

	æq. Time d h ′	Cometa a ☉	Long Comput.	Lat Comput
Novemb. 16.	17. 00	83920	♎ 8.0. 25	0.43.20 aust
18.	21. 34	78020	18.41.50	1.17.30
20.	17. 0	72992	28.1. 45	1.44.30
26.	17. 0	54799	♏ 26.46.30	2.42.30.

If you think fitting to alter any of these Elements I shall not be wanting to adapt the calculus to your Emendations.

The Comet of 1683 does I think agree rather better with a parabolick orb, at least for so much therof as we could observe. its Eleme[n]ts are these

	d h ′	
Perihelion	July 3. 2. 50	in ♊ 25°.29.′½
Latus Rectum	224080.	
Descending Node	♓ 23°.23′	
Inclination	83.11	

But I want time to add the Table of Computed places compared with the Observed, which you shall have by some other opportunity: I find a necessity to derive the places of Hevelius's Comett[7] from the Observations, and not to trust to his computations which Costs me some trouble.

I am Sr

Your most faithfull servt

EDM. HALLEY

I had sent this on Saturday night but could not, now it accompanys your Vlacq, which the Carrier brings you.

NOTES

(1) Add. 3982, fo. 10. The date is surmised. The letter bears no date or postmark, since it was delivered by carrier. In the opening paragraph of his letter dated 15 October (p. 176) Halley states: 'I hope you recd the Vlacq's Tables which I sent you on Munday sennight.' This was 7 October.

There is a copy of this letter in David Gregory's hand in the Library of the Royal Society of London (R.S. Greg. MS. fo. 95). Apart from minor differences in spelling and punctuation, the latter has this note: 'A Correction by Mr Newtons hand. Let the Longitude of the Nodes of this Comet be augmented 9' (or perhaps 10') so that the perihelium be in ♑ 2.2' and let the place & distance of the Comet Jan: 5 remain.'

(2) This letter has not been found.

(3) In a Memorandum, probably by Newton, dated 23 February 1697/8, are the words: 'The Comet of 1607 was probably the same as that of 1682. Its period about 75 years.' See also Letter 532, note (8), p. 173.

(4) See Letter 535.

(5) In the second edition (p. 459) Newton wrote: *Et ex his datis, calculo itidem Arithmetico accurate instituto, loca Cometæ ad observationum tempora computavit, ut sequitur.* (And from these data, by an arithmetical calculation accurately carried out, he [Halley] computed the places of the comets to the times of the observations as follows.) Then follows Halley's table, reproduced above. It contains a few minor alterations. Neither in the second nor the third edition is there any mention of Flamsteed.

(6) See the table in Manuscript 529, note (3), p. 167.

(7) Hevelius communicated his observations on this comet to the Royal Society (see Letter 528, note (3), p. 166).

534 HALLEY TO NEWTON
15 OCTOBER 1695
From the original in the University Library, Cambridge.[1]
In continuation of Letter 533; for answer see Letter 535

London Octob. 15° 1695

Honoured Sr

I hope you rec[eive]d the Vlacq's Tables[2] which I sent you on Munday sennight and with it the Theory of the Comett of 1683, and the Computed places of that of 1680/1. Since then I have with some difficulty mastered that of 1664/5[3] but I was obliged to have recourse to the observations themselves, and to adjust and compare them togather, and recalculate the whole, before I could make them agree with any tollerable exactness, and I suspect that Mr Hevelius, to help his Calculations to agree with the hevens, had added 8 or 9 minutes to the places observed, on the 4th, 5th, and 8th of December, for I find the true places to be as on the other side of this Sheet. And what confirms the whole is that without any forcing, the Last appeearence of this Comett on Mart. 1°. does sufficiently agree with all the rest: so that I think there can be nothing plainer than that Comets do move in orbs about the Sun exceeding near to parabolick;[4] and since in this Comett of 1664, Cassinis Numbers, which supposes the Earth to stand still, differs near two degrees from the observed places, it may be concluded that that Hypothesis is incapable to represent its motion nearer, wheras this of yours traces out its course as exact as the best Astronomicall Tables can any of the planets, notwithstanding that this Comett came so near the earth as to encrease its visible velocity near ten fold. I had mislaid my book wherin I made my Lunar compututions, so could not send it you, but having rummadged for it, and found it, I will send it you this week hoping it may be of some service to you. Next to this, I will examine the Comet of 1682 and send you the result,[5] hoping it will give me no great trouble, because the observations I presume are exact; Sr pray please to command me in any thing you conceive I may be capable to serve you, and you shall be assured that with all readiness I will approve my self

Your most faithfull Servant

EDM. HALLEY

For the Honoured
Mr Isaac Newton
  in Trinity Coll
    Cambridge

Observationes Cometæ Anni 1664/5. Gedani Habitæ.

Temp App					Locus Supputatus
d h '					
Decembr. 3.18:29½	Cometa a Corde Leonis	46.24.20⎞	Long. ♎ 7. 1. 0	7. 1.29	
Londini 17.13.	a Spica ♍	22.52.10⎠	Lat Aust. 21.39. 0	21.38.50 Aus	
				° ' "	
4 18.1¼	Cometa a Corde Leonis	46. 2.45⎞	Long. ♎ 6.15. 0	♎6.16. 5.	
16.45½	a Spica ♍	23.52.40⎠	Lat Aust 22.24	22.24. 0.	
7 17.48	Cometa a Corde Leonis	44.48. 0⎞	Long. ♎ 3. 6	♎3. 7.33.	
16.32	a Spica ♍	27.56.40⎠	Lat Aust 25.22	25.21.40.	
17.14.43½	Cometa a Corde Leonis	53.15.15⎞	Long ♌ 2.56	♌2.56.	
13.27½	ab Humero Orionis Dext	45.43.30⎠	Lat Aust 49.25.	49.25.	
19. 9.25	Cometa a procyone	35.13.50⎞	Long Ⅱ 28.40½	Ⅱ28.43.	
8. 9	Com. a Lucid. Mandib. Coli	52.56. 0⎠	Lat Aust. 45.48	45.46.	
20. 9.53½	Comet a procyone	40.49. 0⎞	Long Ⅱ 13. 3	Ⅱ13. 5.	
8.37½	a Lucid. Mandib Ceti	40. 4. 0⎠	Lat Aust. 39.54	Aust.39.53.	
21. 9. 9½	Cometa ab Hum. dext Orion.	26.21.45⎞	Long Ⅱ 2.16	Ⅱ2.18¼	
7.53½	a Lucid Mandib. Ceti	29.28. 0⎠	Lat Aust 33.41	33.39⅖.	
22. 9. 0	Cometa. a Mandib Ceti	20.29.30⎞	Long. ♉ 24.24	♉24.27.	
7.44	ab Hum. dext Orionis	29.47. 0⎠	Lat Aust 27.45	Aust. 27.46.	
26. 7.58	Cometa a Lucida ♈	23.20. 0⎞	Long ♉ 9. 0	♉9. 2.28	
6.42	ab Aldebaran	26.44. 0⎠	Aust. 12.36	12.34.13.	
27. 6.45	Cometa a Lucida ♈.	20.45. 0⎞	Long ♉ 7. 5.⅔	♉7.08.54	
5.29	ab Aldebaran	28.10. 0⎠	Aust. 10.23	10.23.13	

	28.	7.39	Cometa a Lucida ♈	18.29. 0⎫	Long.	♉.	5.24¾	♉ 5.27.52.
		6.23	a palilicio	29.37. 0⎭	Aust.		8.22.⅝	8.23.37¼
	31.	6.45	Cometa a Cing Androm.	30.48.10⎫	Long.	♉	2. 7.40	♉2. 8.20
		5.29	a palilicio	32.53.30⎭	Aust.		4.13.00	4.16.25
Jan.	7.	7.37½	Cometa a Cing Androm.	25.11. 0⎫	Long	♈	28.24.47	♈28.24. 0
		6.21½	a palilicio	37.12.25⎭	Bor		0.54. 0	B0.53. 0
Cantisii	24.	7.29	Cometa a palilicio	40. 5. 0⎫	Long	♈	26.29.¼	♈26.28.50
Londini		6.13	a Cing. Andromeda	20.32.15⎭	Bor.		5.25.50	B5.26.00
Mart.	1.	8.16	Cometa prope secundam ♈⎫	♈ 29.17.20⎫			♈.29.18.20
		7. 0	observabatur	⎭Bor. 8.37.10⎭ .			B8.26.15

I have not computed as yet to the times of the other observations, but do not at all doubt of their most exact agreement with this Calculus. whose Elements are as follows.

		d	h	′	
Perihelion Cometæ. Novemb.		24	.11	.52	pm T æq.[6] Londini
Nodus Ascendens	♊	21°.13′.55″.			
Inclinatio		21	.18	.40.	
Latus rectum parabolæ		4.10286.			1.02571½
Angulus intra ♌ & perihelion.		49°.27′.30″.			
proinde perihelion in ♌		8	.40	.30.	
Lat Austrina		16	. 1	.45	

And this does exactly answer all the phænomena of this Comett with scarce any difference only about 2 minutes must be sustracted, whilest it was swiftest. which will be done, if we make the angle of the ♌ and perihelion 49°.27′.15″, or 15″ less than before, so that this does agree with the Theory as well as can be wisht for.

NOTES

(1) Add. 3982, fo. 3.

(2) See Letter 532, note (7), p. 172.

(3) See the table of observations which form a part of this letter. For purposes of comparison, the table taken from Cajori, *Principia*, p. 535, is given below. See also Letter 532, note (3), p. 172.

In the second edition of *Principia* (p. 477), following the table, Newton wrote: *Quam probe loca Cometæ in hoc Orbe computata, congruunt cum Observationibus, patebit ex Tabula sequente ab Halleio supputata.* (How accurately the computed places of the comet in this orbit agree with observations will appear from the following table provided by Halley).

(4) See Letter 528, note (2), p. 165.

(5) See Letter 537.

(6) Tempus æquale, i.e. mean time.

Apparent time at Dantzick	The observed distances of the comet from		The observed places	The places computed in the orbit
December d h m		° ′ ″	° ′ ″	° ′ ″
3.18.29½	The Lion's heart	46.24.20	Long. ♎ 7.01.00	♎ 7. 1.29
	The Virgin's spike	22.52.10	Lat. S. 21.39. 0	21.38.50
4.18. 1½	The Lion's heart	46. 2.45	Long. ♎ 6.15. 0	♎ 6.16. 5
	The Virgin's spike	23.52.40	Lat. S. 22.24. 0	22.24. 0
7.17.48	The Lion's heart	44.48. 0	Long. ♎ 3. 6. 0	♎ 3. 7.33
	The Virgin's spike	27.56.40	Lat. S. 25.22. 0	25.21.40
17.14.43	The Lion's heart	53.15.15	Long. ♌ 2.56. 0	♌ 2.56. 0
	Orion's right shoulder	45.43.30	Lat. S. 49.25. 0	49.25. 0
19. 9.25	Procyon	35.13.50	Long. ♊ 28.40.30	♊ 28.43. 0
	Bright star of Whale's jaw	52.56. 0	Lat. S. 45.48. 0	45.46. 0
20. 9.53½	Procyon	40.49. 0	Long. ♊ 13.03. 0	♊ 13. 5. 0
	Bright star of Whale's jaw	40.04. 0	Lat. S. 39.54. 0	39.53. 0
21. 9. 9½	Orion's right shoulder	26.21.25	Long. ♊ 2.16. 0	♊ 2.18.30
	Bright star of Whale's jaw	29.28. 0	Lat. S. 33.41. 0	33.39.40
22. 9. 0	Orion's right shoulder	29.47. 0	Long. ♉ 24.24. 0	♉ 24.27. 0
	Bright star of Whale's jaw	20.29.30	Lat. S. 27.45. 0	27.46. 0
26. 7.58	The bright star of Aries	23.20. 0	Long. ♉ 9. 0. 0	♉ 9. 2.28
	Aldebaran	26.44. 0	Lat. S. 12.36. 0	12.34.13
27. 6.45	The bright star of Aries	20.45. 0	Long. ♉ 7. 5.40	♉ 7. 8.45
	Aldebaran	28.10. 0	Lat. S. 10.23. 0	10.23.13
28. 7.39	The bright star of Aries	18.29. 0	Long. ♉ 5.24.45	♉ 5.27.52
	Palilicium	29.37. 0	Lat. S. 8.22.50	8.23.37
31. 6.45	Andromeda's girdle	30.48.10	Long. ♉ 2. 7.40	♉ 2. 8.20
	Palilicium	32.53.30	Lat. S. 4.13. 0	4.16.25
Jan. 1665 7. 7.37½	Andromeda's girdle	25.11. 0	Long. ♈ 28.24.47	♈ 28.24. 0
	Palilicium	37.12.25	Lat. N. 0.54. 0	0.53. 0
13. 7. 0	Andromeda's head	28. 7.10	Long. ♈ 27. 6.54	♈ 27. 6.39
	Palilicium	38.55.20	Lat. N. 3. 6.50	3. 7.40
24. 7.29	Andromeda's girdle	20.32.15	Long. ♈ 26.29.15	♈ 26.28.50
	Palilicium	40. 5. 0	Lat. N. 5.25.50	5.26. 0
February 7. 8.37			Long. ♈ 27. 4.46	♈ 27.24.55
			Lat. N. 7. 3.29	7. 3.15
22. 8.46			Long. ♈ 28.29.46	♈ 28.29.58
			Lat. N. 8.12.36	8.10.25
March 1. 8.16			Long. ♈ 29.18.15	♈ 29.18.20
			Lat. N. 8.36.26	8.36.12
7. 8.37			Long. ♉ 0. 2.48	♉ 0. 2.42
			Lat. N. 8.56.30	8.56.56

12-2

535   NEWTON TO HALLEY

17 OCTOBER 1695

From the original in the University Library, Cambridge.[1]
In reply to Letter 533; for answer see Letter 537

Sr

I had writ a letter to you last week but stopt it because I had inserted a passage I was uncertain of. Your calculations have satisfied me that the Orb of the Comet of 1680/1 is Elliptical. And by a certain calculation grounded on them upon a supposition that the Latitudes in November as I drew them from ye course Observations of others & printed them, are right; I seem to collect that this Comet rises about 10 or 12 times higher then the orb of Saturn & by consequence revolves in about 400 years. But the Parabolick Orb of this Comet as you have determined it seems to admit of rectification. For the errors in Latitude ought to be double to the errors in longitude in ye months of Febr. & March, supposing the plane of the Orb to be rightly determined. The reason is because the angle of that plane wth the plane of the Ecliptick is about 60 degr. Wherefore considering that ye errors in Longitude (in your calculus) are something more then equal[2] to the errors in Latitude, I diminish the angle wch the plane of ye comets Orb makes wth ye plane of ye Ecliptick by 24⅓′, making it only 60 degr. 56′, & by consequence I place the Nodes in ♋ 2 degr. 2′ & ♑ 2 degr. 2′ or thereabouts; adding about 9′ to ye place we fixt it in before. Also the Latus rectum of the Parabola, wch you make 0|0243 I encrease by 0|002[3] so as to make it 0|0245, & the distance of the comet from ye earth between Decemb. 29 & Jan. 5, I suppose the same as in your calculations, & add 2′ to ye time of the Comets Perihelium so as to make it Decem 8d. 0h. 6′. P.M. By this means the errors in Latitude will I think remain much the same as in your calculations, except that they will in ye end of the Comets motion be decreased about 10″. And the errors in Longitude will be decreased both in the beginning & in the middle, & so much in the end as to remain but about half of the errors in Latitude as they ought to be in a Parabolick Orb. But in my Observations of Feb 25 there is an error of 1′ in Longitude. It was printed as ♉ 26.19′.22″: but should be ♉ 26.18′.22″. I do not think it requisite you should give your self further trouble about this Comet, but if you do, you need only calculate the places of ye Nodes & Perihelium from ye above mentioned data; & (in the Parabolic Orb thus determined) 3 or 4 places of the Comet, suppose Dec. 21. Jan 5 & 25 & Mar 5. For it will then be easy to fill up the other places by interpolation and to assigne such an Elliptick Orb as will answer all the Observations as exactly as can be.

PLATE I. PORTRAIT OF NEWTON (ARTIST UNKNOWN)

You have made ye Orb of the Comet of 1664⁽⁴⁾ answer Observations much beyond my expectation thô wth double pains in calculating all the Observations anew. I can never thank you sufficiently for this assistance & wish it in my way to serve you as much. How far a Comets motion may be disturbed by ♃ & ♄ cannot be affirmed wthout knowing the Orb of ye Comet & times of its transit through ye Orbs of ♃ & ♄. If in its ascent it passes through the Orbe before its heliocentric conjunction with ye Planet, the time will be shortened, if after, it will be lengthened, & the decrease or increase may be a day, a week, a month, a year or more; especially if the Orb be very excentric & ye time of ye revolution long.

I will send you to morrow by Will. Martin the Carrier the Box of brass Rulers & beam compasses & 8s for Vlac's Trigonometry wth many thanks. Those edges of the brass Rulers wch look rough I ground true to one another wth sand.

If it will not give you too much trouble to make an extract of your calculated places of the Moon you need send only those wthout your book, but if you had rather send your book then an extract, it will be the same thing to me. I am

Your most humble Obliged Servant

Is. NEWTON.

Cambridge Oct 17th 1695

For Mr Edmund Halley at
Mr Coxall's in Packer's Court
in Collman-street in
     London

NOTES

(1) Add. 3982, fo. 4.

(2) Newton originally wrote 'double'.

(3) An error for 0·0002.

(4) See Table sent with Letter 534.

536   HALLEY TO NEWTON
21 OCTOBER 1695
From the original in the University Library, Cambridge.
In reply to Letter 535

Octob. 21. 1695

Sr

In your last you have amended the orb of the Comett of 1680/1 from what I could do, and I am glad you concurr with me in the conclusion that it must be Ellipticall, and from your limitations I have recomputed the Comets places as follows.[1]

				Error Long	Lat
Decem. 21°	♒	5 . 7 . 30	21°.42′.49″ Lat.	−0′.8″	−2′.41″
	29	♓	13 . 12 . 22	28 . 10 . 5	+0 .37	+ −
Jan.	5	♈	8 . 49 . 3	26 . 15 . 32	−0 .7	+0 .6
	25	♉	19 . 33 . 40	17 . 58 . 35	−2 .8	+1 .41
Mart.	5	♉	29 . 19 . 19	12 . 6 . 9	−1 .32	+3 .29

By which it appears that you have every where corrected the Longitudes. But for the Latitudes the error both at first and last is about ½ a minute greater, which I hope you will be able to remove when you come to determine the species of the Ellipse.

I have almost finished the Comet of 1682 and the next you shall know whether that of 1607 were not the same, which I see more and more reason to suspect.[2] I am now become so ready at the finding a Cometts orb by Calculation, that since you have not sent the rulers, as you wrote me, I think I can make a shift without them. And I intend as far as I can to limitt the Orbs of all the Comets that have been hitherto observed, of all which I shall duly give you an account. I have sent you the book[3] wherin I did most of my Lunar Computations, and would have made an Extract therof, but am at present busy about the Society's Books,[4] and withall I belive the whole work will be more satisfactory.

There is in the Book of my Lunar Observations a Table of the Nonagesime degree and its altitude, which will be worth your Calculators while to transcribe. I mean that made for the Lat. 51° 30′, which will be of use to find the parallaxes by.

Sr. I am With all imaginable respect

Your most devoted friend and Servant

EDM. HALLEY.

The node of the Comet being ♋ 2°.2′ the angle in the plain of the Orb between the Node and Aphelion is 9°.22′.48″. The Latus rectum 0.0245 ye radius being 1.

NOTES

(1) This table should be compared with the one sent with Letter 533. The complete table extending from 12 December to 5 March appears in the second edition of *Principia*, p. 459. There are slight alterations in it from Halley's table here reproduced.

(2) See Letter 533, note (3), p. 176; also Letter 532.

(3) See final paragraph of Letter 535.

(4) A perusal of the *Record of the Royal Society of London* (1940) makes it abundantly clear that Halley was kept fully occupied by the Society during this period.

537   WALLIS TO WALLER
24 OCTOBER 1695
Extract from the original in the Library of the Royal Society of London

Oxford Octob. 24, 1695.

···   ···   ···

The other thing, is, concerning Mr *Newton*. You publish concerning him, an Extract of a Letter of *De Moivre*,[1] (which is well done, as a publike owning of Mr *Newton's* methode.)[2] Who this *De Moivre* is, I know not. But I have, by me, two large Letters[3] of Mr *Newton's* owne, in the year 1676, written to Mr *Oldenburg* (from whom I had my copy) to be then communicated to *Leibnitz* & *Chirnhause*, with whom he did then correspond. Which are more to the purpose than that of *De Moivre*. If you think fit to publish them in like manner; it may be done in two or three *Transactions* successively. For they are a great deal too much for any one. And it is (I find) not so acceptable (to those who are not Mathematicians) to have too much of this nature in one of them; but rather somewhat of this, & somewhat of other matter. I have (in my *Algebra*) published extracts of these Letters. I am sorry I did not insert the Letters *verbatim*. This is what I thought fit to suggest to you on this occasion, from

Sr

Yours to serve you

JOHN WALLIS.

NOTES

(1) Abraham De Moivre (1667–1754), born at Vitry in Champagne, spent most of his life in England whither he had resorted after the revocation of the Edict of Nantes (1685). Following a study of Newton's *Principia* he rapidly became one of the foremost mathematicians in this country. He was elected F.R.S. in 1697, and was appointed a member of the Commission to report on the claims of Newton and Leibniz regarding priority in the invention of the calculus (1712). His outstanding work was his Doctrine of Chances (*De Mensura Sortis*, 1711) dealing mainly with life annuities. In 1730 he published *Miscellanea Analytica*; this contains his methods of recurring series and complex numbers including the series known by his name.

(2) The work to which Wallis refers in this letter appeared in the *Phil. Trans.* **19** (1695), 52–6, with the title: *Specimina quædam illustria Doctrinæ* Fluxionum *sive exempla quibus Methodi istius Usus & præstantia in solvendis Problematis Geometricis elucidatur, ex Epistola Peritissimi Mathematici D. Ab. de Moivre desumpta*. It deals with the application of the method of fluxions to geometrical problems and is clearly based upon the *Principia*, Prop. 7, Book 2, Lemma 2, pp. 250–3.

(3) See Wallis's letter to Newton (Letter 498), in which he refers to 'the two large letters' of June and August 1676, also his Letters 502, 511 and 518. Newton wrote thanking Wallis for his concern in July 1695 (Letter 512).

The two letters referred to were: (i) the 'Epistola Prior', 13 June 1676 (Letter 165, vol. ii), which contains, *inter alia*, Newton's enunciation of the Binomial Theorem. The letter was read to the Society on 15 June 1676 (Birch, iii, 319), and Wallis, who was a prominent member of the Society, was determined that it should be made public; (ii) the 'Epistola Posterior', 24 October 1676 (Letter 188, vol. ii). Both letters were printed in *C.E.* (1712), the former on p. 49, the latter on p. 67.

538   NEWTON TO HALLEY

Late October 1695

From a draft in the University Library, Cambridge[1]

In the third book of the Principia mathematica pag 459 (Edit 2) there is a table of the Motion of the Comet of 1680 in a Parabolic Orb from Decem 12 to Mar 5.[2] I begg the favour that you would compute its motion in the same Orb for Mar. 9d.8h.38′ true time & send me its distance from the sun in parts whereof 100000 are the mean distance of the center of the earth from the center of the sun & also its Geocentric Longitude & Latitude computed in this orb for the same time Mar. 9d.8h.38′. For I would add to that Table the motion of the Comet computed in the Parabolic Orb for Mar. 9d.8h.38′.

I received from you formerly[3] a Table of the motions of the same Comet in an Elliptic Orb. You put the Node ascendant in ♑ 2gr.2′, the Node descendant in ♋ 2gr.2′.[4] The inclination of the plane of the Orb to the plane of the Ecliptick is 61gr.6′.48″, the perihelium of the Comet in this plane ♐ 22gr 44′ 26″ the equated time of the Perihelium Decem. 7. 23. 9″, the distance of the Perihelium from the ascending Node in the plane of the

Ecliptic 9gr.17′.35″. The axis transversus 1382957 & the Axis Conjugatus 184812 the mean distance of the earth from the Sun being 10000. And in this Orb you computed the places of ye Comet Nov 3, 5 & 10 as follows

1680 Novem	Tempus verum	Long. comp.	Lat comp
	3.16.47	♌ 29.51.22	1.17.32 bor
	5.15.37	♍ 3.24 32	1 6. 9
	10.16.18	15.33. 2	0.25. 7[5]

The first of these three places you have inserted into the Table of the motions of this Comet in an Elliptic orb, wch you have printed in your Astronomical Tables where you treat *De motu Cometarum in Orbibus Ellipticis*. I beg the favour of you to reexamin the two last of them, viz those of Novem 5 15 57 and Novem. 10.16.18.

In the same printed Table you have calculated the place of this Comet upon March 9.8.38 true time: I beg the favour of you to calculate its place in the Parabolic Orb also upon March 9.8h. 38′ true time: For I would add it to the Table of the motion of this Comet in a Parabolic Orb printed in the third book of the Principia Mathematica pag 459, Edit II. Pray send me its computed Long. Lat. & distance from the Sun in parts whereof the radius of the Orbis magnus is 100000.

NOTES

(1) Add. 3965, 14, fo. 605. Although there is no address to the letter, it was clearly intended for Halley.

(2) This table is substantially the same as that which appears in the first edition, following Prop. 41. Halley made a number of corrections; these appear in the second edition. See Letter 529, note (3), p. 167.

(3) Letter 533.

(4) In the second edition (pp. 458–9) Newton declares that Halley, having determined the orbit of the comet more accurately, kept the places of the nodes in Cancer and Virgo 1 gr. 53′, and the inclination of the plane of the orbit to the plane of the ecliptic 61°. 21⅓′, and the time of the comet's perihelion 8 December, 00h. 04m. He found the distance of the perihelion from the ascending node measured in the comet's orbit to be 9°. 20′ and the latus rectum of the parabola 243 parts, assuming the mean distance of the Sun from the Earth to be 10,000 parts. And from these data, by an accurate arithmetical calculation, he computed the places of the comet to the times of the observations as given in the table. For the table which follows see Cajori, *Principia*, p. 513.

(5) In the third edition (Cajori, *Principia*, p. 513) the figures are:
Nov. 3, 17h. 2m., apparent time at London, the comet was in ♌ 29° 51′, with 1° 17′ 45″ latitude north.
Nov. 5, 15h. 58m., the comet was in ♍ 3° 23′, with 1° 6′ latitude north.
Nov. 10, 16h. 31m., the comet was equally distant from two stars in ♌, which are designated σ and τ in Bayer.

539   WALLIS TO HALLEY
11 NOVEMBER 1695
Extract from the original in the Library of the Royal Society of London[1]

Oxford Novemb. 11. 1695

··· ··· ···

Mr Newton's two letters; I am sorry (as I sayd in my last)[2] that I did not print them verbatim, when I published the Extracts out of them. And I would have added them to my first Volume (which was last printed)[3] but that the Volume (without it) was grown too big; which made me omit that, & some other things which I would else have added. I have since written several letters to Mr Newton about it; pressing (with some importunity) the printing of these, & of his Treatise about Light & Colours,[4] (as being neither just to himself, nor kind to the publike, to delay it so long.) As to the Letters, I sent him a fair Transcript, ready for the Press; which if he would print, it might best be done here (& I would take ye care of it,) because we have (most of) the Cuts allready done; & our Printers are allready acquainted with the troublesome Composing, which it will require. But he did not seem forward for either.[5] Perhaps I may find an opportunity (if God give life) to publish those Letters, with some other things of mine. As to that about Light and Colours (for which I am more solicitous) your interest may possibly prevail with him,[6] better than mine, to get it published.[7]

··· ··· ···

Yours to serve you

JOHN WALLIS.

It was suggested to me by one, that he thought you might have a desire to see those two Letters of Mr Newton. I suppose you have them allready in the Royall Society amongst Mr Oldenburgs collections of Papers. For I am sure he had them, & from him I had the Copy from which I transcribed mine, & returned it to him again. But if it be not there (& you desire it) I will spare you mine for a week or fortnight, (upon promise to restore it.) But I am not willing to part with it; because I value it.

These
For Mr Edmund Hally, to
be left with Mr Hunt[8] at
    Gresham-College
        London.[9]

NOTES

(1) W. 2, fo. 55.

(2) See Letter 538.

(3) The first volume appeared in 1695, two years after the second volume.

(4) In the preface to the *Opticks* (1704), Newton says: 'To avoid being engaged in Disputes about these Matters, I have hitherto delayed the Printing, and should still have delayed it, had not the Importunity of Friends prevailed upon me.' It was written in English, as Wallis had so often recommended.

(5) Newton no doubt had not forgotten his protracted quarrels with Lucas and Linus on this subject (vol. II, pp. 183, 189–92, 254–60). In 1676 he had declared: 'I see I have made my self a slave to Philosophy, but if I get free of Mr Linus's buisiness I will resolutely bid adew to it eternally.'

(6) Halley was responsible for seeing the *Principia* through the Press.

(7) Halley replied on 21 November 1695. See Wallis's letter (541), first paragraph.

(8) See Letter 502, note (2), p. 115.

(9) On the reverse of the letter: 'Dr Wallis to Mr Halley desireing him to prevaile wth Mr Newton to publishe his Light & Colours &c. Oxford Novr. 11. 1695.'

540 BENJAMIN SMITH[1] TO NEWTON

18 NOVEMBER 1695

From the original in the University Library, Cambridge[2]

Nov: ye 18*th.* 1695

Sr

I reced your kind Lre dated the 31st of October last, And the fomentation wee applyed as soone as wee could possible; The effects have p[ro]ved very successefull; for the swelling is verry much abated, and the blacknes quite gone. Although att certaine times shee hath still a paine in her brest; after every bathing, shee put the same plaster to her Brest againe, you sent, I could not perswade her to take the Sowes,[3] for since her being wth child almost every thing goes against her Stomach; But shee is resolv'd to try; This wth mine & my wifes service wth our thanks for all your trouble and cost is all from

Your most affectionate

Brother

BEN: SMITH

To
Mr Isaac Newton
Mathematticke Professor
      in
Trinity Colledge in
Cambridge   present

NOTES

(1) Newton's half-brother, born 1650. Two years after the death of Newton's father in 1642, his mother married Barnabas Smith, Rector of North Witham, who died in 1656. There were three children of this marriage, of whom Benjamin was the eldest. He married Sara Bishop of Stickford.

(2) This letter was found among a number of papers relating to the correction of cometary orbits, in the University Library, Cambridge (Add. 3965, 18, fo. 728v).

(3) This word may be a variant of *sowens*, an article of diet made by steeping oatmeal, or rather the husks sifted from it in water until the infusion has become sour, then straining and boiling it to the consistency of light pudding, which is eaten with milk (Joseph Wright, *English Dialect Dictionary* (1898), vol. 5).

541   WALLIS TO HALLEY
26 NOVEMBER 1695
From the original in the Library of the Royal Society of London[1]

Oxford: Nov. 26. 1695.

Sr

In pursuance of yours of Nov 21, I have sent by my Daughter Benson[2] (who is this day gone towards London) the two large Letters[3] of Mr Newton (being a parcell too big to send by the Post.) She intends to lodge at Mrs Pinfold's, over against Hatton-garden in Holborn: And, if you please to send thither for them, you may there receive them; And when you have done with them, you may please to return them to me by some safe hand. And I would willingly see Leibnitz's answer.... I am glad Mr Newton is inclinable to print some of the things he hath by him. So many as he hath on his hands at once, do hinder one another. I am most fond of his book of Light & Colours. His feare of Disputes & Cavils need not trouble him. It will be at his choice, whether or not, to answer them. His Hypothesis will defend itself. We are told here, that he is made Master of the Mint:[4] which, if so, I do congratulate to him. And am, His, and

Your humble servant

JOHN WALLIS

It was Mr Caswell that suggested to me the sending of those Letters. Who doth you no ill offices with me at all, (nor, I think with any body else,) but speaks well of you upon all occasions. I would have you be friendly to one another.

(1) W.2.56. It would seem from this that Halley acceded to the request made by Wallis in the postscript of Letter 539.

(2) Wallis had two daughters, 'handsome young gentlewoemen' (Aubrey, *Lives*, ii, 282); of these, the younger married William Benson (died 1700), of Towcester.

(3) See Letter 537, note (3), p. 184.

(4) There was clearly talk of some vacancy at the Mint and many conflicting rumours were circulating about this time.

By October 1695 it became evident that a recoinage, which had been avoided in the Spring, could no longer be delayed. Clipping of silver coins had become so widespread that the old coin had to be withdrawn, either at face value, or as bullion. The King's Speech at the opening of Parliament on 22 November 1695 announced immediate legislation. The principal Bill passed its third reading in the Commons on 27 December 1695, and received the Royal Assent on 21 January 1695/6 (*Calendar of Treasury Books*, 1696, cxxii). The legislation had been anticipated; new dies for 24 coinage presses were ordered on 23 November 1695 (*ibid.* 1693–6, p. 1415); demonetisation of clipped silver coins was proclaimed on 19 December 1695, and the Mint began recoining on 23 January 1695/6.

The increased work which this involved necessitated a strengthening of the Mint administration. James Hoare, the Comptroller, had borne most of the burden for fifteen years, and was now past it. In February 1695/6 Thomas Hall was appointed to assist the top management at a salary of £400 per annum. The Somers–Shrewsbury letter of 4/14 November 1695 (see *The Private and Original Correspondence of Charles Talbot, Duke of Shrewsbury*, ed. W. Coxe (1821), p. 400) makes it clear that the transfer of the office of Comptroller to Newton, who had been disappointed at not being selected for that office in 1691, was suggested, and it is equally clear that the suggestion came from Montague.

Rumours had already begun to circulate that Newton was being considered for a post at the Mint. It seems clear that the Ministers did not consider the existing officers competent to deal with the problems of recoinage; there were discussions as to how to introduce a suitable man at the Mint, and this may be the explanation of Luttrell's observation (iii, p. 546) on 5 November 1695: 'The King hath been pleased to conferr the place of Master of the mint, vacant by the decease of the late master, upon Dr Newton, Mathematical Professor at Cambridge.' This is clearly incorrect: neither the Comptroller (Hoare) nor the Master (Neale) was dead at this time, though the former's health was certainly failing. Evidently the rumour was revived early the following year; Newton's vigorous denial on 14 March 1695/6 (Letter 544) may have been due to the fact that he had been approached in confidence. In any case, the rumour was no doubt strengthened when Overton, who had been Warden since 1689, was appointed to the office of Commissioner of Customs at a salary of £1200 per annum in place of Sir Patience Ward, who was seriously ill. See Luttrell (iv, p. 34): '24 March 1695/6. Mr Overton is made commissioner of customes in place of sir Patience Ward.' This left the post of Warden vacant, and Montague succeeded in obtaining it for Newton on 19 March 1695/6 (Letter 545). Sir Patience Ward recovered, but Overton was retained in his new office as supernumerary. 'Thursday April 9, 1696. Sir Patience Ward being recovered of his illnesse, continues in his office in the Custome house, and Mr Benjamin Overton is added to them' (Luttrell, iv, p. 42).

Montague's recommendation was probably influenced by his former friendship with Newton when they were at Cambridge; see Letter 545, note (1). There is no evidence that expediting the recoinage was a motive. Newton was in no way responsible for the recoinage, as the work was well under way before the appointment was made; moreover, as Warden he would be responsible for certain disciplines and the oversight of some finance of prosecutions for coinage offences. Nor is there any evidence that Montague transferred Overton simply to make room for Newton. Promotion across departments to lucrative posts on mere merit was unknown. The salary of £1200 per annum would no doubt attract powerful men, and it is possible that Overton had some influence in high quarters which enabled him to obtain such an attractive post. It was this fact which left the Wardenship vacant and Montague seized the opportunity of securing the post for Newton. This seems the most plausible explanation of the sequence of events.

There was no impropriety in Montague's offering, at a time of intense and increasing pressure, a post which did not involve much work, or frequent attendance at the Mint, and there is no reason to think that Montague did not mean exactly what he said, namely, that the post 'has not too much bus'nesse to require more attendance than you can spare'. Newton was still Lucasian Professor of Mathematics at Cambridge, which was a day's journey away. The office had long been a sinecure, and there is no evidence whatever that sinecures in Government service were in any way discreditable. Such posts were openly advertised in the Press, and agencies existed for their purchase and sale. Montague himself held two.

542   HALLEY TO NEWTON
1695/6[1]
From the original in the University Library, Cambridge

Sr.

I had waited on you on Saturday, but I was obliged to go on board my friggat, and besides I could not get time to finish the account of the two Comets I promised you. I find the Inclination Evidently less than the Comet of 1682, in the other of 1607, but some of the observations will allow it the same; they being by no means sufficiently accurate. Your self will best judge whether they may be safely concluded to be the same, as is my present opinion.[2] I will waite on you at your lodgings to morrow morning to discourse the other matter of serving you as your Deputy.

[EDMOND HALLEY]

NOTES

(1) It is difficult to assign a date to this letter. The reference to 'the account of the two Comets I promised you' suggests that it relates to the letters which passed between Newton and Halley towards the end of 1695, particularly during the October of that year; in that case it would probably be early in 1696 when the subject was fresh in the minds of both.

Halley's mention of going aboard his 'friggat' could refer to the *Paramour*. From the records in the National Maritime Museum, we learn that as early as 12 July 1693 'Mr Benjamin Middleton and Mr Edmund Halley petitioned the Admiralty for a vessel of 80 tons to enable them to study the variation of the compass, methods of determing the longitude at sea. and to undertake a voyage of circumnavigation'. This vessel, a pink, was built at Deptford and was launched in April 1694; she was 64 ft. long, 18 ft. broad, 9·7 ft. deep, and of 89 tons.

Halley received his commission in June 1696, and it might well have been in anticipation of this that the above letter was written. Possibly on account of his appointment as Deputy Comptroller of the Chester Mint, further delays occurred. Halley finally received his instructions on 15 October 1698, the *Paramour*, sailed from Deptford on 20 October, and after putting in at Portsmouth for minor repairs, on 22 November the expedition got under way. It is not unlikely that after his return from St Helena in 1678, Halley set off on other maritime adventures. It seems extremely unlikely that he should be given a sea captaincy for such a hazardous enterprise, particularly in view of the frequent disasters at sea, unless he had had some considerable experience in navigation and in methods of determining longitude. See R. T. Gunther, *Early Science in Oxford*, vol. x, p. 108; 'Fry. 22. [March 1688/9] Haly a sayling', and (p. 111) 'We. 3 [April]. Hally returnd'.

(2) See Letter 532, note (8), p. 173.

543   FLAMSTEED TO NEWTON
11 JANUARY 1695/6
From the original in the University Library, Cambridge.[1]
In reply to Letter 530

The Observatory Jan: 11. 1696/5

Sr

Yours of ye 14th of September last told me that you was then newly returnd from a Jorney and undertakeing another. So yt you had not got any time to thinke of the Theory of ye Moon nor should for a moneth. So yt in my Answer to it of ye 17th I said Nothing of her more yn that when you required further observations of her I was ready to supply you. I was then something better of my headach then formerly but it still hung on me till within some few days after being seized with a sharp fit of the stone which gave me 4 or 5 hours strong paines on takeing the usuall remedys for it the Ureters being cleansed & the humors descending in their Naturall Chanells I was freed from it & I bless God for it have felt little of it since but I have had 2 or 3 small fitts of ye stone in ye room of it. Nevertheless I have continued my restitution of ye fixed stars.[2] & finished all the Tables necessary for easeing the labor of Calculation. The Constellations of ♈ and ♉ conteining neare 200 stars visible with ye Naked eye are finished ♊ is on ye stocks, & if God continue my health I hope I may finish the whole Zodiack ere this time 12 moneths. I find some small faults in ye Catalogue I gave you but such as will not much disturbe any of your determinations in the lunar Theory & may be allowed for at last with ease for ought I yet perceave but I dare not be positive till I have gone through the Constellations lyeing on this Circle.

In ye Meane time I have not neglected any opportunity of observeing the moon as often as our much Cloudy weather this yeare permitted: tho I have

191

not deduced her places as yet from my notes. but if they be of more use to you then those of former yeares as you intimated they might be in some of your letters I shall give you such of them as you require wth directions how to deduce her places from them.

But if what I heare be true you will have little need of them for I have been told ever since I came out of Surrey that you have finished ye Theory of ye Moon *on uncontestable principles*[3] that you have discovered 6 severall Inæqualitys not formerly knowne & that nevertheless ye Calculation will not be much more troublesome or difficult then formerly[.] I am heartily glad to hear this & should be more so to have it from your selfe for in truth I suspect you are scarce so forward & I flatter my selfe wth ye opinion yt if you were you would have acquainted me with it as you promised both when I imparted the 3 Synopses of lunar Calculations[4] & observed places to you & in your letters since. Pray let me know how far you are proceeded. You will oblige me, &, if you please, ye true reason *why I have had no letter from you this* 4 *Moneths.*[5]

Some freinds of mine yt live at a distance in ye Country have made New tables for representeing the motions of the two superior planets ♄ & ♃. they Answer all The Tychonick Hevelian & my owne observations of their places within 10 or 12 minutes & those of Bernard,[6] Walter[7] & Mr Horrox nearly within ye same error. So yt now tis evident the faults caused by their inter-mutuall gravitations are not so great as they seemd when these observations were compared with the Rudolphin Tables[.][8] I could acquaint you with more particulars of what has been found out by my freind. but I would not divert you from your own course by *sup*[er]*fluous communications.* if you desire to know more of what is done, you may, whenever you let me know your desires by your letters.

I am often threatned wth ye returne of ye stone & gravell & forced frequently to take my usuall remedies. but I bless God for it I can now when delivered from my feares goe on with what I have under my hands which I could not whilest I had the continuall paines of my head I pray God continue you health & prosper your labors. & am ever

Your affectionate freind & humble Servant

JOHN FLAMSTEED[9]

To
Mr Isaack Newton
fellow of Trinity
Colledge in Cambridge
there   these
Cambridge

NOTES

(1) Add. 3979, fo. 42. At the foot of Newton's letter here referred to, Flamsteed had written: 'Answerd 17 Sept', and on the reverse was a draft of his reply. It is not known if the letter to which the draft refers was ever sent. Flamsteed's reply to Newton's letter of 14 September was dated 19 September (Letter 531); it is clearly the letter to which Flamsteed here refers.

(2) For the many references to this work, which Flamsteed regarded as of paramount importance, see the Index to the present volume.

(3) The italicized passages were underlined in the original.

(4) See Memoranda 468, and the notes which follow it.

(5) Flamsteed had received no letter from Newton since the one referred to above (14 September 1695), nor does there appear to have been a reply to the present letter. See note (9) below.

(6) Edward Bernard (1638–96), astronomer, linguist; Savilian Professor of Astronomy (1673). See vol. I, p. 235, note (2). A comma has been editorially inserted after the word Bernard.

(7) Christiaan Albert Walter, a Danish nobleman who flourished in the seventeenth century. He resided in Paris, and was a frequent visitor to England.

(8) See Flamsteed's letter of 10 December 1694 (Letter 484): 'You requested of me when last here to give you the errors of the Rudolphine Numbers from ye heavens in the planet ♄.'

(9) There appears to have been no further correspondence between the two men for three years. It was revived when Newton wrote to Flamsteed protesting against Flamsteed's use of his name in his correspondence with Wallis (Letter 601).

544   NEWTON TO HALLEY
14 MARCH 1695/6
From the original in private possession

Sr

I understand that a report has been sometime spreading among ye Fellows of ye Royal Society as if I was about ye Longitude at Sea.[1] For putting a stop to that report pray do me the favour to acquaint them (as you have occasion) that I am not about it. And if the rumour of preferment for me in the Mint should hereafter upon the death of Mr Hoar[2] or any other occasion be revived, I pray that you would endeavour to obviate it by acquainting your friends that I neither put in for any place in the Mint nor would meddle wth Mr Hoar's place were it offered me.[3] You will thereby oblige

> Your most humble and
>
> most obedient Servant
>
> Is. NEWTON.

Cambridge. Mar. 14
      1696.

For Mr Edmund Halley at
Mr Coxall's in Packers Court
in Collman-street in
     London.

NOTES

(1) An undated memorandum by Newton, reproduced in Sotheby's *Catalogue of the Newton Papers* (1936), p. 8, states: 'Last Autumn when I was in London Mr Pepys asking me about the possibilities of finding ye Longitude at Sea and desiring my leave that he might say that I thought it possible, I desired him not to mention me about it least it might be a means to engage me in it and reflect upon me if I did not compass it.' Halley had long been interested in the problem of finding the longitude at sea. In a paper to the Royal Society, he 'concluded that it would be scarce possible ever to find the Longitude at Sea sufficient for Sea uses, till such time as the Lunar Theory be fully perfected' (Journal Books of the Royal Society of London, 9 May 1688). This interest was revived when later a prize of £20,000 was offered for a satisfactory solution of the problem. In a letter to Flamsteed, dated 31 July 1714, Sharp wrote: 'I hear Dr Halley puts in for it & indeed so great a prize as 20000 *lib* would animate any one to exert their uttermost endeavours were there any probability' (letter in the possession of F. S. E. Bardsley-Powell). See also Newton's letter to Sir Richard Haddock, 12 August 1699 (Letter 615, note (5), p. 315).

(2) James Hoare. There were two persons of that name, both of whom held offices at the Mint during the Commonwealth. One, a clerk, rose to the position of chief clerk. He died in 1679. The one referred to in this letter became Comptroller before the Restoration and continued in that office till his death about 25 December 1696. Although he showed great zeal in carrying out the duties of his office, he also ran a financial business in Cheapside; it is not to be identified with Hoare's Bank. On his death he was succeeded jointly by Thomas Molyneux and Charles Mason. See Letter 568, note (4), p. 241.

(3) Five days later he was offered the post of Warden of the Mint. See Letter 545, and note (1) thereon.

PLATE II. PORTRAIT OF CHARLES MONTAGUE, FIRST EARL OF HALIFAX,
BY SIR GODFREY KNELLER

545   MONTAGUE[1] TO NEWTON

19 MARCH 1695/6

From the original in the Bank of England

19. *March* 1695

Sr.

I am very glad that at last I can give you a good proof of my friendship, and the esteem the King has of your Merits. Mr Overton[2] the Warden of the Mint is to be made one of the Comrs of the Customes and the King has promised Me to make Mr Newton Warden of the Mint, the Office is the most proper for you 'tis the Chief Officer in the Mint, 'tis worth five or six hundred pounds per An,[3] and has not too much bus'nesse to require more attendance then you may spare. I desire you will come up as soon as you can, and I will take care of your Warrant in the mean time. Pray give my humble service to John Lawton,[4] I am sorry I have not been able to assist him hitherto, but I hope He will be provided for ere long, and tell him that the session is near ending and I expect to have his company when I am able to enjoy it; let me see you as soon as you come to Town, that I may carry you to kiss the Kings hand, I beleive you may have a lodging near Me. I am

Sr

Your most humble Servt

CHAS. MONTAGUE

NOTES

(1) Charles Montague (1661–1715), afterwards Earl of Halifax, K.G., President of the Royal Society (1695–8). On coming up to Trinity College, Cambridge, in 1678, he became intimately acquainted with Newton, nineteen years his senior, and between them they established the Philosophical Club in Cambridge. He sat with Newton in the Convention Parliament (1689) and rapidly rose to positions of great importance: a Commissioner of the Treasury in March 1692; Chancellor of the Exchequer, April 1694 to May 1699; First Commissioner of the Treasury, May 1697 to November 1699; created Baron Halifax, December 1700. His power was waning in 1699.

Newton hoped, through his influence, to secure preferment. Montague did his best, but for some time nothing came of his efforts. As a result Newton became irritated and thought his friends were deserting him; on 26 January 1691/2, he wrote to Locke: 'Being fully convinced that Mr Montague upon an old grudge wch I thought had been worn out, is false to me, I have done with him & intend to sit still unless my Lord Monmouth be still my friend. I have now no fair prospect of seeing you any more unless you will be so kind as to repay that visit I made you last year' (MS. Locke, c. 16, Bodleian Library).

Montague had not forgotten his old friend, and his letter could not have come as a great surprise to Newton (see Letter 541, note (4), p. 189), although he may not have known that it was the Wardenship and not the Mastership which he was being offered until the above letter arrived. The office of Warden was a government sinecure, and it was probably offered to Newton for his greater ease, as Montague's letter suggests. The post was undoubtedly, before Newton's occupation of it, a sinecure in the sense that a conscientious holder supervised enough to see that the work was properly done, while one less conscientious trusted the staff more completely. For fifteen years the Mint was being run by Hall. The letters concerning Mint affairs from now onwards show how little Newton regarded his position as a sinecure. (See E. N. da C. Andrade, *Sir Isaac Newton* (1961), pp. 88 *et seq.*)

Newton was appointed Warden of the Mint on Overton's promotion (see note (2) below) his warrant being dated 13 April 1696 (Number 547).

(2) Benjamin Overton. He became Warden of the Mint in 1689. Seven years later (23 March 1695/6) he was made Commissioner of Customs. See Letter 541, note (4), p. 189, and Craig, *Mint*, pp. 172, 196.

(3) According to Edward Chamberlayne (*Angliæ Notitia*, ed. 1694, pp. 618, 621) each of the seven Commissioners had a salary of £1200 per annum. The Warden of the Mint received a salary of £400 per annum, the Master and Worker £500 plus coinage fees, and the Comptroller £300.

(4) John Lawton, or Laughton, a great personal friend of Montague and of Newton. He became Librarian and Chaplain of Trinity College. Later he was appointed Canon of Lichfield, and he gave to the Library of Trinity a valuable collection of books. See Monk's *Life of Bentley* (1830), pp. 226 and 246, and Brewster, II, 191 note.

546 MEMORANDUM BY NEWTON
MARCH 1695/6
From the original in King's College Library, Cambridge[1]

On Munday March 2d or Tuesday March 3 1695/6, A Londoner acquainted wth Mr Boyle[2] & Dr Dickinson[3] making me a visit, affirmed that in the work of Jodochus a Rhe[4] wth ⊕[5] twas not necessary that the ⊕ should be purified but that the oyle[6] or spirit[7] might be taken as sold in shops wthout so much as rectifying it. That the fire does not destroy the life of the Oyle or spirit in distilling it from the red hot vitriol.[8] That two or three pounds of Oyle or spirit will not afford above half an ounce of fixt salt[9] & that that ye oyle affords more fixt salt then the spirit. That the white spirit is in appearance like rain water only sweet & fragrant, & that Dr Twisdens[10] spirit as I described it to him was genuine. That the white spirit must be rectified seven times from its (?calx) without separating any flegm from it, & that in rectifying, it will endure any heat without losing its life. That the remaining matter for extracting the soul must not be calcined to a red heat; but only well dryed, least the soul[11] fly away. That for extracting the soul the spirit must be digested on this matter not two months but only till it appear well coloured with the extracted soul. That when you draw off the spirit from the soul you must leave the soul not thick like honey or butter but thinner then oyle so that you may pour clean out of your glass like a liquor & that it will keep better in moisture then when too dry & therefore tis safest to err on that hand there being no danger in keeping it too moist. I think he said also that the soul must be volatized by the spirit, but I'm sure tis so in ye Process of Jodochus p 20 & those of Baslius[12] wth $\odot$[13] & other metalls Key 5 p 22, Text p. 123 lin 5 & p 154 l 17, 18, 19, 20 & p 156 lin ult & p 169 l 19, 20, 21, & p 170 l 19, 20, 21, 22, Et Snyders Pharm. Cath[14] p. 15 l. ult & p. 21 l. 25.

He told me also that when all the soul is extracted the remaining matter must be put in a cruciple covered wth a muffle or hollow cup of iron like a bowl inverted & a fire made round about them for an hour wch cannot easily be too hot. Then ye salt extracted wth ye spirit & the matter calcined again & extracted again as before & so on till no more salt can be extracted. That he imbibed this fixt salt always with 1/8th part of the spirit [perhaps 8, 10 or 12 times][15] & that when it was so long imbibed till it became volatile it was not necessary to sublime it. For all is pure, and if in the sublimation any thing should remain below, that would not be a heterogeneous impurity to be rejected but an unripe part of the matter, wch by further imbibition would be all ripened and volatized like the rest. And that if in imbibing you should at any time use too great heat, all the hurt would be only the loss of so much matter as sublimes & dries upon ye upper part of ye vessel. And that in every imbibition he let ye matter imbibed wth 1/8 of ye spirit continue in ye cold for 3 days the better to unite them, and then digested them 4 or 5 days more. And when he had finished ye work wth ye white spirit he imbibed in like manner wth an eighth part of the red soule [perhaps 7 times.][15] And when ye 3 principles were joyned the menstruum became a notable one. It then dissolves and volatizes all metals & gold dissolved & volatized may be digested with it to the end.

When he had finished the imbibitions (whether with both ye white & red spirits, or rather wth ye white alone) he said that the matter flowed wth an easy heat but in cold congealed & grew hard like a stone, & by digestion passed through ye colours black white citrine & red & in the beginning of the decoction it fumed up like a cloud as is described in the process of Jod. Rhe. And that in this decoction if the fire should go out for a while the matter would not thereby lose its life or motion but go on still when ye fire is kindled anew. And that it never putrefied but in the first decoction. Whence I seem to gather that he putrefied wth ye white spirit alone & multiplied only by imbibing wth the red as is described by Jodochus & Basil. This work he fermented by melting with ⊙, & said the whole was finished in 9 months.

NOTES

Newton's very considerable interest in alchemy has received the attention of a number of scholars. None of Newton's published works supports the traditional alchemical doctrines, although he collected books on the subject and left a mass of manuscripts (of which this is one) containing transcripts from printed works on the subject. See D. McKie, in vol. III, p. 443 (addendum to p. 12, note (4)), and *Phil. Mag.* (1942), XXXIII, 847; E. N. da C. Andrade, p. 18 *et seq.*, and Keynes, p. 31, in *Newton Tercentenary Celebrations* (1946); R. J. Forbes, *Chymia* (1949), **2**, 27; F. S. Taylor, *Ambix* (1956), V, 59–84; Lynn Thorndike, *History of Magic and Experimental Science* (1958), VIII, 589–90, 597–9; M. Boas Hall, in *Isaac Newton's Papers and Letters on Natural Philosophy* (1958), pp. 241–8; J. R. Partington, *A History of Chemistry* (1961), II, 468–76, particularly p. 469; 'Some Indications of Newton's Attitude towards Alchemy', by D. Geoghegan (*Ambix*, VI, no. 2, December 1957); Sotheby's *Catalogue of the Newton Papers* (1936), pp. 1–19.

It should be noted that after his appointment at the Mint in 1696 any association of Newton's name with what Professor McKie (*op. cit.*) describes as the 'already discredited pursuit of gold making' might well have been viewed with suspicion.

(1) Keynes MS. 26.

(2) This is probably Robert Boyle, the natural philosopher and chemist, who died in 1691.

(3) Edmund Dickinson (1624–1707), physician and alchemist. Merton College, Oxford; B.A. 1647; M.D. 1656. He practised medicine in Oxford, and later (1684) in London. He was physician in ordinary to the Royal Households of Charles II and James II.

About 1656 he became acquainted with Theodore Mundanus, a French adept in alchemy, from whom he claimed to have learnt the secret of the philosopher's stone. See E. S. de Beer, ed., *The Diary of John Evelyn*, V, 599: 'I went to visite Dr *Dikinson*, the famour Chymist, where we had long Conversation about the Philosophers Elixir, which he believed attainable, & had seen projection himselfe, by one who went under the name of *Mundanus*.' Among his published works is *Epistola ad Theodorus Mundanus, Philosophus Adeptus, de Quintessentia Philosophorum Et de Vera Physiologia*...(1699), a copy of which was in Newton's library (Villamil, p. 74). See J. R. Partington, *op. cit.* pp. 327–8 and *D.N.B.*

(4) Jodocus a Rhe is known mainly by a short treatise which is included in *The Last Will and Testament of Basil Valentine* (London, 1671). In the table of contents (last page) is the entry: 'A Philosophick work by another.' Although this is shown as on p. 17, it is, in fact, on p. 362, which reads as follows:

'A Process upon the Philosophick work of Vitriol Jod. V. R' [Jodocus von Rehe].

In this article Jodocus states that he 'began a new Process at the end of the year 1605. in the City of *Strasburg*'. He continues: 'I took ten pound of Vitriol, dissolv'd it in distilled Rainwater, being warm'd, let it stand for a day and a night, at that time many *feces* were setled, I filtrated the matter, evaporated it gently, *ad cuticulam usque*, I set it on a cool place to crystallize, this onshot Vitriol I exsiccated, dissolved it again in distilled Rain-water, let it shoot again, which work I iterated so long, till the Vitriol got a cœlestial green colour, having no more *feces* about it, and lost all its corrosiveness, and was of a very pleasant taste.'

A copy, in Newton's hand, of 'Rehe (Jodocus a) Opera Chymica, with Transcripts of Letters to Dr. John Twysden from A. C. Faber, and Notes on Fabers Works' is included in Sotheby's *Catalogue of the Newton Papers*, 1936 (p. 14, no. 89).

A manuscript, written in German, containing alchemical treatises and recipes, in the Library of Congress, Washington (MS. Div. acc. 1507(2)), includes a recipe entitled *Universale ex vitriolo* (fo. 74 *a*), apparently ascribed jointly to Basilius Valentinus and Jodocus von Rehe (*Osiris*, 1939, **6**, 528).

That this work by Jodocus has been included in those of Basil Valentine has led to his being confused with the latter.

(5) $\oplus$, a symbol for 'oil of vitriol' (sulphuric acid), in use in the seventeenth century.

(6) The 'oyle' is concentrated acid.

(7) The 'spirit' is dilute sulphuric acid. See also note (11) below.

(8) Vitriol, green vitriol or ferrous sulphate ($FeSO_4,7H_2O$). When heated strongly, the reaction is:

$$2FeSO_4,7H_2O \rightarrow Fe_2O_3 + H_2SO_4 + 13H_2O + SO_2$$

Ferric   Sulphuric   Water   Sulphur
oxide   acid   dioxide

(9) The 'fixed salt' is probably the residue left when the acid is distilled.

(10) John Twysden (1607–88), physician. He was called to the bar of the Inner Temple, 1634; qualified M.D. (Angers) 1646, and F.R.C.P. 1664. He wrote on mathematics, medicine and chemistry. The 'spirit' was probably prepared by a process to which his name was given. See *D.N.B.*

(11) The iatrochemists used a symbolism which included the following parallels:

salt = earth = body

sulphur = air = spirit

mercury = water = soul

(see J. R. Partington, *op. cit.* p. 154).

(12) 'Basil Valentine'. All that is known about him is based upon statements in works attributed to him. He is said to have lived about 1413 in the Benedictine Monastery of St Peter at Erfurt; if this were correct he would have been the most gifted chemist of his age. It is doubtful if he ever existed. The works attributed to him are numerous, the most famous being *The Triumphal Chariot of Antimony*. Like Paracelsus he speaks of spirits inhabiting fire, air, earth and water. He was very familiar with the three mineral acids and he gives recipes for their preparation. See J. R. Partington, *op. cit.* (1961), II, 183–203. According to Villamil, Newton's library contained at least three works by Valentine.

(13) $\odot$ = symbol for gold.

(14) John De Monte Snyders: he wrote 'The Metamorphosis of the Planets, That is A Wonderfull Transmutation of the Planets and Metallique formes into their first Essence'. See Sotheby's *Catalogue of the Newton Papers* (1936), p. 16. The work referred to in the above letter is *Commentatio de Pharmaco Catholico* (Amsterdam, 1666), by John De Monte Snyder. See Ferguson, *Bibliotheca Chemica*, II, 105 and 246.

(15) The square brackets appear in the original.

547   NEWTON'S APPOINTMENT AS WARDEN OF THE MINT[1]

13 APRIL 1696

From a copy in the Royal Mint[2]

Isaac Newton Esqr to be Warden of the Mint

Gulielmus Tertius Dei Gratia Angliæ, Scotiæ, Franciæ et Hiberniæ Rex Fidei Defensor &c. Omnibus ad quos præsentes Litteræ nostræ pervenerint Salutem. Sciatis quod nos pro diversis bonis causis et considerationibus nos ad præsens Specialiter moventibus, de gratiâ nostrâ Speciali ac ex certa Scientiâ et mero motu nostris, dedimus et concessimus, ac per præsentes damus et concedimus, dilecto nobis Isaaco Newton, Armigero, tam officium Custod. Cambij et Monetæ Infra Turrim[3] Nostram London quam Custod. Cunagij[4] Auri et Argenti Infra Turrim nostram London prædict. ac alibi Infra Regnum nostrum Angliæ, ac Ipsum Isaacum Newton tam Custod. Cambij et Monetæ infra Turrim Nostram London Prædict. quam Custod. Cunagij nostri Auri et Argenti infra Turrim prædict et alibi infra Regnum nostrum Angliæ prædict. facimus Ordinamus et constituimus per præsentes in loco Benjamini Overton Armigeri. Habendum et occupandum officium et Custod. illi prædicto Isaaco Newton tam per ipsum quam per Sufficientem Deputatum suum vel deputatos suos sufficientes pro quo vel quibus respondere voluerit, Approbatum vel approbatos prius per Comissionarios pro Thesauro nostro, vel Thesaurarium pro tempore existente quamdiu nobis Placuerit.

...   .

In Cujus Rei Testimonium has Litteras nostras fieri fecimus patentes, Teste meipso apud Westm. Decimo tertio die Aprilis anno Regni nostri Octavo.

1696   Per Breve de privato Sigillo

PIGOTT.

Translation

William the Third, by the Grace of God, King of England, Scotland, France and Ireland, Defender of the Faith etc. To all to whom Our present Letters shall come, greeting. Know ye that, in view of divers good reasons and considerations, which at this time specially influence Us, of Our special grace and of Our sure knowledge and personal initiative, We have given and granted, and by these presents do give and grant, to Our beloved Isaac Newton, Esquire, both the office of Warden of Our Exchange and Mint under Our Tower of London, as well as the office of Warden of Our coinage

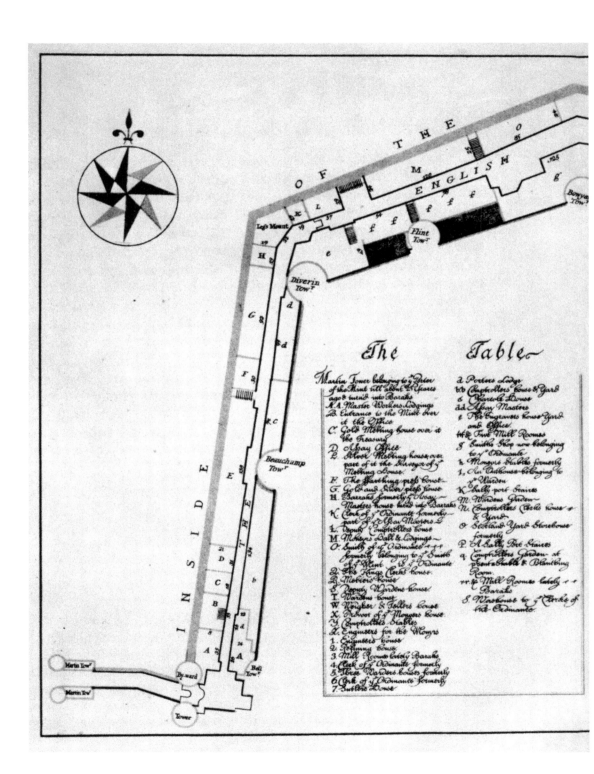

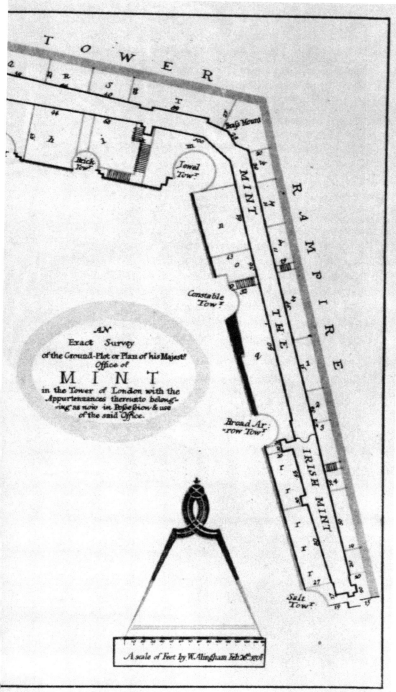

AN
Exact Survey
of the Ground-Plot or Plan of his Majest.ᵉ
Office of
MINT
in the Tower of London with the
Appurtenances thereunto belonging as now in Possession & use
of the said Office.

A scale of Feet by W. Alingham Feb. 26.ᵗʰ 1705.

PLATE III.

THE MINT IN NEWTON'S TIME

of gold and silver under Our aforesaid Tower of London and elsewhere within Our Kingdom of England: and the same Isaac Newton We create, ordain and appoint, by these presents, as Warden of Our Exchange and Mint under Our Tower of London aforesaid as well as Warden of Our coinage of gold and silver under the aforesaid Tower and elsewhere within Our Kingdom of England aforesaid, in succession to Benjamin Overton, Esquire, this office and Wardenship to be held and occupied by the said Isaac Newton either personally, or by his satisfactory deputy or his satisfactory deputies for whom (whether one or several) he has agreed to be answerable, these deputies to be men previously approved by the Commissioners of Our Treasury, or Our Treasuries for the time being, and so long as we shall please.

In witness whereof We have ordained these Letters to be made Patent, in Our presence at Westminster on the thirteenth day of April in the eighth year of Our reign.

NOTES

(1) For fifteen years before Newton's appointment as Warden, the Mint was run by Hoare, the Comptroller. In spite of the view expressed by many historians, Macaulay in particular, the office of Warden was a sinecure, and required only occasional visits to the Mint. Newton however found many tasks to absorb his energies. Thus for criminal work alone there were at least 146 days (possibly many more) when he was at Whitehall or the Mint between July 1696 and the end of 1699 when he became Master. As such, he found many extraneous tasks which were purely voluntary, as, for example, the attempted revaluation of the guinea, the copper coinage, the sale of Queen Anne's tin and the supervision of the Edinburgh Mint.

(2) Royal Mint Record Books, v, 26.

(3) *Infra Turrim*, under the Tower; the Mint lay between the two walls, and beneath them.

(4) *Cunagium*, coining or stamping. The word was used in ancient documents relating to the Mint, and was at one time part of the title of the Master of the Mint, e.g. *magister minte et cunagii* (Rot. Norm. 5 Henry V, m. 9).

548   OATH TAKEN BY NEWTON

2 MAY 1696

From the original in the Public Record Office[1]

You shall swear that you will not reveal or discover to any person or persons whatsoever the new Invention of Rounding the money[2] & making the edges of them with letters or grainings or either of them directly or indirectly. So help you God.

ISAAC NEWTON

Jurat 2do die Maij 1696
coram nobis

STE: FOX[3]
CHAS. MONTAGUE[4]
J SMITH[5]

NOTES

(1) T. 1/37, no. 53. This is in Newton's hand except for the words *Jurat 2do die Maij* 1696 *coram nobis* (sworn in our presence on 2 May 1696). Newton was formally appointed Warden on 13 April 1696.

(2) Up to the end of 1662 coin was 'hammered'. This was replaced by an entirely new process. In order to begin the process of rolling the castings down to fillets horse-drawn mills were used; it is from this fact that the term 'milled' money came to be used. Milled coinage was made current by Proclamation on 27 March 1663.

(3) Sir Stephen Fox (1627–1716) became Paymaster General after the Restoration and he held that office until 1676. He was a Commissioner of the Treasury almost continuously from 1679 to 1702. He was responsible for the founding of Chelsea Hospital.

(4) At this time Montague was Chancellor of the Exchequer. See Letter 545, note (1), p. 195.

(5) John Smith (1655–1723) was a Commissioner of the Treasury (1694–1701); Chancellor of the Exchequer (1699–1701, 1708–10), and Speaker of the House of Commons (1702–8).

549   MINT TO THE TREASURY
6 MAY 1696
From the original in the Public Record Office[1]

> To the Right Honble the Lords
> Commissrs of his Majties[2] Treasury

May it please your Lordships[3]

On the 28th April last Wee laid the following Memorial before your Lordships vizt.

Wee have read the Act to incourage the bringing plate into the Mint to be Coyned and for the further remedying the ill State of the Coyne of the Kingdom.[4]

And find the plate is order'd to be melted down in the Mint to be cast into Ingots (which is a melting more than is usual in the Mint & beyond the agreement with the Melter) but there is no Direction in the Act how this Melting shall be paid for, nor for the refineing of this Silver, the much greatest part of which will be worse than Standard; & some of it exceeding course. & if a great quantity should be brought in it will come to a great Summe of money. which if it were intended should be paid out of the Coynage Duty Wee humbly informe your Lordships that the money arising by that Act is not Sufficient to answer what was to be paid out of it before, much lesse to do this and pay for the presses that people bring in by directions of this Act. Wherefore Wee most humbly pray your Lordships to take it into your Consideration & to give such Orders in it as your Lordships shall think fit. And now Wee further humbly represent to your Lordships the great want of money there is for the uses before mention'd, And also for the paying the Carpenter & other Workmens bills, which amount to 8678*lib*: 00*s*: 00*d* and likewise for paying the Salaries of the Severall Officers & Clerks due at Lady day last amounting to 800*lib*. and the old presses which have been brought in by direction of the Act of Parliament come to about 300*lib*. making altogether 9778*lib*: 00*s*: 00*d*. in part of which there has been received out of the Exchequer.

		lib
On the 28th of March last		400
the 14th of April		700
the 30th of ditto		600
		1700

	lib	
Paid the Workmen above menc̄ond in part of their Bills	1330	
Paid Mr Western for new presses &c	200	1630
Paid for an old presse & charges by Warrant	100	

Which 1630*lib* being Substracted from the 9778*lib* above-mention'd there wants to pay the Workmen & the Salaries and for the old presses, 8148*lib*: but wee are humbly of opinion that the work done by the Carpenter and the rest of the workmen ought to be survay'd & valued before their whole bills are paid off.

All wch is humbly submitted
to your Lordships

ISAAC NEWTON
THO. NEALE[5]
THO. HALL[6]

Mint office 6th May
1696.

NOTES

(1) T. 1/37, no. 57.

(2) 'Majesty's'. Newton also abbreviates it to 'Maties', 'Matys', 'Majts', 'Mats' and 'Maes'.

(3) Newton, in this and other correspondence, variously abbreviates Lordship (Lordships) to Lop (Lops), Lopp (Lopps), Lordp (Lordps), Lordpp (Lordpps). In this and later letters, it is printed in full.

(4) 'Whereas by an Act of Parliament lately made Entituled An Act for the Remedying the ill State of the Coyne of this Kingdom It is enacted that all new Moneys proceeding from the Silver of the clipt moneys thereby directed to be melted downe and recoyned...shall from time to time be brought back into the Receipt of Our Exchequer' (Royal Mint Record Books, VI, 15, dated 18 March 1695/6). The Act came into operation on 1 February 1695/6. See A. E. Feaveaaryear, *The Pound Sterling* (2nd edition), p. 136.

(5) Thomas Neale was Master of the Mint from 1686 till his death in December 1699 when he was succeeded by Newton. He was appointed Groom Porter at the Palace by Charles II. This was an office of the Royal Household, the functions of which were the regulating of all matters connected with gaming within the precincts of the Court. The office was abolished during the reign of George III. Neale was successively Member of Parliament for Petersfield and Ludgershall. He devised and managed public lotteries and general speculations and promoted the National Land Bank. He built the whole of Shadwell in the east end of London, and Seven Dials in Central London. See J. Keith Horsefield, *B.M.E.* pp. 39 and 188.

(6) Thomas Hall had already served as King's clerk, or chief clerk, before he was appointed to assist Neale, the Master and Worker, on 26 February 1695/6 (Royal Mint Record Books, VI, 11). The Committee appointed to inquire into certain Miscarriages of the Mint (see Letter 564, 15 February 1696/7) described him as 'a very careful and diligent worker'. His salary was £400 per annum.

550   MINT TO THE TREASURY

8 JUNE 1696

From the original in the Public Record Office[1]

> To the Right Honble the Lords Commrs of his Majties Treasury.

May it please your Lordships

Wee humbly lay before your Lordships the progress wee have made in the providing of instruments and Officers for the five Mints in the Country.[2]

All the Iron Work & Instruments which cannot be made in the Country are bespoke & in a good forwardness here, & the men that are sent down are to get the rest of the things done there.

And five persons are provided for Deputy Wardens & one of them is already instructing.

And five other persons for Deputy Masters & Workers two of wch are already instructing.

But the Comptroller[3] refuses to appoint any Deputies.

And the Assay Master[4] will find out Deputies, but will not be answerable for them.

The Surveyor of the Meltings will find Deputies.

And the Weigher will find Deputies.

And the King's Clerk has one Deputy already instructed, & three more instructing, & will have the fifth ready before he shall be wanted.

Melters & Refiners, Monyers, & Clerks to the Officers[5] shall be fitted and sent as soon as ye Mints are ready. Several of the persons designed for Deputies and Clerks desire to know what Allowance will be made to them, which wee are not able to answer till your Lordships are pleased to Order what their Allowance shall be.

Is. NEWTON
THO. HALL

Mint Office
8th June 1696.

NOTES

(1) T. 1/38, no. 39.

(2) Country Mints were established in York, Exeter, Bristol, Norwich and Chester following the Recoinage Act of 21 January 1695/6. Their function was the replacing of the clipped coins by milled coins of full weight and value. York and Exeter began coining in August 1696, Bristol and Norwich in September, and Chester in October. They were all closed down in the summer of 1698.

(3) James Hoare. See Letter 544, note (2), p. 194. One whom he eventually appointed was Halley, who went to Chester as Deputy Comptroller.

(4) Daniel Brattle. He was Assayer from 1695 to 1712. Sir John Brattle and his two sons, Daniel and Charles, occupied this post from 1665 to 1723.

(5) For the duties of the several Mint officers see Newton's Memorandum: 'An Account of the Mint in the Tower of London', No. 565, p. 233.

551 NEWTON TO THE TREASURY

JUNE 1696

From the original in the Public Record Office[1]

16 *June* 1696[2]

May it please your Lordships

The Salary of the Warden of his Maties Mint is only 400*lib* per annum with a house of 40*lib* per annum & his perquisites are only 3*lib* 12*s* per annum for sea[3] coales: all which, Taxes being deducted, is so small in respect of the Salaries & Perquisites of the other Officers of the Mint as suffices not to support the authority of his Office.

For the Salary of the Master & Worker is 500*lib* per annum wth a house & very great perquisites & the salaries of the other Officers are proportional, besides their late additional Salaries & allowance of Assistants. For their Salaries upon occasion of the present coynage are at least doubled; & those of the Clerks are encreased almost in the same proportion.

And the Warden is now at much greater expenses & pains then formerly in going constantly to the Mint, overseeing & setting forward the coynage & pixing the money for wch he has need of a Deputy.

Wherefore I most humbly pray your Lordships to take it into your consideration & to give such Orders for the support of the Wardens Office as your Lordships shall think fit

Your Lordships most humble

& most obedient Servant

ISAAC NEWTON

[Reverse]

Read 16 June 96

My Lords at the end of this Coynage will give him a Consideration extr[aordinar]y suitable to his trouble & proporconable to the Increase allowd to the other officers for this extra Coynage. But this to be done out of ye Coynage mo[ney].

NOTES

(1) T. 1/38, no. 48. Newton's request might well have been a matter of principle; he believed that the emoluments were not commensurate with the dignity of the office. When Newton was offered the post of Warden by Montague (see Letter 545) he was assured that the post was 'worth five or six hundred pounds per An. and has not too much bus'ness to require more attendance then you can spare'. Eventually the Treasury offered Newton and Hall £500 'for their extra service in the coinage of the new moneys' (*Calendar of Treasury Books*, vol. XIII, p. 108).

(2) The date, 16 June 1696, has been inserted in an unknown hand to conform with the date of the Treasury Minute above.

(3) Newton wrote 'fee'.

552 MEMORANDUM BY NEWTON

? JUNE 1696

From a draft in the Royal Mint[1]

The State of the Mint

The Mint as appears by the Charter thereof, is a Corporation consisting of the Warden, the Workers, the Moneyers & the other Ministers. The Warden[2] is stiled Keeper of the Exchanges, that is of the Mint wth their revenues & of the Gold & Silver brought thither to be exchanged by coynage. He is also by his Office a Magistrate & the only Magistrate set over the Mints to do Justice amongst the members thereof in all things except in causes of Freehold & causes relating to the Crown. The Workers (one of wch is Master[3] of ye rest) are they who melt refine allay & run the standarded gold & silver into Ingots to be coyned. The Moneyers[4] are they who flat out & coyn those Ingots into moneys. The other Ministers are the Controller,[5] Assaymaster,[6] Surveyor of ye Meltings,[7] Clerk of the Irons,[8] Weigher & Teller,[9] Auditor, Engraver,[10] Porter: whereof the Controller, Assaymaster & Surveyor are, in behalf of the King & his people, a Cheque upon the Master & Worker in his Accounts, Assays, & Meltings. These other Ministers are standing Officers with constant salaries but the workers & Moneyers are no standing Officers nor have salaries but are allowed after a certain rate in the pound weight for all the moneys they make, excepting that the Master & Worker (that is he who is one of the Workers & the Master of the rest) did in ye reign of K. Charles ye 2d become a standing Officer with a salary of 500 *lib* per annum wch is greater then ye Wardens salary, & soon after was appointed by Act of Parliament to receive ye Coynage Duty, the distribution of wch has brought the Mint into his power. For the Seigniorage[11] or Revenue of the Mint was till then received of the King by the Warden. Also ye last winter the Master & Worker was empowered by a clause in another Act of Parliament to incorporate new Moneyers into the Mint without the Wardens consent & soon after formed a Project of erecting Mints in the Country without any Warden at all, & lately has rejected the Wardens judicial power by endeavouring to have differences referred to the Warden & himself. The Moneyers have also shook off the Wardens authority over them by feigning themselves a Corporation, & the Controller (who is the first of the other Ministers) by getting the Office of Master & Worker into his hands (tho an Office inconsistent with his own) hath equalled himself with the Master & Warden & this equality they have ratified by a new Law (no where written but in the Moneyers Charter) that all Orders made by the Master & Controller are binding even to the Warden himself. And thus the Wardens Authority which was designed to keep the three sorts of Ministers in their Duty to ye King & his people, being baffled & rejected & thereby the Government of ye Mint being in a manner dissolved those Ministers act as they please for turning the Mint to their several advantages. Nor do I see any remedy more proper & more easy then by restoring the ancient constitution.

NOTES

(1) Newton MSS. 1, 8–9 (in Newton's hand). This account was a plea to improve the Warden's status.

(2) The Warden, the titular senior (though not the highest paid), was the judicial element; all but the greatest suits at law which affected the Mint or its members had of old been tried by him, but the post had been a sinecure for a hundred years.

(3) The Master Worker was responsible, *inter alia*, for the standard of the coinage. All bullion brought to the Mint, whether ingots, plate, old or foreign coins, was converted into bars of correct fineness by a contractor who hired his own employees.

(4) The conversion of the coinage bars (see note (3) above) into coin was undertaken by the Company of Moneyers. The Moneyers claimed to be an ancient corporation, but the self-appointed designation 'Provost and Company of the Moneyers' does not appear until 1578. The *Report of the Royal Commission on the Royal Mint* (1849) stated that 'their pretensions to be a separate corporation, with legal rights are supported neither by proof nor probability'.

(5) The Comptroller checked the Master by means of the bullion accounts and maintained the buildings; in practice the Comptroller at this time (James Hoare) ran the whole Mint till the start of the Recoinage, and had done so for many years.

(6) The Assay Master 'shall keep a book of bullion brought into the mint, and also of the pot assay, to be made of some ingot.... The master is bound to stand to his report on the trial of gold and silver, in dispute between him and the importers, being made in the presence of the warden, master and comptroller' (Ruding, 1, 38).

(7) The Surveyor of Meltings was a check on the Melter, who was purely a Master's man. This post had already become a sinecure.

(8) Clerk of the Irons. 'It is his duty to keep an account of all the blank dies which are delivered to the graver, and sunk, stamped, and hardened by him. He is also to give an account to the warden, master, and comptroller, when required' (Ruding, 1, 48).

(9) The Weigher and Teller weighed the bullion before it went to the Moneyers, and after it had been coined, to see if the weights agreed.

(10) The Engravers were not permitted to make dies in any other place but the Tower. They were required to deliver each month any faulty dies to the Clerk of the Irons to be defaced in the presence of the Warden, Master, and Comptroller, and not otherwise.

(11) The Seignorage was the Crown's right to a percentage on the bullion brought to the Mint for coining. Until 1666 two prescribed amounts were deducted from new coins by the

Warden. One was handed to the Master for direct costs; this is sometimes distinguished as *Mintage*. The other, after defraying expenses proper to the Crown was handed to the fisc. In 1666 these deductions were abolished.

For a more detailed account of the establishment of the Mint during this period, see Memorandum by Newton, Number 565. For an account of the duties of the various Mint officials, see John Chamberlayne, *Magnæ Britanniæ Notitia*, 1716, pp. 240–1.

553   NEWTON TO THE TREASURY

JULY/AUGUST 1696

From a copy in the Royal Mint[1]

To the Rt Honble the Lords Commissioners of his Majties Treasury.

May it please your Lordships

By the influence (as it seems to me) & for the advantage of George Macy Esqr[2] late Clerk to several successive Wardens of his Majts Mint & a Prosecutor of Clippers & Coyners there hath been granted a Privy Seal to your Lordships & to the said Wardens for impowering the said Wardens & their Deputies (or whom your Lordships should appoint) to seize & gather up the forfeitures[3] of Clippers & Coyners & your Lordships to defray therewith the charges of those that prosecute & reward those that merit in these matters.

But those seizures & forfeitures for want of other rewards are either privately pocketed by my Agents or sold off to them that forfeit or where they come to my knowledge are ballanced & sunk by my Agents Bills & since the ceasing of the clipping trade suffice not to pay for the charge & pains of recovering & gathering them up, & are also a discouragment to Prosecutions by the Offence they give in Courts of Justice for being derived to me out of their ancient channel & by exposing me & my Agents to the censure of prosecuting men for the Estates so that the clipping & coyning trades instead of decreasing did in the time of my Predecessors grow to the greatest height. And the Coyners are now thinned in this City, yet by their flight from hence & by the turning of clippers to coyners, they seem more numerous in other places where I cannot reach them.

And the new reward of forty pounds per head has now made Courts of Justice & Juries so averse from beleiving witnesses & Sheriffs so inclinable to impannel bad Juries that my Agents & Witnesses are discouraged & tired out by the want of success & by the reproach of prosecuting & swearing for money. And this vilifying of my Agents & Witnesses is a reflexion upon me which has gravelled me & must in time impair & perhaps wear out & ruin my credit. Besides that I am exposed to the calumnies of as many Coyners & Newgate Sollicitors as I examin or admit to talk with me if they can but find friends to beleive & encourage them in their false reports & oaths & combinations against me.

I do not find that the prosecuting of Coyners was imposed upon any of my Predecessors (tho some of them have done it) or is consistent with the Privy Seale. For he that gathers up the estates of convicted criminals should not intermeddle with their prosecution. Nor is there any reward or encouragement appointed for my service in these matters, nor am I provided wth any competent assistance to enable me to grapple wth an undertaking so vexatious & dangerous as this must be whenever managed wth diligence & sincerity. And therefore I humbly pray that this duty may not be annexed to the Office of the Warden of his Majts Mint nor enjoyned me any longer, at least not without enabling me to go through it with safety credit & success. All wch is most humbly submitted to your Lordships great wisdome.[4]

Is. NEWTON.

Tis the business of an Attorney & belongs properly to ye Kings Attorney & Sollicitor Gen[era]l, & they are best able to go through it especially wth such assistance as they can procure. And therefore I humbly pray that it may not be imposed upon me any longer. All wch is most humbly submitted to your Lordships great wisdome.

NOTES

(1) This letter (Newton MSS. i, 438–9), which is in Newton's hand, was probably written during the summer of 1696, i.e. after 'the ceasing of the clipping trade'. From May 1696 onwards the clipped coin ceased to be legal tender except by weight.

(2) Up to 1696 coinage offences were dealt with by a succession of Warden's clerks, one of whom was George Macey. After May 1696 there is no further mention of Macey, from which it may be assumed that Newton lost no time in settling down to this task. See articles by Sir John Craig, 'Isaac Newton—Crime Investigator', *Nature*, **182** (19 July 1958), 149, and 'Newton and the Counterfeiters', *Notes and Records of the Royal Society of London*, vol. 18, 1963, p. 136.

(3) The personal effects of persons convicted of coinage offences went to the Warden to pay for the cost of prosecution, but they were often intercepted by other authorities, and the real estate was escheated to the Treasury to reward informers under Privy Seal Warrant of 16 February 1694. The Treasury Papers are replete with similar applications from informers, all of whom should have been (and many were) referred to the Warden. See particularly Letter 571.

(4) These two sentences were crossed out.

554 NEWTON TO PRISON KEEPERS[1]
7 SEPTEMBER 1696
From drafts in the Royal Mint[2]

NEWTON TO THE KEEPER OF THE MARSHALSEA PRISON

Surrey Ss.

To the Keeper of the Marshalsea for the County of Surrey

Whereas you have in your Custody ye body of Charles Eccleston Committed to you by Warr[an]t under the hand & seale of Mr Rich one of his Maties Justices of the peace for ye said County for High Treason in conterfeiting the current coyne of this Kingdom, & forasmuch as it is by[3] their Excellencies the Lords Justices of England referred to me Isaac Newton Esqr one of his Maties Justices of the Peace for the said County & Warden of his Maties Mint to make such person or persons evidences in cases of clipping & coyning as I shall think fit, & I being well satisfied of the good service of the said Charles Ecclestone to ye Government do hereby intend to make him one of his Maties Evidences: These are therefore in his Maties name to require & command you forthwith upon sight hereof to deliver him the said Charles Ecclestone into the safe custody of the Bearer or Bearers in order to give his evidence at Hicks Hall in St John's street against several persons for coyning & for so doing this shall be your Discharge

Given under my hand & seale this seventh day of September Anno Domini 1696

NEWTON TO THE KEEPER OF NEW PRISON

Middlesex Ss To the Keeper of New Prison for ye said County Receive herewith sent you into your safe Custody the Body of Charles Ecclestone being charged upon Oath wth High Treason in Counterfeiting the Current Coyne of this Kingdom, & he being now intended for one of his Maties evidences against several clippers & coyners of false & counterfeit money: These are therefore in his Maties name to require you to keep him in safe custody untill further Order & untill he shall be discharged by due course of Law, & for your so doing this shall be your warrant.

Given under my hand & seale this seventh day of September Anno Domini 1696

Let him not be discharged
without my knowledge.

NEWTON TO THE KEEPER OF THE GATE HOUSE PRISON

Middlesex & } Ss   To the Keeper of the Gate house
City of Westmr }   for ye said City of Westminster

Whereas you have in your Custody the body of Joseph Gregory committed to your Warr[an]t under the hand & seal of Mr Railton one of his Maties Justices of Peace for ye said County & city & forasmuch as it is by their Excellencies the Lords Justices of England referred to me Isaac Newton Esqr one of his Maties Justices of ye Peace for ye said County & Citty & Warden of his Maties Mint to make such person or persons evidences in cases of Clipping & Coyning as I shall think fitt & I being well satisfied of ye good service of the said Joseph Gregory to ye Government do hereby intend to make him one of his Maties Evidences These are therefore in his Maties name to require & command you forthwith upon sight hereof to deliver him the said Joseph Gregory[(5)] into the safe custody of the Bearer or Bearers hereof in order to give his evidence at Hicks Hall in St John's street against several persons for High Treason, & for your so doing this shall be your discharge. Given under my hand & seal this seventh day of September Anno Domini 1696.

NEWTON TO THE KEEPER OF NEW PRISON

Middlesex Ss To the Keeper of New Prison for ye said County. Receive herewith sent you into your safe custody the body of Joseph Gregory being charged upon Oath with High Treason in counterfeiting the Current coyne of this Kingdom & he being now intended for one of his Maties evidences against several clippers & coyners of false money, These are therefore in his Maties name to require you to keep him in safe custody untill further Order & untill he shall be discharged by due course of law & for your so doing this shall be your Warrant. Given under my hand & seale this seventh day of september Anno Domini 1696.

Let him not be discharged wthout my knowledge

NOTES

(1) These four drafts, all dated 7 September 1696, are included because there are no *letters* from Newton on the detail of his criminal work, and also because documents such as the above are exceedingly rare. As Warden, Newton superintended the prosecution of counterfeiters and coiners. For an account of Newton as the scourge of criminals, see the article by Sir John Craig in *Nature*, **182** (19 July 1958), 149, 'Isaac Newton—Crime Investigator'. Newton was the first Warden to do this work; see his own statement in Letter 553, p. 210. See also Hopton Haynes, *Brief Memoires Relateing to the Silver and Gold Coins of England, 1700* (Landsdowne MSS. 801, fos. 54–64).

These four documents relate to two persons, Gregory and Eccleston, committed to prison for currency offences and later released in order to give evidence in further prosecutions. The first got Eccleston out of the Marshalsea; the second got him into New Prison. The third got Gregory out of Gate House Prison, and the fourth got him into New Prison. The two men were moved simply for convenience of safe conveyance to the witness box at Hicks Hall. The Clerkenwell New Prison was built as a relief to the overcrowded Newgate. There were, at this time, a number of persons awaiting trial for felonies or even mere misdemeanours. Since imprisonment was not a usual penalty, the gaol could not have been crowded otherwise. The Marshalsea was for debtors within ten miles of Whitehall (excluding the City of London) and for Surrey felons. Eccleston was a Surrey felon, and was properly there awaiting trial at the Surrey Assizes. He could not properly be tried at Hicks Hall, which was the Sessions House for the County of Middlesex. See note (4) below.

(2) Newton MSS. I, 480 (in Newton's hand).

(3) Newton's signature appears in the margin here, together with a place for his seal.

(4) Hicks Hall, near Smithfield, was the Sessions House for the County of Middlesex. It was a well-known centre for municipal business. It was named after Sir Baptist Hicks, a mercer of Cheapside.

(5) For the case of Gregory, see *Calendar of State Papers*, 25 September 1696, p. 403. In the intervening period (7–25 September), Gregory's evidence got Hollyland sentenced to death. Newton received by the penny post a confession by Gregory of perjury. He confronted Gregory with this and ascertained that the confession was a forgery.

555 HALLEY TO NEWTON
28 NOVEMBER 1696
From the original in the Library of the Royal Society of London[1]

Chester mint Novemb 28° 1696

Honoured Sr

I shall be very sorry if my care to answer in all parts the Instructions given me as Dept Comptroller of this mint,[2] should occasion me to incurr the displeasure of our Superiours. I do not find in the Copy we have of the Indenture of the Mint, that it is prohibited to putt allay to scissell[3] to reduce it to standard, and do not think it necessary to do it when the barrs are formd better. Yet I am given to understand I have erred in so doing, or which is equivalent in putting a Course Ingott into a pot of scissell. Sr My particular instructions are to preserve an exact conformity in the finess and weight of all the Monies to the standard of England nor does our Generall Instructions any where prohibit what we have done: besides I conceived it the Kings Interest not to make the mony too fine for more than one reason: And having observed by the reports of our Assay master often repeated, that the potts sett of exactly standard, proved in the pott and money assays alwais better, I concluded that

213

the silver did refine in the fire, and that the allay did really burn away to
about the quantity of $\frac{1}{4}$ dwt in each fusion: and to be sure of it, I caused an
Ingott of about 1 dwt Worse to be melted down, with the same heat we usually
gave our mettall when we cast barrs, and this without pouring it out we
repeated three times, so that it was about 3 times as long as needed for casting,
and at each heat I took an Assay piece, which our Assay master found to grow
better and better and the last best and very near standard: And the diminution
of wt: was not much more than what was gained in finess; so that it seemed
demonstration to me, that supposing the potts set off standard the barrs would
be about $\frac{1}{4}$ dwt better; and the barrs cast from the scissell of those would be
about $\frac{1}{2}$ dwt better; and allowing accordingly we have found our money with
all the Curiosity[4] I conceive the thing capable of, to answer to the standard of
England: and so I doubt not but it will be found, when our money we have
sent you up, comes to be assayed at the Tower, and our pix[5] at the Exchequer:
However being advertised that this is not the Tower practise I shall forbear it
for the future, unless authorized by your approbation humbly begging your
opinion herin in a line or two by the first opportunity, We have coined about
60 journeys[6] of money and taken in about 94000 ounces of silver.

I am Sr

Your most obedient servant

EDM. HALLEY

To Isaac Newton Esqr.
at his house in Jermyn Street
near St James's
      humbly present
      London

NOTES

(1) Miscellaneous MSS. v, no. 40.

(2) Halley went to the Chester Mint as Deputy Comptroller in 1696 and he remained there
until it closed down in 1698.

(3) The perforated metal sheet left after the blanks have been punched out.

(4) The word was used at this time to signify careful attention to detail. Cf. Evelyn's
Diary, III, 293: 'I dined at Mr *Palmers*...whose curiosity excelled in Clocks & *Pendules*.'

(5) See Manuscript 639, note (2), p. 371.

(6) Journal, or journey-weight. Originally the term meant the quantity of coin which could
be completed in the course of a day, by one gang (Fr. *journée*). A silver journey was fixed in
1663 at 60 lb. troy: the gold journey was fixed in 1670 at 15 lb. troy.

556   THE TREASURY TO NEWTON
7 DECEMBER 1696
From a copy in the Royal Mint[1]

After our hearty commendations Whereas his Maties Warr[ant] under his Royall Signe mannuell beareing date the 17th day of November last past hath been pleased to direct and appoynt the Master & Worker of his Maties Mint for the time being to pay unto you out of any the mony that is and shall be from time to time Imprested to him for the Service of the said Mint the yearly Sallary or allowance of sixty pounds per annum for an addicionall Clerk[2] (over and above your other allowances) from Mich[ae]mas last past Quarterly dureing his Maties pleasure And you haveing represented to us Christopher Ellis as a fitt person to be your addicionall Clerk at ye Said Sallary of Sixty pounds per ann. We approve thereof and direct you to appoynt the said Christopher Ellis to ye said place at ye said Sallary accordingly for wch this shall be your Warrant.

Whitehall Treasury Chambers 7 December 1696

To our loving freind				STE. FOX[3]
Isaac Newton					CHAS MONTAGUE[4]
Esq: Warden of his				J. SMITH[5]
Maties Mint					THOS. LITTLETON[6]

Mr Ellis to be addll Clerk to Mr Newton at 60 *lib* p.an

NOTES

(1) Royal Mint Record Books, VI, 46.

(2) On 26 August 1696 the Lords Justices, who were appointed each year as the board of regents during the King's absence abroad, had recommended that Newton be given an assistant, and, on 17 November, the Treasury sanctioned the revival of an extra clerkship at £60 per annum in consideration of the Warden's extraordinary business in detecting and prosecuting clippers. This authority was antedated to 29 September when the new Clerk, Christopher Ellis, began his duties. See *Calendar of State Papers*, 1696, p. 362.

(3) See No. 548, note (3), p. 202.

(4) At this date Montague was Chancellor of the Exchequer.

(5) See No. 548, note (5), p. 202.

(6) Sir Thomas Littleton, or Lyttleton (1647–1710), 3rd baronet; a Commissioner of the Treasury, 1696–9; Lord of the Admiralty, 1697; Speaker of the House of Commons, 1698–1700; Treasurer of the Navy, 1701–10.

557 MINT TO THE TREASURY
10 DECEMBER 1696
From the original in the Public Record Office[1]

To the Rt Honoble the Lords Commissioners of his Maties Treasury

Wee the Warden and Master Worker of his Majties Mint Having received from your Lordships a Refference dated the 4th day of August last, Upon the Memorial of Mr Jonathan Ambrose[2] hereunto Annex'd wherein he proposes to have Three halfe pence per pound for the First Melting down of the Plate; Wee take this Opertunity humbly to lay before your Lordships That the Master & Worker of the Mint, did alwayes melt down all silver and gold brought thither to be coyned, and did therein Employ such person or persons as he thought fitt for the doeing thereof being bound by the Indenture of the Mint to performe that service, and is answerable for all Loss and Miscarriage that may happen in the doeing thereof, And therefore humbly

Prays that he may Employ the persons that doe it as formerly, and is willing to undertake the severall meltings of the Clipt money at 3 farthings the pound, and on such other Condicõns as are at present allowed for the same, and to undertake the first melting down of the plate at one penny per pound, and the refineing (since the Quantity is like to be great) for whats done in London at 11*d* per pound, but for what's done and to be done in ye Country he humbly does crave the present allowance of 12*d* per pound, he having all along paid the same price, And as to the said Jonathan Ambrose we humbly lay before your Lordships that ye Coynage Duty is not sufficient to Defray the necessaries of the mint at this time, And therefore submitt it to your Lordships Direccõns how and at what rate he shall be paid for the Plate he hath already melted and refined. And we further presume to Represent to your Lordships an Inconvenience we Conceive has arisen by wording ye warrant under his Majties signe Mannual for the Distribucõn of ye 14*d* per pound Allowed by an Act of Last Sessions for making the money, whereby it is Directed that nine pence per pound (part of ye 14*d*) should be paid to ye Corporacõn of monyers, Whereas by the Identure of ye Mint 8*d* per pound and no more is directed to be paid them for their service, And the Additionall penny was only to be paid and allowed them as a bounty from his Matie, So long as ye Warden, Master Worker, and Comptroller or any Two of them whereof ye Warden to be one should perceive the moneys well sized, blanched[3] and Coyned and a due proporcõn of small money made, and the said Service not having by them been well and duely performed, As an Inducemt for ye future, and to make them well size, Blanch and Coyne, the Silver money, and in all things for the Future, to act as they ought, We Conceive it convenient that in ye next Warrant to be signed by his Majtie for Coyning either ye Clipt money or Plate in Pursuance of any Act of this Sessions, that ye giving ye Penny to ye monyers may be Only Condic͠onal, As it is by ye Indenture, And that if they doe not Deserve it, the same money to be saved to ye Crowne.

Is. NEWTON

THO NEALE

December
ye 10*th*
1696.

NOTES

(1) T. 1/41, no. 48. This letter was written by an amanuensis, but is signed and dated by the Warden and the Master Worker.

(2) Jonathan Ambrose was Master's melter from 1681 to 1699.

(3) Blanching was the process by which any discoloration of the coin was removed by the action of acid.

558 NEWTON TO THE TREASURY(?)

1696

From the original in the Royal Mint[1]

Accompt of disbursements from the 6th of Febr 1689 to the 6th Febr 1690 Vizt

For passing their Mats Privy Seal with the Warrant of the Lords Commissrs thereupon to me Concerning Clippers £20:10:10

Charges of being put into the Commission of peace for 7 Countyes. And at the Lords Commissrs of the Broad Seal. And Crown office £14:19:0

The Counties are Middlesex. Westminster, Essex, Surrey, Sussex, Kent & Hartford.[2]

NOTES

(1) Newton MSS. I, 18

(2) Only the last two lines are in Newton's hand; the rest of the document is in another hand.

559 NEWTON TO THE TREASURY

?1696

From the original in the Public Record Office[1]

My Lords

Having lately received Information of some forfeitures[2] that belonged to Clippers & Coiners, and now in the posession of divers persons. To wit.

In the hands of Simon Rolfe of Lyn-Regis[3] Esqr

A Considerable quantity of Clippings Clipt & Counterfeit mony Melted Silver, Tools & Materials & other personal Estate taken upon Edwd Pamphilion (convicted) & other clippers & Coyners tryed at the last Assizes for the County of Norfolk.

In the hands of Joseph Host Esqr Justice of peace in ye sd County

A Considerable quantity of Clippings, Clipt & Counterfeit & Good mony Melted Silver, Tools & Materials &c. taken upon persons tryed at the Said Assize for Clipping & Coyning.

In the hands of the Sheriffe or Under-sheriff of Yorkshire a personal Estate by him Seized that belonged to Clippers & Coiners Convicted & tryed at the Assize there (reported to be) of Considerable Vallue.[4]

It is humbly desired that your Lordships will be pleased to grant your Warrants for the sd Rolfe, Host, & Sheriffe to deliver & pay over the things above Mentioned unto me or my "deputy" pursuant to his Mats Leters of Privy-Seal, As I observe your Lordships have done upon the like Occasion in the time of my Predecessor, perticulerly the 2d of July last & former times when any disputes have bin, as there now is, in this Case, with Sherriffs & others Clayming Right thereunto. Whome to Sue for the Same would prove Very Chargeable, As apears in a Process that hath Comenced 4 years since And still is depending against Two of the London Sherriffs.

All which is humbly Submitted to your Lordships By

My Lords

Your Lordships Most humble

& most obedient Servt

ISAAC NEWTON

NOTES

(1) T. 1/42, no. 50. The letter is in the hand of a copyist, but is signed by Newton.

(2) See Letter 553, note (3), p. 210.

(3) King's Lynn.

(4) See *Calendar of Treasury Papers*, 1557–1696, p. 573.

560   CERTIFICATE BY NEWTON
1696
From a draft in the Royal Mint[1]

To all Mayors Sherriffs Aldermen, Commissioners, Justices of the peace, Constables, Headboroughs, Press-masters & other his Maties Officers whom these may concerne.

Whereas by Ancient Charters from the Crown & confirmed by his present Matie as by several [of] his Progenitors & Predecessors Kings & Queens of England It hath been and is granted unto the Wardens of the Exchanges Monyers Workmen & other Ministers of his Maties Mint deputed unto those things wch concern the said Office That they & every of them are & shall be freed & discharged from being put into any Assizes, Juries, Inquests or Attaints although the same should touch his Maties or his Heires & that none of them against his will should be Mayor Sherriff Bayliffe Escheator Constable Collector Inquisitor or Assessor of Tenths Fifteents Subsidies Talliages[2] Aides or other Impositions nor beare any other Office nor service whatsoever. And that they & every of them should be quit & discharged forever in the City of London & all other Cities Towns & places where they do inhabit & where their estates do lye of all Assessments Taxations Talliages Aids Loanes Contributions Fifteenths Tenths Subsidies & other Impositions whatsoever & every part thereof to the Kings Matie his Heires & Successors by the Commonalty of the Realme of England & otherwise granted & to be granted And likewise from all Imprests Suits Actions Arrests Attaints &c And that all Pleas and complaints (pleas of Freehold & of the Crown only excepted) should be adjudged before the Warden of his Maties Mint. And that they or any of them should not be molested distreined or otherwise grieved in their Lands Tenements Rents Goods or Chattels but should have their due allowance and discharge thereof before the Treasurer & Barons of the Exchequer. And whereas in the Terme of St Michael in the two & thirtith year of the Raign of King Charles the second upon an hearing before the Barons of the Exchequer It was in pursuance to the said Charters ordered that the Collectors of Taxes granted by Parliament or otherwise do upon certificate under the hand & seale of the Warden of his Maties Mint forbeare to make any distress upon their estates goods or chattels or upon their Tenements in any City Town or place where the same do lye but shall return them with the summs assessed upon them insuper upon the accounts of the said Collectors Now I Isaac Newton Esqr Warden of his Maties Mint within the Tower of London do hereby certify that Christopher Priddeth[3] of the Parish of St Leonards Shoreditch is one of my Clerks belonging to his Maties sd Mint, and that he ought to enjoy the Privileges & Immunities before mentioned & all other Privileges that may of right appertein unto him by vertue of the said Charters that he may be free to give his constant attendance unto the service of his Maties said Mint. In witness whereof I have caused the seale of the said Office to be affixed unto these presents Given under my hand at the Mint within the Tower of London this   day of   in the   yeare of the raign of our soveraign Lord King William by the Grace of God of England Scotland France & Ireland Defender of the Faith &c Annoque Domini 1696

NOTES

(1) Newton MSS. III, 405 (in Newton's hand). See also Craig, *Mint*, pp. 50–1 and 223.

(2) Talliage, tallage, tailage, taylage, taliage, etc. An arbitrary tax levied by Norman and Angevin kings upon the town and desmesne lands of the Crown; hence a tax levied on feudal dependants by their superiors; also a toll or custom duty, a grant, levy or imposition (*O.E.D.*).

(3) A Warden's clerk whose service ended about 1696.

561 NEWTON TO MONTAGUE
30 JANUARY 1696/7
From an autograph draft in the Library of the Royal Society of London[1]

Jan. 30. 1696/7

Accepi, Vir Amplissime,[2] hesterno die duo Problematum a Joanne Bernoullo Mathematicorum acutissimo propositorum exemplaria Groningæ edita in hæc verba.

Acutissimis qui toto Orbe[3] florent Mathematicis
S.P.D.[4]
Johannes Bernoulli Math. P.P.[5]

Cum compertum habeamus,[6] [vix quicquam esse quod magis excitet generosa ingenia ad moliendum quod conducit augendis scientiis, quàm difficilium pariter & utilium quæstionum propositionem, quarum enodatione tanquam singulari si qua aliâ via ad nominis claritatem perveniant sibique apud posteritatem æterna extruant monumenta: Sic me nihil gratius Orbi Mathematico facturum speravi quam si imitando exemplum tantorum Virorum Mersenni, Pascalii, Fermatii, præsertim recentis illius Anonymi Ænigmatistæ Florentini[7] aliorumque qui idem ante me fecerunt, præstantissimis hujus ævi Analystis proponerem aliquod problema, quo quasi Lapide Lydio[8] suas methodos examinare, vires intendere & si quid invenirent nobiscum communicare possent, ut quisque suas exinde promeritas laudes à nobis publicè id profitentibus consequeretur.

Factum autem illud est ante semestre in Actis Lips. m. Jun. pag. 269,[9] ubi tale problema proposui cujus utilitatem cum jucunditate conjunctam videbunt omnes qui cum successu ei se applicabunt. Sex mensium spatium à prima publicationis die Geometris concessum est, intra quod si nulla solutio prodiret in lucem, me meam exhibiturum promisi:[10] Sed ecce elapsus est terminus & nihil solutionis comparuit; nisi quod Celeb. Leibnitius de profundiore Geometriâ præclarè meritus me per literas certiorem fecerit, se jam feliciter dissolvisse nodum pulcherrimi hujus uti vocabat & inauditi antea problematis, insimulque humaniter rogavit, ut præstitutum limitem ad proximum pascha extendi paterer, quo interea apud Gallos Italosque idem illud publicari posset nullusque adeo superesset locus ulli de angustiâ termini querelæ; Quam honestam petitionem non solum indulsi, sed ipse hanc prorogationem promulgare decrevi; visurus num qui sint, qui nobilem hanc & arduam quæstionem aggressuri, post longum temporis intervallum tandem Enodationis compotes fierent. Illorum interim in gratiam ad quorum manus Acta Lipsiensia non perveniunt, propositionem hîc repeto.

Problema Mechanico-Geometricum
de Linea Celerrimi descensûs.

Determinare lineam curvam data duo puncta in diversis ab horizonte distantiis & non in eadem rectâ verticali posita connectentem, super qua mobile propriâ gravitate decurrens & à superiori puncto moveri incipiens citissime descendat ad punctum inferius.

Sensus problematis hic est; ex infinitis lineis quæ duo illa data puncta conjungunt, vel ab uno ad alterum duci possunt, eligatur illa, juxta quam si incurvetur lamina tubi canalisve formam habens, ut ipsi impositus globulus & liberè dimissus iter suum ab uno puncto ad alterum emetiatur tempore brevissimo.

Ut vero omnem ambiguitatis ansam præcaveamus, scire B.L.[11] volumus, nos hîc admittere Galilæi hypothesin de cujus veritate sepositâ resistentiâ jam nemo est saniorum Geometrarum qui ambigat, *Velocitates scilicet acquisitas gravium cadentium esse in subduplicata ratione altitudinum emensarum*, quanquam alias nostra solvendi methodus universaliter ad quamvis aliam hypothesin sese extendat.

Cum itaque nihil obscuritatis supersit obnixè rogamus omnes & singulos hujus ævi Geometras, accingant se promptè, tenent[12], discutiant quicquid in extremo suarum methodorum recessu absconditum tenent. Rapiat qui potest præmium quod Solutori paravimus, non quidem auri non argenti summam quo abjecta tantum & mercenaria conducuntur ingenia, à quibus ut nihil laudabile sic nihil quod [sit] scientiis fructuosum expectamus, sed cum virtus sibi ipsi sit merces pulcherrima, atque gloria immensum habeat calcar, offerimus præmium quale convenit ingenui sanguinis Viro, consertum ex honore, laude & plausu, quibus magni nostri Apollinis perspicacitatem publicè & privatim, scriptis & dictis coronabimus, condecorabimus & celebrabimus.

Quod si vero festum paschatis præterierit nemine deprehenso qui quæsitum nostrum solverit, nos quæ ipsi invenimus publico non invidebimus. Incomparabilis enim Leibnitius solutiones tum suam tum nostram ipsi jam pridem commissam protinus, ut spero, in lucem emittet, quas si Geometræ ex penitiori quodam fonte petitas perspexerint, nulli dubitamus quin angustos vulgaris Geometriæ limites agnoscant, nostraque proin inventa tanto pluris faciant, quanto pauciores eximiam nostram quæstionem soluturi extiterint etiam inter illos ipsos qui per singulares quas tantopere commendant methodos, interioris Geometriæ latibula non solum intimè penetrâsse, sed etiam ejus pomœria Theorematis suis aureis, nemini ut putabant cognitis, ab aliis tamen jam longè priùs editis, mirum in modum extendisse gloriantur.

Problema alterum purè Geometricum, quod priori subnectimus & strenæ

loco Eruditis proponimus

Ab Euclidis tempore vel Tyronibus notum est: Ductam utcunque à puncto dato rectam lineam, à circuli peripheriâ ita secari ut rectangulum duorum segmentorum inter punctum datum & utramque peripheriæ partem interceptorum sit eidem constanti perpetuo æquale. Primo ego ostendi in eod. Actor. Jun. p.265 hanc proprietatem infinitis aliis curvis convenire, illamque adeo circulo non esse essentialem: Arrepta hinc occasione, proposui Geometris determinandum curvam vel curvas, in quibus non rectangulum sed solidum sub uno & quadrato alterius segmentorum æquetur semper eidem; sed à nemine hactenus solvendi modus prodiit; exhibebimus cum quandocunque desiderabitur: Quoniam autem non nisi per curvas transcendentes quæsito satisfacimus, en aliud cujus solutio per merè algebricas in nostra est potestate.

Quæritur Curva, ejus proprietatis, ut duo illa segmenta ad quamcunque potentiam datam elevata & simul sumta faciant ubique unam eandemque summam.

Casum simplicissimum existente, sc. numero potentiæ 1, ibidem in actis pag. 266. jam solutum dedimus, generalem verò solutionem quam etiamnum premimus, Analystis eruendam relinquimus.

Dabam Groningæ ipsis Cal.

Jan. 1697]

G R O N I N G Æ

Hactenus Bernoullus: Problematum verò. solutiones sunt hujusmodi.[13]

Probl. I.

Investiganda est curva Linea *ADB* in qua grave a dato quovis puncto *A* ad datum quodvis punctum *B* vi gravitatis suæ citissimè descendet.

Solutio.

A dato puncto *A* ducatur recta infinita *APCZ* horizonti parallela et super eadem recta describatur tum Cyclois quæcun-que *AQP* rectæ *AB* (ductæ et si opus est pro-ductæ) occurrens in puncto *Q*, tum Cyclois alia *ADC* cujus basis et altitudo sit ad prioris basem et altitudinem respectivè ut *AB* ad *AQ*. Et hæc Cyclois novissima transibit per punctum *B* et erit Curva illa Linea in qua grave a puncto *A* ad punctum *B* vi gravitatis suæ citissime perveniet.   Q.E.I.

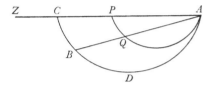

Prob. II

Problema alterum, si recte intellexi, (nam quæ in Actis Lips. ab Auctore citantur ad id spectantia, nondum vidi,) sic proponi potest. Quæritur Curva *KIL* ea lege ut si recta *PKL* a dato quodam puncto *P*, ceu Polo, utcunque ducatur, et eidem Curvæ in punctis duobus *K* et *L* occurrat, potestates duorum ejus segmentorum *PK* et *PL* a dato illo puncto *P* ad occursus illos ductorum, si sint æque altæ (id est vel quadrata, vel cubi vel quadrato-quadrata &c.) datam summam $PK^q + PL^q$ vel $PK^{cub} + PL^{cub}$ &c (in omni rectæ illius positione) conficiant.

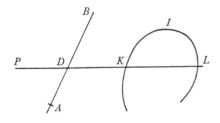

Solutio.

Per datum quodvis punctum *A* ducatur recta quævis infinita positione data *ADB* rectæ mobili *PKL* occurrens in *D*, et nominentur *AD* *x* et *PK* vel *PL* *y*, sintque *Q* et *R* quantitates, ex quantitatibus quibuscunque datis et quantitate *x* quomodocunque constantes et relatio inter *x* et *y* definiatur per hanc æquationem $yy + Qy + R = 0$. Et si *R* sit quantitas data, Rectangulum sub segmentis *PK* et *PL* dabitur. Si *Q* sit quantitas data, summa segmentorum illorum (sub signis propriis conjunctorum) dabitur. Si $QQ-2R$ datur, summa quadratorum $(PK^q + PL^q)$ dabitur. Si $Q^3 - 3QR$ data sit quantitas, summa cuborum $(PK^{cub} + PL^{cub})$ dabitur. Si $Q^4 - 4QQR + 2RR$ data sit quantitas summa quadrato-quadratorum $(PK^{qq} + PL^{qq})$ dabitur. Et sic deinceps in infinitum. Efficiatur itaque ut R, Q, $QQ-2R$, $Q^3 - 3QR$ &c datæ sint quantitates & Problema solvetur.   Q.E.F.

Ad eundem modum Curvæ inveniri possunt quæ tria vel plura abscindent segmenta similes proprietates habentia. Sit æquatio $y^3 + Qyy + Ry + S = 0$, ubi *Q*, *R* et *S* quantitates significant ex quantitatibus quibuscunque datis et quantitate *x* utcunque constantes, et Curva abscindet segmenta tria. Et si *S* data sit quantitas contentum solidum illorum trium dabitur

[Si *Q* sit quantitas data, summa trium illorum dabitur.] Si $QQ-2R$ sit data quantitas, summa quadratorum ex tribus illis dabitur.

Translation

I received yesterday, noble sir, two copies (published at Groningen) of problems proposed by that most acute of mathematicians, Johann Bernoulli, in the following terms:

'To the acutest mathematicians now living throughout the world, greetings from Johann Bernoulli, Professor of Mathematics.'

We[6] are well assured that there is scarcely anything more calculated to rouse noble minds to attempt work conducive to the increase of knowledge than the setting of problems at once difficult and useful, by the solving of which they may attain to personal fame as it were by a specially unique way, and raise for themselves enduring monuments with posterity. For this reason, I hoped I would do nothing more acceptable to the mathematical world than if (following the example of such eminent men as Mersenne, Pascal, Fermat, and above all, the example of that recent nameless Enigmatist of Florence,[7] as well as of the others who before me have acted similarly) I should propose to the most eminent analysts of this age, some problem, by means of which, as though by a touchstone,[8] they might test their own methods, apply their powers, and share with me anything they discovered, in order that each might thereupon receive his due meed of credit when I publicly announced the fact.

This was done six months ago in the *Leipzig Acts* for June, p. 269,[9] where I set forth such a problem, the usefulness and pleasure of which will be recognised by all who successfully apply themselves to it. A period of six months from the first day of this advertisement was allowed[10] to geometers: and, if within that period no solution was published, I promised to reveal my own. But lo and behold, the final date is now past, and no solution has appeared—apart from the fact that the celebrated Leibniz, a man who has rendered pre-eminent service to the more profound geometry, informed me by letter that he had successfully unravelled the knot of what he termed 'this beautiful and hitherto unknown problem'. At the same time he courteously asked me to allow the prescribed time limit to be extended to the following Easter so that meanwhile the same announcement might be made in France and Italy and no ground be left for any complaint about the shortness of the time. This esteemed request I not only granted but I decreed that notice of this extension should be published for I hoped to see whether there existed any who, intending to tackle this lofty and difficult problem, might after a long interval finally master its solution. Meanwhile for the benefit of those into whose hands the *Leipzig Acts* does not come, I here repeat the problem:

Mechanico-geometrical problem
about the line of quickest descent.

To determine the curved line joining two given points, situated at different distances from the horizontal and not in the same vertical line, along which a mobile body, running down by its own weight and starting to move from the upper point, will descend most quickly to the lower point.

The meaning of the problem is this: from the infinite number of lines, joining those two given points, or capable of being drawn from the one point to the other, let there be selected one line of such a nature that, if along it there be bent a metal plate shaped as a tube or groove, a small ball placed therein and released, will pass from one point to the other in the shortest time.

But in order to guard against every possibility of misunderstanding we wish our kindly readers[11] to know that we here admit the hypothesis of Galileo, the truth of which, subject always to the absence of any resistance, no unprejudiced geometer today questions, namely, that *the velocities acquired by falling bodies are in the subduplicate ratio of the heights traversed*: on this occasion we admit the hypothesis, though at other times our method of solving can be applied universally to any other hypothesis whatsoever.

Provided, therefore, no obscurity remains, we earnestly ask the geometers of this age, one and all, forthwith to gird themselves to the task, and to investigate all that they keep hidden in the innermost recesses of their methods. Let him who can, grasp the reward we have prepared for the one who solves, a reward which is not a sum of gold or silver, for that is a reward by which only mean and mercenary minds are hired, minds from which we expect nothing either meritorious or fruitful for knowledge. But since virtue is its own perfect reward and since glory provides an immense stimulus (to action), we offer a prize suited to a man of free birth, a prize composed of the honour, the praise and the applause with which we mean, publicly and privately, in words spoken and written, to crown, to decorate, and to celebrate the perspicacity of our great Apollo-like seer.

If however the festival of Easter shall have passed without our having found any who are capable of solving our noble problem, we shall not grudgingly refuse to the public what we have ourselves discovered. For the peerless Leibniz will forthwith, I hope, publish the solutions, his own and mine, which I entrusted to him some time ago. If geometers carefully examine these solutions, drawn as they are from what may be called a deeper well (of thought) we are in no doubt but that they will recognise the narrow limits of the common geometry, and will value our discoveries so much the more as there are fewer who are likely to solve our excellent problems, aye, the small number even among the very mathematicians who boast that by the remarkable methods they so greatly commend, they have not only penetrated deeply the secret places of esoteric geometry but have also wonderfully extended its bounds by means of the golden theorems which (they thought) were known to no one, but which in fact had long previously been published by others.

A second problem, purely geometrical, which we attach to the first and propose to the learned by way of a gift:

From the time of Euclid, it has been known, even to beginners, that if a straight line is drawn at random from a given point to cut by the circumference of a circle, it cuts it in such a way that the rectangle contained by the segments between the given point and the two points in which it is cut by the circumference, is always equal to the same constant. In the *Leipzig Acts*, June, p. 265, I first showed that this property belongs to an infinite number of other curves and is not peculiar to the circle. Taking this as a starting point, I proposed to geometers the determination of a curve or curves in which, not the rectangle, but the solid under one segment and the square of the other segment is always equal and constant: but by no one, so far, has a method of solving the problem been published. We shall show a solution as soon as it is desired: since, however we can satisfactorily meet this problem only by means of transcendental curves, here is another, the solution of which I can master by purely algebraical means.

The problem is to find a curve with the following property, that the two segments, raised to any given power and taken together, give everywhere one and the same sum.

Of the simplest case, when the index of the power is unity, we have published the solution in the *Leipzig Acts*, p. 266, but the general solution, which at the present moment we suppress, we leave to the analysts to investigate.

Written at Groningen on the first of January 1697

Thus far Bernoulli. The solutions of the problems are as follows[12]

Problem I

It is required to find the curve *ADB* in which a weight, by the force of its gravity, shall descend most swiftly from any given point *A* to any given point *B*.

Solution

From the given point *A* let there be drawn an unlimited straight line *APCZ* parallel to the horizontal, and on it let there be described an arbitrary cycloid *AQP* meeting the straight line *AB* (assumed drawn and produced if necessary) in the point *Q*, and further a second cycloid *ADC* whose base and height are to the base and height of the former as *AB* to *AQ* respectively. This last cycloid will

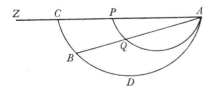

pass through the point *B*, and it will be that curve along which a weight, by the force of its gravity, shall descend most swiftly from the point *A* to the point *B*.   Q.E.F.

Problem II

The second problem, if I have rightly understood it (for what is cited by the author in the *Leipzig Acts* relating to it I have not yet seen), may be thus proposed. There is sought a curve *KIL* so ordered that if the straight line *PKL* be drawn at random from a given point *P* as pole, and it meets this same curve in two points *K* and *L*, the powers of its two segments, *PK* and *PL*, drawn from that given point *P* to their points of intersection with the curve, shall, if they are equally great (that is, both squares, or cubes, or fourth powers, etc.), together make the same sum, $PK^2 + PL^2$, or $PK^3 + PL^3$ etc. for every position of that straight line.

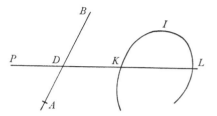

Solution

Through any given point *A* let there be drawn any unlimited straight line *ADB* given in position which meets the movable straight line *PKL* in *D*, and let *AD* be denoted by *x*, and *PK* or *PL* by *y*; let also *Q* and *R* be quantities compounded of any given quantities and the quantity *x* in any way whatever, and let the relation between *x* and *y* be defined by this equation:

$$y^2 + Qy + R = 0.$$

Then if *R* is a given quantity, the product of the segments *PK* and *PL* will be given. If *Q* is a given quantity, the sum of these segments (due consideration being taken of opposite signs in the addition) will be given. If $Q^2 - 2R$ is given, the sum of the squares $(PK^2 + PL^2)$ will be given. If $Q^3 - 3QR$ is a given quantity, the sum of the cubes $(PK^3 + PL^3)$ will be given. If $Q^4 - 4Q^2R + 2R^2$ is a given quantity, the sum of the fourth powers $(PK^4 + PL^4)$ will be given. And so on indefinitely. Hence let it be brought about that *R*, *Q*, $Q^2 - 2R$, $Q^3 - 3QR$ &c are given quantities, and the Problem will be solved. Q.E.F.

In much the same way curves may be found which shall cut off three or more segments having like properties. Let the equation be

$$y^3 + Qy^2 + Ry + S = 0,$$

where *Q*, *R*, and *S* signify quantities compounded in any way whatever of any given constants and the quantity *x*, and the curve will cut off three segments. And if *S* be a given quantity, the solid contained by three of them will be given. [If *Q* is a given quantity the sum of these three segments will be given.] If $Q^2 - 2R$ be a given quantity the sum of the squares of these three will be given.

NOTES

(1) N.1.61*b*. The letter is printed in the Royal Society Classified Papers, I, 30. Above the letter here reproduced are the words in what is probably the hand of Sloane: 'Epistola, præhonorabili viro D. Carolo Mountague Armig. Scaccarij Cancellario & S.R. Præsidi, inscripta qua solvuntur duo problemata Mathematica a Johanne Bernoullo Mathematico celeberrimo proposita.' (A letter written to that most worthy man, Charles Montague, Esq,

Chancellor of the Exchequer, and President of the Royal Society, in which are resolved two mathematical problems proposed by the renowned mathematician John Bernoulli.)

(2) Here the words *ex Gallia* ('from France') are heavily scored out.

(3) 'In the whole world': the contest is now thrown open to all. See note (9) below.

(4) *Salutem plurimam dicit* (greetings).

(5) *Math. P.P.* = *Mathematicæ Professor Publicus*.

(6) The whole of the passage in square brackets is taken from the Classified Papers mentioned above, note (1). At the foot of the passage, in Newton's hand, are the words: *Chartam ex Gallia missam accepi. Jan. 29. 1696/7. Sr Isaac Newton's hand*. Throughout the passage, the punctuation has been slightly altered in order to make clear what the writer intended.

(7) Vincenzo Viviani (1621–1703). Italian mathematician, disciple of Galileo. He was born and died at Florence; see vol. II, p. 100, note (5). For the reference to the 'nameless Enigmatist of Florence', see the Royal Society Classified Papers, I, 29:

Die 4 April 1692

Ænigma Geometricum De Miro Opificio Testudinis Quadrabilis Hemisphæricæ

a

D. Pio Lisci Pusillo Geometra propositum.

The words, *a D. Pio Lisci Pusillo Geometra*, are an anagram upon the words *a postremo Discipulo Galilei*.

(8) The Lydian Stone, a black variety of jasper (basanite) used by jewellers as a touchstone for testing gold and silver.

(9) 'Problema novum ad cujus Solutionem Mathematici invitantur.' This was the 'Problema Mechanico-Geometricum de Linea Celerrimi descensûs' quoted above, and for which a period of six months (June to December 1696) was originally allowed, but this, at Leibniz's request, was extended from December 1696 to Easter 1697. The second problem ('Problema alterum purè Geometricum') was added for good measure (*& strenæ loco Eruditis proponimus*).

The origin of the problem is to be found in a paper by Bernoulli in the *Leipzig Acts* (June 1696, pp. 264–9): *Joh. Bernoulli Supplementum Defectus Geometriæ Cartesianæ circa Inventionem Locorum*, in which he attempted to show how modern analysis was necessary and sufficient in filling the gaps in classical geometry. In particular, he took the product theorem for circle chords ($PA \times PB = k$), and showed that this property did not, conversely, uniquely define a circle through the set of points A, B. He then asked (*casum simplicissimum*) what curve satisfies the relation $PA + PB = k$, and stated an equation (correct but not general) which defined a curve with such a property. It was at the very end of this paper (p. 269) that Bernoulli tacked on the brachistochrone problem as the *Problema Novum ad cujus Solutionem Mathematici invitantur*. This was the problem for the solution of which he set his first time limit of six months. Having received no completely correct solution by the end of the year, he was persuaded by Leibniz to throw it open to the whole world (*toto orbe*), and not merely to readers of the *Leipzig Acts*, and he added the second problem (*en aliud cujus solutio per merè algebricas in nostra est potestate*). Copies were sent to the *Phil. Trans.* and the *Journal des Sçavans*, as well as to Newton and Wallis. Explicit in this paper was a request that the solutions should be sent to Bernoulli; implicit was a second one that the contestants should not publish their solutions

228

before the closing date (Easter 1697). To Bernoulli's consternation, Newton published his solutions in the February issue of the *Phil. Trans.* (**19**, 424): *De Ratione Temporis quo grave labitur per rectam data duo puncta conjungentem, ad Tempus brevissimum quo, vi gravitatis, transit ab horum uno ad alterum per arcum Cycloidis* (Concerning the ratio of the time in which a weight will slide by a straight line joining two given points, and the shortest time in which, by the force of gravity, it will pass from one to another by a cycloidal arc).

(10) *Concessum est...prodiret...promisi*; past tense because Bernoulli is referring to the challenge sent out in June 1696.

(11) *Bonos Lectores*, 'kindly readers'.

(12) *Tenent*. This word is meaningless in its present context. It may well be a dittograph from the next line.

(13) 'The solutions are like this', thus indicating the non-uniqueness of Newton's solutions. The three lines between the words *solutiones* and *sunt hujusmodi* are heavily scored out.

562   NEWTON TO HALLEY
11 FEBRUARY 1696/7
From the original in private possession
For answer see Letter 563

London Feb. 11. 1696/7.

Sr

This morning Collonel Blunt[1] the Kings first Engineer was with me & acquainted me wth a designe ye King had to allow 10*s* per diem for two Masters to teach Engineering (I meane ye Mathematical grounds of it) two hours each day to those of ye Army who will come to hear them publickly, Engineers & Officers & others who shall have ye curiosity & capacity. I proposed you as a fit person to be one of ye two if you should think fit to accept of ye thing. By bringing you acquainted wth ye Officers & making you known to the King it may be a means of making way for something better.[2] The Collonel will call on me 7 or 8 days hence for your answer. I am

Your faithfull friend to
serve you

Is. NEWTON.

I wrote to you ye last Post for
an Engineers place. I question
whether you can have both.

For Mr Edmund Halley
Controller of the Mint at
        Chester

NOTES

(1) Colonel Thomas Blount, or Blunt, a well-known engineer who lived in the latter half of the seventeenth century. He was one of the early Fellows of the Royal Society, having been elected in 1664. He was responsible for many ingenious inventions. He is mentioned four times in the Hooke *Diary*, and was evidently well known to the circle who met regularly at the various coffee-houses during the period 1672–80.

(2) When the Country Mints were established in 1696 (see Letter 550, note (2), p. 205), Halley was installed at Chester as Deputy Comptroller. But he does not appear to have been very happy in this post. In a letter to Sloane, dated Chester, 25 October 1697, he wrote: 'I long to be delivered from the uneasiness I suffer here by ill company in my business, which at best is but drudgery, but as we are in perpetuall feuds is intollerable' (MacPike, p. 103). This 'uneasiness' was aggravated by the disturbances which broke out at Chester between Weddell, the Deputy Warden, and Halley on the one hand, and two clerks, Bowles and Lewis, both of whom had the support of Clark, the Deputy Master, on the other. Halley, alarmed at these dissensions spoke of resigning. Newton, to save him any embarrassment, offered to procure for him 'an Engineers place'. Halley, however, remained at Chester until the Mint closed down in 1698. Shortly afterwards, he was given command of the *Paramour*.

563   HALLEY TO NEWTON
13 FEBRUARY 1696/7
From the original in the Library of the Royal Society of London.[1]
In reply to Letter 562

Chester Feb 13° 1696/7

Honoured Sr

I give you most hearty thanks for your care of my Interest and good inclination to prefer me. As to the proffer you are pleasd to make me of procuring me an Engineers place,[2] I cannot yet tell what to say to it, without I know the occasion or business I should be employed about, for as there is reason to apprehend the war near an end,[3] so unless I were upon a lasting establishment, I suppose I should soon be discarded. I should have been very much obliged to Mr Sam: Newton[4] had he pleased to informe me further: but because loss of time in these cases is dangerous, be pleased to say I will accept of this kind offer, provided Sir Martin Beckman be of opinion that my post is like to be durable, and I am little sollicitous about the service wherever I am employed. Mr Clarke[5] our Master Worker has been at London this two months, and left all his business (which has for this month past kept us all fully employed) to our care, though we know not why we should charge our selves therwith, he not desiring it; but we have been willing to serve the publick by giving a constant attendance and animating all parts of the Mint, so that at this time we have cleard of all yt was imported above 5 weeks since; and have issued about 50000 *li*[*b*] of new Money. Mr Weddall[5] gives you his humble service

I am Sr

Your most obedient servant

EDM. HALLEY

I hope you will be able to
gett us excused this
heavy Capitation tax upon
our Salaries

To Isaac Newton Esqr
at his house in Jermyn Street
              These
                    humbly present

NOTES

(1) Miscellaneous MSS. v, no. 41.

(2) See Letter 562, note (2), p. 230.

(3) The War of the Grand Alliance against France, which began in 1689 and ended with the Treaty of Ryswick in September 1697.

(4) See Letter 496, notes (5) p. 95 and (8), p. 96.

(5) See Letter 562, note (2), p. 230.

564   ARNOLD TO NEWTON
15 FEBRUARY 1696/7
From the original in the Royal Mint[1]

In the Speakers Chamber   *Lune* 15°: *die Feb*: 1696
Mr Arnold in ye Chair
at ye Commee appointed to Examine
into the Abuses of the Mint[2]

Ordered

That the Warden of the Mint doe ag[ains]t Saturday Morning next prepare or Cause to be prepared such Matters and things as the Commee

231

this day directed Mr Chaloner[3] To the End the said Mr Chaloner may make an Experiment before the said Commee in relation to Guineas

ARNOLD

To the Honoured Docr
Newton Warden of his
Majties Mint in the Tower
London

NOTES

(1) Newton MSS. I, 506.

(2) Parliamentary Committee on the Miscarriages of the Mint, 14 January 1696/7.

(3) As a result of certain alleged abuses in the Mint, the House of Commons set up a Committee to inquire into the various charges. Some of these charges were inspired by William Chaloner, who had already informed Montague, the Privy Council, and the public, of these abuses. Not only did he charge the Mint officers with incompetence, he even accused them of actually making counterfeit coin and of allying themselves with counterfeiters. He further offered to submit two inventions which, he claimed, would prevent counterfeiting (see *Calendar of Treasury Papers*, 1697–1701/2, p. 365). Newton having examined these reported to the Committee that they were not practicable. In spite of this the Committee recommended their adoption, and it was resolved:

'That undeniable demonstrations have been given and shewn unto this committee, by Mr William Challoner, that there is a better, securer, and more effectual way, and with very little charge to his majesty, to prevent either casting or counterfeiting of the milled money, both gold and silver, than is now used in the present coinage'. It was further resolved: 'That the house be moved, for leave to bring in a bill, or bills, to prevent the abuses of the officers of the mints, and for the better regulation of the coinage, both in the mint in the Tower, and also in the several mints in the country' (Ruding, II, 53–4). For a Report of the Committee, issued on 8 April 1697, see Ruding, II, 465.

This must have been a severe blow to Newton's pride, and he appears to have been determined that Chaloner should pay the penalty of his crimes, of which Newton had no doubts whatever. There are four long holograph memoranda in the Newton MSS. (I, 496, 499, 503, 504) devoted almost exclusively to Chaloner.

Chaloner was committed to Newgate Prison, 4 September 1697 (see Number 580, note (2), p. 259). He was in and out of prison at least four times before he was sent to the scaffold in March 1699.

565   MEMORANDUM BY NEWTON
Early 1697
From a draft in the Royal Mint[1]

An Account of the Mint in the Tower of London[2]

The Mint or Change is by the Charter thereof a Corporation consisting of the Warden, the Workers the Moneyers & the other Ministers. It may buy & sell, sue & be sued, & tis free from Taxes, arrests & servitude in forreign offices & imployments unless where the Charter is over-ruled by Act of Parliament.

The Warden or Keeper of ye Changes is by the Charter a Magistrate appointed to do right & justice among the Workers, & Moneyers & other Ministers in all their complaints & differences except in causes of freehold & causes relating to ye crown & may call a Court. And hence it is that the Mints are free from arrests. Standing Orders are made by him the Master & Controller (as a Court or Board) or by any two of them whereof he to be one. Hence Letters from ye Treasury are directed to them & sometimes to them by name & to ye rest of ye Officers in generall. In some things (as about ye buildings & assays &c) the Assaymaster may be also called to the Consult. For ye Warden pays the Charges of necessary repairs first avouched by the Master Controller & Assaymaster or any two of them whereof ye Master to be one. He supervises ye whole processe of ye Coynage & pays ye charges thereof & ye salaries of ye Officers except ye Masters salary & wages & such salaries & wages as are appointed to be paid by the Master all wch were received of ye King & paid to the Officers & Workmen by the Warden before ye enacting of the coynage duty, but now the Master retains out of that duty what the Warden should otherwise pay to him. With the consent of ye General of the Mint of Scotland the Warden makes ye standard weights of England & Scotland. He keeps an Indented trial piece to examin the assays by as oft as he shall think convenient, & the Master keeps another to make ye moneys by. He is of late impowered to demand & receive the forfeited estates of abusers of ye coyn for defraying the charges of them that prosecute.[3]

The Workers are the Master & his Assaymaster & Melter & Refiner wth their Assistants Clerks & Underworkmen. The Master upon new occasions of coynage contracts with the King by Indenture & according to ye Indenture by the assistance of his aforesaid servants receives, melts, refines, assays, rules & standards the gold & silver to be coyned, sets it out to ye Potts runs it into standard barrs, & delivers those barrs by weight to ye Moneyers & after coynage receives it from them & delivers it to ye Importer by weight, & by his

233

Clerks enters the accounts thereof in Day-books, Leger-books, Melting books, Pott-books & books of Debtor & Creditor.

The Moneyers draw, cut out, size, blanch,[4] edge & coyne those Barrs into moneys & return the moneys scissel[5] & brokage[6] back to ye Master by weight. They may not pay or distribute any moneys unassayd upon pain of forfeiting their franchises & bodies to prison. They take apprentices & form themselves into a company by electing one of their number to be their Provost. They live in the Country & are bound to attend & do their work whenever summoned by the Warden Master or Controller upon pain of loosing their Franchises & bodies to prison. They are to work in such tasks & so many hours every day (sundays only excepted) as shall be appointed by the Master & for wilfull neglect or refusall the Warden Master & Controller (that is the Court) may expell or otherwise punish them (or any other workmen) as shall seem meet for their Majesties service, & by the same power (as in all other Corporations) may grant freedom to new Moneyers.

The other Ministers are the Controller, Kings Assaymaster Surveyor of the Meltings, Weigher & Teller, Kings Clerk, Clerk of ye Irons &c. These are standing Officers wth set salaries to cheque the Moneyers & Workers & see that they do their duty in working & coyning the gold & silver: the Workers & Moneyers (except the Master since ye reign of Charles II) are not standing Officers nor have salaries but as workmen receive wages after a certain rate in ye pound weight for all the gold & silver they work & coyne.

The Controller is in behalf of the King a cheque upon the Master in his accounts & upon the Assaymasters in their assays. He makes a Controllment Roll upon oath every year of all the bullion molten with its allay & of all the moneys coyned monthly & supervises the whole coynage, & with the Warden & Master locks up the bullion & new coyned moneys & the Pix & Coynage duty: in doing wch the Warden & Controller are a guard upon the Masters bullion & rent moneys & Pix & the Master & Controller are a guard upon the Coynage duty to be issued out by the Warden according to ancient custome

The Assaymaster is in behalf of the King a cheque upon the Master for his assays & keeps a book of all the gold & silver as to quantity & fineness & of all the Pot-assays. He & the Warden & Controller or any two of them chuse ye ingot for the Pot assay; & when the money is coyned he tries it both in weight & fineness before the Warden (who then chuses out the assay-pieces & pix) & the Master & Controller, & if it prove wthout remedy[7] the Master bears ye loss of remelting it. When the Importer & Master disagree about the price of bullion the Assaymaster in the presence of the Warden Master & Controller assays the same & the Master then receives it & stands charged with it according to ye report of ye Assaymaster who is a sworn officer. The assay may be made by the Wardens Indented triall piece.

The Surveyor of the meltings is in behalf of the King a cheque upon the Melter to see that the gold & silver & its allay & nothing else be put into the Melting pott. Whenever the Pot is opened he watches that nothing unfit be put into it & when it is laded out he sees that the ingot for the Pott-assay be duly taken & carried to ye Assay-office. He keeps a book of all the gold & silver molten & of the allay put into it.

The Weigher & Teller weighs all ye gold & silver brought into the Office of Receipt either before or after coynage & when tis requisite he tells it there. He weighs ye silver moneys by Journeys[8] of 60 pound weight & ye gold by journey of 15

The Kings Clerk registers the papers wch pass between the Treasury & the Mint. Also he & the Warden & Controller (by their Clerks) or two of them, in behalf of the King rate & standard all the gold & silver brought into the Mint & examin the Pots set out by the Master, & enter these accounts in Day-books, Leger-books & Melting-books.

The Clerk of the Irons is at present the same person with the Surveyor of the meltings. He keeps an account of all the Dyes made & hardened by the Smith & when they are worn out sees them defaced in the presence of the Warden Master & Controller. He now keeps an account of all the Dyes that are sunck & for that end has one of the keys to ye great Press & to ye box of Puncheons.

There are also two Auditors who yearly examin & allow the accounts of the Warden & Master & the Controllment Roll. a Clerk of the papers who may register Orders, Contracts, Patents Deputations Controllment Rolls, Accounts audited &c. a Porter who removes the Ingots of gold & silver from Office to Office; & an Engraver, two Engineers & a Smith who make & repair the Puncheons, Dyes, Engins & other iron work & are immediately under the Master.

The Warden Master & Controller or any of them as often as need shall require may take up at his Maties price Gravers Smiths Workmen & Labourers & other necessaries for making the irons & moneys & doing all manner of business & therein all Mayors, Sheriffs Bayliffs & other Officers are commanded to assist the said Officers of the Mint.

NOTES

(1) This Memorandum (Newton MSS. 1, 2–3), in Newton's hand, is reproduced, almost verbatim, in the Report of the Parliamentary Committee on the Miscarriages of the Mint, 8 April 1697 (Ruding, II, 465).

(2) See also Memorandum 552.

(3) The Warden was entitled to forfeited chattels of persons convicted of coinage offences. See Letter 553, note (3), p. 210.

(4) See Letter 557, note (3), p. 217.

(5) See Letter 555, note (3), p. 214.

(6) Brokeage (brockage); a coin imperfectly struck, applied particularly to a coin struck on one side only. This was usually caused by a coin jamming in the die, thus interfering with the impression of succeeding coins.

(7) The variation permitted in fineness and weight. See also Letter 736, note (3), p. 512.

(8) See Letter 555, note (6), p. 214.

566 MINT TO THE TREASURY
22 FEBRUARY 1696/7[1]
From the original in the Public Record Office[2]

To the Right Honble the Lords Comrs
of his Maties Treasury.

May it please your Lordships

In obedience to your Lordships commands we have considered the Proposals of Mr Peter Floyer & Mr Charles Shales for receiving the hammered money[3] & delivering the full weight thereof in standard silver into the Mints deducting only for a recompence eight grains out of every ounce Troy, wch is $12\frac{1}{4}d$ $\frac{1}{2}$ farthing in every pound weight. And we most humbly lay before your Lordships that in our opinion the charges of performing that service may amount to upon the pound weight

	d	
For the first melting	$\frac{3}{4}$	
For the wast in melting	2	the sweep being estimated at a farthing
For the loss by the worsness	$2\frac{3}{4}$	
For refining	$\frac{1}{2}$	$\frac{1}{2}$farthing
For incidents	$\frac{1}{2}$	
For the Gentlemens care trouble & hazzard	1	
In the whole	$7\frac{1}{2}$ & $\frac{1}{2}$farthing	

All wch is most humbly submitted to your
Lordships great wisdome

Is. Newton
Tho. Hall

NOTES

(1) This letter is undated, but a draft in the Royal Mint (Newton MSS. I, 197) bears the date 'Feb. 22 1696'.

(2) T. 1/43, no. 53. All this is in Newton's hand except the signature of Thomas Hall, the Assistant Master.

(3) In order to facilitate the recoinage, Floyer and Shales 'offered to undertake the whole charge and risk upon themselves in all and so many mints as their Lordships shall think fit, and to receive all the silver hammered money, and deliver the full weight in standard silver, being allowed eight grains out of every ounce troy-weight' (*Calendar of Treasury Papers 1557–1696*, pp. 568–9). Another document (*ibid.* p. 572) proposed that 'they should deliver the full weight in standard silver into the mint at 9*d* per pound weight to be allowed for waste, worsening, refining, all other charges in receiving and reducing the same to sterling'.

Floyer and Shales were commercial dealers in, and manipulators of copper, tin, and other metals. They had received the Exchequer contract for casting into ingots the hammered coin withdrawn by it in the first half of 1696. Sir Peter Floyer became Prime Warden of the Goldsmiths Company in 1701 and Shales was elected to the Court in 1704. See W. T. Prideaux, *Wardens Court and Livery of the Goldsmiths Company* (published by the Company, 1936). Floyer later received the Mint contract for the internal melting of ingots into coinage bars on 25 February 1702. He was a candidate for the post of Weigher and Teller in 1701. Shales was Melter from 1700 to 1701. For a note on hammered money see Number 548, note (2), p. 202.

567   WALLIS TO NEWTON

1 JULY 1697

From the original in the Library of Stanford University, California

Oxoniæ. Julij 1. 1697.

Clarissime Vir,

Accepi nuper, a D. Leibnitio literas, Hanoveræ datas, Maij 28. 1697.[1] In quibus cum nonnulla sint quæ te quadamtenus spectant, liberem tibi suis verbis exponere; viz. *Si qua esset occasio, D. Newtono, summi ingenii Viro (forte per amicum) salutem officiosissimam a me nunciandi; eumque meo nomine precandi ne se ab edendis præclaris meditationibus diverti pateretur; beneficium hoc a Te petere auderem. Item; Methodum Fluxionum profundissimi Newtoni, cognatam esse Methodo meæ Differentiali, non tantum animadverti, postquam opus ejus ad lucem prodiit; sed etiam professus sum in Actis Eruditorum;*[2] *et alios quoque monui. Id enim candori meo convenire judicavi, non minus quam ipsius merito. Itaque communi nomine designare soleo, Analyseos Infinitesimalis, (quæ latius quam Methodus Tetragonistica patet.) Interim; quemadmodum et Vietana et Cartesiana methodus, Analyseos Speciosæ*[3] *nomine venit; discrimina tamen nonnulla supersunt: Ita fortasse Newtoniana et mea differunt in nonnullis.*[4] Hæc ea verbatim transcripsi, ex Nobilissimi Leibnitij,

237

literis; ut videas, id ab Externis etiam desiderari, quod ego non tantum petij, sed obtestatus sum aliquoties, aliique mecum, nec tamen obtinuimus; ut, quæ apud te [im]primis desideratissima, ederentur. Quippe dum hoc aut negas, aut differs; non tantum tuæ famæ, sed et bono publico deesse videris. Duas illas Epistolas tuas (longiusculas et refertissimas) Anno 1676 scriptas, (unde ego Excerpta quædam antehac edidi,) curabo ego (nisi tu id vetes) subjungi Volumini cuidam meo[5] (jam aliquandiu sub prælo) quamprimum per præli moras licebit. Tuam de Lumine et Coloribus Hypothesin novam[6] (cujus aliquot specimina jam ante multos annos dederis) quam per annos (si recte conjicio) plus Triginta, apud te supprimere dictum est; oro ut propediem edendam cures; ut quam ego insignem fore Naturali Philosophiæ accessionem, jamdudum existimavi; et publico deberi: (Quam et Prælo fuisse diu paratam audio.) Idem dixerim de pluribus quæ apud te latent, quorum ego non sum conscius. Hæc interim raptim monenda duxi,

Tuus ad officia,

JOHANNES WALLIS.

[At the foot of the page Wallis has added]

I put it into this form, that, if you think it proper, you may desire Dr Sloan[7] to insert it in the *Transactions*.

For Mr Isaac Newton, Controller
of the Mint, at the Tower
        London.

Translation

Most Distinguished Sir,

    Lately I received from Mr. Leibniz a letter dated Hanover, May 28, 1697.[1] As it contains matters which to a certain extent concern you, I should like to set them before you in his words: 'If you have an opportunity (perhaps through a friend) of conveying to Mr. Newton, that man of great mind, my most devoted greeting, and of begging him in my name not to let himself be dissuaded from publishing his distinguished meditations, I dare to ask this favour of you. There is another matter, not only did I recognise, after his work was published, that the most profound Newton's Method of Fluxions was like my differential method, but I said so in the *Acta Eruditorum*[2], and I also informed others. I judged this course suited both to my honesty and his deserts. I am therefore accustomed to designate both by the same name, Infinitesimal Analysis (which is wider in scope than the Tetragon method). However, just as the methods of Vieta and of Descartes have both come to be termed *Analysis Speciosa*,[3] though there are still some points of difference, so perhaps Newton's method and mine differ in some points'. I have transcribed these words verbatim from the noble Leibniz's letter, so that you can

238

see that foreigners also want something I have not only sought, but begged for several times[4] (and others with me) but which we have not yet obtained, namely that the much desired material which you now suppress might be published. For while you refuse this, or postpone it, you show yourself neglectful of your own reputation and the public good. Those two letters of yours (rather long and full of matter) written in 1676 (excerpts from which I have already published). I shall take care (unless you forbid it) to subjoin to a certain book of mine[5] (now for quite a long time in the press) as soon as printing delays allow. I beg you will very soon publish your new hypothesis of light and colours[6] (of which you gave some examples many years ago) which (if I reckon rightly) you are said to be suppressing for more than thirty years: inasmuch as I have now for a long time thought it will prove a notable accession to philosophy, and it is owed to the public: and I hear that it has long been ready for the press. I should say the same about many things you keep hidden, of which I am not yet aware. In the meantime, I have thought it worth while to give you this hurried reminder.

Yours devotedly,

JOHN WALLIS.

NOTES

(1) Wallis published Leibniz's letter in the third volume of his *Opera* (p. 680). The passage beginning 'Si qua esset occasio' occurs with very slight alterations in a postscript to the letter which Wallis published. In a lengthy reply to Leibniz, sent on 30 July 1697, Wallis says: 'Quæ Newtonum spectant, ad eum scripsi tuis verbis, simulque obtestatus sum meo nomine, ut imprimi curet quæ supprimit scripta. Quod & sæpe ante feceram, sed hactenus in cassum.' (Concerning those matters which pertain to Newton I have written to him in your words and at the same time I have urged him on my own account to see to it that the writings which he is suppressing should be printed. I had done this often before, but so far with no effect.)

(2) October 1684.

(3) That is, analysis (or algebra) performed with letters instead of numbers.

(4) See Letter 502 in which Wallis complains to Waller that 'what he [Newton] there delivers, passeth now with great applause, under the name of Leibnitz's *Calculus Differentialis*'. See *The Early Mathematical Manuscripts of Leibniz*, translated by J. M. Child (1920).

(5) Volume III of the *Opera*, which was published in 1699.

(6) See Newton's 'Advertisement' to *Opticks*, 1 April 1704 (Number 672).

(7) Sir Hans Sloane (1660–1753). Physician and naturalist. Elected F.R.S. 1684; Secretary to the Society, 1693–1712; President, 1727–41. He accompanied the Duke of Albemarle to Jamaica where he made a valuable collection of natural history specimens. He became Physician General to the Army in 1716; President of the Royal College of Physicians, 1727. Founded the Botanical Gardens, Chelsea, 1721. Contributed many papers to the *Philosophical Transactions*. See *Record of the Royal Society of London* (1940).

568   MINT TO THE TREASURY
6 JULY 1697
From the original in the Public Record Office[1]

To the Rt Honble the Lords Commers of his
Majties Treasury.

May it please your Lordships

We have according to your Lordships directions considered the petition of
Mr James Roettiers[2] hereunto annexed, & do most humbly represent that tis
very true that the Dyes & Puncheons in his custody were seized by direction of
a Committee of ye House of Commons backt with a vote of the House, & that
by your Lordships order he was afterwards removed from the imploymt of
Engraver to his Maties Mint, & his working-rooms & Tools were seized by the
Warden & Master & Worker. That the Warden soon after restored almost all
his tools to his Father, & told him that before his sons departure to Flanders
the rest should be restored together wth such Dyes & Puncheons as were his.
That we know not how far Mr Harris may have engaged himself to your
Lordships to make new Puncheons as good as Mr Roettiers, but we have
hitherto endeavoured to engage him (contrary to his mind) to copy after
Mr Roettier's Puncheons that the money may be all alike. And that Mr Neale
intends to pay him his demand of sixty pounds & tenn shillings, & with your
Lordships approbation to give him fifty pounds more (being ye summ he
desired) for ye five hundred pair of Dyes for the Country Mints.[3]

And we most humbly submit it to your Lordships great wisdome whether he
shall be any more imployed as Engraver in his Maties Mint or allowed a
maintenance here till Mr Harris shall shew such specimens of his Art as he may
have promised to your Lordships or return with his family to Brussels his native
country. And we also most humbly desire your Lordship's direction about the
Medal-Dyes & Puncheons whether they shall be restored to him, & about his
father whether he shall still be allowed an habitation in the Mint.

Mint Office   Is. NEWTON
July 6.—97   THO. NEALE
                                                          THO. MOLYNEUX[4]
                                                                        1697

NOTES

(1) T. 1/46, no. 43. This letter is in Newton's hand.

(2) The three brothers, John, Joseph and Philip Roettiers received a joint salary of £325 a year on appointment as engravers in 1662. In 1669 they received an additional grant from the King of £450 a year for life as engravers of medals subject to the condition that, after the death of one of them, the two survivors would receive £350 a year and, after the death of the second one, the survivor would receive £250 for life. Joseph went abroad in 1679, and Philip left in 1685 to take up an appointment in the Brussels Mint. John's two sons, James and Norbert filled their places. According to a Treasury Report (2 July 1689), John 'hath for some time past lost the use of his right hand by the shrinking of the tendents, and is not able to work any more'. See *Calendar of State Papers*, 1668–9, p. 270, and *Calendar of Treasury Papers*, 1670, p. 53.

When John's disability became known, the post of Chief Engraver, with a salary of £325 was given to George Bowers. Bowers died on 1 March 1690. Although John Roettiers and his two sons had been suspended in the post of Chief Engraver, they continued to work in the Mint and to enjoy the life grant of £450 a year made in 1669. The two sons, James and Norbert, applied unsuccessfully for the vacancy caused by the death of Bowers, claiming that they had been 'engravers to ye Mint in the last two King's reignes and did make for the present Majesty the Coronation Medalls and Puncheons for the Guynyes and halfe Crowns and supplied Dyes to the Mint to Coyne with untill the Place was given to one Mr George Bowers who is since dead'. The appointment was given to Henry Harris. Harris was quite unskilled in coinage work, and on 22 March 1690, three days after his appointment as Chief Engraver, an agreement was made with James and Norbert Roettiers under which they would act as his assistants provided £175 of the salary of £325 a year payable to Harris was paid to them, and that the allowance of £450 payable out of the Exchequer be constantly paid to them as well as the fee, or fees, payable out of the Tower (*Calendar of Treasury Papers*, vol. VII (1689–90), p. 108).

When the brothers Roettiers were discharged from coin work in 1697 they were succeeded by John Croker who had been employed by Harris since 1696. Croker was paid £175 of Harris's salary of £325 'as Mr. Rotier has and had' (*Treasury Minutes*, T. 29, vol. 10, p. 28).

(3) York, Exeter, Bristol, Norwich and Chester. See Letter 550, note (2), p. 205.

(4) Thomas Molyneux, M.P., was appointed, along with Charles Mason, M.P., Comptroller of the Mint on 27 January 1696/7 (Royal Mint Record Books, v, fo. 26v). Both men were convicted of corruption and were discharged in April 1701. They were succeeded by John Ellis the following month.

569   NEWTON TO THE TREASURY
JULY 1697
From a draft in the Royal Mint[1]

To the Rt Honble ye Lds Commrs of his
Maties Treasury.

May it please your Lordships

We ye Warden Master Worker & Controller of his Maties Mints do most
humbly represent yt by our Charter confirmed by many Kings for above five
hundred years past all the Officers Workers Moneyers & other Ministers of the
Mints are incorporated throughout all England into one body under ye
Government of a Warden with freedome from Arrests & other great Privileges
& without subjection of our Corporation to ye particular Government or
Jurisdiction of ye Tower of London or its liberties. And that the Kings of
England have given us ye ground & houses between the Lines in the said
Tower out of their royal favour & for our greater security & not to deprive us
of those their other royal grants above mentioned & bring us into subjection
trouble & danger. And also that his Maty by the Indenture of the Mint doth
will & streightly charge all officers of the Tower that all ye Officers of ye Mint
& their servants & all importers of gold & silver shall have free ingress &
regress & issue by the Gates & through ye same Tower & Franchises thereof
inward & outward at all times without any arresting disturbance letting or
gainsaying of the chief Governour Constable Lievetenant or the Porter or any
other Officer or Person whatsoever he be for any manner of debt matter or
cause whatsoever it be & without any thing given to any of them or to any
other to have such entry or issue.

And we further represent to your Lordships that we have at all times in
every thing shewed due respect to my Lord Lucas[2] as chief Governour of ye
Tower [& are not conscious to our selves that any under Officer or Workman
within our jurisdiction has any ways failed therein.][3] And therefore we are
concerned that we should be now forced to represent to your Lordships that
on Satturday the third Instant, Philip Atherton[4] one of our Labourers was
apprehended in our Melting house without the licence or knowledge of our
Warden by Roger Bayly a Constable by virtue of a Warrant of Robert
Bateman Esqr one of his Maties Justices of ye Peace for ye Liberties of ye
Tower backt with another Warr[an]t of my Lord Lucas & is now detained in
New Prison from ye Kings service. And that on the same day one of the
Warders, to let out of the Mint some people wch he had brought in, seized the

242

Gate of ye Mint & took ye Porters son by the throat; whence arose a fray between them wch caused such a tumultuous concourse of people as rendred ye money unsafe wch was then coming down the street of the Mint in Trays & yt ye said Warder threatens revenge upon the Porters son wherever he meets him & my Lord Lucas himself threatens further (notwithstanding his Maties will & command to ye contrary exprest in ye Charter & Indenture of the Mint) to sue our Porter for the same.

And we further represent to your Lordships that the Centinals lately appointed to guard the door of the Engravers working room where the Dies & Puncheons were seized & kept by the Warden had orders to permit entrance as well to my Lord Lucas as to ye Warden. And that our Engraver Mr Harris[5] having sollicited my Lord to turn Mr Roettiers family out of the Mint & my Lord having divers times prest the Warden to do it & threatned to apprehend Mr James Roettiers[6] & at length an Information being made that King James[7] was seen in Mr Roettiers House, My Lord did thereupon seize the Gates of the Mint & search Mr Roettiers house two several days without giving notice to any Officer of the Mint of his intended search, & now demands a list of the names of all persons belonging to ye Mint, in order to search their houses at pleasure, under pretence that they entertein such lodgers as belong not to ye Mint & are dangerous to ye Tower. Where as his Maty by your Indenture of the Mint places in us the care of enquiring after such as live or inhabit in ye Mint without our licence or other right & impowers us to turn them out, & military searches without us have not hitherto been used so far as we know, nor are necessary to ye safety of the Tower the Mint being shut up on all sides by the Lines & we being ready to search either alone or wth his Lordship upon notice of any danger; & if the military searches without us should be now allowed & Constables let in upon us at pleasure we cannot undertake any longer the charge of the Dyes & Puncheons & Marking Engins & other coyning Tools & of the Gold & Silver wch lyes scattered about in all the rooms apperteining to ye coynage, nor will Merchants & other Importers think their estates secure in the Mint. We were placed in a Garrison that ye Exchange & Treasury of ye Nation might not be invaded by our Guards but guarded in our custody from all manner of invasion.

And we further represent that ye Centinals begin to be rather a grievance then security to us. The Press-room about a fortnight ago was in danger of being robbed one of ye Locks of the door being broken, wch makes a suspicion that ye Centinal was at least privy to it, so that the Moneyers are now fain to lodge two Labourers in ye Press room for their security which they never thought necessary to do before. And about the same time two of our houses being greatly disturbed at Midnight by a drunken Officer upon the Guards who drew his sword brake ye window & would have forced himself into one of ye houses & running after one of our servants made a pass at him & ran him through ye cloaths & a Centinal being called either to come to their assistance or to call his Corporall with a File of Musketeers, the Centinal would not stir nor call. And when a Corporal & File of Musketeers were called by another person, the [drunken officer] endeavoured to perswade a by-stander to beare false witness for turning the blame upon our men. And my Lord Lucas instead of being told the truth of such matters is only informed by the soldiers that our men get drunk & affront the Centinals, & has upon such informations ordered the Centinals to fire at us. Whereas heretofore the Centinals used not to be set singly as now, but to walk ye Mint-street all night in a body strong enough to apprehend any drunken or disorderly person without firing at him. For why should we lose a good Artificer upon pretence of his being drunk, when ye best are most addicted to yt crime & it was never yet made death? Or why should every Centinal be impowered under any feigned pretence to shoot his enemy or any other man that complains, if such bloody discipline may safely be avoyded? Or why should ye people who live in ye Mint be so terrified as to leave their habitations in it to ye neglect of ye Kings service & insecurity of the treasure.

And we are further constrained to represent that the Warden on thursday last telling my Lord Lucas the misdemeanours committed in the Mint by this drunken Officer, his Lordship & the Deputy Governour together did the next day being Friday give strict charge to his Warders not to suffer either meat or drink to be brought into ye Tower to workmen in the Melting-house or Press-room, & that if any were sent for, they should take it away from the messengers & give it to the soldiers, his Lordship pretending that the workmen abuse the same to excess & sell it to others, (which we never heard of) & yt his Warders have an unreasonable trouble in examining persons that go in & out. And this order being strictly put in execution gave so great discontent to ye work-men, that but for ye perswasion of the Officers of the Mint they would have left off their work that day & the two next days & instead of nine potts were not able to melt above six & say they can work no more unless they may have speedy redress.

And we are further to acquaint your Lordships that my Lord Lucas repre-sents that he doth not invade the Privileges of the Mint but that ye power wch he useth over us is asserted by ancient Records & Court Rolls preserved in the Tower,[8]

All wch we most humbly desire your Lordships according to your great wisdomes to take into serious consideration in order to our releife & pray that the Records & Court Rolls alledged by his Lordship may be produced &

compared with our Charter & Indenture & the rights between the Tower &
the Mint thence adjusted stated & limited so that both parties may know their
Duties Privileges & Powers in respect of each other for conserving a good
correspondence & friendship between them for the future & that what has
hitherto by mistake been done against us may not be drawn into precedent
hereafter.

NOTES

(1) Newton MSS. III, 409–10 (in Newton's hand). Newton drew this account of the
privileges of the principal officers of the Mint from old records. It does not represent the
conditions which obtained in his own day.

(2) Lord Lucas was Lieutenant of the Tower. His observations upon the complaint referred
to in this letter were sent to the Treasury on 27 July 1697; a copy is among the Newton MSS.
(III, 386–9). Lucas's answer to what he called the 'untrue Callumnies' occupies seven pages
and is supported by eleven affidavits from members of his command. His account of the search
of Roettiers's house is as follows:

'Whereas upon information that ye late K. James was seen in Mr Roteer's house, wch was
about 7: a Clock in the Evening. I stay'd 2: hours in which time I sent thrice to speake wth
any of the Officers of the Mint, but Could find none, nor the porter or any of his fammily,
whereupon about 9: a Clock, my selfe, the Deputy Governr and a Justice of the peace, went
into and search'd Roteer's house, & 3: or 4: days after, by order of theire Excellencies ye Lords
Justices, I went again to Roteer's house, & took the examinations of all his family, wch is all
the Millitary or other searches that were made.'

The letter concludes:

'And now (My Lords) haveing fully answer'd and disprov'd that Libell (wch I think) had it
not been deliver'd to your Lordships, deserved no Answer, in this Manner, And I do submit
to Your Lordships whether I am obliged upon such a Libell (so full of untruths and scandalls)
to set forth any of the Records Rolls or Jurisdictions of the Tower, Which are as antient as any
in the kingdome.'

(3) The square brackets are Newton's.

(4) Atherton was arrested for assaulting a head constable in order to extricate his wife from
custody.

(5) In March 1690, Henry Harris succeeded to the post of Chief Engraver on the death of
George Bowers (see Letter 568, note (2), p. 241). John Croker was first employed by Harris
about 1696 (see Letter 630). As a seal engraver Harris engraved in incuse whereas coinage
tools needed work in relief and subtler modelling. The coin engraver worked in three stages:
  (i)   an incuse matrix for the principal part of the design;
  (ii) with this he 'raised' a punch in relief, also cutting other punches for letters, detail,
etc.;
  (iii) he then 'sank' the working dies by using all these punches.
Further hand work was done in stages (ii) and (iii).

245

(6) See Letter 568, note (2), p. 241.

(7) James the Second fled to France on the Revolution of 1688.

(8) For a full account of these disturbances at the Mint see Craig, *Newton*, pp. 15–17, and *Mint*, p. 195.

570   HALLEY TO NEWTON
2 AUGUST 1697
From the original in the Library of the Royal Society of London[1]

Chester August 2° 1697

Ever Honoured Sr

Our Adversary Clark[2] went on Saturday afternoon for London with a resolution to loade the Warden and my self, with all the Calumnys he can: You see what weapons they fight with, and stick at nothing to compass their ends: he has taken Bowles along with him and Brown, who are I suppose to serve him on all occasions, as far as their oaths will go. If need be, I begg you would interpose your protection, till we can be informed of any sort of accusation, and that we may be heard before we are in any case judged. I hope your potent friend Mr Montague will not forgett me if their should be occasion, but as I am conscious to my self of no transgression, so I doubt not but to acquitt my self of any imputations their malice can invent: Mr Clark being gone, we begin to pay again, there having been but 3 days intermission.

I am

Honoured Sr

Your most faithfull Servt

EDM. HALLEY

To the Honoured
Isaac Newton Esqre
at his house in Jermyn Street
near St. James's Square
            These
                humbly present

NOTES

(1) Miscellaneous MSS. v, no. 42.

(2) See Letter 562, note (2), p. 230.

571   THE TREASURY TO NEWTON
12 AUGUST 1697
From the original in the Royal Mint.[1]
For answer see Letter 573

Whitehall Treasury Chambers
12th August 1697

The Lords Commrs of his Mats Treasury are pleased to referr this Peticion to Isaac Newton Esqr Warden of his Mats Mint who is to consider the same and Certify their Lordships a true State of the matter therein contained together with his Opinion what is fitt to be done therein.[2]

WM LOWNDES[3]

Fran: Moore ref: to Warden of the Mint

To the Rt Honoble the Lords Commissioners of his Majts Treasury.
The humble Peticion of Francis Moore Gent Sheweth
That whereas Wm Hawkins[4] late of London was Executed at the Citty of Lincoln on ye 26th day of July now last past for Counterfiting the Current Coin of this Kingdom whereby his Estate both reall and Personall is become forfeited unto his Majty. And for as much as Your Petr hath been att expence and trouble in travelling and makeing diligent enquiry to find out his Estate hath found some Estate with writeings of Considerable Vallue which did belong to the said Wm Hawkins

Your Petr therefore humbly Prays Your Lordships will be pleased to Order and impower Your Petr to receive & recover such Papers, Moneys or other Estate reall or Personall for his Majts use as have belonged to the said deceased, And for ye faith & full performance whereof Your Petr is ready to give such Security as may be to Your Honors Satisfaccion. And for his Encouragemt therein and defraying his Expences, Your Petr is willing to Submitt himself to Your Lordships pleasure & bounty as in such Cases have been usuall, or in this may seem meet to Your Lordships.

And Your petr shall ever Pray &c

NOTES

(1) Newton MSS. 1, 485.

(2) See Letter 553, note (3), p. 210, regarding the disposal of the effects of persons convicted of coinage offences. Some of these had been appropriated by Sheriffs of London, Middlesex, and other authorities. Newton's assertion of Mint claims was successful; the Treasury compensated the Sheriffs. See Letter 559.

(3) William Lowndes, M.P. for the borough of Seaford. He joined the Treasury in 1675, and was promoted secretary in April 1695, and continued in that office till his death in 1724. He compiled *A Report containing an Essay for the Amendment of the Silver Coins* (1695). See *D.N.B.*

(4) William Hawkins was executed at Lincoln for counterfeiting on 26 July 1697. See Letter 573 for a draft of Newton's answer to the petition.

572   NOAH NEAL TO NEWTON
15 AUGUST 1697
From a copy in the Royal Mint[1]

Noah Neal Esqr his Letter to Isaac Newton Esqr   Warden of the Mint

Stamford Augt the 15*th* : 97

I have sent you more work as you will see by the enclosed information, the method I have to propose to you to take them is this, Thorpe writes to Colsons for some money he hath sent him a bill for £10 upon his correspondent one Coreye, a gooldsmith in fleet street near fleet bridge and hath told him the matter that when Thorpe comes for the money that he secures him & whoever comes with him, describing Bacon to be a coiner man with a scar in his face: Thorpe writes to Colsons to direct his letter to be Left for him at Mr Serjeants at the Mityr in pye corner (And to his bro[the]r in Law who also hath seen the Stamps but is unwilling to appear yet) he orders him to direct to be Left at Mr Boswell at ye Scotch armes in High Holbourne: for that I think you had best see who comes for yt Money, it may be both may come, if only Thorpe; then you might send immediatly & Search for Bacon at these two places; for otherwise we know not where to find him: but I must beg your pardon for presumeing to direct you in this matter who knows soe much better then I can direct. Your Deputy hath been wth me; I hope the Lds of the Treasy will grant my request about these two Little houses. for I am sure I have earned them....

I am Sr your most humble servt   N.N.

NOTE

(1) Royal Mint Depositions, 280. This is one of many letters concerning Newton's efforts to secure the conviction of criminals.

573   NEWTON TO THE TREASURY
AUGUST 1697
From a draft in the University Library, Cambridge.[1]
In answer to Letter 571

My Lord,[2]

I have considered the annexed Petition & find that ye Petitioner did apprehend Will Hawkins & James Harrison & by taking a journey of 36 miles to Lincoln Assizes and staying there a week wth his witnesses did prosecute them to conviction & the estate he petitions for amounts to about 10 *lib* per annum & this is not yet desired for his Ma[jesti]es use & I beleive that suitable rewards such as may be a lease of this estate for as many years as your Lordships shall think fit may tend to encourage Gentlemen in the County to serve his Maty more readily in these matters

NOTES

(1) Add. 3964. The above is a draft in Newton's hand on the back of Letter 571. It is without signature, address, or date, but clearly refers to the petition by Francis Moore.

(2) Probably should read: My Lords.

574   FLAMSTEED TO NEWTON
4 SEPTEMBER 1697
From the original in the University Library, Cambridge[1]

The Observatory Sept 4. 1697

Sr.

Tho I got from London hither on Thursday last yet I had not time to examine the lunar observations till this morneing when turneing over my papers I found that on Dec 11. 1694 at 4h.31′.10″. I had stated ye Moons Right Ascension 338°.38′.50″ her distance from ye pole 94°.18′.00″ correct onely by the allowance of 1′.20″ for refraction. but Dec 13. at 6h.03′:06″ her Right Ascension 3°.52′.10″ distance from ye Pole onely correct by the allowance of 55″ Refraction 83.18.40.[2] So yt as I affirmed there was no allowance made in ye first for parallax nor in ye 2d yt I know of till I came to peruse my calculation & compare it with the Note I took from my letter whereby I found I had given you the distance from ye Pole correct by parallax 82°.37′.50″. which I assure was no willfull mistake as you might well apprehend by my asserting my sincerity in it & had you given me notice of it when you found it first I had imediately given you the same correction of it that this brings you. how I came to commit this mistake will be easily apprehended when you see ye papers from whence I copied those notes & till then I must intreat you to lay by all your apprehensions of any Intended practise for I assure [you] I had none whatever you may suppose or hath been suggested to you to ye Contrary. You know how ready I have been to rectifie all mistakes when they have hapned & since I have none to assist me yt is skillfull tis impossible for me wholly to avoyd them.

I doe not find that I have many more observations about ye latter end of Dec 1694. & begining of 95 then I have imparted to you. but when you come hither you shall see the book of observations & I will take care to have those you select examined & calculated for your service

There was no great need of this letter at this time but to let you see how ready I am to correct any mistakes, to serve my freinds, & to prevent their mis-apprehensions. & that I am ever   Sr

Your obliged humble Servant

JOHN FLAMSTEED

NOTES

(1) Add. 3979, no. 36. The letter to which this is a reply has not been found.

(2) At this point the words 'there was no fault in your first' have been crossed out.

575   NEWTON TO THE TREASURY
19 OCTOBER 1697
From the original in the Public Record Office[1]

To the Rt Honble the Lords Commrs of
his Maties Treasury.

May it please your Lordships

According to your Lordships direction I have made & adjusted three setts of money weights to be standard weights for money current in the Kingdome of Ireland.[2] And they are ready to be delivered when & where your Lordships shall please to order. I am

Your Lordships most humble

& most obedient Servant

Mint Office
Oct. 19. 1697

IS. NEWTON

NOTES

(1) T. 1/48, no. 49. This letter is in Newton's own hand.

(2) Attested weights may conceivably have been wanted by the Irish Exchequer to criticize its receipts of coin. Oversight of coinage weights fell in the Warden's sphere.

From almost the end of the fifteenth century specifically Irish silver was confined to the Tower of London. The weights were probably intended for English coin current in Ireland. See Craig, *Mint*, ch. XXI.

576 FLAMSTEED TO NEWTON

10 DECEMBER 1697

From the original in the University Library, Cambridge[1]

The Observatory: Dec 10. 1697

Sr.

It was not till last night yt I examined ye Calculation wherein you advised me of an error It was easily found in ye 4th line by compareing it wth ye 8th. & I saw it had propagated through ye Whole worke, repeateing it I find

The Vis Centrifugat at ye ☽	‚0041208[2]
Gravity answering it at ye Earth	14 ‚49 feet
Gravity by Mr Huygens expermt is	15 ‚083 f.
Distance of ye ☽ yt answers this gravity of ye Earth	60 ‚5 semidiameters[3]
Or yt ye ☽ may be retained at ye distance of	59 ‚3
semidiameters from ye Earth if wee allow to a degree. English Measure	70 ‚514 Miles
The French Measure Makes	69 ‚125
Mr Norwoods	69 ‚5[4]

Tis not impossible that either or both of them might erre a mile betwixt their Measures on ye earth & observations in the heavens considering that their Instruments were loose: & not large enough for so nice an Experiment

But I doe not thinke it probable yt they both did so I thinke I have reason rather to suspect, what I have often told you of late, yt some other cause heightens the simple gravity in these Northerne Countries. Had wee, or could wee procure, some Nice Measures of the length of ye Pendulum yt vibrates seconds under ye Equator, they would determine whether there were such a Cause or Not

I am fully satisfied yt all ye Celestiall bodyes are moved by the laws of Gravity. & yt ye ☽s motions are subject to them but since lunar Eclipses seem not to allow the Parallaxes to be dim[in]ished. & wthout a true determination of them it will be impossible to determine her true place from any Measures taken in ye heavens I would gladly see this point first firmly setled before you proceed any further. for an error committed here will propagate as far as yt did in its progress of which I have given you the Correction above

The ill weather permitts me not to stir abroad yet I blesse God I enjoy my health better then I have done of late yeares by much. & proceed on my business as my affaires & slender assistance will permit. I have materialls enow by me but I want hands to worke ym up. I heare nothing of my Lord Rumny at present. Nor doe I expect till the businesse Is done & past any remedy. I intend to waite on him in the holidays haveing some business that will Introduce me. I may then call at your house to pay you my respects. & thankes for your Endeavors to secure us here which I wish may prove effectuall I am Sr

Your obliged Servant

JOHN FLAMSTEED

To Isaack Newton Esqr
Warden of ye Mint
at his house in German
Street. neare
St James's

NOTES

(1) Add. 3979, fo. 37.

(2) That is, 0·0041208. Oughtred, in his *Clavis Mathematicæ*, employed this method of writing decimal fractions.

(3) Taking Huygens's figure of 15·083, and the Moon's distance from the Earth as 60·5 semidiameters, and assuming the inverse square law to hold, the '*Vis Centrifugat* at ye ☽' would be 0·004120.

(4) The various estimates of the length of a degree along a meridian which were available at this time were:

Snell (1617)	67 miles	
Picard (1679)	69·945 miles	(Latitude 49°.22′ N)
Cassini (1685)	69·545 miles	(Latitude 53°.15′ N)
Norwood (1635)	69·54375 miles	

Richard Norwood (1590–1675) measured the distance between London and York, and published his estimate of the length of a degree along this meridian in his *Seaman's Calendar* (1637), where he stated that 'one degree of a Great Circle measured on the Earth is 367200 feet or 69½ miles and 14 poles', a result which is correct to within half a mile.

577   DAVID GREGORY TO NEWTON

23 DECEMBER 1697

From the original in the University Library, Cambridge[1]

Much Honoured Sir

The last time I waited on you, you was pleased to give me a favourable answer to my proposal of having some establishment in London that is consistent with what I have here. The present settlement of the Duke of Glocesters family gives me occasion to mention this to you again. No doubt it will be thought necessary to instruct that young Prince in Mathematicks, and to take your advice in the choice of a tutor therin.[2] As this would exceedingly fitt my humour and circumstances, it is such wherin I would have all probability of success: And I hope Sir you will allow me your assistance in it as you shall find reason and occasion. I would have waited on you at London now but the circumstances of my family make it difficult for some weeks yet. In the mean time I beg you will give me your opinion and advice about this, and whether you find that such ane imployment will soon be thought necessary. I am in all duty

Oxon. 23 Dec.   Honoured Sir
    1697.
                    Your most oblidged and most humble Servant

                                        D. GREGORY

        For
    The Much Honoured
Isaac Newton Esquire
Warden of the Mint
at his House in Germin-street
        St James's
        London.

NOTES

(1) Add. 3980, no. 9.

(2) Both David Gregory and Flamsteed were anxious to secure the post of tutor to William, Duke of Gloucester, son of Queen Anne, who at the date of this letter was $8\frac{1}{2}$ years old. Neither secured the post. See Letter 600, note (9), p. 296. The boy died at the end of July 1700, a few days after reaching his eleventh birthday.

578   HALLEY TO NEWTON
30 DECEMBER 1697
From the original in the Library of the Royal Society of London[1]

Honoured Sr

The parliament having this day voted the Continuance of all the Country Mints, I should be very unwilling to leave Lewis and Clark[2] to enterprett my resignation to be any other than a voluntary cession, as they will most certainly do, unless I prosecute, as I have already begun, the undue preferences by them made. Abbom Grays affaire I value not, as being what I hope may be justifiable on many accounts, should the Lords belive me consenting to it, but the Mint at Chester I assure you cannot subsist as it ought, whilst Lewis governs Clark as he does, and Mr Neale supports both. Wherfore I begg that Lewis may appear face to face with me, before the Lords, there to answer to his throwing the standish[3] at Mr Weddell, the giving the undue preference to Pulford, and some other accusations of that nature I am prepared to lay before their Lordships. I came to town purposely to charge that proud insolent fellow, whom I humbly begg you to belive the principall Author of all the disturbance we have had at our Mint, whom if you please to see removed, all will be easie: and on that condition I am content to submitt to all you shall prescribe me. Nevertheless, as I have often wrote you, I would urge you to nothing, but what your great prudence shall think proper, since it is to your particular favour I owe this post which is my chiefest ambition to maintain worthily; and next to that to approve my self in all things   Your most faithfull & Obedient Servt

Decemb 30°   EDM. HALLEY
1697

To Isaac Newton Esqre
    Humbly present
        These

NOTES

(1) Miscellaneous MSS., v, no. 43.

(2) See Letter 562, note (2), p. 230.

(3) A stand containing ink, pens, and other writing materials.

579 A MANUSCRIPT BY NEWTON
1697
From the original in the Royal Mint[1]

Observations concerning the Mint.

Of the Assays.

The Assaymasters weights are 1, 2, 3, 6, 11, 12 & represent so many ounces. The weight 12 is about 16 or 20 grains more or less as he pleases to have his weights made. With this he weighs the silver into the fire & recconning a wast answering to two penny weight he weighs it out of the fire by the weight 11 to see if it be standard, & if it be heavier or lighter he adds in the lighter scale penny weights & if need be an half penny weight & grains to see how much better or wors. His scales turn wth ye 128th part of a grain, that is wth ye 2560th part of ye weight 12 wch answers to less then ye 10th part of a penny weight. They are fenced about wth glass windows to keep them from ye motion of ye air & have in them little thin brass platters to take away the weights by wthout handling the scales.

He cuts off from every Ingot a piece of about a drachm for two assays beats it out into a thin plate, scrapes it clean & cutts it into the ballance &c. In assaying the money he clips a little off from severall pieces of money & assays them together. The Assay drops of the money & of the pott-assays (but not of the Ingots) are his fee. He makes two assays of every ingot, puts 13 Coppels[2] at once into the furnace uses the poorest lead assayd & run into bullets. A bullet is twice the weight of ye silver. He foliates a bullet wth ye hammer; tears it in two, wraps up the silver in one half, & adds a whole bullet to it, so that the lead is 3 to one. He lets the fire cool gradually till ye silver set least by cooling too quickly the silver spring & the assay thereby make the silver seem wors then it is.[3] When ye lead is blown off the silver looks very bright. The Assay Furnace is of copper plates luted half an inch thick within. It is about 18 inches square 10 inches high to ye grate (wch is of iron barrs) & about 15 inches above ye grate. The muffle stands upon ye grate & ye coppels are set in wth a pair of tongues upon the floor of ye grate through a round hole in ye side of ye Furnace wch is afterwards filled wth live charcoal. In a quarter of an hour the lead fumes away & the operation is done. The King pays for the muffles coppels & furnaces. Pottern ore[4] is the poorest of silver &[5] the poorest sorts of ores are ye richest in silver. ....[6] commends ye Lead of Villach[7] as best for Assays because poorest in silver.

Of the Melting

The Melter runns from 600 or 700 to 800 lb of late 1000 lb weight of silver in a pot & melts 3 potts a day in each furnace within the space of about 12 or 14 hours. The first pot is about 5 hours on ye fire ye two next about 4 hours a piece. When ye silver is molten he puts in the allay. For each melting (including fire, pots, Hoops tongues shovels ladles ingot molds sand & wages of melters & mould makers) he is allowed three farthings per pound weight & for wast five farthings & as much for melting the scissel[8] & for its wast, that is in all 4d per lb wt (vizt 2d for Ingots & 2d for ye scissel). Formerly he had only $3\frac{1}{2}$ per lb wt[9] for ☽ & 1s 1d per lb wt for ☉. The sweep he has into ye bargain & makes it up for himself at his own charges. A pot for 800 lb weighs about 500 lb wt & cost 20d per lb, and lasts about 6 weeks or two months more or less that is about 120 meltings so that pots cost about $\frac{1}{8}d$ per lb wt of silver melted in them & if hoops ingots molds & other utensils be added they cost less then $\frac{1}{6}d$ per lb wt. A pot in three meltings each day spends about 25 bushells of coales per diem, & imploys about 10 men at 20d per diem each in making molds, looking to the fire & filling & lading out the potts. The mens wages & coals at 6d per bushel coming to $\frac{1}{8}$ per lb wt or something less. The sweep amounts to   [10] & the charge of making it up to   [10] per lb wt. And the coales at 6d per Bushel to about $\frac{1}{16}$ of a penny per lb wt. The Pots shrink in the fire by long use so that a Pott which when new holds 800 lb wt, when it has been used a month or six weeks will hold but 700 or 650 lb wt or perhaps less.

The Scissel if the Pot is crouded full & well charged a 2d & 3d time wasts as little (or wthout a sensible difference) as if it be filled wth Ingots, & the three meltings (if the pot be not quite so full) are done in ye same time or within a little.

The hammered money was melted last year at ye Exchequer with a blast in small potts of 50 lb weight a piece, 75 lb weight of money in a pott, about 12 pot fulls each day. The potts cost 8 pence a pound & last about 30 or 35 meltings or potfulls a piece. So that ye potts cost $\frac{1}{8}$ of a penny per lb wt of silver melted in them. But ye blast makes quicker dispatch this way with perhaps less then half ye expence of fire then in ye other way wth great pots The little pots are best for coarse silver to be refined, the great ones for standard silver because they alter the fineness least & make least wast for the melter. Mr Floyer & Mr Shales[11] were payd $\frac{3}{4}d$ per lb wt for melting at ye Excheqr this Winter besides potts (wch weighed about 50 lb per pott, cost 8d per lb of iron or $\frac{1}{7}d$ per lb wt of money melted in them) & Refitting of Ingots Mittens for workmen, earthen potts, sandover,[12] baskets cartage of potts &c (wch cost about $\frac{1}{70}$ of a penny per lb wt or $\frac{1}{10}$th of ye potts) but the potts &c should be included in ye $\frac{3}{4}d$ for melting. Every pot each day takes up a bushel of coals or above in the first melting each morning & half a bushel or less in ye rest, that is about $\frac{7}{12}$ of a bushel [at ea]ch[13] melting at a mean rate, that is if the coale be 6d a [bus]hel, about $\frac{1}{20}$th of a penny per lb wt. The wast [at] the first melting of hammered money wth the blast in these little potts is recconned at 2d (or $\frac{2}{3}$ dwt) per lb wt, the sweep being allowed for in this recconning & estimated at a farthing per lb wt. The Plate taken in at Chester last May proved generally about 5 dwt or 6 dwt (per lb wt) worse then standard (by reason of the soader)[14] with a wast of about 5 ounces per $\oplus$ wt[15] or 1 dwt per lb wt

Of the making the Moneys

Sixteen ounces Troy of sixpenny Blanks were blancht[16] in 6 minutes & lost of their weight in blanching the first experimt 8 gr the next 10 gr ye next 7 gr the next 9 gr & at a second blanching for 7 minutes of time one grain more[;] at a middle recconning they lost at one blanching $8\frac{1}{2}$ gr. Whence a pound Troy loses about $6\frac{1}{3}$ gr. & a pound Troy of crown blancks 3 gr of $\frac{1}{2}$ crown blancks 4 gr & of shilling blanks 5 gr. By experimt I found that a pound Troy of $\frac{1}{2}$ crown blancks lost $3\frac{1}{2}$ gr.

A sixpenny barr weighing 16 ounces Troy in Nealing[17] three times, got 3 grains in weight ye first time, lost $\frac{1}{2}$ a grain ye second time & got $1\frac{1}{2}$ grain the third time, that is in all the three nealings it grew heavier by 4 grains. A shilling barr of 15 ounces Troy in one nealing grew heavier by $1\frac{1}{2}$ grain. So that Nealing increases ye weight of a shilling barr of a pound weight Troy by about 1 gr or $1\frac{1}{4}$ gr & of a sixpenny barr by about $1\frac{1}{2}$ or 2 gr. And Nealing & blanching together decrease the weight of a pound weight of sixpenny blancks by about 5 gr, of shilling blancks by 4 gr, of $\frac{1}{2}$ crown blancks 3 gr of crown blancks $2\frac{1}{3}$ gr. And if the sixpenny, twelvepenny, half crown & crown blancks be taken in common in ye proportion of 1, 4, 3, 2 the nealing & blanching together decrease the weight of a lb wt by $\frac{5+16+9+4\frac{2}{3}}{10}[=]\frac{34\frac{2}{3}}{10}$ or $3\frac{1}{2}$ gr. If the blancks be not well nealed they will not blanch well.

The Moneyers melt their limel[18] per se without any mixture to make it run & in melting it grows better 2 dwt 3 dwt or 4 dwt & loses 1, 2 or 3 lb wt [?dwt] of its weight The limel is not above the $\frac{1}{100}$th part of ye money. And if the loss in the limel be $\frac{1}{80}$th part thereof by scattering & $\frac{1}{80}$th by melting, the wast by the limel will be $\frac{1}{4000}$th of the money that is $\frac{3}{16}$ of a penny per lb wt

There is also a wast in the milling by the dropping off of sand with some particles of silver & in the nealing by some blanks falling out of the pan upon the hearth & lying there till they be half consumed by the fire and in shreds of silver scattered up and down the rooms & lost in ye [dus]t or by sticking to the workmens shoes: all wch cannot amount to $\frac{1}{4}$ of a penny per lb wt. So that the whole wast in the making of the moneys by the Moneyers comes not to 1d per lb wt.

Two Mills with 4 Millers, 12 horses two Horskeepers, 3 Cutters, 2 Flatters, 8 sizers one Nealer, thre Blanchers, two Markers, two Presses with fourteen labourers to pull at them can coyn after the rate of a thousand weight or 3000 lib of money per diem And if for the horses & labourers one with another be allowed after ye rates of 22d per diem it comes to about 3 lib per diem, that is three farthings per lb wt.

So that the whole charge of coynage besides the allowance to the moneyers for their hazzard & pains comes only to $1\frac{1}{2}$ $\frac{1}{8}d$.

NOTES

(1) Newton MSS. I, 10–11; an account of the methods and economics of minting.

(2) Coppel; the spelling 'cupel' is more usual. It is a shallow crucible, made of bone-ash, for assaying gold or silver with lead, in the process called 'cupellation'.

(3) A fused metal, if cooled too quickly, decrepitates ('spits').

(4) Pottern ore: *O.E.D.* gives *pottern*: as 'of or pertaining to potters'. Pottern-ore, or pottern-lead-ore, as the above letter notes, was used by potters as a glaze for earthen vessels because it vitrifies easily. Lead ores were long known to contain traces of silver.

(5) Here Newton has inserted the words: 'steel ore & other'.

(6) Newton has left a blank here. Clearly, Agricola (Georg Bauer, 1494–1555) is meant. Newton was familiar with Agricola's work (*De Re Metallica*, Basel, 1556) for according to R. de Villamil (*Newton: the Man*, p. 63) he had a copy of the later edition which was published in Basel in 1621, and he used this when he was at Cambridge. Moreover, Newton was experimenting on alloys for use as mirrors in his reflecting telescope from 1672, and this confirms his acquaintance with Agricola's work. See 'Some Notes on Newton's Chemical Philosophy', by Douglas McKie (*Phil. Mag.* **33** (1942), 847–70, especially pp. 850–2).

The relevant passage from *De Re Metallica* (pp. 187–8) is as follows:

'Plumbum autem omnino omni argento careat, quale est Villacense. Quod si ejusmodi non fuerit in promptu, seorsum plumbum est experiendum, ut certò sciatur, quantam argenti portionem in se contineat, utque calculis subductis de vena rectè judicetur; nisi enim tale plumbum fuerit, experimentum erit falsum & fallax.' (The lead, however, must be quite free from silver, such as is that known as *Villacense*. But if lead of this sort is not obtainable, the lead must be separately tested, in order to determine what proportion of silver it contains, so that after an exact assessment a correct judgment of the vein may be made, for unless such lead is used, the trial will be false and misleading.)

(7) A translation of *De Re Metallica* was published by H. C. and L. H. Hoover (London, 1912) and a footnote to the above passage (p. 239) reads: 'The lead free from silver, called *villacense*, was probably from Blayberg, not from Villach, in Upper Austria, this locality having been for centuries celebrated for its pure lead.' These mines were worked prior to, and long after, Agricola's time.

(8) See Letter 555, note (3), p. 214.

(9) Here, and elsewhere, Newton has abbreviated 'lb wt' (pound weight) to 'lwt'.

(10) Gaps have been left here, and the figures have not been inserted.

(11) See Letter 566, note (3), p. 237.

(12) Sandover (=sandiver): this is the vitreous scum floating on top of the molten metal, covering, and presumably refining, the silver. It is referred to by Agricola (see note (6) above), p. 235, note 17.

(13) A corner of the sheet has been torn off. Words which are missing are surmised; these are indicated in square brackets.

(14) Soader, or sowder, which later came to be written 'solder'.

(15) Hundredweight (cwt).

(16) Blanching is a process for removing, by the action of acid, any discoloration from the face of the metal.

(17) That is, annealing.

(18) Limel: the metal dust from filing the blanks in order to reduce them to their correct weight (cf. *limaille*, filings).

580 PETITION BY CHALONER

Late 1697

From a copy in the Royal Mint.[1]
For Newton's answer see Memorandum 581

To the Honble the Knights Citizens & Burgesses in Parliamt assembled

the humble Petition of William Chaloner[2] Gent. Sheweth That your Petitioner[3] did in the last sessions of Parliamt discover several abuses committed in the Mint & shewed by what methods false money was coyned, that laws might more effectually be made to prevent the same & severall Acts were accordingly made against it, & then some of the Mint threatned by some means to prosecute him & take away his life before the next sessions of Parliament, telling him that this Honble House had no power to meddle with the affairs of the Mint, therefore they would not obey the Order made by a Committee of this Honble House

That ye said Committee[4] promised your Petitioner he should suffer no dammage for his discoveries about the Mint & by ye directions of this Honble House a member thereof did represent his case to ye King & his Majty was pleased to say that he should suffer no dammage for the said matters, yet they committed him to Newgate & kept him in Irons for seven weeks alledging that he had abused the Mint in Parliamt & they did falsly & illegally preferr a bill of Indictmt against him but could bring no evidence against him to prove it altho they used strange methods to procure something. For many witnesses have made oath before a Judge of the Kings Bench & some Justices of peace, that some of the Mint have imployed & given Privilege to several persons to coyn false money (who put it away among the subjects for good) all which was done with an intent to draw him into coyning to take his life away & the better to effect the same they allowed the said persons money from time to time to buy tools & to carry on the said conspiracy against your Petitioners life. But all their endeavours could not get him to be concerned in coyning, but on the contrary he hath made it his business for a considerable time past to find out the Treasons & Conspiracies against the King & Kingdome & thereby hath discovered many who have been convicted for the same, & was this year writing a book of the present state of the Mint & the defects thereof (as he promised the said Commee) wch he hoped would have been of service to the publick. But the Mint having notice of it they committed him to Prison & so prevented him from doing it. All wch can be proved if this honble House is pleased to require it.

That your Petitioner is utterly ruined by his endeavours to serve the King & Kingdom & by his discoveries against the Mint to this Honble House

Your Petitioner therefore most humbly prays that this Honble House will be pleased to consider his great sufferings & ruined condition as being incapable of providing for himself & family by what he intended for the service of the Publick & grant him such redress as shall seem best in your Honours great Wisdom & Justice

And your Petitioner shall ever pray &c

WM CHALONER

NOTES

(1) Newton MSS. 1, 497. This is a copy which Newton made in his own hand.

(2) Chaloner was arrested after he had forged Exchequer Bills, a new type of Government security introduced by Montague in 1696. This, however, did not fall within the province of Newton, who was concerned with currency offences. He believed Chaloner to be the most accomplished counterfeiter in the country, and it was on the suspicion that he was planning further counterfeiting offences that Newton had him arrested. Chaloner was committed to Newgate Prison on 4 September 1697, but he obtained his acquittal by suborning the prosecution's witnesses. The reference to 'seven weeks' in the second paragraph helps to fix the date of the petition which was probably late 1697.

(3) 'Petitioner' is throughout abbreviated to 'Petr.'.

(4) The Committee set up to inquire into certain miscarriages of the Mint. See Letter 564, note (3), p. 232.

581   MEMORANDUM BY NEWTON

EARLY 1698

From a draft in the Royal Mint.[1]
In reply to Letter 580

An Answer to Mr Chaloner's Petition.

Mr Chaloner before a Committee of the last Sessions of Parliamt laboured to accuse & vilify the Mint & approve himself a more skilfull coyner then they that he might be made their supervisor & then supply Tho. Holloway[2] with Tools out of the Tower to counterfeit his own milled money by a way wch he then concealed from that Honble Committee, boasting secretly that he would fun[3] the Parliamt as he had done the King & Bank before. And while he was upon this designe I gave the Chairman of that Honble Committee minutes for a new Act of Parliamt against Coyners & the Act was afterwards drawn up by some of the Judges & Officers of ye Mint & brought into the House of Lords without Mr Chaloners assistance.

About the end of that Sessions Mr Neale[4] (as I have heard) moved the House to give leave that Mr Chaloner might be prosecuted for taxing the Mint with coyning great quantities of false money but I do not know that there was any designe or menace to prosecute him for anything else then that calumny & much less to take away his life. Nor did I ever hear any one pretend that the Honble House of Commons had no power to meddle with ye Affairs of the Mint or that their Committee wch then sat had not all the power wch the House thought fit to give them: but when the Committee sent an order to me about preparing an Edger for Mr Chaloner to grove new money, I told Mr Chaloner that in regard of an Oath wch I had taken[5] I could not safely carry him to the Workmen of the Mint about it, but if he would give me directions I would take care of the matter; but Mr Chaloner refusing to give me directions (as he ought to have done by that Order) I directed the workmen (without him) to grove some half crowns shillings & sixpences & carried them to the Committee.

That his Maty by order of this Honble House was now moved in behalf of Mr Chaloner I did not hear till after I had committed him wch I was moved to do because he stood charged wth new designes of counterfeiting Bills & Money forreign & domestick. And particularly upon the rising of the last sessions of Parliament he advised Tho. Holloway to take a house in the Country convenient for coyning & agreed with him that he should find materials & teach Holloway to coyn & Holloway & his brother John should coyn together at that house & they three should share the money so made. And accordingly

Tho Holloway did take a house at Egham in Surrey & Chaloner did make some progress in teaching him & his brother & in preparing coyning tools, & for doing this & having coyned very great summs of Gold & not for offending the Mint he was committed & prosecuted. But the principal witness fled into Scotland & the Indictment being drawn wrong two others of the Witnesses were afterwards drawn off to swear against it & against the Mint, altho by swearing on both sides their credit fell & to draw off the Kings witnesses & swear them against the King gravells prosecutions & renders it dangerous for any man to prosecute & is therefore accounted a misdemeanour.

There are divers witnesses that Mr Chaloner last spring & Summer was forward to coyn & I do not know or beleive that any privilege or direction was given by any of the Mint to draw him or his confederates in or that any conspiracy was made against him or any money given to buy coyning tools. Neither can I find that he did ever make it his business to find out any treasons or conspiracies against the King & Kingdom but what were of his own con-triving, as in the case of the Printers. And as for his intended book about the Mint I heard nothing of it till I saw it mentioned in his printed case & therefore did not commit him for that book. When I had committed him he told me he had written a book about preventing the counterfeiting of Bills & offered to let me see it, but said nothing of any book intended by him about ye Mint.

If therefore he be ruined tis by his endeavouring not to serve the King & Government as he pretends but to coyn false money. And if he would but let the money & Government alone & return to his trade of Japanning,[6] he is not so far ruined but that he may still live as well as he did seven years ago when he left of that trade & raised himself by coyning.

NOTES

(1) MSS. 1, 497–8. In Newton's hand. The date is probably early 1698. See p. 260, note (2).

(2) Thomas Holloway, or Holliday, till now a confederate of Chaloner, was the 'principal witness' mentioned in the third paragraph above. Chaloner later cheated Holloway and thus turned his wife, Sarah Holloway, into his bitterest enemy. Holloway fled to Scotland where he was eventually hanged. See Hopton Haynes, *Brief Memoires Relateing to the Gold and Silver Coins of England*, 1700 (Lansdowne MSS.), p. 176.

(3) Possibly a dialectical pronunciation of *fon*, a verb now obsolete. Formerly it meant to befool, hoax.

(4) Besides being Master of the Mint, Neale was a member of the House of Commons.

(5) Taken on 2 May 1696 (see p. 201), shortly after his appointment as Warden.

(6) In a letter (Newton MSS. 1, 501), Newton, on one of the rare occasions when he broke out into the picturesque, described Chaloner as 'A japanner in clothes threadbare, ragged, and daubed with colours' who 'turned coiner and in a short time put on the habit of a gentleman'. See Craig, *Newton*, p. 124.

582   ROBINSON[1] TO NEWTON
5 FEBRUARY 1697/8
From the original in the University Library, Cambridge[2]

Chester 5 Feb: 97/8

Honrd Sr

I am sorry so short a time should afford
an occation of complaint agst Mr Weddall,[3] it is wha[t]
could scarcely be expected from any man but hi[m;]
he hath writt a most impertinent Reflecting Let[ter]
to Mr Fosbrooke[3] our Surveyour, telling him that B[......][4]
and his man Robinson wer coming down; a[nd]
I think Sr I ought not to suffer such base reflex[ions]
from one that hath all along endeavourd my ru[in]
you were pleasd' to tell me, that you would undertak
he should not any more disturb or abuse me, but by
this letter men[t]ioned I find the same turbulent
Spiritt, so that I fear our animositys will not be
abated wch is no small trouble to me, (who I can say[)]
was never accounted a fomentor of differences; however
I am under no obligačon to Lye under the Lash of his
tongue, & tho for peace sake I was willin[g]
to forgett former injurys, yett I am not to be Hectord'
over att pleasure by one (Laying aside his deputačon)
who (without vanity) is in no way my Supe[rior;]
Sr I heartily ask your pardon for this trouble wishing
I may not be compelld' by your Deputys[5] behaviour t[o]
take an effectuall remedy to make him quiett, or
bridle his pen & tongue; Sr I am still opposd' [to]
our Moneyers Agent here, in makeing use of th[is]
place for erecting my furnace, wch you were all[ways]
pleased to allow me to have, there being no o[ther]
convenient place for that purpose; the bellowes [wch you]
procur'd to sett up there, the Gentlemen cannot [now]

spare; & if he taks them away (as he says he in[tends)]
I cannot make up my Sweep; so that I beg a fur[ther]
order for the same that no further obstru[ction be]
given to the buissness of the Mint, but t[hat I may]
wthout disturbance goe on & hope to app[ear]
Honrd Sr

Your humble Servtt

THO: ROBINSON

To
The Honrd Isaac Newton, Esqr
   in German Street in
      London
These

NOTES

(1) Thomas Robinson (see note (3) below).

(2) Add. 3964.8, fo. 19. The sheets are badly damaged and many letters and words have had to be surmised. These are indicated by square brackets.

(3) The Deputies at the Chester Mint at this time were as follows:
Warden: Robert Weddell.
Master: Thomas Clark (Clarke).
Comptroller: Edmund Halley.
Assayer: Peter Pemberton.
King's Clerk: John Collins.
Surveyor of Meltings: Francis Farbrook.
Porter: Edmund Webb.
Master's Assistant: Thomas Robinson, who took charge when Clark came to London.
Comptroller's Clerk: Lewis, who was superseded (22 July 1697) by Greenall, and later by Smith (25 October 1697).

See *Calendar of Treasury Papers*, XLVIII, 52, 9 November 1697, 'Memorial of the Officers of the Mint, to the Lords of the Treasury, presenting them with a list of officers and clerks employed at the five Country Mints, together with their salaries, praying their Lordships to pay the same'.

(4) Only the initial letter is decipherable, but it probably refers to Bowles, a clerk (see Letter 562, note (2), p. 230).

(5) The text of the letter points to Weddell, who as Deputy Warden was officially Newton's deputy.

583   JOHN NEWTON[1] TO NEWTON
5 FEBRUARY 1697/8
From the original in the University Library, Cambridge[2]

Most hond Sr

The Czar[3] intends to be here to morrow before 12 and I thought my self obliged to signify to You hee likewise expects to see You there. I have taken all possible care to have things in readiness & have not time to add more but that I am

Most hond Sr

Your most humble & obediant Servant

J NEWTON

Fryday Feb 5th. 1697/8[4]

NOTES

(1) Newton's family tree shows several persons with the name John Newton. The sender of this letter might have been Sir John Newton, of Barrs Court, Gloucestershire, and later of Thorpe, in Lincolnshire, who was reputed to be a kinsman of Isaac, both having been descended from John Newton, of Westby, in Lincolnshire. See Newton's letter to Sir John Newton, April 1707 (Letter 719), note (1).

(2) Add. 3966.15, fo. 310.

(3) The Czar, who was born in 1672, and who became Peter the Great, conceived the idea of establishing a Russian Navy, and realizing the superiority of English shipbuilding paid a visit to this country early in 1698. He stayed at the house of John Evelyn, and during the day worked as a labourer in the shipyards at Deptford. He also visited Greenwich Observatory (see Weld, I, 257). Sir John Craig observes: 'It may have been a pleasant change when the Czar of all the Russias, Peter the Great, took 3 February 1698 off from Deptford dockyard to be shown round the Mint by the Chancellor of the Exchequer' (*Mint*, p. 195).

(4) The date is not correct. 5 February 1697/8 fell on a Saturday. The available evidence is that the Czar visited Deptford on 6 February.

584   MEMORANDA BY DAVID GREGORY
20 FEBRUARY 1697/8
From the original in the University Library, Edinburgh[1]

20 febrii 1697/8.

Newtonus hoc Problema solvit. Chordam flexilem ita in diversis sui punctis onerare ut in datam curvam pondere suo incurvetur. Et vicissim Fili[2] dato modo onusti figuram determinare. Credit rationem Catenariæ ad

265

suam abscissam, esse geometrice et facile exprimibilem, Catenariæ ad suum applicatum aliquanto difficilius.

Planetæ ex eodem loco Decidui, et per reflexionem vel aliter in tangentem orbitæ cujusvis respective actæ, si velocitas cujusvis minuatur in ratione $\sqrt{2}$ ad 1, vel si Sol duplo major deveniat instanti quo quisque orbitam propriam attingit, planeta eam describeret orbitam quam nunc describit.

Joannes Bartolus Lucinianensis anno 1611 scripsit *Tractatum de Radiis Lucis et Visus, in vitris perspectivis et Iride*, in 4to. In quo Iridis colores prorsus similiter explicat quo Cartesius. Hunc modum Irid[em] explicandi habuit ab affini suo Marco Antonio de Dominis Archi[epis]copo[3] Spalatrensi, qui illam longo ante tempore obtinuit, ut in præfatione asserit. Deerat Titulus huic libello, quem vidi apud Newtonum. Postero tamen die apud Sloanum vidi illum cum titulo, Marci Antonij de Dominis Tractatus &c, editus per Jo: Bartolum, Venetiis 1611, cujus exemplar dono mihi dedit D. Gray M.D.

Newtonus vicissim propositurus erat Bernoullio et Leibnitio Problema de via projecti cum resistentia est in duplicata ratione velocitatis, quod perperam solverat Leibnitius, in *Actis Lipsiæ*.

Vix credit Marchionem Hospitalium potuisse per se invenire curvam celerrimi descensus.

In condenda tabula refractionis stellarum Atmosphæræ Altitud[inem] non considerat ultra 40 aut 50 milliaria provectam; illam in 10 [aut] 12 partes dividit et methodo approximandi Geometris familiari hoc schemate contento rem perficit.

Vidi Ego Lemma primum Lib. 3 saltem ejus vicarium, quatuor titulis absolutum. Vidi etiam Egyptiorum [Græ]corum &c opiniones de Mundi systemate, in religione adu[mbratas]. Vidi etiam 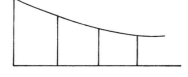 hæc Problemata aliter quam in libro Princip. soluta. Inven[ire] vim Solis (et Lunæ) ad mare movendum.

Diaphana magis sulphurea licet leviora radios magis refringunt, ut oleum (vel spiritus vini) magis quam aqua fontana.

Vidi apud Sloanum Historiam debellatæ Americæ per D. Oviedo Hispanum conscriptam, librum rarissimum.

D. C. Wrennus ait se esse compotem Methodi Gravitatem explicandi mechanice. Ridet [D.] Newtonum credentem illam per Mechanicam non fieri, sed a Creatore primitus il[latam].

Nihil scripsit Wrennus de via Planetarum aut motibus cœlestibus.

Translation

Newton solves this problem: so to load a pliable rope at different points on it that it is bent by its own weight into a given curve, and contrariwise, to determine the shape of a string loaded in a given manner. He believes that the relation of a catenary to its abscissa is expressible geometrically without much difficulty, but that of a catenary to its ordinate is somewhat more difficult to express. [See figure.]

Planets falling from the same place and diverted by turning aside or otherwise into the tangent of any orbit drawn respectively for each: if the velocity of any such planet is diminished in the ratio of $\sqrt{2}$ to 1, or if the Sun were to become twice as large at the instant when each planet reaches its appropriate orbit, the planet would describe that orbit which it now describes.

Giovanni Bartolo of Luciniano, in the year 1611, wrote a treatise *De Radiis Lucis et Visus, in Vitris perspectivis et Iride*, in quarto. In this he explains the colours of the rainbow in a way quite like that of Descartes. He obtained this method of explaining the rainbow from his relative by marriage, Marco Antonio de Dominis,[3] Archbishop of Spalato, who reached it a long time earlier, as he claims in the preface. The title-page was missing from the copy of this book which I saw at Newton's house. The next day, however, I saw it complete with the title page in Sloane's house. *The treatise of Marco Antonio de Dominis* edited by Jo: Bartolus at Venice 1611. Mr Gray, M.D., made me a present of a copy of it.

Newton had meant in his turn to propound to Bernoulli and Leibniz a problem about the path of a projectile when the resistance varies as the square of the speed, a problem which Leibniz had solved in a way not entirely satisfactorily, in the *Leipzig Acts*.

He hardly believes that the Marquis de l'Hôpital could have discovered the curve of quickest descent unaided.

In drawing up the table of refraction of the stars he does not consider that the height of the atmosphere extends further than 40 or 50 miles. He divides it into 10 or 12 parts, and by the method of approximation familiar to geometers achieves his result by drawing this figure.

I saw the first lemma of Book Three, which is certainly a substitute for this problem, completed under four headings. I saw also the notions of the Egyptians, Greeks, etc., about the shape of the universe foreshadowed in religious form.

I also saw these problems solved in a different way from that used in the book, the *Principia*; to find the force of the Sun (and the Moon) to move the sea.

Transparent bodies of a yellower colour, although lighter, refract rays to a greater extent, as for instance oil or spirits of wine more than spring water.

At Sloane's house I saw a history of the Conquest of America written by Senor Oviedo, a Spaniard, a very rare book.

Mr C. Wren says that he is in possession of a method of explaining gravity mechanically. He smiles at Mr Newton's belief that it does not occur by mechanical means, but was introduced originally by the Creator. Wren wrote nothing about the path of the planets, or the motions of heavenly bodies.

NOTES

(1) A. 90. The sheet is torn along the right-hand edge and many words have had to be surmised.

(2) The word was originally written *filum*, then altered to *fili*; similarly *onustum* was altered to *onusti*, to agree with *fili*.

(3) Marco Antonio de Dominis (1566–1624), Professor of Mathematics at Padua; Bishop of Segni, 1600, Archbishop of Spalato (the modern Split) and Primate of Dalmatia, 1602. He came into conflict with the Church of Rome, and was sent to prison, where he died. In the work here referred to (*De Radiis Visus et Lucis*, Venice, 1611) he came very near the true explanation of the formation of the rainbow, showing that in the primary bow one reflexion and two refractions were involved. Descartes was no doubt familiar with this work; his own explanation was fuller than that of de Dominis inasmuch as he gave an explanation of the formation of the secondary bow, showing that in this two reflexions as well as two refractions were involved. See J. F. Scott, *The Scientific Work of René Descartes*, pp. 72–4, 202.

585   JACQUES CASSINI[1] TO NEWTON
6 APRIL 1698
From the original in the Library of King's College, Cambridge[2]

Clarissimo Viro Domino Isak Newton
Jacobus Cassini S.P.D.

Cum e Londino, reversurus in Galliam, huc pervenissem, accepi a patre meo[3] epistolam una[m] cum maximis Satellitum Saturni digressionibus, quas a me expostulaveras. Has tibi mandare et gratitudinem meam tuorum erga me beneficiorum simul exhibere, mihi liceat. Tuam domum adivi ut te inviserem, sed mala usus sum fortunâ cum tunc abfuisses. Vale, Vir Clarissime, et sic habeas me tibi semper esse addictissimum. Dover die 6a April 1698. St. N.

Digressiones maximas Satellitum Saturni[4] ab ejus centro comparavimus cum diametro ipsius annuli ad quam earum ratio per accessum recessumque Saturni a Terrâ non variatur et tamen hoc modo comparatæ digressiones non nihil variabiles deprehensæ sunt.[5]

Primus sive intimus visus [est]distare a Saturni centro una diametro annuli minus particula variabili intra 1/20 et 1/40. Secundus ut plurimum diametro una cum quadrante. Tertius diametro una cum tribus quadrantibus. Quartus diametris tribus cum dimidio.

Quintus diametris undecim cum tribus quadrantibus.

Revolutiones	Dies	Horæ	Minuti
Primi	1	21	19
Secundi	2	17	43
Tertii	4	12	27
Quartii	15	23	19
Quinti	79	22	0

Translation

To that distinguished man, Isaac Newton, greetings from Jacques Cassini

When, being about to return to France from London, I had reached this point of my journey [i.e. Dover], I received a letter from my father[3] together with an account of the greatest deviations of the satellites of Saturn which you had requested of me. Allow me to convey these to you and at the same time express my gratitude for your kindnesses to me. I went to your house in order to pay you a visit, but it was my ill fortune to find you away. Farewell, noble Sir, and be assured that I am always deeply devoted to you.

Dover 6 *April* 1698 *N.S.*

We have compared the maximum deviations of the satellites of Saturn[4] from its centre with the diameter of its ring. The ratio of these [deviations] to the diameter by reason of the approach and recession of Saturn from the Earth does not vary, nevertheless, when compared in this way, the deviations were found variable to some extent.[5]

The first or innermost was seen to be distant from the centre of Saturn by the diameter of the ring, diminished by a fraction varying between $\frac{1}{20}$ and $\frac{1}{40}$, the second by $1\frac{1}{4}$ diameters, the third by $1\frac{3}{4}$ diameters, the fourth by $3\frac{1}{2}$ diameters, and the fifth by $11\frac{3}{4}$ diameters.

NOTES

(1) Jacques Cassini (1677–1715), son of G. D. Cassini (see note (3) below). With his father he measured the length of the arc of the meridian between Dunkirk and Perpignan, and concluded that the Earth was elongated at the Poles.

(2) KMS. 93.

(3) Giovanni Domenico (Jean Dominique) Cassini (1625–1712). Appointed first Director of the Paris Observatory, 1699. Besides the four satellites noted below (note (4)) he discovered the division in Saturn's ring which is known by his name. A letter from him, printed in the *Phil. Trans.* **16** (1687), 299, is entitled: 'A Letter of Monsieur *Cassini* to the Publisher, giving his Corrections of the Theory of the five *Satellites* of *Saturn*; with Tables of the Motions of those *Satellites*, adapted to the Meridian of *London* and the *Julian* Account. Paris le 10 Octobre 1686.'

(4) The five satellites of Saturn which were known at this time, with their periods, were:

	Period		
	d.	h.	m.
Tethys	1	21	19
Dione	2	17	41
Rhea	4	12	25
Titan	15	22	41
Japetus	79	7	56

Titan was discovered by Huygens in 1655; the other four were discovered by G. D. Cassini between 1671 and 1684. See Spencer Jones, p. 255.

(5) What follows is best understood from the following extracts from the letter of G. D. Cassini referred to above (note (3)):

'La distance du premier Satellite au centre de Saturne m'a paru variable, & son mouvement sensiblement inegal....J'ay dernierement determiné sa moyenne distance de 39/40 du diametre de l'anneau de Saturne....

'La distance du second Satellite du centre de Saturne m'a paru plus uniforme. Je l'ay determinée d'un diametre de l'anneau & $\frac{1}{4}$...

'La distance du Troisieme du centre de Saturne paroit d'un diametre de l'anneau & $\frac{3}{4}$...

'La distance du Quatrieme Satellite au centre de Saturne paroit de 4 diametres de l'anneau.

'La distance du cinquiesme Satellite au centre de Saturne de 12 diametres de l'anneau.'

At the conclusion of the letter (ibid. p. 306) Cassini goes on to say:

'I shall only add, That the Proportion of the Squares of the times of the Periods, to the Cubes of the Distances, (which is proposed as probable by *Kepler*, but now demonstratively found true by Mr. *Newton*,) gives us nicely the Proportion of the Distances of these Planets from the Center of *Saturn*; and supposing the Satellite of *Hugens* four Diameters of *Saturn's* Ring distant from him, we shall find by the Periods, the Distances, as follows.

	Periodus			Distantia
	d.	h.	′	
Intimi	1	21	$18\frac{1}{2}$	0·964
Penintimi	2	17	$41\frac{1}{2}$	1·235
Medii	4	13	$47\frac{1}{4}$	1·740
Penextimi	15	22	41	4·000
Extimi	79	7	54	11·621

'These Distances may be used, as more accurate than those obtained by Observation, which yet differ but little therefrom.'

The meaning is this. If we assume the distance of the fourth satellite in terms of the diameter of the ring to be 4, then the distances of the remaining satellites would be represented by the corresponding numbers in the last column.

Assuming the outermost diameter of Saturn's ring to be 171,000 miles, and the distances of the five satellites from the centre of the planet to be respectively 183,000, 234,000, 327,000, 749,000, 2,210,000 miles, Cassini's results are found to agree closely with modern values. See *Journal des Sçavans*, 22 April 1686, p. 77.

586   NEWTON TO ALL COUNTRY MINTS
16 APRIL 1698
From the original in the Royal Mint[1]

Mint Office in the Tower of L.
Apr. 16. 1698.

Sr

The business of the Country Mints now drawing towards an end,[2] the Lds of
ye Treasury are minded that we should lay your Accounts before them & for
that end I desire you wth the assistance of the Master & Comptroller to fill up
the blanks in the inclosed paper wth all the care & exactness you can & then
to return it to me or a fair copy thereof signed by you all that we may rely
upon it.

Where nothing was imported (as perhaps in the branch of wrought Plate
between Nov 4 96 & Jul. 1 97, or in that of publick monies since Mar. 1) you
are to fill up the blanks with cyphers. And so in the two last columns cyphers
are to be put where nothing remained due to Importers, or nothing was
reserved for melting refining & coyning. And in the two columns under
Ingots & Sweep if in any Case you did not add the sweep to ye Ingots, in that
case set down the weight of the Ingots alone wth the letter I before it.

If you do not know the tale of the new monies made out of every branch
imported, set down the whole tale or summ of the monies made out of all the
branches together from time to time, that is from the beginning of the coynage
to Nov 4 96, thence to Jul. 1, 97, thence to Nov. 4, 97 & thence to Mar. 1 or
to ye day of your signing the paper. Over against the word summ in ye
inclosed paper you have blanks for this purpose. But if you do not know these
summs nor can collect them out of your books then send me in a Letter the
weight & tale of all ye new monies coyned till such & such periods of time
when you did both weigh & tell all the moneys then coyned, dividing the
whole coynage into five or six periods.

Let ye ablest of your Clerks be imployed in consulting & comparing your
books & computing what is necessary out of them, but trust not the computation of a single Clerk nor any other eyes then your own. And let the inclosed
or a fair copy thereof filled up & signed by the Officers be sent back in a week
or 10 days after the receipt hereof or sooner if you can. And if there remain
any blanks wch you cannot fill up in that time, you may keep a copy of the
Paper you send me, to be filled up more completely afterwards & sent to me
as soon as you have finished it. I have sent the same Papers to all the Mints
that your Accompts may be conformable to one another & thereby the better

fitted for the view of the Lords & put in a method for Auditing. Tis by
Mr Neals consent & desire as well as by the suggestion of the Lords that I take
care of this matter & therefore I doubt not but you will have all the assistance
his Deputy & the Controller can give you & the free inspection of their
Books. I am

Your loving Friend

Is. NEWTON.

NOTES

(1) Newton MSS. II, 251. This is in Newton's hand.

(2) Norwich closed down in April 1698; by September the rest had ceased to operate.   See
Craig, *Mint*, p. 192.

587   HARINGTON[1] TO NEWTON

22 MAY [1698?][2]

From *Nugæ Antiquæ*, II, 104.[3]
For answer see Letter 588

A letter from JOHN HARINGTON to Mr. NEWTON, afterwards Sir ISAAC;
with a Scheme of the Harmonic Ratios

Sir

At your request I have sent you my scheme of the Harmonick Ratios
adapted to the Pythagorean proposition, which seems better to express the
modern improvements, as the ancients were not acquainted with the sesqui-
alteral divisions, which appears strange. Ptolemy's *Helicon* does not express
these intervals so essential in the modern system; nor does the scheme of four
triangles, or three, express so clearly as the squares of this proposition. What I
was mentioning concerning the similitude of ratios, as constituted in the sacred
architecture, was my amusement at my leisure hours, but am not master
enough to say much on these curious subjects. The given ratios in the dimensions of Noahs ark, being 300, 50, and 30, do certainly fall in with what I
observed; the reduction to their lowest terms comes out 6 to 1, which produces
the quadruple sexqialteral ratio; and 5 to 3 is the inverse of 6 to 5, which is
one of the ratios resulting from the division of the sexquialteral ratio; the
extremes are as 10 to 1, which produce by reduction 5 to 4, the other ratio
produced by the division of the sexquialteral ratio. Thus are produced the
four prime harmonical ratios, exclusive of the diapason, or duple ratio. I have

conjectured that the other most general established architectural ratios owe their beauty to their approximation to the harmonic ratios; that the several forms of members are more or less agreeable to the eye, as they suggest the ideas of figures composed of such ratios. I tremble to suggest my crude notions to your judgment, but have the sanction of your own desire, and kind promise of assistance to rectify my errors. I am sensible these matters have been touched upon before, but my attempts were to reduce matters to some farther certainty as to the simplicity and origin of the pleasures affecting our different senses; and try, by comparison of those pleasures, which affect one sense from objects whose principles are known as the ratios of sound, if other affections, agreeable to other of our senses, were owing to similar causes. You will pardon my presumption, as I am sensible neither my years nor my learning permit me to speak with propriety herein; but, as you signified your pleasure of knowing what I was about, have thus ventured to communicate my undigested sentiments, and am, Sir

Your obedient servant

JOHN HARINGTON[4]

Wadham College,
May 22d. 1698

NOTES

(1) John Harington (1679–1725), of Kelston, in Somerset, was admitted to Wadham College, Oxford, in June 1696. See Edleston, p. 302: 'Mr John Harington (of the family of "Ariosto" Harington and "Oceana" Harington), an undergraduate of Oxford, seems to have had some conversation with Newton upon a method which had occurred to him of representing musical intervals by the addition of numbers proportional to the sides (3, 4, 5) of a right-angled triangle, and to have alluded to the bearing of the subject upon the principles of architectural beauty.'

(2) The date is clearly printed 22 May 1693. But as Edleston points out (p. 304, note) Harington was only 17 when he was admitted to Wadham College in 1696, i.e. three years after the alleged date of the letter. It is more than likely that the figure 3 in the date has been printed in error for 8.

(3) Sir John Harington, *Nugæ Antiquæ* (1792).

(4) At the end of the letter Harington has added: 'A Scheme for all the Harmonical Intervals demonstrated from the Golden Proposition of Pythagoras. Euclid 47.'

588   NEWTON TO HARINGTON

30 MAY 1698

From *Nugæ Antiquæ*, II, 108.
In reply to Letter 587

Mr NEWTON's Answer to Mr John HARINGTON, 1693[1]

Sir,

By the hands of your friend, Mr Consel, I was favoured with your Demonstration of the Harmonic Ratios,[2] from the Ordinances of the 47th of Euclid. I think it very explicit and more perfect than the Helicon of Ptolemy, as given by the learned Doctor Wallis.[3] Your observations hereon are very just, and afford me some hints which, when time allows, I would pursue and gladly assist you with any thing I can, to encourage your curiosity and labours in these matters. I see you have reduced, from this wonderful proposition, the inharmonics as well as the coincidences of agreement, all resulting from the given lines three, four, and five.[4] You observe that the multiples hereof furnish those ratios that afford pleasure to the eye in architectural designs: I have, in former considerations, examined these things, and wish my further employments would permit my further noticing thereon, as it deserves much our strict scrutiny, and tends to exemplify the simplicity in all the works of the Creator; however I shall not cease to give my thoughts towards this subject at my leisure. I beg you to pursue these ingenious speculations, as your genius seems to incline you to mathematical researches. Your remarks that the ideas of beauty in surveying objects arises from *their* respective approximations to the simple constructions, and that the pleasure is more or less, as the approaches are nearer to the harmonic ratios. I believe you are right; portions of circles are more or less agreeable, as the segments give the idea of the perfect figure from whence they are derived. Your examinations of the sides of polygons with rectangles certainly quadrate with the harmonic ratios. I doubt some of them do not; but then they are not such as give pleasure in the formation or use. These matters you must excuse my being exact in, during your inquiries, till more leisure gives me room to say with more certainty hereon.[5] I presume you have consulted Kepler,[6] Mersenne,[7] and other writers on the construction of figures. What you observe of the ancients not being acquainted with a division of the sexquialteral ratio is very right; it is very strange that geniuses of their great talents, especially in such mathematical considerations, should not consider that although the ratio of three to two was not divisible under that very denomination, yet its duple members six to four easily pointed out the ditone[8] four to five, and the minor tierce[9] six to five, which are the chief perfections of the diatonic system,[10] and without which the ancient system was doubtless very imperfect. It appears strange, that those whose nice scrutinies carried them so far as to produce the small limmas,[11] should not have been more particular in examining the greater intervals, as they now appear so serviceable when thus divided. In fine, I am inclined to believe some general laws of the Creator prevailed with respect to the agreeable or unpleasing affections of all our *senses*; at least the supposition does not derogate from the wisdom or power of God, and seems highly consonant to the macrocosm in general. Whatever else your ingenious labours may produce I shall attentively consider, but have such matters on my mind, that I am unable to give you more satisfaction at this time; however, I beg your modesty will not be a means of preventing my hearing from you, as you proceed in these curious researches; and be assured of the best services in the power of

Your humble Servant,

Is. NEWTON

[*Jermyn Street*]
May 30, 1698

NOTES

(1) See Letter 587, note (2), p. 273.

(2) The harmonic ratio between two notes is the ratio of their frequencies. If this ratio, or *interval* as it is usually called, can be expressed in small integers (e.g. 1:2 as in the case of a note and its octave, or 2:3 as with a note and its dominant) the two notes are acoustically consonant. To find the *sum* of two intervals, or frequency ratios, these ratios are multiplied together; similarly their difference is the ratio of one ratio to the other. The interval (frequency ratio) between one note and another may be determined by multiplying the frequency ratio of the first and an intermediate one by the frequency ratio between that intermediate one and the second. See examples below.

(3) *Claudii Ptolemæi Harmonicorum libri tres. Cum Appendice: De Veterum Harmonica ad Hodiernam Comparata.* Wallis, *Opera*, vol. III.

(4) As strings under the same tension whose lengths are arithmetically proportional produce notes whose frequencies are harmonically proportional; thus such strings whose lengths are proportional to 5, 4, 3 produce notes whose frequencies are proportional to 12, 15, 20. It was the discovery of the wonderful harmonic progression in the notes of the musical scale that led Pythagoras to the view that in numbers lay the elements of all things. 'All things are numbers.'

(5) Newton's duties at the Mint were fully absorbing his time.

(6) Kepler, *Harmonices Mundi libri v* (1619), p. 126. Following the ideas of Pythagoras, Kepler sought the underlying harmony in the universe, believing that there was a relation between musical intervals and the distances of the planets from the Sun.

(7) Marin Mersenne (1588–1648), a Franciscan friar and friend of Descartes. As a mathematician he wrote on the theory of numbers, although his best-known work was in physics. He wrote *Cogitata Physico-Mathematica* (1644), a somewhat unequal work, and *Les Preludes de L'Harmonie Universelle* (1634).

(8) Ditone, sometimes called the Pythagorean third. It is the interval between two notes whose frequency ratio is $81:64$. It is the *sum* (see note (2) above) of two major tones (frequency ratio $9:8$) and is the excess of four perfect fifths (frequency ratio $3:2$) over two octaves, i.e. $(9/8)^2$, or $(3/2)^4 \div 2^2$.

(9) Minor tierce, or minor third (frequency ratio $6:5$) is the amount by which a perfect fifth exceeds a major third (frequency ratio $5:4$), i.e. $3/2 \div 5/4 = 6/5$.

(10) The diatonic scale on any note has frequencies proportional to the numbers 24, 27, 30, 32, 36, 40, 45, 48. There is one kind of semitone, its frequency ratio being $16/15$ but there are two kinds of tones, their frequency ratios being $9/8$ and $10/9$. The diatonic scale differs from the 'tempered scale' on a modern pianoforte in which the octave is divided into twelve equal semitones each with a frequency ratio of $\sqrt[12]{2}$.

(11) Limmas. The great limma is the amount by which a minor third exceeds a minor tone, this latter being the amount by which a major third exceeds a major tone (frequency ratio $9:8$). The frequency ratio of the minor tone is $10/9$, or $5/4 \div 9/8$. The frequency ratio of the great limma is $27:25$, or $6/5 \div 10/9$. The small limma is the amount by which a major tone exceeds a diatonic semitone (frequency ratio $16/15$). Its frequency ratio is $135:128$, or $9/8 \div 16/15$. See Grove's *Dictionary of Music and Musicians*, edited Eric Blom (1954).

589   MEMORANDA BY DAVID GREGORY
? JULY 1698
From the original in the University Library, Edinburgh[1]

Adnotata Phys: et Math: ex Newtono 1698 particulatim de Refractione

Diametri Ellipseos Jovis sunt ut 9 ad 8[2] non ut 40 ad 39 (quæ est insensibilis) nam in Corol deficit coefficiens tam in numeratore quam denom: (quæ ultima est 17).

Theoria Lunæ non absolvetur ob Flamstedij Iracundiam nec de Flamstedio erit sermo, ad 4 tamen minuta rem complebit, quam complevisset ad 2, si Flamst: observationes subpeditasset.

Saturni Orbem definiet, Aliorum Orbes ob observationum penuriam non definiturus.

2 aut 3 Cometarum orbitas definiet, quorum unus est anni 1664.[3] has Halleius faciet et protrahet.

                 ...   .

Omnem Cometarum theoriam de novo instituit, et Precessionem Equinoctiorum. Sect: 4 et 5 eximit.

Tab: Refractionem[4] Aeris 6 Septimanis universo Calculo condidit et Flamst: communicabit. hæc fit in vero Systemate tam densitatis aeris in diversa altitudine, quam refractionis per totum corpus aeris. Refractio Horizontalis est 34' 45": 3 gr [est] 13' 20"; 10 gr est 4': 52". Calor Aeris prope terram minuit refractionem, foret enim 37' absque hoc. Tab: hæc ab zenith ad 3 gr elev: non potest errare ultra 2".

Cometa prope ☿ transiens ad orientem, mutavit istum locum in perihelium, ut Luna Aquas attrahendo diluvium fecit.

...   .

Sunt 16 genera Curvarum secundi generis,[5] et 76 Curvæ Newtonus conscripsit tractatum de illis quem mihi impertietur ut eum edam.

...   .

Translation

Physical and Mathematical Notes from Newton, 1698, particularly about Refraction

The diameters of Jupiter's elliptical orbit are as 9 to 8,[2] and not as 40 to 39 (which is not perceptible), for in the Corollary a coefficient is missing in the numerator as well as in the denominator (which last is 17).

On account of Flamsteed's irascibility the theory of the Moon will not be brought to a conclusion, nor will there be any mention of Flamsteed, nevertheless he [Newton] will complete to within four minutes what he would have completed to two, had Flamsteed supplied his observations.

He will determine the orbit of Saturn, but not those of the other planets because of the lack of observations.

He will determine the orbits of two or three comets, one of which is that of 1664:[3] these Halley will work out and chart.

...   .

He constructs afresh the whole theory of comets, and the precession of the equinoxes. He leaves out sections 4 and 5.

He has drawn up in a table the refractions[4] of the air for six weeks by a general reckoning and he will communicate it to Flamsteed. On a true system this [table] relates as much to the density of the air at various heights as for the refraction through the whole body of the air. The horizontal refraction is 34'.45": at 3 degrees it is 13'.20"; 10 degrees at 4'.52". The heat of the air near the Earth diminishes the refraction. For apart from this it would be 37'. This table from the zenith to three degrees of elevation cannot have an error of more than 2".

A comet passing near the Earth to the east has altered its course in perihelium just as the Moon by attracting the waters caused a deluge.

...   .

There are sixteen species of curves of the second class,[5] and 76 curves. Newton has written a treatise on them, which he will communicate to me for me to publish.

NOTES

(1) Greg. MSS. C. 62.

(2) The eccentricity of Jupiter's orbit is 0·0483, which is about ten times less than this ratio indicates.

(3) See Halley's letter to Newton (Letter 534).

(4) *Tab. Refractionem.* This must be a slip for *Tab[ulam] Refractionum*, in which case the meaning would be: 'He has drawn up a table of refractions.' Otherwise, if *Tab.* is an abbreviation for *Tabula* (ablative case) we should expect it to be preceded by *in*.

(5) *Enumeratio linearum tertii Ordinis*, published as an Appendix to the first edition of the *Opticks* (1704), contains theorems on the theory of curves. Newton divides these into 72 species.

590   WALLIS TO FLAMSTEED
13 AUGUST 1698
From the original in the Corpus Christi College Library, Oxford[1]

Oxford Aug. 13. 1698.

Sir,

I have understood from Mr Caswell, a good while since, that you had very considerable Observations (for divers years together) to prove a Parallax of the fixed stars to ye Earth's Annual Orb: I have desired Mr Caswell to press you, to let us have an account of them so as to have them published. I do again request it of you, & hope you will not refuse it. I am now drawing toward an end of publishing a third volume[2] of Mathematical Tracts, wherein will be a Collection of Letters relating to such matters. If you will do me the favour to draw up these observations in form of a Letter to me (or to whom else you please) & let me have it: I will (with your leave) publish it amongst those Letters. The thing will be an honour to you, & to our nation. I know your hands are full of other work; but I hope you may spare so much time as to draw up such a letter (in Latine;)[3] & I will take upon my self the trouble of seeing it printed here; & you shall command from me the imploying of as much time to serve you in what way you shal propose to me. But pray let it be done quickly, lest it come out too late. 'Tis pitty you should loose the honour of being the first who hath made such a discovery.

I am   Sr

Yours to serve you,

JOHN WALLIS.

These
For Mr Flamsted, at the
Observatory in
    Greenwich.

NOTES

(1) C. 361, no. 125. The letters which passed between Flamsteed and Wallis about this time are here included on account of the impact they made upon the relations between Newton and Flamsteed.

(2) The final volume of Wallis's *Opera*, which appeared in 1699, was now nearing completion, and Wallis had planned to include in it a number of letters from eminent men of science of the day (*Epistolarum quarundam Collectio Rem Mathematicam Spectantium*). Believing that Flamsteed had detected a parallax of the pole star, and fearing that his country might be deprived of the honour of the discovery, he was anxious to include an account of it in the above work. 'You will find in the collection of letters', he wrote to Samuel Pepys (Add. MS. no. 113, Bodleian) 'some remarkable observations of Parallax of the Earth's Annual Orb by Mr Flamsteed (in confirmation of the Copernican Hypothesis) a thing long sought for, and never observed till now.' Although Flamsteed at once replied, Wallis had to repeat his request on 26 August and 8 November. See Letters 592 and 596.

It should be noted that Flamsteed with the instruments at his disposal could not possibly have detected the parallax, since it amounts to no more than 0″.106, and quite possibly much less. The earliest discovery of a parallax of a fixed star is due to Bessel who in 1838 announced that he had discovered a parallax of 0″.54 in the star 61 Cygni. A few months later, Thomas Henderson discovered a parallax of 0″.98 in Alpha Centauri. The displacement which Flamsteed had attributed to parallax was shown by Bradley to be due to aberration. He announced his discovery in a letter to Halley (1 January 1728/9); an account of it had already appeared in *Phil. Trans.* 35 (1697/8), 637. See Letter 592, note (2), p. 281.

(3) Wallis, having satisfied himself that the letter contained nothing 'but what is fit to be published', translated it into Latin. See Letter 597.

591   MINT TO THE TREASURY
18 AUGUST 1698
From the original in the Public Record Office[1]

To the Right Honble the Lords Commissioners of his Majts Treasury

May it please your Lordships

In examining the Annexed Petition we find that ye three brothers John, Joseph, and Phillip Roettiers[2] were Gravers to the Mint from ye year 1661 Att an allowance of 325£ yearly and in ye year 1669 by a new Grant were further allowed 450£ yearly out of the Exchequer during their joynt lives as Gravers for ye Coyn & Medalls and after ye death of one of them the two survivors were to receive 350£ yearly during their joynt lives & the survivor of these to receive 250£ during his life

About ye year 1680 Joseph Roettier went off but the whole payment continued in Consideraçon (as we suppose that the Petitioner James Roettier ye son of John Roettier (the other Petitioner) supplied the place

Phillip Roettier another of ye Patentees went off in Feb 1684 But John Roettier haveing another son Norbertus the whole allowance of 450£ still continued & has been paid till March 25. 1697 Norbertus went off three years ago And the Petition is for ye Arrear and continuance of it

This Pension of 450£ per annum was alwayes paid by a speciall warrant from ye Lord high Treasurer or Lords of ye Treasury and never came into the Mint Accompt

As to ye Petitioners wants we believe the Petition may be true and that James Roettier is very capable of makeing medals and on that score deserving his Majts favour

All wch we humbly submit to your Lordships great wisdom

Mint Office
Aug. 18 1698

Is NEWTON.
THO: NEALE.

NOTES

(1) T. 1/55, no. 62. Only the signature, the address and the date are in Newton's hand.

(2) See also Letter 568, note (2), p. 241.

592   WALLIS TO FLAMSTEED
26 AUGUST 1698
From the original in Corpus Christi College Library, Oxford.[1]
In continuation of Letter 590

Oxford Aug. 26. 1698.

Sr

Yours of Aug. 23. I received this morning. I am glad you are inclinable to draw up those Observations of yours concerning the Parallax of ye Earths Annual Orb.[2] The time you mention will I beleive be soone inough; for I find our Printers more slow than I could wish. I saw your letter to Mr Caswell, with which I was well pleased. I think so much of it as concerns the Rectifying of your Instrument, may be spared. It will be sufficient to give us the Observations as they would have appeared if ye Instrument had been rectified: &, as to this rectifying of ye Instrument, we may trust you. Nor will it be necessary to give a large account of the form of your Instrument; it will be sufficient to say it is a Mural Quadrant, or larger Arch, fixt to a Wall, in ye plain of ye Meridian, & furnished with Telescopick Sights, such as to distinguish a very smal Arch of a few seconds. (For, I think this is ye case.) And then, yt (the Position & Sights of ye Instrument being duly Rectifyed,) the Observations were such as there set down. A more large account of particulars may afterwards be done at leisure; wth more Observations to be henceforth made. But I am willing that at lest a short account be given of ye Observations you have; to preserve ye memory and reputation of it to yourself; as the first who have effectively discovered it. The Letter to Mr Caswell (if I do not mis-remember) gives but the Observations of one year; but intimates more to have been made in confirmation of it. It is not necessary that ye Parallax of other stars[3] should justly agree to that of ye Polar; for we are not to presume that they be all at an equal distance from us. Some may possibly have a discernable Parrallax & others not. The greater stars may be reasonably thought (but we are not sure) to be nearest; & those nearest to ye Pole of ye Zodiack most liable to a Parallax. But, if it can be discovered in any; it is a sufficient Demonstration of the Earths motion.[4]

I am

Sr

Yours to serve you

JOHN WALLIS.

These
For Mr John Flamsted
at the Observatory at
    Greenwich.

NOTES

(1) C. 361, no. 131.

(2) Though it appears from this letter that Flamsteed readily acceded to Wallis's request, either his health or his preoccupation with other matters prevented him from giving the required information at once, for on 8 November following Wallis wrote again requesting it. To this Flamsteed replied at length some time in November; a Latin translation of his reply appeared in the third volume of Wallis's *Opera* (p. 701) under the date 20 December 1698: *Epistola D.* Johannis Flamsteed (*Mathematici Regii* Grenovici) *ad D.* Wallisium, *Dec.* 20. 1698. *De* Parallaxi Orbis Annui Telluris *Observata.*

(3) The parallaxes of 'other stars' and of Polaris, and their distances from the Earth are:

Star	Parallax	Distance from Earth
α Centauri	0″·98	210,000*r*
61 Cygni	0″·54	382,000*r*
α Lyræ	0″·26	793,000*r*
Sirius	0″·15	1,375,000*r*
Arcturus	0″·127	1,624,000*r*
Polaris	0″·106	1,946,000*r*

r being the radius of the Earth's orbit (H. Godfray, *Treatise on Astronomy* (1886), p. 203).

(4) On the reverse of the above letter is a rough draft, in Flamsteed's hand, of Letter 595.

281

593   NEWTON TO LOCKE[1]
19 SEPTEMBER 1698
From the original in Corpus Christi College Library, Cambridge[2]

Jermin Street. September: 19*th*: 1698

Sr

I have enquired the weight and finesse of the peices of money mentioned in your Letter[3] and they are as follows.

A Holland Ducat[4] weighs 2 dwt: 4 gr: and is 1 car 2 gr: better than our Standard

A three guilder peice[5] weighs 20 dwt: 6 gr: and is 3 dwt worse than standard

A Spanish Pistol[6] weighs 4 dwt: 8 gr: and is $\frac{1}{4}$ dwt worse.

A peice of eight[7] weighs 17 dwt: 12 gr: and varies in finess: The Pillar peice[8] is 2 dwt better: the Mexican 1 dwt worse, the Peruvian 16 dwt worse.

The French Lewis d'or[9] weighs 4 dwt: 8 gr: and is $\frac{1}{4}$ dwt worse.

The French Crown[9] weighs 17 dwt: 12 gr: and is 06 dwt worse

The Cross Doller[10] weighs 18 dwt and is $12\frac{1}{2}$ dwt worse.

The Jacobus peice[11] coin'd for 20 shillings is the 41th: part of a pound Troy, and a Carolus 20*s* peice of the same weight. But a Broad Jacobus (as I finde by weighing some of them) is the 38th part of a pound Troy.

The whole number of guineas coined in the three last Reigns (recconing $33\frac{1}{2}$ Guineas to a pound weight Troy) is 7983739.[12]

Bank money at present is at about 5 per cent discount in Holland and Gold at $6\frac{1}{4}$ or $6\frac{1}{2}$ per cent discount. So that a Guinea is worth 20*s* 8*d* or 20*s* $8\frac{1}{2}d$ of our Milld Money at present in Holland.

The paper which I left with you I have received, but the Goldsmiths MS I cannot yet procure. If there be any thing further in which I can serve the Lords Commissioners of the Council of Trade, you may command

Your most humble & obedt Servt

Is: Newton

Memdm: The grains of the penny weights are different from the grains of the Carats. For 24 grains of the first sort make a penny weight and 4 grains of the other sort make a carat.

Memdm: Mr: Newton said that a Carat is sometimes taken for the 24th part of an ounce, and sometimes for the 24th. part of a pound.

NOTES

(1) John Locke (1632–1704). See vol. III, p. 76, note (1). During this period Locke was a Commissioner at the Board of Trade (with Pollexfen, q.v.). He wrote extensively on matters affecting the finances of the nation, e.g. *Some Considerations of the Consequences of the Lowering of Interest, and Raising the Value of Money* (1692); *Further Considerations Concerning Raising the Value of Money* (1695), etc. See Horsefield, *B.M.E.* pp. 57–60.

The correspondence between Newton and Locke concerns the attempts to stabilize the relation between silver, which was the standard currency, and gold. Originally, the guinea (first coined in 1663) was equivalent to 20 shillings; by 1695 its value rose temporarily almost to 30 shillings. Its weight was 129 39/89 grains, or about 5 dwt. $9\frac{1}{2}$ grains. It was struck from 22 carat gold. The crown, mentioned below, was a silver coin of weight 464 16/31 grains, or about 19·3 dwt.

(2) MS. Locke, b 3, fo. 127.

(3) This letter has not been found.

(4) The 'Holland Ducat', which corresponded to our half guinea, was the standard coin of the Netherlands.

(5) The 'three guilder peice' was the Netherlands equivalent to our silver crown.

(6) The 'Spanish Pistol' (pistole) was a Spanish gold coin. The name 'pistole' was likewise applied to the French *louis d'or*, as well as similar coins circulating in Italy, Germany, etc.

(7) This was the Spanish dollar. It was a coin of eight *reales* (*real* = *nummus regalis*, or royal money). The name 'piece of eight' was derived from the figure /8 by the side of the coat of arms. It corresponded to our crown.

(8) The 'Pillar peice' was a variety of the 'piece of eight'. The further references here are to Spanish colonial coins struck in slightly inferior alloys.

(9) The 'French Lewis d'or' (or *louis d'or*) was a gold piece corresponding to our guinea. It was first struck in 1640 by Louis XIII, and was worth at that time about 17s. 6d. The French crown (*écu*) corresponded to our crown.

(10) The 'Cross Doller', corresponding roughly to our crown, was struck in the Spanish Netherlands.

(11) The 'Jacobus peice coin'd for 20 shillings' was the fifth issue (1619–25) *unite* of James I. The 'Broad Jacobus' was a hammered gold coin struck in the reign of James I in 1604, and originally worth 20s. With this coin was introduced the name *unite*, from the inscription it bore: *Faciam eos in gentem unam* (I will make them one nation), an allusion to the Union of England and Scotland. The *unite* of James I was also known as a *Jacobus*, that of Charles as a *Carolus*. All these coins were *hammered*, i.e., struck manually by hammers, and they were all generically *broad*. The first issue, during the reign of James I, retained the weight of that of the last issue of Elizabeth I, namely $33\frac{1}{2}$ to the pound troy, or $171\frac{63}{67}$ grains. It was superseded in November 1604 (second issue) by a somewhat lighter 20s piece at 37·2 to the pound troy, or $154\frac{26}{31}$ grains: this is the Broad Jacobus. The value of this piece was raised to 22s. in 1611, without change of weight or design. This was followed in July 1619 by a lighter piece, of reduced diameter, 41 to the pound troy: this is the 'Jacobus peice' to which Newton refers. Charles I, after trying for

a few weeks 44 pieces of 20*s*. to the pound, returned to the 41 pieces to the pound of James I: this is Newton's 'Carolus'. Charles II continued the same weight; after an intermediate rise in value he substituted 20*s*. pieces ('guineas') in 1663 at $44\frac{1}{2}$ to the pound, $129\frac{32}{89}$ grains.

(12) '$33\frac{1}{2}$ Guineas to a pound': the Mint's obligation from 1663 onwards was to get $44\frac{1}{2}$ pieces from the pound troy. This covers the reigns of Charles II (except the first 3 years), James II, and William and Mary.

The number of guineas coined in 1694 was 57,000; in 1695 the number rose to 717,000. In 1696 the number was 138,000, the coinage having been suspended from 2 March to September of that year. See Horsefield, *B.M.E.*, p. 76.

594   FLAMSTEED TO COLSON[1]

10 OCTOBER 1698

Extracts from a copy in the Royal Greenwich Observatory[2]

The Observatory ☽ *Octob*. 10. 98

Sr

...he [my servant James] brought in a report from you, he told me that both you and Collonell Bruce told him that Mr Newton had perfected the Theory of ye Moon *from Mr Halleys Observations* and imparted it to him wth Leave to publish it and that Mr Halley would publish it in a short time. Sr I can scarce beleive that Mr Halley, however Indiscreet, could be the Author of this report; Since he has seen a Synopsis of 152 Observed places of the Moon wth her calculated places, and the Ellements of the Calculation all done by my own hand, and knows I Imparted them wth as many more as made them above 200 to Mr Newton and that their is a very Fair correspondence keept between us for this purpose: and some have been given him very lately. But to clear you wholly and take of all Occasions of your Injureing either Mr Newton or me by Spreading this or the like false storyes for the future, I must Acquaint you that Mr Newton assures me he has not imparted his Lunar Theory to Mr Halley (so that all he knowes of it must be onely collected from discourse he has had wth him) *nor made use of one of Mr Halleys Observations in rectyfying of it....*[3]

I would not Injure Mr Halley either wth Mr Neuton (on whom I know he has dependance) nor the Collonel (by whom he may make some advantage): therefore when I found Mr Newton concerned at the report (wch I gave in near as few words as I have wrote it) I added no more, but that I wondred why or by whom it should be spread. Nor would I write [to the] Collonel, nor had to you but that I find my servant discorsed to you on his own head and omitted w[ha]t I cheafly injoyned him wch I have marked before wth a Line underneath...

284

Mr Newtons Theory when perfected must needs agree wth my Observations since tis built, as he freely owns[,] upon them and his doctrine of Gravitation: and the one wthout the other will not doe the buissness; but both togeather will, as he says himself. Mr Halleys could be of no use to him, because he used the Tychonic places of the Fixed Starrs to rectifye and state the Moons by: the Hevelian were not extant; and had they been published, they were got wth Plain Sights.

I shall make a new Table of Refractions for the Collonell, in a day or two or a True Theory. I have not leave to impart Mr Newtons; I beleive he will see cause to wth draw it. I intend to be in London, god willing, on fryday next but the days being short I shall not have leasure to See the Collonel at your house. I shall be at Garways⁽⁴⁾ betwixt one and two If you come down hither in the mean time let it not be on Wednesday for I have company that day: at any other you shall be wellcome to

Your Friend & Servant

To Mr Colson Teacher of the   J.F. MR.
Mathematickes⁽⁵⁾ at his house
in Goodmans Feilds   London⁽⁶⁾

NOTES

(1) John Colson (1680/1–1760), English mathematician; entered Christ Church, Oxford, 1699. Elected F.R.S. in 1713, and became Lucasian Professor of Mathematics in 1739. He was Vicar of Chalk, Kent, from 1724 to 1740, and later was Rector of Lockington, Yorkshire. He published many mathematical treatises and translations, notably *The Method of Fluxions and Infinite Series: with its Application to the Geometry of Curve-Lines. By the Inventor Sir Isaac Newton, Kt.... Translated from the Author's Latin Original not yet made publick.* By John Colson, M.A. and F.R.S. (1736).

(2) Vol. 33, fos. 11–12.

(3) See the assurance given by Newton (Letter 480): 'If you can have a little patience wth me till I have satisfied my self about these things & made the Theory fit to be communicated wthout danger of error I do intend that you shall be the first man to whom I will communicate it'; also (Letter 494): 'As for your Observations you know I cannot communicate them to any body & much less publish them wthout your consent.'

(4) Garways, or Garraways, a noted coffee-house in Change Alley, Cornhill, named after Thomas Garway, who first sold tea there, which he recommended for disorders of all kinds. It was one of the leading auction rooms in the city. See H. B. Wheatley, *London Past and Present*, **2**, 85–6, where it is described as 'a celebrated place for sandwiches, sherry, pale ale, and punch'. It was rebuilt after the Great Fire (1666) and was finally closed down on 11 August 1866. It is frequently mentioned in the Hooke *Diary*.

(5) Colson became the first master of Sir Joseph Williamson's Mathematical School at Rochester.

(6) Only the address is in Flamsteed's hand.

595   FLAMSTEED TO NEWTON
10 OCTOBER 1698
From a draft in Corpus Christi College Library, Oxford[1]

To Mr Newton Octob 10. 1698

Sr

The right ascensions of the $\mathbb{)}$s center which my [servant] being diligent to show you sent you in my absence are all too small by different quantitys betwixt 3 & 5 minutes by reason of a small æquation he forgott to apply in them, & which I had applyed in those I got when you were here. I am busy about other things at present bear with me one forthnight & they shall be repeated & corrected & sent you by Sr

Yrs.

NOTE

(1) It is not known if the letter, of which the above is a draft, was ever sent. It was written on the reverse of a letter from Wallis to Flamsteed (Letter 592).

596   WALLIS TO FLAMSTEED
8 NOVEMBER 1698
From the original in the Corpus Christi College Library, Oxford.[1]
In continuation of Letter 592

Oxford Nov. 8. 1698

Sr

I am now entering upon the Printing of some *Latine* Letters, in an Appendix to a Third Volume of mine, which hath been now in the Press for some years: Amongst which I would be glad of yours concerning the Parallaxis of the Pole-star.[2] Mr Caswell tells me, it will contain your Observations thereof for seven years. I desire I may have it by the end of this Month; that it may not come too late. For we are now drawing to an End. I think you need not encumber it with those particulars which relate to the *Rectifying of your Instrument*; but give us the Observations as they would appear, supposing the Instrument so rectifyed. It will, I think, be not to your disadvantage to have it there. And it will be to the Reputation of our Nation, to be the first that have been able to make-out that Parallax. I am

Sir,

Yours to serve you,

JOHN WALLIS.

For Mr John Flamsteed,
at the Observatory,
    in
        Greenwich.

NOTES

(1) C. 361, no. 133. Through his eagerness to obtain information regarding Flamsteed's 'discovery' of the parallax, Wallis unwittingly became the cause of an aggravation of the ill-feeling which had been growing between Newton and Flamsteed for some time. Flamsteed eventually sent the required information, and in the course of his letter he referred, perhaps incautiously, to the observations with which he had supplied Newton. Wallis acknowledged Flamsteed's letter on 10 December (Letter 597), and finding 'nothing of it but what is fit to be published' passed it on to David Gregory, who in turn acquainted Newton with its contents. Newton took the strongest exception to the use of his name, and at once wrote to Gregory directing him to request Wallis to suppress the paragraph containing the reference to himself. Wallis, having learned of the attitude adopted by Newton, wrote to Flamsteed on 28 December (Letter 598) suggesting a modification of the offending paragraph. Meanwhile, on his return to Greenwich, having been absent for a few days, Flamsteed wrote a long letter to Newton on 2 January (Letter 600) defending his action, and asking Newton to state whether he had given such orders to Gregory; and, not receiving an answer at once, wrote to Wallis on 7 January (Letter 602) requesting him to let the paragraph stand. Very soon afterwards Newton's letter, dated 6 January (Letter 601), did arrive, having been delayed, whereupon Flamsteed again wrote to Wallis on 10 January (Letter 605) desiring him to 'alter ye Offensive Innocent Paragraph as you intimated'.

It is to Flamsteed's credit that notwithstanding these differences he continued to visit Newton when he was in London and to procure for him such observations as he still required.

(2) See Letter 590, note (2), p. 279, and Letter 592, note (2), p. 281.

597   WALLIS TO FLAMSTEED

10 DECEMBER 1698

From the original in the Corpus Christi College Library, Oxford.[1]
In continuation of Letters 592 and 596

Oxford Dec. 10. 1698.

Sir,

I have finished the Translation of your letter into Latin,[2] some while since. I find nothing of it but what is fit to be published, & therefore leave out nothing of it. I sent you the two first sheets of it by Dr Gregory,[3] who sayd he should

287

see you, & would give them to you. If you desire I should send you ye third
sheet, it shal be done. I am well satisfied with it; And think it proper, that ye
English Letter be published in the *Transactions*, (for it well deserves a place
there.) And to that end I shall return you the English Letter when you give
me order so to do. But, I would advise that it be printed at Oxford (as in
such cases is sometimes done) that I may see to its being correctly printed.
For I find the Correctors of ye Press in London, are apt to mis-take in things
Mathematical. Mean while I desire you will look carefully to the numbers,
that they be truely written, & then return me the two sheets I have sent you.
Particularly in the 10th Collation (if I do not mis-remember) where, I think,
you subtract 5″ instead of Adding it.

I am

Yours to serve you

JOHN WALLIS.

For Mr John Flamsteed
at the Observatory in
      Greenwich.

NOTES

(1) C. 361, no. 135.

(2) Wallis had requested Flamsteed 'to draw up such a letter (in Latine)'; Flamsteed,
however, writing in great haste, had written in English and had forwarded the letter to Wallis
some time in November. Wallis's translation of the letter into Latin was published in the third
volume of his *Opera* (p. 701). See Letter 592, note (2), p. 281.

(3) Gregory evidently communicated the contents of this letter to Newton; Flamsteed
maintained that by so doing he had rendered a great disservice, for it was his 'officious
flattery' that was the cause of Newton's violent letter of 6 January 1698/9 (Letter 601).

598   WALLIS TO FLAMSTEED

28 DECEMBER 1698

From the original in the Corpus Christi College Library, Oxford.[1]
For answer see Letters 602 and 605

Oxford Dec. 28. 1698.

Sir,

I received on Munday last, Dec. 26. your letter of Dec. 24. and the paquet
of papers directed to me, of which I shal take care. And at the same time I
received another letter from one in London,[2] which desires me not to print

any Paragraph of your letter which speaks of your giving to Mr Newton Observations of the Moon. He is a friend to both of you; but he doth not give me his Reasons why. I thought best to acquaint you with it, & desire your advise upon it. If you order me to leave out that Paragraph (& the next which follows, about your two friends in the North, of a like nature:) instead thereof, I was thinking (if you like it) to put this, [Aliaque intervenerunt negotia (quæ nunc narrare nec est opus) quibus distinebar ne potuerim isti rei vacare: Atque (inter alia) ut amicis aliquot, id flagitantibus, exhiberem Planetarum (Saturni, Jovis, et Lunæ) loca plurima, ex Observatis meis Calculo deprompta. Quæ moneo ne putes negligentiæ Desidiæve meæ imputandum, quod non citius huic operi me accinxerim.][3] which I think may serve your turn or wish.

. . .   .

I doe not think of making any other considerable variation from your copy as it is now sent (otherwise than as to prepare it for the Printer;) but (because you omit the time in the Date) I shall supply Dec. 2. 1698,[4] or what other time you shall appoint. I am

Sr

Yours to serve you

JOHN WALLIS.

For Mr John Flamsteed
at the Observatory in
     Greenwich.

[On the cover of this letter, Flamsteed has added]

You say Dr Gregory is a freind to both of us. I much doubt it. had hee been my freind he would have sent me word that paragraph would displease Mr Newton. a letter would have come hither and an answer have gone back in allmost as little time as one goes from London to Oxford. tis much to be suspected he is onely Mr Newtons freind for Mr Montagues sake. since his Countrymen gave out formerly that he had found abundant errors in the *Princip.* Now that Mr Newton gave them to him. to deale plainely with you his freinds resort to Hindmarshes[5] Shop in Cornehill & Who they are you may easily be informed even at Oxford.

    That I was at London on the ♀day[6] they arrived and the following. that I wrote to Mr [N]ewton on monday & sent him then account of What Dr Gregory wrote to Dr Wallis as also the paragraph of my letter which Dr Gregory would suppress. yt receaveing no Answer by thursday morneing I then wrote to him againe for one. that since he takes no notice of my letters

I conclude I need take no notice of Dr Gregorys. nor you neither & therefore thinke you need not alter yt paragraph at all.

Dr Gregory is a freind of Mr Halleys tho he was his competitor[7] but I perceive by this transaction he is no freind of mine tho I shewd him more freindship then he could reasonably expect on yt occasion & Mr Halley as much enmity but he thinkes Mr Halley has an Interest in Mr Newton & therefore is become his freind, & takes the same courses Hawley [*sic*] did to Ingratiate with him whose favour may be of use to him wth Mr Montague[8]

NOTES

(1) C. 361, no. 137.

(2) David Gregory.

(3) 'Other matters cropped up (which now there is no need to explain) which hindered me from being able to devote myself to this matter, and one of these matters which intervened was that I should show a number of friends who were pressing me to do so, several places of the planets, Saturn, Jupiter, and the Moon, obtained by calculation from my observations. And I beg you not to attribute it to negligence or inactivity on my part that I have not more speedily devoted myself to this work.' The square brackets are Wallis's.

(4) In Wallis' *Opera* the date is given as December 20 1698. See Letter 592, note (2), p. 281.

(5) Joseph Hindmarsh was active as a publisher, opposite the Royal Exchange in Cornhill, between 1678 and 1696, after which the business was conducted by H. Hindmarsh, possibly his brother. From *The Term Catalogues* (ed. E. Arber, 1906) Joseph appears to have been a prominent Tory and a high Anglican. See E. G. R. Taylor, *Mathematical Practitioners*, p. 279.

(6) See Letter 600.

(7) Halley and Gregory were candidates for the Savilian professorship of astronomy in 1691 when Gregory was appointed.

(8) See Letter 545, note (1), p. 195.

599   FLAMSTEED TO NEWTON
29 DECEMBER 1698
From the original in the Royal Greenwich Observatory[1]

The Observatory Dec 29. 1698

Sr

I have examined the times of the determined R. Ascent & D a P. of the ☽s limbes which I gave you when you were last here,[2] & find them all just save ye 2d. April 25 1695 which make 18h.38'.07" & they are all fit for use

In a letter of yours of April 23. 1695[3] you gave me a New Table of the æquations of the Apoge wherein you made

Ye greatest æquation near	$12°.10\frac{1}{2}'$
Ye greatest Excentricity	66850
least	43566
The greatest horizontall parallax of ye ☽	$61'.37''$
least	55 .05
The diminution of this parallax in ye ☐ in ye	
least dist	46
in ye greatest	53

In ye same letter you direct to Add $12'.00''$ to my place of ye Apoge
In some others you make ye greatest Physicall partes $13'$.
The meane Variation $35\frac{1}{2}$
You add $2'$. nearly to the ☽s Meane Motions which I find the Observations in ye Synopses require

I give you this breife of your Communications already made least you should have forgot what I had in my hands already & put your self to a needless trouble of Causeing them to be copied over againe If you have made any alterations in these or any additions to the Theory yt will make ye Numbers answer my observations better you will oblige me if at your leasure you will please to impart them to Sr

Your affectionate & humble servant

JOHN FLAMSTEED MR

PS

In your letter you say these corrections will Answer all my observations within 10 minutes Mr Halley boasts that those you have given him will represent them within 2 or 3 or Nearer

I wish you many happy yeares. J F

To Mr Isaack Newton Warden
of ye Mint at his house in Jermin
Street near St Jamess
            London.

NOTES

(1) Vol. 33, fo. 12. There is a copy in the University Library, Cambridge (Add. 3979, no. 38).

(2) Flamsteed and Newton had met several times during December. In particular Newton visited Flamsteed on 4 December to obtain from him twelve computed places of the Moon on

dates which he had previously specified. On Flamsteed's instructions, James Hodgson (Flamsteed's assistant) had calculated these places during Flamsteed's absence in Derbyshire and had sent them on 8 September 1698. Flamsteed, on examining these on 11 November, 'found them all false' and computed them afresh. The amended results were communicated to Newton on the above date (4 December). A further correction was sent with this letter. In the margin of the letter quoted above Flamsteed has written:

'Mr Newton came to see me ⊙ December ye 4th in the time of evening service I imparted too him the Right Ascentions and distances from ye Pole of the ☽s Limbes Recalculated in my 5th book of Calculations pag 184, but not haveing examined the times I told him I could not assure they were true stated by my servant James. haveing since examined them I wrote this letter on yt occasion & sent it Dec 29th.

'Since this was wrote I find 45″ ought to be added to ye distances of ye ☽s limbs from the Pole which I was not then aware of: I acquainted him there was a further fault in ym. When I was last wth him [4 December], he is reserved to me contrary to his promise I lie under no obligation to be open to him.'

By this time Newton was aware of Flamsteed's mention of his name in his letter to Wallis (Letter 602), and that may explain his 'reserve'.

(3) Letter 500.

600   FLAMSTEED TO NEWTON

2 JANUARY 1698/9

From the original in the University Library, Cambridge.[1]
For answer see Letter 601

The Observatory Jan. 2d. 1698/9

Sr

I was in your neighbourhood on Saturday last[2] but thought it too early to disturbe you with a visit, when I had nothing to offer you but my respects, & ye usuall wishes of many happy yeares,[3] which you should accept by this: I had not troubled you now but that on my returne I receaved a letter from Dr Wallis[4] in which he writes thus. *I have receaved your packet,* (yt is my letter concerneing ye parallax of ye Pole Star) *and at the same time I received another letter from one in London*[5] *which desires me not to print any paragraph of ye letter which speakes of your giveing* <u>Mr Newton observations of ye moon.</u> *he is a freind to both of you, but he does not give his reasons why, I thought best to acquaint you with it, & desire your advice upon it* Sr. I wrote my letter to Dr Wallis on this subject in great hast, & when I had much other business on my hands, in November last; & to silence some busy people yt are allwayes askeing, *why I did not print?* I tooke occasion to let them know, that since ye yeare 1689 when I was first fitted for it I have been layeing in a stock of observations to rectifie the places of ye fixed stars; that in 1694 I rectified my solar tables, & layd a foundation for ye rectification of the fixed stars, yt in 1695 I furnished you with 150 observed places of ye ☽, with ye places also calculated from my Tables, in order to the Correction & restitution of her Theory; that I had Tables for abridging ye labor (usually imployd in calculateing ye stars places from my *data*) under my hands; & others to make ye Catalogue more usefull, & I wrote my letter in English. & the good Dr haveing promised me a weeks worke as a recompense for my paines I sent him word I would excuse yt, if he would save me the labor of putting it into Latin: It was then but 3 sheetes which (he accepting ye condition) I sent him & thereby gained time to copy six moneths observations from my first books & furnish my country calculators wth ye R Ascentions &c of ye stars in the Southerne constellations, to calculate their Longs & latitudes from: In a forthnights time I receaved 2 of the 3 sheetes from ye Dr. loose in a Wrapper from Dr Gregory, with directions to Leave them when perused (for him to returne) at Mr. Hindmarshes,[6] a booksellers shop, *where ye non Jurats*[7] *resort*, in Corn hill, the third sheet soon followed but on perusall of them I found it was requisite to adde almost another, to explaine some places, where I had been too short, or where the Dr, not haveing thoroughly understood my meaneing, (by reason he had not seen my Instrumts nor was acquainted with my methods) had not exprest it as I would my selfe: This took me up more time then I expected, which made me to send my packet by the post least Dr Gregory should not convey it so soon to the Dr [Wallis] as I desired. however I gave Dr Gregory notice that I had returnd it, & he was as diligent to Write to Dr Wallis as above, for what occasion I know not. I shall give you the Whole paragraph. Wherein I have mentiond my accommodateing you with Materialls, & I assure you, I have not mentiond you on that account any where besides in my letter, onely [in] ye booke I have where I shew yt if wee allow ye Nutation. this Parallax must be greater as much as it is. My words are these:– "Contraxeram etiam cum Do Newtono (doctissimo tunc temporis in Academia Cantabridgiensi Mat[hes]eos Professore) necessitudinem, cui Lunæ loca ab observationibus meis ante habitis, deducta 150 dederam, cum locis simul e tabulis meis ad earum tempora supputatis, tum similia in posterum prout assequerer promiseram, cum elementis calculi mei, in ordine ad emendationem Theoriæ lunaris Horroccianæ qua in re spero eum successus consecuturum expectationi suæ pares".[8]

Sr. This is the paragraph & *all of it* I think there is not neare so much in it as you acknowledg to my selfe, & (I have heard from other worthy Gentlemen) you have acknowledged to them & therefore I cannot thinke that it was from any intimation of yours (tho hee says it would be *displeaseing to you if it were printed*) but out of a designe, to ingratiate with you, that he puts an arrest upon this paragraph. I thinke ye Word *Horroccianæ* may be omitted. tho I put it in because you allow yt Theory as far as it goes, you found the faults of it by ye differences from my observations. Hee was our Countryman. & tho your Theory will be new in that (tho you give us ye reasons & derive it from Naturall Causes) yet he gave ye groundplot & it will be an honor both to you & me to doe him justice.

Sr My observations lie ye King & Nation in at least 5000 *lib*, I have spent above 1000 *lib*. out of my own pocket in building Instrumts & hireing a servant to assist me now neare 24 yeares; tis time for me (& I am now ready for it) to let the World see I have done something that may answer this expence. & therefore I hope you will not envy me the honor of haveing said I have been usefull to you in your attempts to restore the Theory of the Moon I might have added the Observations of the Comets places given you formerly of the superior planets & refractions at ye same time wth ye ☽s. but this I thought would look like boasting & therefore forbore it.

I desire you would please to let me know by a line whether Dr Gregory ever shewed you my letter. I meane Dr Wallis his translation of it, which I thinke I have altered in ye paragraph above from what it was. but can not say in what words because I returnd ye Dr [Wallis] his Copy, with my transcript of it enlarged & altered; to geather but whenever tis printed you will find it agree wth ye copy above exactly.

Sr. I am told Dr Gregory is s[c]hem[in]g to be Tutor in Mathematicks to the Duke of Gloscester.[9] *Which place I was told some Moneths agone* (when ye setling of his household was first discourst of) *was designed for me.* To make a Variance betwixt you & me & Dr Wallis & to engage you to procure him the favor of Mr Montague[10] I am apt to beleive he concernes himself in this business. he thinkes perhaps it will depreciate me & keep me from being his Competitor. let him not trouble himselfe. I have an Interest much beyond his when ever I please to move yt way. but I doe not thinke the Duke yet fit for a Mathematicall tutor or that he will be this four or five yeares I hate flattery & shall not goe to Court on this account till I am sent for. or have notice yt I am desired That place indeed might afford me ye opportunity of procureing help for my assistance for I could defray ye charges out of pay. but I feare It would be as prejudiciall to me other ways & therefore shall not move to traverse the Drs designe, except he force me to it by his *treacherous behaviour.*

Sr I beg an Answer to this letter speedily.[11] & you need tell me no more but that you have seen the paragraph before writen or not seen it, That you gave such orders to Dr Gregory or not that I may return an Answer to Dr Wallis. and hereafter if any such flatterers as he come to say any thinge to you that may tend to make a difference betwixt us pray tell ym you will informe me &

you will forthwith be rid of them: I shall allwayes use the same course towards you whereby our freindship that began early may continue long & be happy to boath of us which through Gods blessing I hope it may at least I shall allwayes endeavor it, being ever Sr

Your most affectionate freind &

humble servant

JOHN FLAMSTEED MR

Pray enquire what company Dr Gregory keeps yt you my not be deceaved in his Character. our Scotch thinke to carry all before ym by the helpe of ye Bp of Salisbury.(12) whom I esteeme (next ye Bp of Woster)(13) above ye rest of our Clergy but I cannot thinke him wise in placing his Countrymen about ye young Duke.

To Mr Isaack Newton.
Warden of the Mint
at his house in German
Street Near St James's
London these
present

NOTES

(1) Add. 3979.40. The words in italics were underlined by Flamsteed.

(2) 31 December 1698. In the correspondence which follows, dates are parti cularly im portant.

(3) Either a New Year's Day greeting or a reference to Newton's 56th birthday, 25 December.

(4) Letter 598.

(5) David Gregory.

(6) See Letter 598, note (5), p. 290.

(7) The non-jurors were those clergy of the Church of England who, adhering to the belief in the divine right of kings, refused to take the oath of allegiance to William III and Mary after the Revolution of 1688. Archbishop Sancroft was the principal of them.

(8) 'I had become closely associated with Mr Newton, at that time the learned Professor of Mathematics at the University of Cambridge, to whom I had given 150 places of the Moon, deduced from my observations, previously made, and at the time of these observations, her places as computed from my tables, and I had promised him similar ones for the future as I obtained them, together with the elements of my calculation in due order, for the improvement of the Horroccian theory of the Moon, in which matter I hope he will have the success comparable with his expectations.'

295

(9) William, Duke of Gloucester, son of Queen Anne. He died at the end of July 1700, shortly after reaching his eleventh birthday. See Letter 577, note (2), p. 253.

(10) At this time Montague was First Commissioner of the Treasury. See Letter 545, note (1), p. 195.

(11) Newton replied on 6 January following (Letter 601).

(12) Gilbert Burnet (1643–1715) succeeded Seth Ward as Bishop of Salisbury in 1689; of Scottish origin. F.R.S. 1663 (apparently never admitted). Professor of Theology at Glasgow, 1669–74. He was appointed preceptor to the Duke of Gloucester, 1698. Author of *The History of the Reformation* (3 vols. 1678–1714); *History of my own Time*. See T. E. S. Clarke and H. C. Foxcroft, *Life of Gilbert Burnet Bishop of Salisbury* (1907).

(13) Edward Stillingfleet (1635–99) became Dean of St Paul's (1678) and Bishop of Worcester (1689). See *D.N.B.*

601   NEWTON TO FLAMSTEED

6 JANUARY 1698/9

From the original in the Corpus Christi College Library, Oxford.[1]
In reply to Letter 600; for answer see Letter 604

Jermin street Jan. 6. 1698/9

Sr

Upon hearing occasionally that you had sent a letter to Dr Wallis about ye Parallax of ye fixt starrs to be printed & that you had mentioned me therein with respect to ye Theory of ye Moon[2] I was concerned to be publickly brought upon ye stage about what perhaps will never be fitted for ye publick & thereby the world* *put into an expectation of what perhaps they are never like to have.* I do not love to be printed upon every occasion much less to be dunned & teezed by forreigners about Mathematical things or to be thought by our own people to be*² *trifling* away my time about them when I should be about ye Kings business.[3] And therefore I desired Dr Gregory to write to Dr Wallis against printing that clause wch related to that Theory & mentioned me about it.*³ You may let the world know if you please how well you are stored wth observations of all sorts & what calculations you have made towards rectifying the Theories of ye heavenly motions: But there may*⁴ *be cases* wherein your friends should not be published without their leave. And therefore I hope you will so order the matter that I may not on this occasion be brought upon the stage. I am

Your humble servant

Is. NEWTON

[Added by Flamsteed]

* When Mr Halley boast tis done & given him as a secret tells ye Society so & forreigners see Mr Colsons letter[4] to me.

2* Was Mr Newton a trifler when he read Mathematicks for a sallery at Cambridge. Surely ye Astronomy is of some good use tho his place be more beneficiall

*3 I know what I have to doe without telling.

*4 where persons thinke too well of themselves to acknowledge they are beholden to those who have furnisht them with ye feathers they pride themselves in when they have great fr[iends] &c.

NOTES

(1) C. 361, no. 97. It is not easy to understand Newton's attitude as revealed in this letter. We learn from Flamsteed's *Diary* (R.G.O. vol. 35) that Flamsteed frequently visited Newton at the Mint, and also at his house in Jermyn Street whither he had moved in August 1696. Already there were signs that the rift between the two men was gradually becoming wider. On Sunday evening, 4 December 1698 (see Flamsteed's letter to Newton, Letter 599, note (2), p. 291), Newton called upon Flamsteed at the Observatory in order to obtain twelve more computed places of the Moon which he needed to establish his lunar theory, and which were obtainable from no other source. Wallis, it should be noted, having repeatedly asked Flamsteed for particulars of his 'discovery' of the parallax of the pole star (see his letters 13 and 26 August, and 8 November 1698), had found nothing in Flamsteed's reply to which objection might be raised. Wallis returned the first two sheets of his translation of Flamsteed's letter to Flamsteed by David Gregory, through whom the contents of the letter became known to Newton, who lost no time in writing the above letter.

(2) See the passage in Flamsteed's letter to Newton (Letter 600), beginning 'Contraxeram etiam cum Do Newtono'.

(3) The 'King's business' was the management of detectives and informers, the handling of witnesses who were often themselves criminals, and the conviction of counterfeiters. The Mint had little else to occupy the Warden. The great recoinage was complete, and the Warden need only attend on Wednesday and Saturdays when the Mint was actually coining. The pressure on Newton was due to his zeal in carrying out his many extraneous tasks.

(4) The contents of this letter may be surmised from Flamsteed's letter to Colson (Letter 594). See also More, p. 478.

602   FLAMSTEED TO WALLIS
7 JANUARY 1698/9
From the original in the Royal Greenwich Observatory.[1]
In reply to Letter 598

The Observatory.   Jan. 7. 1698/9 ♄

Hond. Sr.

Tho Yours of ye 28th arrived here on fryday was sevennight. Yet being then at London I received it not till my returne home on Saturday night.[2] I wondered to find by it that Dr Gregory should concerne himselfe any farther wth my letter about ye Parallax of ye Pole star then onely to transmitt it, had he been my friend as you suppose him, he might as easily have wrote to me to advise me to alter that Paragraph[3] as to you to suppresse it. The Penny Post comes something sooner here then the Generall Post does to Oxford. I fear you mistake him much: his friends are neither friends to you nor me: they resort commonly to Mr Hindmarches, a Booksellers shop in Cornehill; and who they are you may Learn at Oxford. the truth is,

The Dr is sueing for ye Mathematicall Tutorship to the Young Duke of Gloscester[4] who will not have occasion for a Tutor in Mathematicks this 4 or 5 Years: he knowes I was Named for that Employ when the seteling of his houshold was first discoursed of and that I have an Interest tho I doe not look after it, for reasons not to be recited in this Letter. he hopes to Gain it by his Interest in ye Bp of Salisbury[5] and that Mr Newton may be of good use to him by procureing him the favour of Mr Montague: for this reason he has taken this occasion to ingratiate wth Mr Newton by suggesting I have worte [*sic*] something that may derogate from him but I am apt to beleave that he will rather Injure then help himselfe by this peice of flattery.

For Mr Newton owns not onely the 150 Lunar Observations[6] I fitted him wth to examine whether the moons moc̄ons answerd those he thought she ought to have by ye Laws of Gravitation; but moreover that he has *made use of No Observations but mine in rectifieing of her Theory.* he vindicates me from ye suggestions of those of Dr Gregorys friends and party, and does me Justice when ever any occasion offers so that I am apt to beleave Dr Gregorys letter was his own contrivance wthout the knowledge of Mr Newton who cannot be Offended at the Mention of 150 Observations imparted to him.

Since I have accomodated him wth as many as would make them up 300 togeather wth 100 at least of ye Superior Planets ♄ & ♃ and aboute 100 of refractions: besides my observation of Comets and the Diameters of ye Planets of wch Nothing is said in that Epistle least I should seem to boast.

This I have said to Mr Newton in a letter I wrote to him last monday morning.[7] I expected an Answer on Thursday and none comeing wrote to him then againe to desire him to let me know. whether wt Dr Gregory had wrote to you was by his *direction or not*. and haveing noe returne conclude he thinks not fit to take notice of it or that he is not in Town. I think it concernes not Dr Gregory to have been thus bussy and *that neither you nor I ought to take any more notice of it then Mr Newton does and therfore you may please to let that Paragraph and the next stand as it is* without alteration. but as for wt you think fitt to add at the Begining[8] I am obliged to you for ye Intimation you may add it If you please. I approve it.

Onely I desire you that hereafter you acquaint not Dr Gregory wth any thing that passes betwixt you and me: that so hee may have no oppertunity of makeing friends *against* me at my cost. Mr Caswell is a very honest as well as a very ingenuous[9] person and scornes flattery and Basness. wee have been long friends: you need not be so reserved on his account. You may impart your mind to him in any thing that concernes me: he will write to me and save you ye Labor.

I begg your Pardon for the Length of this Letter. I have only to add that my Observations lie the King and Nation in more then 5000*lib* and my self in 1000*lib*. out of my own Pockett to build my Instruments and hire assistance. I have suffered much in my health by my night Labors the paines I have imployed in Calculations have been unconceavable: all the recompence I expect is acknowledgemt of my Industry; wch those that would deprive me of at the same time are unjust to our Nation as well as injurious to me. The Dr I am apt to think is not so much displeased at that Paragraph, as at the whole Letter; wch he fears may contribute to undeceive some people that had taken up false notions of me from ye misrepresentations of his party; and may doe me to great heed to ye prejudice of his pretensions. I have never yet opposed him: but If he takes these wayes of making friends, he must expect that I shall take Notice of it. be he wt he will I shall ever be wt I ought, that is Sr.

Your Obliged & Most humble

Servt

JOHN FLAMSTEED M R

To ye Reverend Dr Wallis at his house in Oxford

NOTES

(1) Vol. 33, fo. 13. This letter was written by an amanuensis, probably Hodgson. The passages in italics were underlined, most likely by Flamsteed. Only the address, at the foot, is in Flamsteed's hand. See Letter 598.

(2) 31 December 1698.

(3) The paragraph commencing *Contraxeram etiam cum Do Newtono* in Letter 600.

(4) See Letter 600, note (9), p. 296.

(5) Gilbert Burnet (see Letter 600, note (12), p. 296).

(6) See Memoranda 468 and 516.

(7) That is, 2 January. Actually Newton replied on Friday, 6 January; Flamsteed received this letter on the Monday following (9 January), and wrote to Wallis the next day, desiring him to 'alter the offensive innocent paragraph' in the way intimated by Wallis in his letter of 28 December. Meanwhile, Wallis had written to Newton desiring to know his wishes regarding the paragraph.

(8) This was Wallis's suggested alteration. It refers to the passage in his letter of 28 December beginning *Aliaque intervenerunt negotia.*

(9) Flamsteed frequently wrote *ingenuous* when he meant *ingenious.*

603   WALLIS TO NEWTON

9 JANUARY 1698/9

From the original in the University Library, Cambridge[1]

Oxford, Jan. 9. 1698/9.

SIR,

I had lately an intimation from Dr Gregory, as if it were a desire of yours, That (in a letter of Mr. Flamsteads's[2] concerning the parallax of the Earth's Annual Orb, which I am about to print) I would omit a paragraph wherein you are mentioned: It is, wherein, when I had pressed his communicating to me and the publike his observations concerning this matter; he excuseth his delay of gratifying me therein, from his diversions by other Business; & (among the rest) this for one: 'Contraxeram etiam cum Do. Newtono doctissimo tunc temporis in Academia Cantabrigiensi Professore necessitudinem, cui Lunæ loca ab observationibus meis ante habitis deducta 150 dederam, cum locis simul e Tabulis meis ad earum tempora supputatis, tum similia in posterum prout assequerer promiseram, cum elementis calculi mei in ordine ad emendationem Theoriæ Lunaris Horroccianæ qua in re spero eum successus consecuturum harum expectationi suæ pares, hinc [suscepi][3] cum ingeniosis duobus viris (in Septentrionalibus Angliae partibus) ut ipsis impertirem loca duorum superiorum planetarum (Saturni et Jovis) in ordine ad restituendum illorum motus: quam rem ipsi tunc moliebantur.'[4] Upon which I wrote to Mr Flamsted to have his opinion in it; For I was not to alter his letter without his order.

300

I have since an answer from him, to this purpose: That it was (he supposeth) rather a suggestion of Dr. Gregory's own, than any order from you, to have it left out; for that you do readly acknowledge in all companies, that you had from him such Observations; and that upon his now writing to you on this occasion, he hath received no answer from you to the contrary, & therefore he thinks that you do acquiesce, as to the publishing that paragraph with ye rest, & seems yet to adhere to it.

For my part, I am willing to serve you both, and not willing to displease either: And therefore desire you will please to accommodate that busyness between yourselves. And I shall readly comply with what you two agree upon therein. And I desire I may have an account from you, as soon as may be; because ye Press will quickly be ready for it.

> I am,
>
> Sir,
>
> Yours to serve you,
>
> JOHN WALLIS[5]

I do not apprehend any prejudice to you in printing it, (being merely true matter of fact): And it seems of concernment to him to satisfy the world (from this & other things mentioned,) that he is not idle: though he be not yet in a readyness to publish that whole of his Observations (for which he is frequently called upon,) it being a great work.

For the Worshipfull Mr
Isaac Newton, Master of the Mint,
at the Tower.
> London

NOTES

(1) Add. 3977, fo. 18.

(2) Letter 598.

(3) Some such word is missing here.

(4) 'I had also become closely associated with Mr. Newton, at that time the most learned Professor in the University of Cambridge, to whom I had given 150 places of the Moon, deduced from my observations made previously, together with her places computed from my tables at the times of these observations, and I promised him similar ones for the future, according as I obtained them, along with the elements of my calculation, in order for the improvement of the Horroccian theory of the Moon, in which matter I hope he will have all the success comparable with his expectation therefrom. Hence I arranged with two ingenious

men in the northern parts of England that I should communicate to them the places of the two superior planets (Saturn and Jupiter) in due order with a view to re-establishing their motions. This is the problem on which they themselves were at this time working.'

(5) It is not known if Newton replied to this letter.

604   FLAMSTEED TO NEWTON
10 JANUARY 1698/9
From the original in Corpus Christi College Library, Oxford.[1]
In reply to Letter 601

Sr

Yours dated Jermin Street Jan: 6, arived here Last night the 9th, with the Generall post-mark & Charge upon it, as if it had come from some place less yn 80 milles remote from London; I waited for it from ye 2d to ye 7th Instant Saturday night, & then wrote to Dr Wallis, yt *I thought he needed not take any notice of Dr Gregorys Letter to him, to forbear printing that clause in mine wherein I had mentioned you, since you tooke no notice of two of mine* I had wrote to you that weeke conserning it, *wch made me think, you thought it not worth your while to consern your selfe about it*; Now I find *you did desire Dr Gregrey to write soe to him*, I shall write to him my selfe to alter that passage, soe as he first advised, and soe as I belive you will find no just cause of offence in it, my Letter goes to him this night, the altered paragraph you have at the foot of this Letter[2]

I did not think I cou'd have disobliged you, by telling the world know[3] yt the Kings Observatory had firnished you with 150 places of the Moon,[4] derived from Observations here made, and compared with tables; in order to correct her Theory: since (not to seeme to boast) I said nothing of what more it has *furnished you*, freely with, *as I had Leasure* & Mr Halley, has not stuck to tell it abroad, both at ye Society, & else where, *that you had compleated her Theory & given it to him as a Secret.*[5] I cou'd not think you wou'd be unwilling our Nation shou'd have the honour of furnishing you with soe many and good observations for this work, as were not (I speak it without boasting) to be had else where: or that it shou'd be said you were about a new work which others said you had perfected. I thought not it cou'd be any deminution to you, since you pretend not to be an observer your selfe. I thought it might give some people a better notion of what was doeing here then had bin impressed upon them by others whom God forgive. You will pardon me this freedom & excuse me when I tell you if forreigners donn[6] & troubl you tis not my fault but those who think to recommend themselves to you by advanceing ye fame of your works, as much as they possibley can; I have somtimes told some

ingenious men that more time and observations are required to perfect the Theory but I found it was represented as a little peice of detraction *which I hate* & therefore was forced to be silent. I wonder that *hints* shoud drop from your pen, as if you Lookt on my business as *trifling* You thought it not soe surely when you resided at Cambridge it's property is not altered; I think it has produced somthing considerable allready & may doe more if I can but procure help to work up the observations I have under my hands *which is one of the designs of my Letter to Dr Wallis was to move for*, I doubt not but it will be of some use to our ingenious travellers and saylers; and other persons, that come after me, will think their time as Little mispent in these Studyes, as those did that have gon before me. The works of the Eternall Providence I hope will be a little better understood through your Labours & mine, then they were formerly. think me not proud for this expression, I Look on pride as the worst of sins, Humility as the greatest virtues, this makes me excuse small faults in all mankind bear great injurys without resentment & resolve to maintain a reale friendship with ingenious men to assist them what Lies in my power without ye regard of any interest but that of doeing good by obliging them
To Mr Newton Jan: 10 ♂ 1698/9[7]

NOTES

(1) C. 361, no. 99.

(2) See Flamsteed's postscript to Letter 605 for 'the altered paragraph'.

(3) Flamsteed has inserted subsequently, but unnecessarily, the word 'know'.

(4) See Memoranda 468 and 516.

(5) According to a story related by a servant and quoted by More (p. 478), Flamsteed's friend Colson had stated 'that Mr Newton had perfected the theory of the Moon *from Mr Halley's observations*, and had imparted it to him, with leave to publish it; and that Mr Halley would publish it in a short time'.

(6) See Letter 601. 'I do not love to be printed upon every occasion much less to be dunned & teezed by forreigners.'

(7) Only this line is in Flamsteed's hand. The passages in italics are underlined in the original, probably by Flamsteed.

605   FLAMSTEED TO WALLIS
10 JANUARY 1698/9
From the original in the Royal Greenwich Observatory.[1]
In continuation of Letter 602

To ye Revernd Dr Wallis at his house
in Oxford

The Observatory Jan. 10. 1698/9

Hond Sr.

Yesterday I received a very Artificiall Letter[2] from Mr Newton. it had been 3 days in comeing from Jermin Street by ye Generall Post; I am sorry it arrived not sooner, for then it had saved me the Labor of contradicting mine ye Last Saturday.[3] To obleige Dr Gregory, Mr Newton will not approve of that Paragraph,[4] and gives me reasons for it such as they are. I have answered him this Afternoon: but, at the same time, desire you to alter ye Offensive Innocent Paragraph as you intimated;[5] so as you thinke it will give no offence, and as you wrote it in your Last of December ye 28th. I should be glad to know of you (or Mr Caswell by your Order) what forwardness ye Volume of Letters is in, and when wee may expect it extant. I shall send him a Copy both of Mr Newtons letter, and my answer, by ye next post, or next but one. my Servt is absent; so I have no body to transcribe it. I hope now all the trouble Dr Gregorys officious flattery has caused, is over; and that you will remember hereafter not to commit any thing to his hands for me, but it shall be sealed up in a Cover. Excuse me and assure your selfe I am allways

Hond Sr

Your affectionate servant

J FLAMSTEED

I think your Alteration is thus:

aliaque intervenerunt negotia (quæ nunc narrare, nec est opus) quibus distinebar ne potuerim isti rei vacare atque inter alia (ut amicis aliquot id efflagitantibus exhiberem) Planetarum Saturni Jovis Martis loca plurima tum ex Observationibus meis cum e Tabulis calculo deprompta. Quæ moneo ne putes Negligentiæ desidiæve meæ imputandum quod non citius huic operi me accinxerim.[5]

(1) Vol. 33, fo. 15. The letter is in the hand of a copyist. Only the inscription above the date is in the hand of Flamsteed.

(2) Letter 601. According to *O.E.D.*, one meaning of 'artificial', current in the sixteenth century, was 'merely made up, factitious; hence, feigned, fictitious'.

(3) Letter 602.

(4) See the first paragraph of Letter 600.

(5) See Letter 598 for Wallis's suggested alteration, and translation.

6o6   CHALONER TO NEWTON
20–24 JANUARY 1698/9
From a copy in the Royal Mint[1]

William Chaloners Letter to the Warden of the Mint

[Newgate Gaol]

Sr,

I have been close Prisoner 11 weeks[2] and no friend sufferd to come near me but my little child I am not guilty of any Crime, and why am I so strictly confined I do not know I doubt Sr You are greatly displeased with me abo[u]t the late business in Parliamt[3] but if you knew the truth you would not be angry with me for it was brought in by some persons agt my desire
Sr, I presume you are satisfied what ill men Peers and your Holloway[4] are who wrongfully brought me into a great deal of trouble to excuse their Villanys Wherefore I begg you will not continue your displeasure agt me for I have sufferd very much so I wholy throw my self upon your great Goodness I am

Sr

Your most humble & obedient servant

W. CHALONER

NOTES

(1) Royal Mint Depositions, 133. This is the earlier of two letters which Chaloner sent to Newton after his arrest. See note (2) below.

(2) A warrant for the arrest of Chaloner on a suspicion of forging malt tickets was issued on 6 October 1698 (*Calendar of State Papers*, 1698, p. 400), and he was committed to Newgate Prison during the first week in November, say 3–7 November. If the eleven weeks ran from his committal to prison, then the date of the above letter would be between 20 and 24 January

1698/9. In the second letter, which was sent shortly after the above, Chaloner protested that he had not the skill to forge malt tickets. 'Pray Sr consider that these persons that accuse me are those formerly convicted for Crimes. . . . Do not take notions and suggestions for truths. . . . For Gods sake. . . do not let me go murthered out of this world' (*ibid.*).

Malt lottery tickets were a second series of Government Premium Bonds, which, following the success of the Million Pound Lottery of 1693–4, were authorized by Act of Parliament. They carried fixed interest, with additional prizes awarded by periodical lotteries. Both payments were secured by an increase in the duty on malt, hence these bonds were known as Mault Tickets, or Mault Lottery Tickets. They passed from hand to hand and could easily be reproduced by a skilful forger. Chaloner had engraved a copper plate from which one hundred copies were printed. This charge, however, was dropped, and Chaloner was proceeded against on the more serious charge of forging coin of the realm. 'He was committed by Tho: Railton. . . on oath for High Treason in counterfeiting the current coyn' (Newgate Roll in the Middlesex Registry, 21 January 1698, O.S.). It was upon this charge that Chaloner was convicted.

There are seven depositions taken by Newton during February 1698/9, bearing directly on Chaloner. These are mainly from former confederates.

(3) Letter 564, notes (2) and (3), p. 232.

(4) See Memorandum by Newton, No. 581 and note (2), p. 262.

607 MINT TO THE TREASURY
JANUARY/MARCH 1698/9
From the original in the Public Record Office[1]

> To the Rht Honble the Lords Commissrs of his Majtys Treasury
> A Report of the Officers of His Majtys Mint

May it please Your Lordships

In obedience to Your Lordships Reference of the 17th Instant, Wee have considerd the annexed Petition of Daniel Crichlow Cittizen & Button-maker of London; And do find that the Press of Gerrard Bovey[2] was not licensed by Your Lordships as is alleged by Crichlow, but only permitted him by some of the Officers of the Mint, upon his giving security that ye Press should not be used any other ways, but in making of buttons only; and to deliver it back into the Mint upon demand.

That this was done before the late Act of Parliamnt[3] which makes it Treason to have a Press for Coining; and therefore Wee have called it in, conceiving that such a precedent may be dangerous for increasing ye Number of such Presses, til they become so common, as to be lyable to great Abuse

That Your Lordships have Authority by the late Act of Parliament aforesaid to Licence such Presses: But in what cases such Licence may be granted, without danger of abuse; Wee humbly submitt to Your Lordships Wisdom to determine

Tower Mint Office
1698/9.

Is. NEWTON
THO: NEALE
THO: MOLYNEUX[4]

1698

NOTES

(1) T. 1/60, no. 35 *b*. The date given in the *Calendar of Treasury Papers* is '?Jan/Mar 25 1698/9'.

(2) On 4 April 1699, Gerard Bovey, button-maker, petitioned for the restoration to him of a certain engine (invented and long used by himself for the manufacture of buttons) which was detained by the officers of the Mint as an engine that might be used for the coining of money (Sotheby, *Catalogue of the Newton Papers* (1936), p. 107). There is a similar petition by one Jacob Wallis for the restoration of a hand wedge press long since invented and used by him for the manufacture of 'hooks and chains for watches' (*ibid.* p. 108).

(3) The Act stated: 'That all persons, whose professions require such-like tools or engines as may be made use of for coining or clipping, be obliged to register their names and places of abode.... That no presses, such as are used for coining, be in any other place than his majesty's mint' (Ruding, II, 36/7). The Act received the Royal Assent on 3 May 1695.

(4) Thomas Molyneux was joint Comptroller with Mason after Hoare. See Letter 568, note (4), p. 241.

608   CHALONER TO NEWTON

c. 6 MARCH 1698/9

From a copy in the Royal Mint[1]

William Chaloners Letter to Isaac Newton Esq

Most mercifull Sr

I am going to be murtherd allthough perhaps you may think not but tis true I shall be murdered the worst of all murders that is in the face of Justice unless I am rescued by your mercifull hands

Sr pray considr my unprecedented Tryall & That no person swore they ever saw me actually coyn but yt I should own it 2ly that Holloway, Abbott Coffee Hitchcock Peck & Hanwell what they swore were facts done in the City & Surrey which is directly agt Law to be taken before a Jury in Middx yet their evidence was given in charge agt me to the Jury I do not arraign none for the same nor the Court but such a thing tho agt Law has been done agt me and I like to loose my life thereby 3ly that all the evidence (except that of Jurd Carters wife) Swore to 8 years ago yet the Indictmt sets forth only the 24th of Augt last now that was not proved nor could be for I was yn out of Towne not coyning in her house as she falsly swore. 4ly That all the Evidence agt me was perfect Mallice I haveing convicted some and others given evidence agt them 5ly That there is no president of a person being convicted of Coyning without some Circumstances to corroborate the evidences testimony nor indeed can a man be convicted of any other Crime whatever without some Circumstances to prove the fact besides the evidence bare Swearing of it Some body must have lost something to prove the Thiefe Some person robbd to prove the highwayman besides Evidence only High Treason agt the King must have But poor I am Convicted by Malitious evidence without any Circumstances at all 6ly It was hard for me to be taken out of my 5 weeks Sick Bedd the last 3 weeks light headed

So yt I was not provided for a tryall nor in my Senses wn tryd 7ly Wt Mrs Carter swore agt may appear direct mallice I having 3 yeares before Convicted her husband of Forgery and discoverd where he and she were coyning for wch he is now in Newgate But I desire God allmighty may Damne my Soul to eternity if every word was not false that Mrs Carter and her Maid swore agt me abo[u]t coyning and Mault Tickets[2] for I never had any thing to do with her in coyning nor ever intended to be con[c]erned in Mault Tick[e]ts nor ever spoke to her abo[u]t any such thing Mrs Holloway[3] swore false agt me or I desire never to see the Great God and I desire the same if Abbot did not swear false agt me so yt I am murderd O God Allmighty knows I am murderd Therefore I humbly begg your Wor[shi]p will considr these Reasons and yt I am convicted without Precedents and be pleased to speak to Mr Chancellr to save me from being murthered O Dear Sr do this mercifull deed O my offending you has brought this upon me O for Gods sake if not mine Keep me from being murderd O dear Sr no body can save me but you O God my God I shall be murderd unless you save me O I hope God will move your heart with mercy and pitty to do this thing for me I am

Your near murderd humble Servant

W. CHALONER

NOTES

(1) Royal Mint Depositions, 205. The evidence submitted was irrelevant to the indictment. Chaloner was certainly innocent of some of the specific offences with which he was charged, and the evidence for the others was probably perjured. Letters from other prisoners whom Newton employed to spy on Chaloner are to be found in the Mint Depositions under numbers 131, 132 and 134.

It is probable that if this letter had moved Newton, Chaloner's life would have been spared. But Newton, being convinced of the enormity of Chaloner's many crimes, seems to have passed the appeal on to his clerk to be entered in the book of depositions, for there is no reference to it apart from this.

Chaloner was executed on 22 March 1699. Luttrell (IV, 489) reports: 'Yesterday Challoner the coiner was convicted of high treason at the Old Baily, having follow'd that trade for several years, for which he was often in Newgate, and had been hang'd e're now, but that he became evidence against others of his gang, who were executed.'

(2) See Letter 606, note (2), p. 306.

(3) Sarah Holloway, mother of Thomas Holloway; see Memorandum 581, note (2), p. 262.

609 MINT TO THE TREASURY
8 APRIL 1699
From the original in the Royal Mint[1]

To the Rt Honble the Lords Commrs of his Majties Treasury.

May it please your Lordships

We have considered the annexed Petition and believe that Mr Hoare the Petitioners Father did for neare forty yeares with great skill & diligence serve the Crown in the Office of Comptroller of the Mint & during the late Recoynage being infirm with old age imployed two Deputies (his Nephew to inspect the whole Mint & one Mr Wells to inspect the Melting House) & also came sometimes in his own person to the Mint. And we find that in consideration of the extraordinary coynage & for rent of his Garden he was allowed an additional salery after the rate of 300 *lib* per annum till the time of his death. And what he may further deserve to equall him with the other Officers we most humbly submit to your Lordships great judgment & justice.

Mint office Apr. 8
1699

Is. NEWTON.
THO: NEALE:[2]
THO MOLYNEUX

NOTES

(1) Newton MSS. II, 478. This, which is in Newton's hand, is a reply to an application for remuneration for the grandson of James Hoare, who employed, besides the two deputies referred to, a grandson (Henry Hoare) from 8 July 1695 to work in the Mint without pay in order to qualify him for employment later. As the latter did not succeed to any office in the Mint, a belated application was made by his father for his services.

(2) Neale's signature is crossed out.

610   MINT TO THE TREASURY
8 APRIL 1699
From the original in the Royal Mint[1]

To the Rt Honble the Lords Commrs of his
Majts Treasury

May it please your Lordships

We have considered the annexed Memorial & Bill of the Moniers amounting to 8032 *lib*. 13*s*. 6*d* and humbly conceive that the 2083 *lb* demanded for coyning 47566 lwt [pounds weight] of silver at 10½*d* per lwt, and the 733 *lib* lent to carry on the coynage, should be deducted out of that summ, being allowed in other accompts; and that the 510 *lb*. 9*s*. 0*d* demanded for Horses & Carriage, and the 153 *lib*. 8*s*. 0*d* for Smiths' Tools should not be allowed. These summs being deducted there remains 4512 *lib*. 16*s*. 6*d* due for coyning Tools for erecting the new Mints. This recconing the Moniers give in according to a bargain made between them & the Master & Worker & approved by your Lordships for furnishing the severall new Mints wth Tools at certain rates. But notwithstanding this bargain we have prevailed wth them to abate 512 *lb*. 16*s*. 6*d* besides the carriage of the Tools into the country, and after this abatement their Bill comes to 4000 *lib*. And what is further to be done in this matter we most humbly submit to your Lordships great wisdome.[2]

Is. NEWTON
THO: NEALE

Mint Office   Apr 8
        1699

NOTES

(1) Newton MSS. i, 214. This is in Newton's hand.

(2) The settlement should have been the Master's business, not the Warden's. The letter shows Newton to be a shrewd bargainer; the tools were sold to the Moneyers at double the scrap value. See Letters 612 and 613.

611   JAMES GREGORY[1] TO COLIN CAMPBELL[2]
29 MAY 1699
Extract from the original in the University Library, Edinburgh

Edr 29 *May* 1699

...   .

Mr Neuton is lately made a member of the royal Academy of Paris,[3] and has 1500 livres of Pension. and Viviani[4] is made of the Royal Society of London. Mr Neuton is more backward from printing then ever being now the Chief officer of the mint of England. So that we begin to dispair of any of his things, more then what we have, in his lifetime. Mr Hally has gott a ship from the government,[5] in which he has sett sail to goe round the globe on new discoverys, and the rectifying of geography.... Mr Flamsted has rectifyed above 3000 fixed stars: but is so perversly wicked that he will neither publish nor communicat his observations....

I am

Reverend Sir

Your most humble Servant

JAS GREGORY

NOTES

(1) James Gregory, brother to David Gregory, whom he succeeded as Professor of Mathematics in Edinburgh.

(2) The Reverend Colin Campbell (1644–1726) of Achnaba (Argyllshire) is described by C. P. Finlayson (*Edinb. Med. Journal*, 1953, p. 52) as being of 'a self-effacing, other-worldly disposition, which found its main outlet in mathematical speculation'. See *D.N.B.*

(3) Newton was one of the eight foreign associates elected to the remodelled French Academy of Science on 21 February 1698/9.

(4) Vincenzo Viviani (1621–1703). He was elected F.R.S. on 29 April 1696. See Letter 561, note (7), p. 228, and vol. II, p. 100, note (5).

(5) The *Paramour*, which sailed in November 1698. See Letter 542, note (1), p. 190.

612   MINT TO THE TREASURY
14 JUNE 1699
From the copy in the Royal Mint.[1]
For answer see Letter 613

To the Rt Honble the Lords Comrs of his Maties Treasury
May it please your Lordships

In Obedience to your Lordships directions wee have Ordred the Corporaċon of the Monyrs to weigh and Value the Tooles and Utensils that were made on Acct of the late great Coynage both for the mint in the Tower and the Mints in the Country and finde by their report that all the cast Iron Weights 9 Tun: 17 lb: 2 qr: 4 oz and all the wrought Iron: 9 Tun: 9 lb: 0 qr: 0 oz: the value whereof is Inconsiderable being 23*lib* 0*s*: for the Cast Iron at one farthing per pound weight and 88*lib*: 4*s* for the wrought Iron at one penny per pound weight But the Monyers are willing to allow double that Value for them rather than they should be demolished and will lay up soe many of them as will keep wthout spoiling and are not of Dangerous consequence to be kept that his Matie may have the Benifitt of them if Occasion should require and for support of their owne Tooles yt They are Obleidged to Maintain to ye King All wch is most humbly submitted &c

June 14 1699

IS NEWTON
THO: NEALE
THO MOLYNEUX

NOTE

(1) Royal Mint Record Books, VI, 65. The Moneyers were prepared to buy the tools used in the recoinage at double the scrap value. The Treasury approved a figure of £500. See Letter 613.

613   THE TREASURY TO MINT
21 JUNE 1699
From the copy in the Royal Mint.[1]
In reply to Letter 612

Gentlemen

The Lords Comrs of his Maties Treasury haveing considered your report relating to the Tooles & Utensils to be returned to the Corporation of Monyrs

Doe direct you to abate for them upon payment of their Bill the sume of Five hundred pounds in Consideration of the said Tooles & Utensils being returned to them

I am

Gentlemen

Your most humble Servt

Treary Chambers   WM LOWNDES[2]
21 *June* 1699

Officers Mint

NOTES

(1) Royal Mint Record Books, VI, 65.

(2) William Lowndes, Secretary to the Treasury. See Letter 571, note (3), p. 248.

614 MEMORANDUM BY NEWTON

3 AUGUST 1699

From the original in the Royal Mint[1]

Tower Gold in the Pix[2] Aug 3. 1699

Guineas in ye Pix 882½. The whole weight 236 oz. 18dwt. 5gr, making in ye pound weight 44½ ⅕ or 44 *lib*. 14*s*. 0*d*. Told out to the fire 1 five pound piece, 35 Guineas & 9 half Guineas amounting to 44½ Guineas. Weight to ye fire 11oz. 18dwt. 23gr
from the fire 11oz. 18dwt. 15gr. wast 8gr.
Upon ye assay it proved standard.

Tower silver in the Pix.

The tale thereof 348*l*. 2*s*. 8*d* The weight 111lwt. 10oz. 3dwt. making by the pound weight. 3*lib*. 2*s*. 3*d*. Better ob[3] wt
  Bristoll made 3*lib* 2*s* 2*d* Exeter 3. 2. 3.
  Chester 3. 2. 3 York 3. 2. 3.
  Norwich 3*lib*. 2*s*. 6*d* in ye pound weight.

All of them standard except Bristoll wch was ob wt worse.

313

NOTES

(1) Newton MSS. I, 239.

(2) For Newton's account of the Trial of the Pyx, see Manuscript 639; also Craig, *Mint*, pp. 394–407.

(3) One pound troy of standard (sterling) silver contained:

$$
\begin{array}{lll}
\text{silver} & 11\text{oz} & 2\text{dwt} \\
\text{copper} & \underline{18} & \text{,,} \\
& 12\text{,,} & 0\text{,,}
\end{array}
$$

'Better ob' means 11oz 2½dwt of silver per lb; 'ob weight worse' means 11oz 1½dwt of silver; 'ob' is an abbreviated form of *obol*, an ancient Greek silver coin.

615 NEWTON TO HADDOCK[1]

12 AUGUST 1699

From the original in Trinity College Library, Cambridge[2]

Mint Office Aug. 12 99

Sr Richard

I received just now a letter from Mr Burden[3] your Secretary whereby I am desired to meet you & Sr Cloudsly Shovell[4] for examining a proposal of Monsr Burden about finding the Longitude[5] & am ready to wait upon you when & where you shall please to appoint. I am

Your most humble Servant

Is. NEWTON

For the Rt Honble Sr Richard Haddock
Ld Commr of the Admiralty

[Added at the foot of the letter, presumably by Haddock]

Saty. Morning. answd. 18th. d: to ye Mint

NOTES

(1) Sir Richard Haddock (1629–1715), knighted, 1675; he took part in several actions from 1666 to 1672; Admiral and joint Commander-in-Chief, 1690; later, Comptroller of the Navy, 1690–1715. See *D.N.B.*

(2) R. 16.38, fo. 436.

(3) Josiah Burchett who served as Secretary of the Admiralty from 24 June 1681 to 20 May 1702, and who is frequently mentioned in the *Calender of Treasury Papers*, 1697, etc. Newton

first wrote Burdett, then scored it out and substituted 'Burden'. It is possible that Burden was the name of an assistant, but it seems more likely that Newton made a slip here. See *Bulletin of Institute of Historical Research*, XIV (1936–7), 169 and 183; also *D.N.B.*

(4) Sir Cloudsley, or Clowdesley, Shovel (1650–1707). He is one of the greatest English Admirals and he took part in many successful naval operations. See Trevelyan, *Blenheim* and *D.N.B.*

In 1707 there was published:

Neptune's Rival; or Great Britains Hero:
BEING

A Full and True ACCOUNT of the Life and Glorious Actions, and much lamented Death, of Sir Cloudsly Shovell, Vice-Admiral of GREAT BRITAIN, who was unfortunately cast away on the Rocks of Scilly, the 22nd of October at Eight a Clock at Night in Her Majesty's Shipp the Association; together with an ELEGY upon him.

(5) The loss of the *Association* (see note (4) above) was undoubtedly due to unsatisfactory methods of determining the longitude at sea, and the loss of this ship focused attention upon the important problem of rectifying this matter. Four days after the date of the above letter Newton exhibited at the Royal Society an improved form of his sextant: 'Mr Newton shewed a new instrumt contrived by him, for observing the Morn-stars, the longitude at Sea, being ye old instrumt mended of some faults, with which notwithstanding Mr Hally had found ye Longitude, better yn ye Seamen by other methods' (Journal Books of the Royal Society of London, x, 145).

For reference to the necessity of providing methods of determining the longitude at sea, see Letter 544, note (1), p 194, and Brewster, ii, 257.

616   HOCKETT[1] TO NEWTON
14 SEPTEMBER 1699
From the original in the Stanford University Library, California

Revd Sr.

Your poor servant, cloisterd in a cell, as unknown to the Great, as he is unfitt to converse with Them, tho (as Mr Hopkins[2]: of Trin: Coll: assures him) happy in the favor of the Seniority, wants yet an interest in the Master: not so much perhaps in regard to the disposition of his Patron, as with Respect to the Quality of His Person, wch makes courtship a debt. That he might therefore neither presumtuously depend, nor unmannerly approach Him: he hath sollicited my Ld: Bp: of St Asaph,[3] that by Your Rev[ere]nces mediation, he might be commended to His favor: This was promisd him, but lest a pressure of private troubles, heartily deplored, should have weighd down such a weak engagemt, he humbly begs Your generous and candid interpretation,

315

that You will accept the intendment of his Lordships intercession, as if He had actually made it. Upon this foundation, but especially on the benignity of Your own nature, one formerly Fellow of Trin: Coll: not otherwise known to You, than for what the greatness of Your minde will not suffer You to remember to his prejudice, penitently implores Your favor for his son, this next Election candidat of a Fellowship, of whom his Tutor speakes very well; that You will be pleas'd graciously to commend him to our Hond: Master. As Your internal accomplishmts have raisd You to a greatness wch Your minde yet transcends, with an humble confidence he hopes, that to Your intellectual and moral perfections, You will add this instance of charity, a grace so much the more Divine, by how much less it is deserv'd, a virtue immediately linkd to the throne of the Highest, at wch for the continuance of Your present prosperity and complemt of Your everlasting happiness, (as all that he can pay) You have the sincere and fervent addresses of

Revd: Sr

Your most devoted client

and humble servant

JO. HOCKETT[4]

East Dean
Sepbr: 14*th*: 99.

For the Reverend
    Mr Isaac Newton.

NOTES

(1) John Hockett (or Hacket), admitted Trinity College, Cambridge, 1659; B.A., 1663; Fellow of Trinity, 1664. Probably Rector of Motson, Hants. See J. and J. A. Venn, *Alumni Cantabrigienses*, Part I (to 1751). His son, John, was admitted at Trinity in 1693; scholar, 1695; B.A., 1697. He does not appear to have been successful in obtaining a fellowship of his college.

(2) Probably Daniel Hopkins, Fellow of Trinity College, Cambridge, 1682; D.D., 1707.

(3) Edward Jones (1641–1703), educated at Westminster and Trinity College, Cambridge; Fellow, 1667; M.A., 1668. He taught at Kilkenny School, where Dean Swift was one of his pupils. Dean of Lismore, 1678; Bishop of Cloyne, 1683–92, of St Asaph, 1692–1700. He retained his see till his death, but was suspended for a time on charges of simony and maladministration. See *D.N.B.*

(4) On the same sheet of the above letter Newton has written down some notes on the Chronology of the Ancient Kingdoms. See *The Chronology of Ancient Kingdoms Amended* (1728) by Newton.

617   NEWTON TO THE TREASURY
1 OCTOBER 1699
From the original in the Public Record Office[1]

To the Rt Honble the Lords Commrs
of his Majts Treasury.

May it please your Lordships

The prosecution of Coyners during the three last years having put me to various small expenses in coach-hire & at Taverns & Prisons & other places of all wch it is not possible for me to make accompt on oath, your Lordships were pleased to give me hopes of an allowance for the same wch I humbly pray may be an hundred & twenty pounds[2] and that your Lordships will please to direct an Order to the Auditor Mr Bridges[3] for allowing the same in my Accompts. All wch is most humbly submitted to your Lordships great wisdome

Is. NEWTON

Mint Office Oct. 1
1699.

NOTES

(1) T. 1/63, no. 45. This is in Newton's hand.

(2) Newton recovered the expenses he had incurred in the work of prosecuting offenders. The repayment of the sum here mentioned was sanctioned by the Treasury on 9 November following (*Calendar of Treasury Papers*, 1699, p. 200).

The total Mint expenditure was between £200 and £300 a year. Much larger sums in rewards came from other sources. The reason for putting this particular sum separately is that Newton had not kept an itemized account.

The charge was perfectly proper, though probably understated, since Newton did not keep records. This was not unnatural because Newton only got an advance of £300 for three years' expenses and found the rest temporarily from his own money. Annual settlements for this service were not usual.

(3) Probably Brook Bridges, one of the two Auditors of the Imprest (in the Court of the Exchequer).

618   BLACKBORNE[1] TO NEWTON
30 DECEMBER 1699
From the original in the University Library, Cambridge[2]

Honour'd Sr

I herewith present you with a Clause in the Compa[ny']s Charter empowering them to coyn Moneys in India[3] which I hope will fully satisfy the

Enquiry you were pleased to make the last night relating to ye Companys
Petition for sending Mills & Presses for India. I am with due respects

Sr

Your most humble Serv.

Ro. BLACKBORNE

East India house
30th Decr. 1699
    Sr
        For your Self.

NOTES

(1) Robert Blackborne (born 1652) was Secretary to the East India Company, 1696. See
Sir William Foster, *The East India House* (1924).

(2) Add. 3966, fo. 117.

(3) The Mint on being asked for coining tools enquired what right the Company had to
use them. As the above letter indicates the Company was able to produce the necessary
authority, given in the Charters of 1677 and 1686. Earlier, Queen Elizabeth had ordered the
Mint to make special coins for the Company. See Craig, *Mint*, pp. 374–5.

619   STACY TO NEWTON

?1699

From the original in the University Library, Cambridge[1]

Sir,

    I being before recommended to your favour for ye place of Keep[e]r of
New Prison,[2] ye Keep[e]r being now dead, & I understand yt ye Electtion
for that place will be Thursday next being the Twentysixth of this Instant att
Hicks[3] Hall where I humbly desire your Wor[shi]ps presence vote &
Interest for that place for

Sr

Your humble Sert

to Command.

JOHN: STACY

To Sr Isaac Newton Knt[4]
in Jarmin Street
    Humbly
        Present

318

(1) Add. 4005.15, fo. 82. This letter is undated, but was probably sent during Newton's Wardenship of the Mint (1696–9), or soon after.

(2) For Newton's association with the New Prison see Letter 554.

(3) This word, which is difficult to decipher, is probably Hicks Hall, near Smithfield, the Sessions House of the County of Middlesex. See Letter 554, note (4), p. 213.

(4) The text suggests that this letter was sent during Newton's tenure of office as Warden of the Mint. But Newton did not become 'Sr Isaac Newton Knt' until 1705. Possibly Stacy wrote to Newton after the latter had been knighted, knowing that he had formerly been associated with the prison. In any case it seems doubtful that Newton's position at the Mint would entitle him to a vote for the keepership.

620 MEMORANDUM BY NEWTON

c. 18 JANUARY 1699/1700

From the original in the Royal Mint[1]

The Lordship of Twyford[2] is in the Parish of Colsterworth[2] & accordingly hath time out of mind paid to Coulsterworth all Assessments to Church & poor & for the High ways & by a statute of the 14th of Cha II ch 6 & 3 & 4 W & M ch 12 all lands liable to Church & Poor ought to pay to ye Constables Assessments (vide the statute). Coulsterworth is in the soak of Grantham & Twyford is not in the soak of Grantham but in ye hundred of Beltisloe[2] & therefore the Constable of the neighbouring town of Northwitham[2] wch is also in ye hundred of Beltisloe, hath collected ye Kings Taxes at Twyford & been allowed by the Constable of Coulsterworth the sessions Charges And now because the Constable of Witham collects the Kings Taxes at Twyford the Justices at their Quarter Sessions held for ye parts of Kesteven in the County of Lincoln Jan 8 1699 have ordered that Twyford shall thence forward pay (not ye Church & Poor & Highways, but only) Constables Levies to Northwitham.

Q. 1. Whether notwithstanding that Order Parish Assesmts to be raised by the said statutes ought not to be paid to Coulsterworth, the several statutes in such cases limiting them to be raised by the Churchwardens & Constables of the Parish

2. If the Constable & town of Witham levy an Assesmt upon Twyford & distrein what will be the properest way to try the matter at Common law by Replevin,[3] or action for taking & deteining the goods quousque &c

3 Whether may not the Constable of Coulsterworth make a Deputy at Twyford & by his Deputy if not by himself act in Bentisloe [Beltisloe] hundred without giving the Constable of Northwitham any further trouble.

3 Whether is not ye Constable of Coulsterworth constable of the whole Parish including Twyford, wth power to act in Bentisloe Hundred either by himself or a Deputy wthout giving the Constable of Witham any further trouble.

NOTES

(1) Newton MSS., II, 628 (in Newton's hand). Newton does not question the Church, Poor and Highway Rates. These have been paid, and apparently are to continue to be paid to Colsterworth.

The King's Taxes for Twyford have been paid to Northwitham. From the first paragraph it appears that now the Justices have ordered the Constable's levies to be paid to Northwitham. It is this that Newton questions. For reasons which are not clear, Newton appears to prefer to pay to Colsterworth rather than to Northwitham.

The order for the levy was made on 18 January 1699/1700, and this helps to fix the date of the Memorandum. See Craig, *Newton*, pp. 39–40.

On the reverse of this Memorandum is a draft of a reply to Pollexfen's book, *A Discourse of Trade, Coyn and Paper Credit* (1697).

(2) Colsterworth is on the River Witham, $7\frac{1}{2}$ miles south of Grantham. Twyford is about half a mile to the north of Colsterworth, and North Witham is about a mile south of Colsterworth. Beltisloe still survives, in the deanery of that name, in the diocese of Lincoln. See A. Ferguson, 'Newton and the Principia', *Phil. Mag.* **33** (1942), 871, and Turnor, *Grantham*.

(3) A legal term, meaning the restoration to, or recovery by, a person of goods and chattels distrained or taken from him, upon his giving security to have the matter tried in a court of justice, and to return the goods if the case is decided against him (*O.E.D.*).

621   NEWTON'S APPOINTMENT AS MASTER OF THE MINT
3 FEBRUARY 1699/1700
From a copy in the Royal Mint[1]

Isaac Newton Esqr. To be Master and Worker of the Mint 3d Feb. 1699/1700.

William the Third by the Grace of God King of England, Scottland, France, & Ireland Defender of the Faith etc. To All to whom these presents shall come greeting. Know yee that wee for divers good causes and considerations us thereunto moving of our especial grace, certain knowledge, and meer motion, have given and granted, and by these presents do give and grant unto Our trusty and Well beloved Subject Isaac Newton Esqr. the office of Master and Worker of all our Moneys both Gold and Silver within our Mint in our Tower of London and elsewhere in our Kingdom of England, in the Roome & place of Thomas Neal Esqr.[2] lately deceased, together with the fee or yearly pension of five hundred pounds, to be quarterly paid to him from the five and twentieth day of December now Last past, to witt at or upon the five and twentieth day of March, the four and Twentieth day of June, the Nine and Twentieth day of September and the five and twentieth day of December in every year by even and equall portions out of Such Money as by any Act or Acts or Parliament already made for encouraging of Coinage are allowed to be issued yearly out of Our Exchequer for the fees and Sallaryes of the Officers of the Mint or Mints, and towards the providing maintaining and repairing the houses, Offices, and buildings and other uses in the said Act or Acts.

...   .

And know yee also that wee for the considerations aforesaid have given and granted, and by these presents do give and grant unto the said Isaac Newton all edifices, buildings, Gardens, and other fees, allowances, proffitts, privileges, franchises and immunities belonging to the aforesaid Office, To Have and to Hold the said Office of Master and Worker unto the said Isaac Newton to be exercised by himself or his sufficient deputy or deputies for whom he will answer, and also the said fee or yearly pension of five hundred pounds Sterling together with the said houses, edifices, buildings, garden and other fees, allowances, profits, advantages, priviledges, franchises, Liberties and immunities for and during Our pleasure in as full and ample manner to all intents and purposes as Sr Ralph Freeman Knight[3] deceased, Henry Slingsby Esq[4] or the sd Thomas Neale or any others heretofore holding the said Office have had, held, or enjoyed or ought to have had, held, or enjoyed the same, Provided always and under this Condition and Our Will and pleasure is that the said Isaac Newton after he is in possession of the said Office, by virtue of these presents shall within two Months next after signification of our pleasure in that behalf by the Commissioners of Our Treasury, or high Treasurer of England, now or for the time being make and execute on his part Such Indenture and Agrements for & concerning the making of the Several Sorts of Money Gold and Silver, and for payment of the several Officers of the Mints, and other matters and things relating thereunto and to the execution of the Said Office, as were formerly made, with alterations and additions if any shall be thought fitt and directed by us for the better management and carrying on of that Service. In Witnesse whereof wee have caused these our Letters to be made patents, Witnesse ourselfe at Westminster the third day of February in the Eleaventh Year of Our Reigne

By Writt of Privy Seal

COCKS[5]

NOTES

(1) Royal Mint Record Books, v, 29.

Newton was appointed Master on 26 December 1699, on the death of Neale. On 10 January following he was authorized by Royal Warrant to act under the Indentures of the previous Master. His own Indentures were concluded on 23 December 1700 (*Calendar of Treasury Papers*, 1699–1700). Fresh Indentures were also required on a change of Sovereign. See Craig, *Mint*, pp. 224–5.

(2) See Letter 549, note (5), p. 204.

(3) Sir Ralph Freeman served as Master of the Mint jointly with Thomas Aylesbury from 1635 to 1643. He was reappointed at the Restoration, and an indenture was made with him on 20 July 1660. Under a patent of 30 December 1662 the post of Master was held jointly by Sir Ralph Freeman and Henry Slingsby, and following the death of the former the latter became sole Master.

(4) Henry Slingsby was suspended from his post of Master on 15 July 1680, owing to certain irregularities, and he resigned on 20 April 1686. He died in 1690 (see *Commons Journal*, 8 April 1697, vol. xi, p. 775). Evelyn, on 12 January 1687–8, wrote: 'Mr. Slingsby, Master of the Mint, being under very deplorable circumstances on account of his creditors, and especially with the King, I did my best endeavours with the Lords of the Treasury to be favourable to him.'

When it became apparent that Slingsby had seriously abused his trust, a Royal Warrant of 16 August 1678 granted the office of Master of the Mint to Thomas Neale 'after the death, surrender, forfeiture or other determination of the estates and interest of Henry Slingsby, then in possession of the Office'. See Craig, *Mint*, pp. 176–8.

(5) Probably Charles Cox, who was appointed Clerk of the Patents in 1699 (Luttrell, iv, 507).

622   A MANUSCRIPT BY NEWTON
27 FEBRUARY 1699/1700
From the original in the University Library, Cambridge[1]

The Observatory at Greenwich is more westward then ye Observatory at Paris by 2gr. 19′, then Dantzick by 18gr. 48′, then Uraniburg[2] by 12gr. 51′. 30″, then Roma by [12° 30′].[3]

The mean motions of the Sun & Moon from ye Vernal Equinox in the meridian of Greenwich, I put as follows. vizt anno 1680 upon ye last day of

$$-20$$

December at noon st[ilo] v[eteri] the mean motion of the Sun, 9s. 20gr. 34′. 46″ of his Apogee[4] 3s. 7gr. 23′. 30″, of the Moon 6s. 01gr. 45′. 45″ of her Apogee[4]

$$-15$$

8s. 4gr. 28′. 5″ of her Node ascendent 5s. 24gr. 14′. 35″. And anno 1700 upon the last day of December at noon st.v. the mean motion of the Sun 9s. 20gr.

−20

43'.50" of his Apogee[4] 3s. 7gr. 44'.30" of the Moon 10s. 15gr. 19' 50" of her

−15

Apogee[4] 11s. 8gr. 18' 20" of her node ascendent 4s. 27gr. 24'.20". For in 20 twenty Julian years or 7305 days the Sun moves 20 revolutions, 00sig. 00gr. 9'.4" his Apogee[4] 21'.00", the Moon 247 revol. 4s. 13gr. 34'.5", her Apogee[4] 2rev. 03sig. 03gr. 50'.15". Her Nodes 1rev. 09sig. 26gr. 50'.15". These motions are in respect of the Vernal equinox. If from them you subduct the motion of the Equinox (wch in 20 Julian years is about 16'.40" backwards)[5] there will remain the motions in respect of the fixt stars in 20 Julian years vizt of the Sun 19 rev. 11s. 29gr. 52'.24". Of his Apoge[4] 4'.20". of the Moon 247 rev. 4s. 13gr. 17'.25" of her Apogee[4] 2 rev. 3s. 3gr. 33'.35". of her Node 1 rev. 00s. 27gr. 6'.55".

According to this recconing the tropical solar year consists of 365 days 5 hours 48 minutes & 57", & the sidereal of 365d. 6h. 9' 14½".[6]

The mean motions above described are affected wth several inequalities. And first there are annual equations of the mean motions of ye Sun and Moon & of her Apoge & Nodes. The annual equation of the Suns motion arises from ye excentricity of his Orb wch is $16\frac{11}{12}$ supposing the radius of that Orb to be 1000. Tis thence called Æquatio centri & when greatest amounts to 1gr. 56'.20".[7] The greatest annual equation of the Moons mean motion is 11'.49", that of her Apoge 20' that of her Node 9'.30". And these four annual equations are constantly proportional to one another so that when any one of them is greatest they are all greatest, & when any one of them is less all the others are less in the same proportion. Whence from the Sun's annual equation may be readily computed the other three annual equations & therefore a table of the Suns annual equation is sufficient Let the Suns annual equation found by that table at any given time be called P. Make $\frac{1}{10}P=Q$, & let $Q+\frac{1}{60}Q=R$. $\frac{1}{6}P=D$. & $D+\frac{1}{30}D=E, D-\frac{1}{60}D=F$[8] and the annual equation of ye Moon at ye same time shall be R, that of her Apoge E, & that of her Node $\frac{1}{2}F$. And note that when the Sun's equation is added that of the Moon must be subducted; that of her Apoge added & that of her Node subducted. And on the contrary when the Suns equation is subducted that of the Moon must be added that of her Apoge subducted & that of her Node added.

There is another equation of the Moons mean motion wch depends upon the position of her Apoge to ye Sun & is greatest when the Moons Apoge is in the Octants of the Moons Orb & vanishes when it is in the Syzygies & Quadratures. This equation when greatest is 3' 56" if the Sun be in his Perige or 3' 34" if he be in his Apoge & in other positions of ye Sun it is reciprocally proportional to ye cube of the distance of the Sun from the earth. And when ye Moons Apoge is not in ye Octants of her Orb this equation becomes less, & is to ye greatest equation in the same position of the Syzygies as the sine of the doubled distance of the Moon from the nearest Syzygy or Quadrature is to ye Radius. It is added to the Moons motion in the passage of ye Apoge backwards from ye Quadratures to ye Syzygies & subducted from it in ye passage of ye Apoge from the Syzygies to ye Quadratures.

There is a third equation of ye Moons motion wch depends upon the position of the Moons Nodes to ye Sun & is greatest when ye Nodes are in the Octants & vanishes when they are in the Syzygies & Quadratures. This equation is proportional to ye sine of ye doubled distance of the Node from the nearest Syzygy or Quadrature & when greatest amounts to 47″ & is added in the passage of ye Nodes from ye Syzygies to ye Quadratures, & subducted in their passage from ye Quadratures to ye Syzygies.

From the Sun's true place subduct the mean motion of ye Apoge equated as above, & the remainder will be the Annual Argument of her Apoge. Thence compute her excentricity & the second equation of her Apoge as follows. Let T represent the earth, TS a line drawn from the earth towards the Sun, $TACB$ a line drawn from the earth towards the mean place of ye Moons Apoge equated as above, STA the annual Argument of her Apoge, TA her least excentricity, & TB her greatest excentricity. Bisect AB in C & with ye center C describe a circle AFB. Take ye angle BCF equal to ye doubled annual Argument & TF shall be her excentricity & BTF the second equation of her Apoge. For determining these things let ye mean distance of ye Moon from ye Earth or radius of her Orb be 100000, her greatest excentricity TB 66782 and her least excentricity TA 43319, so that the greatest equation of her Orb when her Apogee is in ye Syzygies be 7gr. 39′.30″ & when it is in

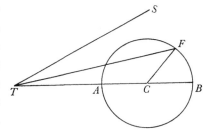

$$+30$$

ye Quadratures 5gr. 57′.56″, & that the greatest equation of her Apogee be 12gr. 15′.4″.

Having by these principles made a Table of Equations & Excentricities to every degree of ye Annual Argument & thereby upon any occasion found the Excentricity TF & the angle BTF wch is the second & principal equation of her Apoge, add that equation to ye equated place of her Apoge if ye Annual Argument be less then 90gr or more then 180gr & less then 270gr, otherwise subduct it & the summ or difference will be the place of the Moons Apoge twice equated wch subducted from ye place of the Moon thrice equated as above leaves her mean Anomaly & this Anomaly with her Excentricity (by a Table of Equations made to every degree of her mean Anomaly & to ye Excentricities 45000, 50000, 55000, 60000 & 65000) will give you her Prostaphæresis[9] or Æquatio centri wch is her 4th equation, & subducted if her mean Anomaly be less than 180gr or added if greater gives you her place four times equated.

The greatest Variation or Variation of ye Moon in the Octants of her Orb is almost reciprocally proportional to the cube of ye earths distance from the Sun. Make it 37′ 25″ when the Sun is in his Perigee & 35′ 4″ when the Sun is in his Apogee. Make the differences of ye Variation in the Octants reciprocally proportional to ye differences of the cubes of ye earths distances from ye Sun & thereby let a Table be made of ye Variation in ye Octants (or of its Logarithm) to every 10th or to every 6t or 5t degree of the Suns mean Anomaly, & then say. As ye radius to ye sine of the doubled distance of ye Moon from the nearest Syzygy or Quadrature so is the Variation in ye Octant to ye Variation required. And this Variation added in ye 1st or 3d Quadrant of the Moons monthly course & subducted in the second or 4th Quadrant will give you the Moons place five times equated.

The summ of the distances of the Moon from the Sun, & of the Moons Apoge from the Suns Apoge is ye Argumt of ye sixt Equation & as the Radius is to the sine of that Argument so is 2′.10″ to ye 6t Equation. Subduct this equation if that Argument be less then 180gr. add it if greater.

Say also As the Radius to the sine of the distance of the Moon from the Sun so is 1′.30″±31 to the seventh equations, & subduct this equation if the Moon be in the increase, add it if in the decrease. So have you the Moons place seven times equated. This is her place in her Orb.

If the sixt & seventh equations be increased & decreased in reciprocal proportion to ye Moons distance from us, that is, in direct proportion to her horizontal parallax, they will become exacter And this may be readily done by Tables made to every minute of that Parallax and to every 5 or 6 degrees of ye aforesaid Argument for ye sixt equation & of ye Moons distance from ye Sun for the seventh.

From the Suns true place subduct the mean motion of ye Node ascendt equated as above, & the remainder will be the Annual Argument of her Node. Thence compute the second equation of her Node as follows. In the figure above described let TA be to AB as 56 to 3 or $18\frac{2}{3}$ to 1. Bisect BD[10] in C & wth ye center C & radius CA or CB describe a circle AFB. Take the angle BCF double to ye annual Argument of ye Node (found as above) & ye angle BTF shall be ye second equation of ye Node to be added when ye Node is in its passage from ye Quadratures to ye Syzygies, otherwise subducted. So have you the place of the Moons Node ascendent. And thence is to be computed (by Tables made for that purpose after ye ordinary manner) the Moons Latitude and the Reduction of her longitude in her Orb to ye Ecliptic, supposing the inclination of her Orb to ye Ecliptic to be 4gr. 59′.35″ when her Nodes are in ye Quadratures & 5gr. 17′ 20″ when they are in ye Syzygies. When you have thus found the Moons Longitude & Latitude the Obliquity of the Ecliptic 23gr. 29′ will give you her right ascention & Declination.

When the Moon is in her mean distance from us in the Syzygies I put her horizontal Parallax 57′.30″, her hourly motion 33′.32″.32‴ & her apparent diameter 31′.30″. And when she is in her mean distance from us in ye Quadratures, I put her horizontal Parallax 56′.40″ her hourly motion 32′ 12″ 2‴ & her apparent diameter 31′.3″. In her mean distance in the Octants you may reccon that her center is distant from the center of ye earth sixty semidiameters of the earth & two ninth parts of a semidiameter.

The Suns horizontal parallax I put 10″ & his apparent diameter in his mean distance 32′.15″.

In Lunar Eclipses the Atmosphere of the Earth by refracting & scattering the Sun's light casts a shadow as if it were opake or dusky to ye altitude of at least 40 or 50 geographical miles. A geographical mile I call ye sixtith part of a degree of a great circle upon the earth & these miles upon ye body of the Moon in her mean distance from us answer to seconds. whence ye semidiameter of the earths shadow upon ye Moon is to be enlarged about 50 seconds, or (if you please) the Moons horizontal parallax in Lunar Eclipses is to be enlarged in ye proportion of about 69 to 70.[11]

NOTES

(1) Add. 3966. There is a copy, in David Gregory's hand, in the Library of the Royal Society of London (R.S. Greg. MS. fo. 15). This latter is headed: 'A Theory of the Moon by Mr Neuton', and it bears the date 27 February 1699/1700. It differs in minor respects (spelling, punctuation, etc.) from the original here reproduced, and in the use of the word *aphelium* throughout the second paragraph for what Newton, on second thoughts, wrote *apogee*. It seems that Newton corrected his own version (see note (4) below) after Gregory had made his copy during one of his frequent visits to Newton.

There is a Latin translation in Horsley (III, 245), commencing, 'Observatorium *Grenovicense* occidentalius est *Parisiensi* 2gr. 19′; *Uraniburgo* 12gr. 51′. 30″ & *Gedano* 18gr. 48′.'

It is significant that neither in Newton's original, nor in the copy made by Gregory, is there any mention of Flamsteed, although Gregory, in his *Astronomiæ Physicæ & Geometricæ Elementa* (1702), states (p. 332): 'sicuti per plurima Lunæ Loca à Cl. D. Flamstedio observata expertus est' (as is clear from the many places of the Moon observed by the distinguished Mr Flamsteed). See Manuscript 623, note (1), p. 328.

(2) Tycho Brahe's Observatory. See Letter 709, note (3), p. 476.

(3) Newton left a blank after the word 'by'.

(4) In these places Newton first wrote 'aphelium', then crossed it out and substituted 'apogee'. See note (1) above.

(5) The annual motion of the equinox is $50''\cdot 26$.

(6) The Julian Year is $365\frac{1}{4}$ days. It errs by about 3 days in 400 years. The tropical year is 365 days 5 hours 48 minutes 46 seconds; the sidereal year is 365 days 6 hours 9 minutes 10 seconds.

(7) See Letter 470, note (11), p. 15, where the figures are $1° 56' 26''$ and $16\frac{15}{16}$.

(8) This should read
$$D + \tfrac{1}{30}D = E,$$
$$D - \tfrac{1}{60}D = 2F.$$

It is corrected in Gregory, *Astronomiæ Physicæ & Geometricæ Elementa* (1702), p. 333, and Horsley, iii, 246.

(9) Prosthaphæresis, see Letter 497, note (3), p. 99. It was frequently written erroneously by Newton and David Gregory.

(10) This should read *AB*.

(11) At the foot of his copy, Gregory has written:
'25 Martij 1700. Equatio maxima ☽ Centri sive Prosthaphæresis Solis aut Terræ est 6' aut 7' minor quam vulgo perhibetur (adeoque excentricitas ♄ minor) sive loco $2°:3'$ est tantum $1°:57$.
                                                                                                                    56
Declinatio Eclipticæ est exactissime $23°:29$.

In 25 locis ☽ in quadrato ☉, Newtoni locus computatus non differt ab observata ultra 2', præterquam semel 9', et iterum 4' quæ duo procul dubie sunt observata.

Oportet aliquando recedere a vera philosophia ad commodam Tabl. [the next word is indecipherable].

Mr Newton suspects this to change according as the ☽ apogee is in ♋ or ♑ for he says that in 4 years it must change.'

623   A MANUSCRIPT BY NEWTON

(Undated)

From the original in the University Library, Cambridge[1]

The Theory of the Sun & Moon

Compute the mean motion of the Sun by Flamsteeds Tables, subducting 10'' or 20'' from it & making his greatest Equation $1°.56'.20''$ (the Radius being 1000 & excentricity $16\frac{11}{12}$) & adding $36'.10''$ to the longitude of his Aphelium A.C. 1701 *ineunte*,[2] & making his Aphelium to move forwar[d]s 16' or $18'.36''$ in 100 years. And call the Sun's Prosthaphæresis[3] with its proper signe $\pm P$ & with its signe changed $\mp P$.

Compute the Moon's meane motion by Flamsteeds Tables adding $2'.10''$ or $2'.18''$ to it, wch call M & $18'$ or $18'\ 9''$ to the Longitude of her Aphelium wch call A & taking $2'\ 20''$ from the longitude of her Node, wch call N.

Make $\mp\frac{1}{10}P = Q$ & $Q + \frac{1}{60}Q = R$ the annual Equation of the Moons mean motion. For the greatest annual Equation is $11'.49''$.

Make $\pm\frac{1}{6}P = D$, and $D \pm \frac{1}{40}D = E$, & E the annual Equation of the Apoge.

$$\begin{array}{ccc} & 2F & -F \\ & & \end{array} \qquad \text{Node.}$$

For the greatest annual Equation of the Apoge is 19.52.

$$\text{Node} \qquad 9.27.$$

Subduct $A + E$ from the Sun's true place & the remainder will be the

$$N - F$$

Annual Argument of her Apoge

Node

NOTES

(1) Add. 3966 (10), fo. 74. It is undated and unsigned, but it is in Newton's hand, and it clearly relates to the previous manuscript (622), with which it should be compared. It is noteworthy that whereas the earlier manuscript makes no mention of Flamsteed (see p. 326, note (1)) the present one contains two references to Flamsteed's tables.

(2) 'In the beginning of the year 1701, anno Christi.'

(3) See Letter 497, note (3), p. 99.

624   MEMORANDUM BY NEWTON

25 APRIL 1700

From the original in the Library of the Royal Society of London[1]

Elementa motuum Solis et Lunæ ab Æquinoctio verno.[2]

Tempus æquabile, quod verum dici solet, diurnæ non solis sed Fixarum revolutioni proportionale est et inde condendæ sunt Tabulæ pro æquatione Temporis.

In Observatorio Regio Grenovicensi, Anno Christi 1701 ineunte ad meridiem Kal. Jan. stylo veteri, erit medius motus Solis 9s. 21gr. $42'.38''$. Apogæi ejus 3s. 07gr. $44'.30''$, Lunæ 10s. 28gr. $30'.12''$ & Apogæi ejus 11s. 08gr. $25'.14''$.

Uraniburgum est orientalius Observatorio Regio Parisiensi 00h. $42'.10''$ & hoc Observatorium est orientalius Grenovicensi 00h. $09'.15''$, et inde per reductionem habentur motus illi medij eodem die et hora ad meridianum Uraniburgi, vizt. Solis 9s. 21gr. $40'.32''$ Apogæi ejus 3s. 07gr. $44'.30''$ Lunæ

10s. 28gr. 01′ 58″ & Apogæi 11s. 8gr. 25′.00″. Et ante undecim dies seu meridie Kal. Jan. stylo novo erit motus medius Solis 9s. 11gr. 50′.00″ Apogæi ejus 3s. 7gr. 44′.32″, Lunæ 6s. 03gr. 05′.33″ & Apogæi ejus 11s. 07gr. 11′.28″.

Maxima Solis Prostaphæresis[3] quæ Keplero est plusquam 2gr. 3′ debet esse tantum 1gr. 56′.20″.

Ubi hæc æquatio additur vel subducitur medio motui Solis debet ejus pars decima e contra subduci vel addi medio motui Lunæ. Nam medius motus Lunæ non est unformis sed per vices tardescit et acceleratur propterea quod orbis Lunæ dilatatur in perigæo Solis et contrahitur in ejus Apogæo.

Postquam motus medius Lunæ sic correctus habetur, reliqua peragenda sunt per Tabulas Kepleri: et Æquinoctium vernum incidet semper in diem horam et minutum ubi longitudo Solis per hoc computum prodit 00s. 00gr. 00′.00″.

Translation

Elements of the Motions of the Sun and the Moon from the vernal Equinox[2]

Mean time, which is usually called true time, is proportional to the daily revolution, not of the Sun, but of the fixed stars, whence tables are to be constructed for the equation of time.

In the Royal Observatory at Greenwich in the beginning of the year 1701, at midday the first of January, Old Style, the mean place of the Sun will be 9s. 21gr. 42′.38″; of his apogee 3s. 07gr. 44′.30″, that of the Moon 10s. 28gr. 30′.12″, and of her apogee 11s. 08gr. 25′.14″.

Uraniborg is east of the Royal Observatory at Paris by 00h. 42′.10″, and this Observatory is east of that of Greenwich by 00h. 09′.15″, and hence by reduction are obtained those mean places on the same day and hour referred to the meridian of Uraniborg, namely of the Sun 9s. 21gr. 40′.32″, of his apogee 3s. 07gr. 44′.30″, of the Moon 10s. 28gr. 01′.58″, and her apogee 11s. 8gr. 25′.00″. And eleven days before at midday on the first of January, New Style, the mean place of the Sun will be 9s. 11gr. 50′.00″, of his apogee 3s. 7gr. 44′.32″, of the Moon 6s. 03gr. 05′.33″, and her apogee 11s. 07gr. 11′.28″.

The maximum prosthaphæresis[3] of the Sun, which according to Kepler is more than 2gr. 3′, should be only 1gr. 56′.20″.

When this correction is added to, or subtracted from the mean place of the Sun, one tenth part of it should be subtracted from, or added to the mean place of the Moon. For the mean motion of the Moon is not uniform but by turns is retarded and accelerated because the Moon's orbit is expanded in the Sun's perigee and contracted in his apogee.

After the mean place of the Moon thus corrected is obtained, what remains is to be done by Kepler's tables, and the vernal equinox will always fall on the same day, hour and minute where the longitude of the Sun by this computation comes to 00s. 00gr. 00′.00″.

NOTES

(1) Early Letter Book of the Royal Society of London, N. 1, fo. 63.

(2) See Edleston, p. 304, and Montucla (*Nouvelle Edition, An.* 7), IV, p. 325. The decree of the German Diet (Ratisbon, 23 September 1699) reforming the Julian Calendar, ordained that:

(i) In the year following, the day after 18 February 1700 should be named 1 March, the intervening days being suppressed.

(ii) Easter should be determined by astronomical calculation (i.e. the exact time of the vernal equinox and the full Moon following it). The Proclamation which followed some days afterwards gave rise to considerable discussion amongst theologians and men of science, in which this country took part, as is clear from the following extracts from the Journal Books of the Royal Society of London: 21 February, 1699/1700. 'A letter from Hanover to Dr. Sloane from Mr Leibnitz was read, concerning the changing of the Stile.

'Dr Sloane said, he heard, Mr Newton had made a very good calculation of the Year, and that the Settling yt affair might be helpt by yt. Dr Sloane was ordered to wait on Mr Newton about it.'

25 April 1700. 'The Dr [Sloane] read an answer to Mr Leibnitz's letter containing Mr. Newton's opinion concerning the alteration of the stile. [This is the Memorandum reproduced above.] The V.P. [Sir Robert Southwell] said yt his opinion was yt this paper be sent to Mr Leibnitz, and in ye mean time that he procure Mr. Flamsteds and Dr Wallis's opinions, and send to him: also yt a copy of this be kept.'

On 11 May, Wallis wrote the following letter to Sloane. The letter was read to the Society on the 22 May following:

'Sir

'I thank you for your letter of May 7. with the Papers inclosed. As to the Gregorian Correction (as they call it) of the Kalendar: I think it was very unhappyly introduced at first. Not onely as not perfectly True: but principally, as having caused a Confusion of Stile throughout Christendome which, from time to time, hath continued hitherto; nor do I see, when it is like to end.

'The Julian Civil Year, is a much Better form. . . .' (Early Letter Book of the Royal Society of London, W. 2, fo. 66.)

Sloane sent Newton's Paper, as revised by Flamsteed, to Leibniz on 4 July 1700. The letter concludes: 'The Royall Society have laboured to gett his [Newton's] Theory of the Moon, Book of Colours &c. printed, but his excessive modesty has hitherto hinder'd him, but the Society will do what further they can with him.' (Early Letter Book of the Royal Society of London, S. 2, fo. 14.)

(3) See Letter 497, note (3), p. 99.

625   NEWTON TO BRAINT[1]
6 MAY 1700
From the original in the Burndy Library, New York

Mr Provost

You are to take care that the five pound pieces & forty shillings pieces of Gold & the Crowns & half crowns be made with ANNO REGNI DUODECIMO on the edges till further order

Is. NEWTON

To Mr John Braint
Provost of the Moniers
Mint Office *May* 6t
1700

[On the reverse]

Order for altering ye Edging ye Money
1700

NOTE

(1) John Braint, one of the moneyers, elected Provost 12 June 1694. See Letter 633, where he is described as 'Engineer of the Mint'.

626   FLAMSTEED TO LOWTHORP[1]
10 MAY 1700
Extracts from a copy of a letter in the Royal Greenwich Observatory[2]

The Observatory May 10. 1700:

My Friend

I esteem it ye Duty of all Christians as much as in them lies to have peace with all men & it was ye sole Consideracõn of this Duty and a Tenderness for you yt induc'd me to advise You to be carefull of Your Behaviour towards Mr Newton...he is so possest with Prejudices against me by some People's suggestions whom You know very well yt I can have no free discourse with him:

...   .

I Look't upon his imparting wt he had deduced from ym [my observations of the stars and the planets] to Dr Gregory & your Captaine [Halley] as a

331

Greater breach of Promise[3] yn if he had imparted ye Observations them-selves...my Work was like ye building of St Paul's. I had hew'd ye Materials out of ye Rock, brought ym together & formed ym but yt hands & Time were to be allowed to perfect ye building & Cover it...wn he comes hither I shall not be averse to impart more Lunar Observations to him provided yt he withdraw wt he has imparted to Others or stop their reflecting discourses & own before Sr Chris: wt he has already receiv'd & What I then imparted to him...I believe him to be a good Man at ye Bottom. but through his Naturall temper suspitious & too easy to be possest with calumnies especially such as are imprest with Raillery....

P.S. 'Tis given out at Oxford yt Mr Newton has improv'd his doctrine of Gravity so far yt he can answer all my Lunar Observations exceeding Nearly; & yt there is now Little Need of them since all ye Inequalitys of ye Moons Motions may be discovered by ye sole Laws of Gravitation without ym...I had imparted above 200 of her observ'd Places to him, which one would think should be sufficient to Limit any Theory by...but still he is more beholden to them for it yn he is to his speculations about gravity, wch had misled him....

[Flamsteed has written at the foot] Supscribed to Mr John Lowthorp

NOTES

(1) John Lowthorp (1659–1724), of Holderness, Yorkshire. Admitted sizar at St John's College, Cambridge; M.A., 1683. Ordained priest, March 1683/4, and was probably a naval chaplain. He wrote *An Experiment of the Refraction of the Air made at the Command of the Royal Society* (28 March 1699), *Phil. Trans.* **21** (1699), 339. He was elected F.R.S. in 1702. In 1705 he published an abridged edition of the *Philosophical Transactions* with the title: *The Philosophical Transactions and Collections to the End of the Year 1700. Abridg'd and Dispos'd under General Heads. By John Lowthorp, M.A. and F.R.S.* This work received the *imprimatur* of the Royal Society on 12 May 1705. See Venn, *Alumni Cantabrigienses*, Part I, vol. III, p. 104.

(2) This letter (vol. 33, fo. 32) is in the hand of a copyist.

(3) See Memoranda 468, note (2), p. 7.

PLATE IV. MURAL ARC, GREENWICH OBSERVATORY. FROM 'HISTORIA
CŒLESTIS BRITANNICA'

627 FLAMSTEED TO NEWTON
18 JUNE 1700
From the original in the Royal Greenwich Observatory[1]

135 A Letter. To Mr Is. Newton. Warden of ye Mint

Shewing what have been ye Errors of ye Murall Arch.

Sr

♂ *Junij* 18 1700[2]

That the Earths Axis is not allwayes inclined at the Same Angle to the plane of ye Ecliptick is a discovery wholly oweing to you & Strongly proved in ye 4th book[3] of your *prin. Phil. Nat Math.* how much the alteration of this Angle or ye Nutation of the Axis[4] ought to be you have not yet Shewed: and whether you have yet determined or no I know not. but haveing found it sensible in the observations made with the Murall arch described in my letter to Dr Wallis (wherein the parallax[5] of the Earths Orbe at ye Pole Star is proved by eight yeares continued observations) & not doubteing but an account of the observations that shew it would be gratefull to you I have resolved to impart them to you as to one who not onely delights in these things but are able to judge of them, & has a particular interest in this which proves the gravitation of the parts of Matter & the truth of your Theory of our Earth.

When I wrote that letter to ye Dr I told him that to determine the Parallax[5] and this Nutation to some competent exactness, wee ought to have Instruments of 15 or 20 foot Radius for takeing the Meridionall distances of ye Stars from ye Vertex I then had employed onely such of these as I thought most proper for yt particular discovery & no more yn was just necessary by reason of the hast of ye press. which forced that letter out of my hands before I had time to examine such observations as I foresaw would most probably make the

137 Nutation sensible But haveing now had occasion to consider & compare a competent Number of them for another Use & finding this Nutation sensible whenever proper observations are compared: Since No great favor of a science so usefull to our Nation appeares who might by the easy Charge of a large Instrument at once oblige all Ingenuous men. & bring certeine honor to himself and Country thereby. Since my cost & paines already bestowd have met wth quite other rewards then they deserved I shall lay by the hopes of seeing any such formed till I may have some setled station where I may build a larger then I proposed at my own expence and from my present stock of observations examine what may be the utmost possible quantity of this Nuta-

tion. and this I the rather chuse to doe because when I acquainted you wth my discovery of the Parallax of ye orbe at ye Pole-Star you suspected yt ye Nutation had caused those alterations in its Zenith distances which I tooke for the Parallax, which makes me thinke you esteeme this Nutation to be something More then it really is, as you perhaps judged the Parallax less then it was really found.

I determined the least quantity of ye Parallax of ye Orbe at the Pole Star, that my observations would allow & shewed that what ever this Nutation shall be found it must be added to ye parallax determined which was diminished by it, this was needfull yt you might not mistake it for the Nutation. it will be necessary now to shew What is the greatest quantity of the Nutation yt you may not esteem it greater yn really it is which your opinion of the Parallax makes me thinke you may

Had the wall to which my instrumt is fixed continued stable & unmoved, had it continued fixed both the Parallax of the Orbe with the Nutation of the earths Axis would have been discovered & determined as exactly as ye small

139 smallness of my Arch would permit in ye compass of one yeares observations whereas Now I find my selfe obliged to examine all I have made to come to some near guesse at it & to confirme what is derived from ye first years observations by ye consent of ye following

June ye 21 ♀   But since it does not retaine its first position but sinkes every yeare these
1700:   cannot be found without the errors caused by ye graduall sinkeing of the Arch being first known. I suspected some such thinge might happen when I first began to use it & therefore concluded

That frequent observations of the Meridionall Zenith distances of the stars in ye foot of Castor would shew the errors soonest & most easily. both because these stars lyeing very neare the Solstitiall Colure[6] alter their distances from ye pole insensibly. as also because they culminate within less yn 30 degrees of our Vertex so yt they were not liable to Uncerteine & Variable refractions.

Whereas those stars of ♐ yt lie neare ye winter solstice & opposite points of ye Ecliptick pass ye Meridian low & so may be intangled with refractions suspected to be variable

I considered further that these stars in the foot of the twins were the more proper for this enquiry because their latitudes from ye Ecliptick being small the parallax of the Earths Orbe must vanish & become insensible of them.

I was sensible yt the Nutation of the Earths Axis (if so much as to become sensible) must be perceaved in these stars because tis made directly towards & from them but I esteemed it then so small that it would not be sensible in my observations

But however that it would not affect those stars yt lie neare ye æquinoctiall

334

points & Colure[6] because their distances from the pole of the World are not alterd by it as those of ye stars in ye foot of ye Twins are

141   Onely in these Stars which lie in ye Constellations of ♍ & ♓ the declinations or distances from ye pole vary fastest (about 20″ per yeare) which is no obstacle to ye inquiry, the variation being readily found by Tables prepared for that Purpose & easily allowed for.

Tis necessary to have the errors of the Instrumt truely stated in order to find the true distances of the stars & planets from our Vertex which are Corrupted (as they are copied from ye Instrumt) by ye sinking of the Wall on which tis fixed I shall therefore first seek out these by stars both neare ye solstitiall & æquinoctiall colure. but I must advise you yt those nearest the æquinoctiall points & colures are most to be relied on & of them such as have least Latitude from ye Ecliptick are cheifly to be regarded because those which are farther distant from ye Ecliptick may be suspected to be corrupted by ye parallax of the orbe, & those yt are furthest removed from ye Equinoctiall points to be intangled wth ye Nutation

Some little effect & some small errors both these may have caused in those many places of ye Moon & planets I have imparted to you but so small that it needs not be regarded I onely mention them yt you may not thinke I was not aware of them.

1689   July 15. 16. 17. the begining of the divisions on the Murall Arch was found & determined by my selfe & Mr Sharp[7] yn my servant by ye Transits of ye bright ✱ in ye Dragons head near our Zenith both on ye Instrumt & Westerne plaine.

	its distance from ye Vertex
By ye transit of a star in ye Swan over the Meridian observed on ye Westerne plane Oct: 5 following	Bor 22′ 05″
but on ye Instrumt Septem. 19 & Octob. 8 following it was	23. 55
Difference	1. 50
Error	0. 55

by ye revolves of the screw on ye edge these distances were about 10 seconds less. I propose therefore the error at present 1′.00″

Whereby all the distances taken to ye South are too much those towards ye North or on ye Pole Stars side too little.

I thought not of the sinking of the Wall & therefore doubted not but this
143 fault came by some stroke or Injury the Arch might have receaved since its

first verification & division but to be certeine of it No: 15 following I took ye distance of ζ in Cassiopea from our zenith & found it

	° ′ ″	Revs	Cents	
on ye Inst	0.42.00 = 15	78	B	1
Decemb. 15	0.41.50 = 15.	69		2
but on ye Western plane Dec 13	0.44.20 = 16.	71		3
againe Dec 16	0.43.55 = 15.	72		4

Compareing ye first & 3 of these togeather the difference (2′.20″)

	′ ″
halved gives ye error of the Instrument	1.10
ye 2d & 3d	1.15
ye 1st & 4th	1.10
ye 2d & 4th	1.02½
The Meane is near by	1.10* * *

Againe. Dec. 16. on ye Western plane Persei τ = 0.00.25 = 0.14

17 on ye Instrumt   Aust 0.01.50 = 0.39::

halfe their summe makes ye error   1.07 *

Dec 16 ye Shoulder of Perseus γ

	° ′ ″	re cents
on ye western plane	Aust 0.48.10 = 18.14	
17 on ye Instrumt	0.45.55	

their difference 2′.15″ halved }
gives the error of ye Instrumt }   01.07 *

whence ye error of the Instrument may be concluded at this time some little less yn 1′.10″ I allow it tho In copying ye Observations from ye first notes into my faire Journall I have made it onely 1′.00″[8] & used it so all the following year not suspecting yn the subsideing of the wall & not thinkeing it worth regardeing when I first began to perceive yt ye error caused by it was some little augmented.

Both from these experiments & ye observations of the following year 1690 compared with those of Nov: 1689 I conclude ye error of the Instrument in the November observations to have been onely 1′.10″[8] as in December which I make use of every where in reduceing them to ye truth in ye next page.[9]

Wherein for finding the error of the Instrument in ye following yeares I have transcribed ye Meridionall Zenith distances of such stars as will be convenient for this purpose & some others yt have been frequently observed such as ye stars in the head of Aries,[10] ye lions heart, ye Virgin's Spike, ye Stars of Scorpius Sagittarius & Capricornus, yt may be employed for ye same purpose wth due allowances.

June 24. 25. 26
absent at
Hampton
court &
Windsor

336

Distances fr. ye Vertex

1689			on ye Inst ° ' "	correc ° ' "	Var ' "
1689 No 28 ♒		λ	60.41.00	60.39.50	22.26
		δ	68.54.40	68.53.30	22.28
	♓	β	49.19.00	49.17.50	22.48
		γ	49.52.40	49.51.30	22.56
		b	47.46.50	47.45.40	23.15
		κ	51.54.45	51.53.35	23.24
		.	52.03.10	52.02.00	23.24
		θ	46.47.45	46.46.35	23.26
		ι	47.31.40	47.30.30	23.38
		λ	51.24.05	51.22.55	23.40
		ω	46.19.50	46.18.40	23.51
		1c	44.42.50	44.41.40)	23.53
		2c	44.14.50	44.13.40)	
		d	45.00.55	44.59.45	23.54
		δ	45.35.30	45.34.20	23.38
		ε	45.16.10	45.15.00	23.22
		e	47.28.45	47.27.35	23.20
		ζ	45.33.20	45.32.20	23.06
		f	49.30.35	49.29.25	22.59
		μ	46.56.50	46.55.40	22.37
		η	37.45.10	37.44.00	22.35
		π	40.56.25	40.55.15	22.22
		ν	47.34.30	47.33.20	22.12
Dec 1	♍	α	60.59.40	60.58.30	22.46
9	♉	α	35.38.$^{15}_{20}$	35.37.$^{05}_{10}$	10.06
12		γ	36.38.35	36.37.25	11.34
	1.θ		36.20.40	36.19.30	10.46
	2.		36.14.55	36.13.45	
		α	35.38.$^{20}_{25}$	35.37.$^{10}_{15}$	10.06
16	♉	α	35.38.$^{2}_{5}$	35.37.10	10.06
16	Π	η	28.56.00	28.54.50	0.12
					+ *
		μ	28.51.30	28.50.20	0.40
		ν	31.07.25	31.06.15	1.19
		γ	34.51.55	34.50.45	2-16
					- *
17	♈	γ	33.43.35	33.42.25	21.43
		β	32.12.40	32.11.30	21.01
		α	29.30.45	29.29.35	21.07
Dec 26	♓	η	37.45.15	37.44.05	22.35
Nov 16	♉ Plei	η	28.22.20	28.21.10	14.27
1690 Jan. 6	♉ Ao.	α	35.38.25	35.37.15	10.06
		β	23.11.20	23.10.10	5.16
		ζ	30.34.35	30.33.25	4.01
	Π προπ		28.15.00	28.13.50	
	10.δ		28.55.55	28.54.45	
	10.μ		28.51.30	28.50.20	
23		δ	28.58.40	28.57.30	6+34
	♋	μ	29.02.40	29.01.30	11.11
		ζ	32.56.30	32.55.20	11.38
		β	41.22.40	41.21.30	12.07

Distances from ye Vertex

1690			observed ° ' "	correct ° ' "	Var ' "
Feb. 7	♉	α	35.38.30	35 37.20	06
	Π	H	28.15.20	28.13.50	1-22
		η	28.56.25	28.54.55	0+12
		μ	28.51.50	28.50.20	0.40
		ν	31.07.50	31.06.20	1.19
		γ	34.52.20	34.50.50	2.16
		ξ	38.18.00	38.16.30	3.06
		λ	34.26.00	34.24.30	6.26
	1 δ		28.58.40	28.57.10	6.34
	♋	δ	28.56.25	28.54.55	14.28
10	♌	α	38.01.$^{45}_{40}$	38.00.$^{15}_{10}$	20.22
14	♉	α	35.38.$^{30}_{5}$	35.37.$^{00}_{5}$	
	Π	γ	34.52.15	34.51.45	
	♋	γ	28.56.30	28.55.05	14.20
		δ	32.13.45	32.12.15	14.28
	♌is	α	38.01.45	38.00.15	20.22
	♍	γ	51.13.50	51.12.20	23.44
		α	61.00.10	60.58.04	22.46
	♎♋	α	66.11.15	66.09.45	18.35
		β	59.41.$^{10}_{15}$	59.39.$^{40}_{5}$	16.44
19	♉	α	35.38.40	35.37.10	
	♏is	π	76.36.55	76.35.25	13.31
		δ	73.09.00	73.07.30	13.21
		β	70.22.25	70.20.55	12.52
		σ	76.14.15	76.12.45	11.32
		α	77.07.50	77.06.20	10.47
	{1 2ω		71.14.30	71.13.00	12.44
			71.26.35	71.25.05	
		ν	70.04.15	70.02.45	12.18
Mar. 25	♍is	ω	41.38.40	41.37.10	23.34
		ξ	41.30.35	41.29.05	23.43
		ν	43.13.30	43.12.00	23.43
		2ζ	41.31.15	41.29.45	23.50
		β	47.58.50	47.57.20	23.46
		A	41.19.15	41.17.45	
		b	46.06.30	46.05.00	23.51
		π	43.08.50	43.07.20	23.52
		o	41.02.05	41.00.35	23.54
		η	50.25.45	50.24.15	23.54
		q	59.12.45	59.11.15	23.49
		χ	57.45.50	57.44.20	23.45
		γ	51.13.55	51.12.25	23.44
		δ	46.23.$^{40}_{5}$	46.22.$^{10}_{5}$	23.30
		ε	38.51.35	38.50.05	23.22
Junij 5	♍	α	61.00.15	60.58.45	22.46
	♎	α	66.11.25	66.09.55	
		β	59.41.10	59.39.40	
	♏is	π	76.37.10	76.35.40	
		α	77.07.55	77.06.25	
	Ophi.	θ	76.04.40	76.03.10	

♀ Junij 28 1700

144

Dist from ye Vert

145

1690			observed ° ′ ″	correct ° ′ ″	Var.	1690						′ + ″
Junij 5	♐	ξ	72.28.00	72.26.30	4.17	Junij 2	♍	π		76.36.50	76.35.20	13.31
		ξ	72.54.50	72.53.20			♏	δ		73.09.00	73.07.30	13.21
		o	73.35.40	73.34.10	5.06			β		70.22.30	70.21.00	12.52
		π	72.55.20	72.53.$^{40}_{30}$	5.31		1⎫	⎱ω		71.14.35	71.13.05	12.45
30	♑	α	64.53.20	64.51.50	12.12		2⎭			71.26.40	71.25.10	12.43
		2dα	64.55.$^{35}_{40}$	64.54.$^{5}_{10}$	12.12			ν		70.04.20	70.02.50	12.18
		β	67.10.55	67.09.25	12.28							
		η	72.29.45	72.28.15	16.03	Sept 16	♓ mfκ	λ		54.58.10	54.56.40	23.27
		θ	69.53.05	69.52.35	16.11					54.16.10	54.14.40	23.30
		ι	69.34.40	69.33.10	17.24					54.25.40	54.24.10	23.32
		γ	69.29.10	69.27.40	18.38							
		δ	68.57.10	68.55.40	19.05	Sep 20	♓ ante	λ	54.			
Sept. 15	♑	γ	69.29.30	69.28.00				λ	51.24.00	51.22.30	23.40
		λ	64.14.05	64.12.35	19.04				49.42.30	49.41.00	23.44
		μ	66.26.50	66.25.20	19.29				52.07.10	52.05.40	23.46
									50.16.05	50.14.35	23.47
	♒	θ	60.46.30	60.45.00	20.49							
	♓	ε	45.16.10	45.14.40	23.22							
		e	47.28.40	47.27.10								
		ζ	45.33.10	45.31.40								
		f	49.30.25	49.28.55								
	Π *	μ	28.51.45	28.50.15								
16	♓	β	49.19.00	49.17.30								
		γ	49.52.55	49.51.25								
		κ	51.54.45	51.53.15								
19	Π *	η	28.56.25									
20		*μ	28.51.50									
	♑	θ	69.53.20	69.51.50								
	♓	β	49.19.05	49.17.35								
		γ	49.53.00	49.51.30								
		κ	51.54.45	51.53.15								
		λ	51.24.00	51.22.30								
		ω	46.19.45	46.18.30								
		d	45.00.30	44.59.00								
30	♑	γ	69.29.20	69.27.50								
		δ	68.57.45	68.56.15								
No. 14	♍	β	47.59.00	47.57.30								
17 & 19	♓	δ	45.35.$^{25}_{30}$	45.$^{33.55}_{4.00}$								
		ε	45.16.$^{20}_{10}$	45.14.$^{50}_{45}$								
28	♓	κ	51.54.45	51.53.15								
		λ	51.14.00	51.22.30								
Dec. 13	♈	α	29.30.$^{0}_{5}$	29.29.$^{10}_{15}$								
14 & 15	♍	α	61.00.20	60.58.50								
16	♉	α	35.38.30	35.37.00								
1690												
Mar 14	Π	μ	28.51.50	28.50.20								
		γ	34.52.25	34.50.55								

♀ Junij 28
1700

Before I enquire what Argumts for ye Nutation of the Earths Axis my observations afford me, it will [be] neccessary to enquire what were the errors of the Instrument both before the Middle of Dec 1689, when it was determined by Experimt & after & how it stood in ye whole yeare 1690 following

1

1689 Dec 16 & 17. α Tauri a Vertice in Meridiano.		35.38.20/25
1690 Dec 16	eadem	35.38.15

The annuall accesse of this star to ye pole is 10″. therefore if the Instrumt had not sunk the star's Meridionall Zenith distance yt was 35.38.20/25 in Dec 89 would have been on Dec 16 (90) 35.38.15/10 but it was then found 35°.38′.35″ whence tis evident yt ye error is increased since the 16 of Dec. 89 neare 20″. & because then it was found 1′.10″ in Dec 90 it will be 1′.30″

2

In like manner Dec 17 1689. γ Tauri. its Zen Mer dist   36.38.45
Cor[rected] by its Annuall access to ye pole 10″. gives its
dist 1690   =36.38.35

Which was then found by ye Instrumt	36.38.50
more yn it ought by	0.15
adde ye error of Dec 89 =	1.10
it Makes ye error Dec 90 to be	1.25

These observations I have pitched upon for the triall because both being made on ye same days of ye yeare, were equally affected with ye Nutation & parallax of the orbe if wee suppose them sensible at this star, & therefore ye difference of ye observed zenith distances after they are correct[ed] by the annuall access to ye pole can proceed from nothing but ye sinking of the Instrumt

June 29 ♄
1700

1689 No: 28 η Piscium 37°.45′.10″	γ [Piscium] 47°.34′.30″		⎱
Dec 26 37 .45 .15	47 .34 .30		⎰

hence tis evident yt ye error was the same

3

on ye 28 of November yt it was on ye 26 of December yt is 1′.10″ in both, or but 5″ bigger at ye latter.

4

1689 Dec 16 η Geminorum 28.56.00	μ [Geminorum] 28.51.30	
1690 Jan 6 & 10 28.55.55	28.51.30	

whence tis evident againe that the same error continued to ye 10th of Jan. following for ye alteration of ye Nutation could cause no sensible alteration of these stars zenith distances in this place, & at this small Intervall of time

To avoyd all suspition of change of ye Meridionall Zenith distances by ye Nutation I shall next enquire the encrease of the error of ye Instrumt by such

149

stars on which it could have little or no influence such are they as lye neare the

20

Equinoctiall Colure. but the Sun approaching the stars of Pisces I cannot longer find them to employ & therefore make use of ye Virgo's Spike which tho it may be a little can not be much affected by it

5   1689 Dec. 1. this stars Meridionall Zenith distance is   60.59.40
         but 1690 Feb. 14 I find it noted   61.00.10
         its annuall access from ye pole is 20″ & therefore ought to be now   61.00.15

er Dec 1′.10″   whence it appeares yt the error in Feb 1690 is about 30″ more yn it was in
    +20   Dec. 89 I state it 1′.30″
     5
Feb14th‾1‾.‾30‾   1690. Apr 11 & 12. α Virginis   61.00.05/10
                          26   61.00.10
                  May 12   61.00.15
                  June 5   61.00.15::

from all which compared togeather tis manifest that the error was ye same from Feb: 14 to June the 5th 1690: yt is 1′.30″

That it continued the same all this Summer will be evident by compareing observations of the same solstitiall star taken at both ye æquinoxes, when ye Earths Equator librated widest from ye Ecliptick & the effects of ye Nutation were ye same & therefore not to be regarded.

                                          °   ′   ″
6   1690 March 7. μ Geminorum   28.51.45
         Sept ye 15   28.51.45
                 19   28.51.50

Therefore ye Instrument continued immoved from March 7. to Sept 19 & ye error 1′.30″

In September ye stars of Pisces yt had been observed in November last became againe observable. I will examine therefore by them how much the error of ye Instrument is increased since No. 28 (89)

)ᵃᵉ July 1.   Which being done on ye other side[11] tis evident that on Nov: 28. 1689 the
  1700   error of the Instrumt was about 1′.10″. as it was found by ye Experimts of December 13. 16 & following.

The errors being thus found from No: 28. 1689 to Jan 10 1690 to be 1′.10. & from Feb 14 to Dec 16. 1690 to be 1′.30″ it remains to be enquired what it was from Sept the 13 to No: 28 & When it was that it encreased from 1.10 to 1.30 betwixt ye 10 January & 14 Feb: for the first

7   1689 Sept 13 γ Capricorni   69.29.10   δ [Capricorni]   68.57.30/5
        Octob 26   69.29.10   68.57.25
     1690 Sept 15 & Oct 29   69.29.10   Sep 25)   68.57.55

The annuall access of thes stars to the pole is 15″ which being applyed the error will be found in Sept 1689 to be 45″[12] or 50″ as on ye left hand page[13]

Novemb. ye 15 λ Piscium	51°.23′.50″	η [Piscium]	37°.44′.50″: 1689		
Novemb ye 28	51 .24 .05		37 .45 .10		
	15		20		

whence tis apparent yt the error was greater by about 15″ or 20″ on ye 28th of November yn on ye 15 or 16. & since on ye 28 it was 1.10 on ye 15 it will be 0′.50″ or 55″ as it had been found by experimts on ye 5th & 8th of October 1690: Jan 10 & 13 ye Meridionall distances of ye Bulls south Eye from our Vertex were ye same 35°.38′.25″=807.72: so yt ye same error still continues

				Rev cents
			° ′ ″	Rev cents
January ye 23	I find it		35.38.25	=807.75
			30	
Feb:	2	. . . .	35.38.30	=807.77
	5	. . . .	35.38.30	=807.76
	8	. . .	35.38.35	=807.79
	10	. . . .	35.38.35	=807.80
	14	. . . .	35.38.35	=807.75::
	18	. . .	35.38.35	=807.80
	19	. . .	35.38.40	=807.79
	22	. . . .	35.38.40	=807.79
Mar	13	. . . .	35.38.40	=807.78
			35	

If wee allow the parallax of the Orbe sensible at this star the earth recedeing from it, its latitude must become less & consequently its Zenith distance must be diminished on this account.

And if we admit the Nutation perceptible in it, the pole of our Globe approaches it on this account & its zenith distance is still More diminisht by ye Nutation.

But for all these causes workeing the same Way to make the Zenith distances less they are found encreaseing continually from ye 23 of January to ye 22 of Feb: not accounting any thing for its access to the pole on ye account of ye Recess of ye æquinoxes because altogeather insensible in a moneths time at this star.

Hence I conclude the error of the Instrumt encreased 20″ betwixt ye 10th of Jan: & ye 22 of February. that haveing stated it on Jan 10 onely 1′.10″ on ye 2th of Feb. it was encreased to 1′.20″ & on ye 7th or 8 to 1′.30″ which error continued till that yeare following to December ye 16. on which day the observations end the threds being broke. Before the begining of ye New Yeare 1691. the Instrument againe sunke & ye error found bigger afterwards. D G (July 1. 1700)

153   July 2 ♂
1700
From these collations I have stated the Errors of the Instrumt from the 13 of Sept. 89 to ye 16 of December (90) as in ye top of page 150.[14]

And hereby corrected the Zenith distances obs[erv]ed & copied from ye Instrument of a good Number of remarkable fixed Stars, chosen conveniently for finding the Error of the Instrumt in future yeares & discovering ye Nutation.

To each of the Stars when first observed I have added the Variations of their distances from ye Pole for one degree increase of their longitudes whereby their true distances from ye Vertex or Pole may be gained for any time to come within an age & compareing them (correct by ye Variation) with ye observed the error of ye Instrument & Nutation (if sensible) discovered & determined:

July 3. 1700
☿
In ye end of year books I have inquired the Errors of the Instrument by severall stars not haveing any regard either to the parallax of ye Orbe or Nutation which *their agreemt shows to be small*, but here I shall pitch onely on such as may serve to determine the Errors more accurately & give the Nutation if sensible

And therefore I shall make use of those stars yt lie nearest the æquinoctiall colure employeing also ye Solstiall Stars as often as I find them observ[ed] at ye times of ye equinoxes

Had I been aware of the Meridionall Zenith distances being corrupted by ye Parallaxes & Nutation when I first began to employ the Murall Arch I had been as carefull to forecast for these as I was for the Pole Star on another account: Since I was not, it can not be suspected that any of them are wrested to shew what they would not afford I give the observations simply as they were copied from ye Instrumt my Reader if skillfull will see whether they are justly applyed or no

155
I must onely acquaint him that when the zenith distances of any stars have been observed severall nights togeather one after another and there is some small difference betwixt them I make use of that which is biggest because that when through hast due care has not been taken to cleane ye edge of ye Index the little dirt and filth adhereing to it sometimes makes ye distance numbred on ye diagonalls 10 or 15 seconds less yn it ought to be or really was and would have been numbred had the dust been wiped off:

I find by compareing my observations of the Suns Meridionall distances from ye vertex at ye Solstices & the latitude thence deduced wth the latitude found by ye pole star that some such fault has been committed as requires the allowance *of about ¼ of a minute to be added to all the Zenith distan[ces] observed,*

* which might happen by ye stretching of ye feet or bending of ye beam compasses when the points of 60 and thirty degrees were layd of. & *that this must
be applyed in all the measures taken whether ye stars past the Meridian to ye North or
South of our Vertex or rather 20″ when above 40 degrees South or North 10″ when less*

When therefore these observations come to be applyed either 15″ must be
deducted from ye errors [(]which are allways to be substracted from ye Zenith
distances) in ye southern part of ye Arch or the Zenith distances correct by the
simple errors must be augmented 15 seconds.

I shall Copy but a few observations of those many I have employed at ye
end of the year books or Diarys of my Observations where I have sought out
the errors, but rather excerpe such observations as I have not yet made use of
in the Inquiry of ye Errors & which I thinke most proper on all accounts
whereby the Errors I have formerly determined will either be confirmed or
corrected

And herein I shall take care to compare Observations of Stars made at
ye same time of ye year cheifly for finding ye errors because then neither ye
Nutation nor parallax of ye orbe can affect them, but for enquireing ye
Nutation it selfe I must compare observations of stars lieing neare ye Solstitiall
Colure & taken about the Solstices with other observations of the same stars
got near ye æquinoxes. or of ye stars of Virgo & Pisces taken at ye same
times

Hence I conclude ye error of ye Instrumt

1691		′ ″	
January.	20	2.05	per stars in Gemini
February	20	2.10	per [stars in] Gemini & γ Tauri & Virginis
Martij	10	2.15	per ν & γ Virginis
Ap:			
May			
June		2.15	per α Leonis & α Virginis
July			
August			
Sept: Ineunte		2.20	per β γ κ λ Piscium
Octobris		2.25	
Novembris. 2.		2.25	per α Leonis
Novem	22.	2.35:	
	23.	2.35	per stars in Gemini
	24.	2.30	
Decembris.	1.	2.30	
	9.	2.35:40 Statuatur 2′.40″ vel 2′.35″	

157

343

It may be an Argument for ye Nutation of ye Earths Axis that the Errors of Instrument are found greater by about 15″ or 20″ seconds by the stars in Gemini then by the Stars of Pisces in December. for . admitting ye Nutation to be about ¼ or ⅓ of a minute

July 8 ☽
1700

In ye moneths of March & September the stars in ye foot of Gemini are nearest ye vertex & ye error will be [the] same yt it is found by ye stars of Pisces & Virgo

But in June & December ye said stars of Gemini with those yt lie near ye opposite partts of ye same Colure will be remotest from our Pole & Vertex, & therefore the Error of the Instrumt greater by them yn by ye stars of Virgo & Pisces by about ¼ or ⅓ of a minute.

July 9 ♂
1700

159

I find but few observations of ye Meridionall Zenith distances of ye stars of Virgo & Pisces taken this yeare because most of my paines was employed in constellations remote from Eccliptick & therefore can onely determine the Error of the Instrument from other stars as I have done in this page. but from some few I had of them Arguments may be drawn for a Nutation of ye Earths Axis

1692		′ ″	
	1.	2.40 per α Libræ	
Januarij	23 Error	2.45 per α Leonis & η μ Geminorum	
Febr.	16	2 50 per stars in Gemini	
Martij	4	2.50 per γ Geminorum	
Apr.	22	2.50 per α Virginis & α Leonis	
Maij	4	3.00 per α Virginis	
	16	3.00 per α Virginis	
	& 19	*3.20 25 per δ & β Scorpii	
Julij & Aug 15		*3.30 per π Sagittarii vid year book	
Sept	14	3.05 per β γ Piscium ⎫	Arg Nutationis si conferetur
	28	3.15 vel 20 per π ρ o: ⎬	cum observationibus
		η μ Geminorum⎭	Decemb
No:	27	3.20 per α Virginis	
Dec.	5 & 11	⎰3.20 per β γ ε Piscium ⎫	Argumentum
		⎱3.35 per h η μ Geminorum⎭	Nutationis

July ☿ 10
1700

NB. the errors come bigger by ye stars of Scorpio & Sagittarius in June & July then α Virginis before

Sept 28 rather bigger by ye stars of Gemini then of Pisces

Dec 5 & 11 certeinly bigger. an Argument of a very small Nutation for it was encreaseing Sept 28.

* The errors in March & September may be taken from ye observations of ye stars of Pisces Gemini & Virgo indifferently but in June & december onely from ye stars of Virgo & Pisces which are good all the yeare

1693

Martij	4 Error	3.40 per star η or μ Geminorum
	18	3.40 per stars in Virgo
May	5	3.55 per stars in Virgo
July	21	4.00 per star in Sagittarius
Sept	25	4.00 per star η μ Geminorum
	26	4.00 per π Sagittarii & β γ ε Piscium
No	20	4.00 per γ κ λ Piscium
Dec	19	4.25 per stars in Gemini
NB Sept 4 Error		3.35 per star in Pisces

NB this yeare the errors are found less by ye stars of Pisces on ye 4th of Sept then by ye stars of Gemini & Sagittarius by ¼ or ⅓ of a minute as they were ye last yeare

161

1694

		′ ″	
Januarij	5 Er[ror]	4.20/5	per stars in Virgo
		4.20	per stars in Gemini
Feb:	21	4.25	per stars in Gemini & α Leonis
Mar	14	4.25	per α Leonis & γ Virginis
	29	4.25	per η γ α Virginis
Apr	25	4.30	α Leonis & β η α Virginis & α Leonis
Junij	15	4.25/30	per stars in Sagittarius
Aug	24	4.20	per κ λ Piscium
Sept	21	4.15	per β κ Piscium
	22	4.20	per β κ λ ε Piscium
	23	4.30	per h η μ Geminorum
	30	4.25	per μ ν γ ξ Geminorum
Oct	27	4.30/35	per α Leonis & β γ δ ε Piscium
Dec	13	4.30/35	per η Piscium, γ α Virginis

In this yeare the errors found by ye stars in Gemini & Virgo agree very well togeather those found by ye stars of Sagittarius agree wth them. but when ye stars of Pisces come observable in September ye errors are found less yn by ye stars of Gemini whereas, admitting the Nutation, they ought to be the same & onely lesse yn those of December found b[y] ye stars of Gemini

345

This yeare I find no observations of the stars of Gemini taken in December. whereby I might have resolved this doubt. The errors seeme encreased not above ¼ of a minute[15] betwixt ye 5th of January & ye end of [the] yeare.

Last yeare I wanted observations of the stars of Virgo & Pisces in December. The reason of these defects is because at ye times I was busy about getting a stock of observations for rectifieing stars of ye Northerne & other constellations & therefore thought not of takeing any observations of these stars for determineing ye Nutation. those I use are such as I had taken with a different view & their agreemt shews ye excellency of observations made wth Telescopicall sights & what exactness may be expected from Instrumts of a bigger Radius firmely fixed.

I was ill all this yeare till Michaelmas with ye headach which ended in a fit of the stone. afterwards most of the Headach & stone [are] bettr except when I get great Colds[16]

NOTES

(1) Vol. 39, pp. 135–61. This letter was left incomplete; it is not known if it was ever sent.

(2) In the date, 'Junij' is overwritten 'Julij'; the letter was continued into the latter month.

(3) There is no 4th Book of the *Principia*. Flamsteed was referring to the 3rd Book, the *De Mundi Systemate*.

(4) Nutation. The pole of the Earth does not describe a circle (see 'Precession' in Glossary), but it moves in a slightly wavy curve the mean distance of which from the pole of the ecliptic remains constant. See Spencer Jones, p. 57.

(5) See Letter 493, note (6), p. 86, and Letter 590, note (2), p. 279.

(6) The colures are the two great circles of the celestial sphere which intersect at right angles at the poles, and divide the equinoctial and the ecliptic into four equal parts. The equinoctial colure is the great circle passing through the celestial poles and the equinoctial points. The solstitial colure is the great circle passing through the celestial and ecliptic poles and through the solstitial points.

(7) See Letter 704, note (1), p. 464.

(8) Flamsteed originally wrote: 1°.00′.

(9) This is sheet 145 of Flamsteed's letter (p. 338).

(10) Here and subsequently, the names of the stars will be written in place of the astronomical symbols for them, which Flamsteed used; e.g. where he wrote ♉α, 'α Tauri' will be printed. The sign * has been replaced by 'star'.

(11) On sheet 148, Flamsteed wrote:

observata dist a Vert

		° ′ ″	° ′ ″		
1690 Sept 15 ✕	ε 45.16.10	e 47.28.40	ζ 45.33.10	f 45.30.25	**A**
& 89 No 28 esset	45.15.50	47.28.25	45.33.00	45.30.15	
reperitur	−0.20	−0.15	−0.10	−0.10	
Sep. 15	1.30	1.30	1.30	1.30 Sept 15. 90 Error	
fit er Sept 15(90)	1.10	1.15	1.20	1.20 Err No: 28. 1689	

1690 Sept 16 ✕	β 49.19.00	γ 49.52.55	κ 51.54.40	com κ 52.03.15	**B**
esset	49.18.40	49.52.20	51.54.25	52.02.50	
	−0.20	0.35	0.15	0.25	
error Sep 28	1.30	1.30	1.30	1.30	
Error (1690)	1.10	0.55	1.15	1.05 Err No. 28. 1689	

1690 Dec 3 } ✕	χ 51.54.$\frac{45}{50}$ com.52.03.05	λ 51.24$\frac{05}{10}$	**D**	
et 11 }				
No 28.89	51.54.25	52.02.50	51.23.45	
	−0.20	−0.15	−0.20	
Error Sept 90	1.30	1.30	1.30	
Error No: 28 (89)	1.10	1.15	1.10	Error No: 28. 1689

1690 Sept 20: ✕	λ 51.24.00	ω 46.19.45	d 45.00.30
esset	51.23.45	46.19.30	45
	−0.15	−0.15	
error Sept 20	1.30	1.30	**C**
1689. No: 28. Error	1.15	1.15	

These stars all alter their declination 20″ per Annum. I have allowed so much in these Collations the alteration of 2 moneths being one 3½″ which is scarce sensible on the limb see my letter to Dr Wallis

(12) Flamsteed wrote: 45′.

(13) On sheet 148, Flamsteed wrote:

1689 Sept 13 γ Capricorni	69.29.10	δ	68.57.30
adfer Anni accessus	15		15
diff esset Sept 90	68.28.55		68.57.15
observata	68.29.40		68.57.55
erroris augmtum	0.45		0.40
err Sept 90	1.30		1.30
error Sept 89	0.45		0 50

(14) On sheet 150, Flamsteed wrote:

1689 Sept 13.	Error 0'.40" or 1.00 per coll 7	July 15. 0.00 per Ex
Octob 26	0 55 per collat 7 or 1.00	Octob 5. 8. 0.55
No: 15 & 16	0 .55 or 55 per Coll 8	No: 15 & Dec 15 0.55
Novemb: 28	1 .10 per Collat 3	Dec 16 & 17 1.10
Dec 16	1 .10 per Exper	
1690 Jan. 10	1.10 per collat 4:	
Feb: 2	1.20 per collat 9	
Feb: 7	1.30 per Collat 5	
continues 8		
to Decemb 16.	1.30 per collat 1. & 2	

(15) Flamsteed has written 'mitude'.

(16) At this point the lengthy letter was brought abruptly to a close.

628   NEWTON TO THE TREASURY
21 JUNE 1700
From a draft in the Royal Mint[1]

May it please your Lordships

It has been usual for the Masters & Workers of his Majts Mint to give 2000 *lib* security to the King before the sealing of the Indenture of ye Mint and to mention the same in ye Indenture. Thus did Sr Robt Harley[2] in the second year of King Charles the first, Sr Ralph Freeman[3] in ye twelfth year of King Charles the second and Mr Slingsby[4] in ye 22th year of the same King. Mr Neale having spent a large estate and being far in debt made his way into ye place by granting half the profits thereof to other Officers of the Mint & giving 15000 *lib* security to the King and had time given him to find this security after the sealing of his Indenture. But his Melter Mr Ambrose[5] (through whose hands all ye coynage was to pass) gave but 2000 *lib* security to Mr Neale. The Coynage being now much less then in the time of any of the Masters & Workers above mentioned, if your Lordships please to accept of the usual security of 2000 *lib* I am ready to give it that it may be entred in the Indenture now to be sealed.[6] All wch is most humbly submitted to your Lordships great wisdome

Is. NEWTON

Mint Office
June 21. 1700.

NOTES

(1) Newton MSS. 1, 64. This is in Newton's hand. There is another version MSS. 1, 66.

(2) Sir Robert Harley (1579–1656), Parliamentarian; Oriel College, Oxford; B.A., 1603; M.P. for Radnor and Herefordshire. He was appointed Master of the Mint in 1626 and he held that office until 1635. By an Ordinance of Parliament, he was reappointed in 1643, from which year he continued to coin with the King's dies. He was discharged from his office in 1649 for refusing 'to stamp any coin with any other stamp than formerly' (Ruding, 1, 408, note 6). His grandson, Robert (1661–1724), became the first Earl of Oxford in 1711.

(3) Sir Ralph Freeman (*fl. c.* 1650). See Letter 621, note (3), p. 322.

(4) Henry Slingsby. See Letter 621, note (4), p. 322.

(5) Jonathan Ambrose. See Letter 557, note (2), p. 217.

(6) Fresh indentures were required on a change of the Master.

629   NEWTON TO CATHERINE BARTON[1]
5 AUGUST 1700
From the original in the Royal Mint[2]

[London] Aug. 5. 1700

[Dear Niece]

I had your two letters & am glad ye air agrees wth you & th[ough the] fever is loath to leave you yet I hope it abates, & yt ye [re]mains of ye small pox are dropping off apace. Sr Joseph [Tily][3] is leaving Mr Tolls house & its probable I may succeed him[. I] intend to send you some wine by the next Carrier wch [I] beg the favour of Mr Gyre & his Lady to accept of. My Lady Norris[4] thinks you forget your promis of writing to her, & wants [a] letter from you. Pray let me know by your next how your f[ace is] and if your fevour be going. Perhaps warm milk from ye Cow may [help] to abate it.

I am

Your very loving Unkle,

Is. Newton[5]

For Mris. Catherine Barton
   at Mr Gyre's at Pudlicot neare
      Woodstock in
         Oxfordshire
By Chipping-Norton Bagg.

349

NOTES

(1) Newton's mother had three children by her second husband, Barnabas Smith, one of whom had married Robert Barton. Catherine (born 1679) was the daughter of this marriage. She was reputed to be a lady of singular charm and beauty, and she had many admirers, among whom were Lord Halifax and Dean Swift. She lived with her uncle for nearly twenty years. In August 1717 she married John Conduitt, Member of Parliament for Whitchurch in Hampshire and later for Southampton. He was elected F.R.S. in 1717 and was much esteemed by Newton. On Newton's death, Conduitt succeeded to his post at the Mint. There was one daughter, Catherine Conduitt (born 1718), who married the Hon. John Wallop, afterwards Viscount Lymington. She died in 1740, leaving one daughter and four sons, from the eldest of whom the Portsmouth family are descended.

Catherine Barton's relationship with Halifax, who became a widower in 1698, has been the subject of much speculation. There is no doubt that she lived in his house, though in what capacity it is impossible to establish with certainty. Although Brewster (II, 270–81) had previously suggested the opposite view, A. De Morgan, as the result of careful investigation (*Newton: His Friend and His Niece*, 1885), has come to the conclusion that she was his wife, but the marriage, if there was one, was never publicly acknowledged. What is certain is that in a will, dated 10 April 1706, Halifax left her an annuity of £200 per annum which was purchased in Newton's name and in a codicil, dated 1 February 1712, he 'did appoint his Executor to assigne the said Annuity to the said Catherine Barton' (MS. 127, King's College Library, Cambridge).

(2) Newton MSS. II, 30. The original is almost undecipherable, and many words have had to be surmised. The letter is also printed in More, pp. 470–1, and Brewster, II, 213.

(3) Sir Joseph Tily, attorney-at-law; knighted 1696. See his letter to Newton, 24 November 1707.

(4) Lady Norris, *née* Elizabeth Read, daughter of Robert Read of Bristol. She married (1) Isaac Meynell, goldsmith, (2) Nicholas Pollexfen, merchant, and (3) Sir William Norris, of Speke, near Liverpool. Norris was a fellow of Trinity College, Cambridge. He represented Liverpool in the third, fourth, and fifth parliaments of William III. He was created baronet in 1698. He was seized with an attack of dysentery on a journey from Mauritius to St Helena, and died on 10 October 1702.

(5) The reverse of this letter is covered with Newton's calculations of the value of foreign coins.

630   MEMORANDUM BY NEWTON
[1700]
From the original in the Royal Mint[1]

About 2 years & an half ago Mr Harris[2] meeting wth Mr Croker[3] a Jeweller & finding yt he could emboss desired him to practise engraving & for his encouragemt promised him that as soon as he could grave well enough for ye service of the Mint he should have a place & an house in the mint & an hundred pounds per annum for his service there & thereupon Mr Croker left his own imployment & practised graving & supported himself at his own charge except that he received of Mr Harris by severall little parcells during the first two years until the displacing of Mr Roettiers[4] about 80*lib*. And since ye displacing of Mr Roettiers he hath received 26*lib* wthout any lodgings or further encouragement. And about whitsonday was a twelvemonth When Mr Harris took Mr Fitches[5] house to work in, he promised Mr Croker that he should have Mr Roettiers Salary. And about 3 months ago when Mr Croker complained to Mr Harris that he had no encouragement to go on & prest Mr Harris for a settlement saying that he could not otherwise work a stroke more

Mr Harris replied that he had promised him an hundred a year & that if he would go off he might when he pleased. And on thursday was a senight Mr Croker presenting Mr Harris wth a paper wherein he represented to him that he had promised him Mr Roettiers Salary & a house for graving for the coyn besides what he did for medalls [that he was ordered to grave new puncheons for a five *lib* piece & double Guinea & for ye small money but had no encouragment][6] & desired a settlement to depend upon. To wch Mr Harris (wthout reading the paper) answered in general that he would make a settlement but 3 or 4 days after when Mr Croker spake to him again about it & told him that he was ordered to grave...but had no encouragemt he replied that he would do nothing till he was in the other house & then he would make a settlement.

NB. Mr Harris cannot emboss nor punch, nor draw. All ye assistance Mr Croker had from him was to be told of wt Mr Harris thought was faulty & to be supplied wth pieces of Mr Roettiers & others to copy after.

NOTES

(1) Newton MSS. I, 174. The date is surmised from the contents of the Memorandum.

(2) In 1690 Henry Harris succeeded to the post of Chief Engraver on the death of George Bowers, and against the applications of James and Norbert Roettiers (see Letter 568, note (2), p. 241). The two latter were engaged to engrave tools for coinage. See Letter 569, note (5), p. 245.

(3) John Croker (1670–1741) was born in Dresden, the son of a wood carver. Owing to the early death of his father, his godfather, a goldsmith, took him as an apprentice and he was brought up a jeweller, but whilst he was still young he embraced the career of a die sinker. After visiting various places in Germany and Holland he came to England in 1691, and was appointed as assistant to Henry Harris, the Chief Engraver, about 1696. After the death of Harris in 1704 Croker was appointed his successor.

351

He engraved all the dies for the gold and silver currency during the reigns of Anne and George I. From 1702 he was constantly engaged in medal engraving. Among the medals for which he was responsible were: Queen Anne's Bounty (1704), Union of England and Scotland (1707), Sir Isaac Newton (1726). All bear his signature, 'I.C.' See Craig, *Mint*, p. 202; G. Duveen and H. G. Stride, *The History of the Gold Sovereign* (1962), p. 57; and *D.N.B.*

(4) In March 1697 (see Letter 568, note (2), p. 241).

(5) Thomas Fitch, Weigher and Teller from 1695 until 1701, when he was succeeded by Hopton Haynes.

(6) The square brackets are Newton's.

631   NEWTON TO THE TREASURY
20 JANUARY 1700/1
From a draft in the Royal Mint[1]

To the Rt Honble the Lords Commrs of his Majts Treasury

May it please your Lordships.

The great value put upon French & Spanish Pistoles[2] in England has made them of late flow plentifully hither above all other sorts of Gold, especially the French Pistoles wch are better sized and coyned and less liable to be counterfeited & by consequence of more credit then the Spanish. For Pistoles pass amongst us for 17s. 6d a piece whereas one with another they are worth but about 17s. $0\frac{1}{2}d$ or 17s. 1d at the rate that Guineas of due weight and allay are worth 21s. 6d. And tho allowance be made for the lightness of our Silver monies by wearing yet Pistoles will be worth but between 17s. 2d and 17s. 3d.[3]

About four years ago by the English putting too great a value upon Scotch money the Northern borders of England were filled with that money and Scotland with ours the Scots makeing about 8 or 9 per cent profit by the Exchange untill your Lordships were pleased to put a stop to the mischief. The case being now the same (but of much greater consequence) in the reputed par of the Exchange between the English money and Pistoles, wch runs 3d or 4d in a Pistole too high to the Nations loss in the course of Exchange, we thought it our duty humbly to represent it to your Lordships in order to such a remedy as your Lordships shall think fit.

We presume also to lay before your Lordships that by reason of the great demand of silver for Exportation in Trade, the price of Bullion exceeds that of silver monies 3d or 4d and sometimes 6d or 7d per ounce whereas monies ought to be of great or greater value then Bullion by reason of the workmanship and certainty of the Standard. And this high price of Bullion has not onely put an end to the coynage of Silver, but is a great occasion of melting down and Exporting what has been already coyned. All wch is most humbly submitted to your Lordships consideration and great wisdom

NOTES

(1) Newton MSS. II, 139. The letter, which is unsigned, is in the hand of a copyist, but it is clearly the work of Newton.

(2) See Letter 593, note (6), p. 283.

(3) The circulation of gold coins in 1701 was about $9\frac{1}{4}$ millions. This included foreign coins, mainly louis d'or and Spanish pistoles which had been imported in large quantities and were circulating at 17s. 6d. each. As a result of fresh assays Newton calculated that worn coins were worth 17s. $0\frac{1}{2}d.$, or 17s. 1d. in comparison with the perfect standard guineas, and he advised an official valuation of 17s. The Government accepted his recommendation by a Proclamation dated 5 February 1700/1. The coins were promptly sent to the Mint to be turned into guineas. See Craig, *Mint*, p. 215, S. Dana Horton, *The Silver Pound* (1887), p. 261, also Letter 593, note (1), p. 283.

632   KEMP[1] TO NEWTON

15 FEBRUARY 1700/1

From the original in the Royal Mint[2]

Tower London Feb. 15th 1700.

Worthy Sr

I Desire ye favour of knowing your pleasure Concerning ye Tryall of John Crossly[3] whether you please to be at it, and whether John Sutcliff[3] is bound to Appeare against the said Crossly, I being fearfull Sutcliffe will not except he be allready tide to it

An Answer per bearer will Infinitely Oblige

Your Humble Servt

THO: KEMP

To
Isaac Newton Esqr Master
& Worker of his Majesties
Mint in ye Tower of
     London

NOTES

(1) Thomas Kemp, one of the Mint moneyers.

(2) Newton MSS. II, 640. The letter is addressed to Newton instead of to Sir John Stanley, the Warden and proper authority (see Letter 635, note (6), p. 357).

(3) John Crossly, probably a counterfeiter; John Sutcliffe, a reluctant witness for the Crown.

633 NEWTON TO BRAINT
26 FEBRUARY 1700/1
From the original in Trinity College Library, Cambridge[1]

Feb. 26. 1701

Sr

You are directed to mark the five pound & two pound pieces & the Crowns & Half crowns with ANNO REGNI DECIMO TERTIO on the edges[2] after Decus et Tutamen[3]

Is. NEWTON

To Mr John Braint
Engineer of the Mint.[4]

NOTES

(1) R. 16. 38, fo. 436. There is no address at the head of this letter.

(2) The gold coins struck whilst Newton was at the Mint were the 20s. piece, together with, in smaller numbers, 10s., £2 and £5 pieces. Silver coins were crowns, half-crowns, shillings and sixpences. Other silver coins were groats, threepences, twopences and pence. The Mint Indenture prescribed '18 ounces in groats, threepences, twopences and pence upon every hundredweight'. The impressing of larger coins with a lettered edge was introduced in 1662, it having been tried unsuccessfully in the time of Elizabeth I.

(3) 'An ornament and a safeguard.' The quotation is from Virgil, *Æneid*, v, 262.

(4) See also Newton's instruction to Braint, Letter 625, where Braint is referred to as 'Provost of the Moniers'.

634 MEMORANDA BY DAVID GREGORY
21 MAY 1701
From the original in the University Library, Edinburgh[1]

1. To discourse with Mr Newton about the change of the Inclination of the Orbite of a Satellite; the draught of my objection is in a paper apart.

2. To gett the æquatiuncula in the Theory of the Moon[2] of Mr Newton spoke to Mr Hally, as Mr Hally wrote to me.

3. To know what Mr Newton & Mr Halley mean by desiring me to leave out somewhat I have of Mr Halley concerning the Comet 1680.

4. To talk with Mr Newton concerning my doubt about inserting some things in my Astronomy. these things are among the Index's &c torn from the end of my Astronomy.

5. To endeavour to gett Mr Newtons table of Refractions.[3]

6. To consult Mr Newton about a preface, & upon the whole.

7. To talk about Euclid especially the data;[4] & if I should write a preface, & what instances put in it.

8. To endeavour to gett his book of Light & Colours,[5] & to have it transcrib'd if possible.

9. To see to gett at least his book *de Curvis secundi generis*.

10. To see if he has any design of reprinting his *Principia Mathematica*[6] or any other thing.

11. To ask Mr Newton about Cassini's figure of a Planets Orbit: & its reconcileing Ward and Keplers hypotheses.[7]

[overleaf] things to discuss w. Halley[8]
3 about Mr Newtons designs, & his own observations of the Variation of the Needles declination[9]

NOTES

(1) Greg. A 68$_2$. There is a copy, in Gregory's hand, in the Library of the Royal Society of London (R.S. Greg. MS. fo. 74).

(2) See Manuscript by Newton (622).

(3) Sent with Letter 496.

(4) Euclid's *Data* consists of a collection of 94 exercises. It contains a number of important theorems omitted from the *Elements*.

(5) The work eventually appeared in 1704. See *Advertisement to Opticks* (672).

(6) Newton appears to have entertained ideas of bringing out a second edition of the *Principia* about 1694 when he began to approach Flamsteed for observations which would enable him to perfect his lunar theory which the new edition should contain. 'I desire only such Observations,' he wrote (Letter 478), 'as tend to perfecting the Theory of the Planets in order to a second edition of my book.'
In the previous and the following pages there are many allusions to the proposed new edition. Newton's duties at the Mint, however, were absorbing much of his time, and it became clear that he would never have leisure to complete the work himself.
We gather from Bentley's letter to Newton (Letter 742) that Newton had agreed to let Bentley act as editor. Bentley however soon afterwards abandoned the task, and eventually Cotes was persuaded to undertake the work. The second edition appeared in 1713 under the editorship of Cotes.

(7) See J. C. Adams, 'On Newton's Solution of Kepler's Problem' (*Monthly Notices of the Royal Astronomical Society*, vol. 43, 1882, pp. 43–9).

(8) Halley returned to London in August 1700. His corrected Chart of the Variation was shown at the Royal Society (see *Phil. Trans.* (1700), **22**, 725).

(9) On the reverse of the Royal Society copy, Gregory has written:

    (1) To talk with Mr Halley about the whole of my Astronomy

    (2) About the frugall part of it, & about a cutter [of] figures in wood, about buying paper &c.

    (3) About Mr Newtons design, & his own observations of the Variation of the Needles declination.

    (4) If he intends any further voyages.

635   MINT TO THE TREASURY
21 MAY 1701
From the original in the Public Record Office[1]

> To the Rt Honble the Lords
> Commrs of his Majties Treasury

May it please your Lordships

In Obedience to your Lordships Order of Reference we have examined the annexed Petition of the Tellers of the Bank praying a reward for telling all the new moneys proceeding from the twelve Generall Remaines[2] at the request of the Officers of the Exchequer; And we find by your Lordships warrants to Mr Neale[3] that all the new Monyes proceeding from those Remaines were told at the Mint; and the tale as well as the weight reported to your Lordships by your Order. We find also that the Tellers of the Exchequer were at first sent to the Mint to tell the same; but when the Coynage encreased so that a sufficient number of Tellers could not be spared from the Exchequer, they were assisted by the Tellers of the Bank. For they sent their Porter William Dumford from time to time to summon the Tellers of the Bank to attend at the Mint, as appears by the Affidavit of the said William Dumford hereunto annexed. We find also by the schedule hereunto annexed, & long since signed by Mr Neale and his Deputy Mr Fauquier,[4] & by Mr Fitch[5] our Weigher and Teller, and Mr Turner who supervised the Tellers and reported the tale, that ye severall Tellers of the Bank were imployed the several dayes certified by the said schedule. And by the annexed Affidavit of the said Tellers we find that they have not yet received any reward for the sd service. But what reward they were promised, or what part of those twelve Remaines was carried from the Mint to other places than the Exchequer we do not find.

All which is most humbly submitted

J STANLEY[6]

Is. NEWTON

Mint Office May 21
1701.

NOTES

(1) T. 1/74, no. 24. This is written by an amanuensis, except the last line of the letter, the address, the date and his signature, which are in Newton's hand. See also note (6) below.

(2) 'twelve Generall Remaines'. Exchequer receipts of clipped coin to June 1696 were melted down in twelve (fortnightly) instalments; the ingots from each were separately delivered to the Mint.

(3) Neale died in December 1699, when he was succeeded by Newton as Master.

(4) John Francis Fauquiere, or Fauquier (d. 1726), was a director of the Bank of England when Neale made him his deputy. He was appointed Deputy Master by Newton at a salary of £60 per annum. He took over a share of the Master's duties from about 1716, when Newton's health showed signs of failing. See Craig, *Newton*, pp. 34 and 116–17.

(5) Thomas Fitch. He was appointed Weigher and Teller in 1695. See Memorandum by Newton (630), note (5), p. 352.

(6) Sir John Stanley succeeded Newton as Warden of the Mint (December 1699). As Warden he took formal precedence at the Mint Board, and signed before the Master their communications with the Treasury. F.R.S., 1698.

636   A MANUSCRIPT BY NEWTON

28 MAY 1701

From the original in the Library of the Royal Society of London[1]

Tabula Quantitatum et Graduum Caloris

Caloris partes æquales	Caloris gradus	Calorum Descriptiones et signa
0	0	Calor aeris hyberni ubi aqua incipit gelu rigescere. Innotescit hic calor accurate locando Thermometrum in nive compressa quo tempore gelu solvitur
0. 1. 2		Calores aeris hyberni

357

2. 3. 4		Calores aeris verni et autumnalis
4. 5. 6		Calores aeris æstivi
6		Calores aeris meridiani circa mensem Julium
12	1	Calor maximus quem Thermometer ad contactum corporis humani concipit. Idem circiter est calor avis ova incubantis
$14\frac{3}{11}$	$1\frac{1}{4}$	Calor balnei prope maximus quem quis manu immersa et constanter agitata diutius perferre potest. Idem fere est calor sanguinis recens effusi.
17	$1\frac{1}{2}$	Calor balnei maximus quem quis manu immersa et immobili manente diutius perferre potest.
$20\frac{2}{11}$	$1\frac{3}{4}$	Calor balnei quo cera innatans & liquefacta defervendo rigescit et diaphaneitatatem amittit.
24	2	Calor balnei quo cera innatans incalescendo liquescit & in continuo fluxu sine ebullitione conservatur.
$28\frac{6}{11}$	$2\frac{1}{4}$	Calor mediocris inter calores quo cera liquescit & aqua ebullit.
34	$2\frac{1}{2}$	Calor quo aqua vehementer ebullit & mistura duarum partium plumbi trium partium stanni & quinque partium bismuti defervendo rigescit. Incipit aqua ebullire calore partium 33 et calorem partium plusquam $34\frac{1}{2}$ ebulliendo vix concipit. Ferrum vero defervescens calore partium 35 vel 36, ubi aqua calida & 37 ubi frigida in ipsum guttatim incidit, desinit ebullitionem excitare.
$40\frac{4}{11}$	$2\frac{3}{4}$	Calor minimus quo mistura unius partis Plumbi quatuor partium Stanni & quinque partium Bismuti incalescendo liquescit, [& in continuo fluxu conservatur.][2]
48	3	Calor minimus quo mistura æqualium partium stanni et bismuti liquescit. Hæc mistura calore partium 47 defervendo coagulatur.
57	$3\frac{1}{4}$	Calor quo mistura duarum partium stanni et unius partis bismuti funditur, ut et mistura trium partium stanni et duarum plumbi, sed mistura quinque partium stanni et duarum partium bismuti in hoc calore defervendo rigescit. Et idem facit mistura æqualium partium plumbi et bismuti.

68	$3\frac{1}{2}$	Calor minimus quo mistura unius partis bismuti & octo partium stanni funditur. Stannum per se funditur calore partium 72 & defervendo rigescit calore partium 70.
81	$3\frac{3}{4}$	Calor quo bismutum funditur ut et mistura quatuor partium plumbi et unius partis stanni. Sed mistura quinque partium plumbi et unius partis stanni ubi fusa est et defervet in hoc calore rigescit.
96	4	Calor minimus quo plumbum funditur. Plumbum incalescendo funditur calore partium 96 vel 97 & defervendo rigescit calore partium 95.
114	$4\frac{1}{4}$	Calor quo corpora ignita defervendo penitus desinunt in tenebris nocturnis lucere, & vicissim incalescendo incipiunt in iisdem tenebris lucere sed luce tenuissima quæ sentiri vix possit. Hoc calore liquescit mistura æqualium partium Stanni et Reguli martis,[3 et mistura septem partium bismuti & quatuor partium ejusdem Reguli defervendo rigescit.
136	$4\frac{1}{2}$	Calor quo corpora ignita in tenebris nocturnis candent, in crepusculo vero neutiquam. Hoc calore tum mistura duarum partium reguli martis & unius partis [Bismuti tum etiam mistura quinque partium reguli martis & unius partis] Stanni defervendo rigescit. Regulus per se rigescit calore partium 146, & mistura duarum partium [reguli et unius partis cupri rigescit calore partium] $140\frac{1}{2}$
161	$4\frac{3}{4}$	Calor quo corpora ignita in crepusculo proxime ante ortum solis vel post occasum ejus manifesto candent in clara vero diei luce neutiquam, aut non nisi perobscure.
192	5	Calor prunarum in igne parvo culinari ex carbonibus fossilibus bituminosis constructo & absque usu follium ardente. Idem est calor ferri in tali igne quantum potest candentis. Ignis parvi culinaris qui ex lignis constat calor paulo major est nempe partium 200 vel 210. Et ignis magni major adhuc est calor, præsertim si follibus cieatur.

In hujus Tabulæ columna prima habentur veræ caloris quantitates computatæ [in proportione arithmetica][4] inchoando a calore quo aqua incipit gelu rigescere tanquam ab infimo caloris gradu seu communi termino caloris et frigoris, et ponendo calorem externum corporis humani esse partium duodecim. In secunda columna habentur gradus caloris in ratione geometrica sic ut secundus gradus sit duplo major primo, tertius item secundo, et quartus tertio, & primus sit calor externus corporis humani sensibus æquatus. Patet autem per hanc Tabulam quod calor aquæ bullientis sit fere triplo major quam calor corporis humani, et quod calor stanni liquescentis sit sextuplo major & calor plumbi liquescentis octuplo major & calor Reguli liquescentis duodecuplo major & calor ordinarius ignis culinaris sexdecim vel septemdecim vicibus major quam calor idem corporis humani.

Constructa fuit hæc Tabula ope T[h]ermometri & ferri candentis. Per Thermometrum inveni mensuram calorum omnium usque ad calorem quo stannum funditur & per ferrum candens inveni mensuram reliquorum. Nam calor quem ferrum calefactum corporibus frigidis sibi contiguis dato tempore communicat hoc est calor quem ferrum dato tempore amittit est ut calor totus ferri.[5] Ideoque si tempora refrigerij sumantur æqualia calores erunt in ratione geometrica, & propterea per tabulam logarithmorum facile inveniri possunt.

Primum igitur per Thermometrum ex oleo lini constructum inveni quod si oleum ubi Thermometer in nive liquescente locabatur occupabat spatium partium 10000, idem oleum calore primi gradus seu corporis humani rarefactum occupabat spatium 10256 & calore [aquæ jamjam ebullire incipientis spatium 10705 & calore aquæ vehementer ebullientis spatium 10725, et calore][2] stanni liquefacti defervientis ubi incipit rigescere et consistentiam amalgamatis induere spatium 11516 & ubi omninò rigescit spatium 11496. Igitur oleum rarefactum fuit ac dilatatum in ratione 40 ad 39 per calorem corporis humani, in ratione 15 ad 14 per calorem aquæ bullientis, in ratione 15 ad 13 per calorem stanni defervientis ubi incipit coagulari et rigescere et in ratione 23 ad 20 per calorem quo stannum deserviens omnino rigescit. Rarefactio aeris æquali calore fuit decuplo major quam rarefactio olei, & rarefatio olei quasi quindecim vicibus major quam rarefactio spiritus vini. Et ex his inventis ponendo calores olei ipsius rarefactioni proportionales et pro calore corporis humani scribendo partes 12 prodiit calor aquæ ubi incipit ebullire partium 33 et ubi vehementius ebullit partium 34, & calor stanni ubi vel liquescit vel deserviendo incipit rigescere et consistentiam amalgamatis induere prodiit partium 72, & ubi deserviendo rigescit et induratur partium 70.

His cognitis ut reliqua investigarem calefeci ferrum satis crassum donec satis canderet et ex igne cum forcipe candente exemptum locavi statim in loco frigido ubi ventus constanter spirabat & huic imponendo particulas diversorum metallorum et aliorum corporum liquabilium notavi tempora refrigerij donec particulæ omnes amissa fluidate rigiscerent & calor ferri æquaretur calori corporis humani.[6]

Deinde ponendo quod excessus calorum ferri et particularum rigescentium supra calorem atmosphæræ Thermometro inventum essent in progressione geometrica ubi tempora sunt in progressione arithmetica, calores omnes innotuere. Locavi autem ferrum, non in aere tranquillo sed in vento uniformiter spirante ut aer a ferro calefactus semper abriperetur a vento et aer frigidus in locum ejus uniformi cum motu succederet. Sic enim aeris partes æquales æqualibus temporibus calefactæ sunt & calorem conceperunt calori ferri proportionalem.

Calores autem sic inventi eandem habuerunt rationem inter se cum caloribus per Thermometrum inventis & propterea rarefactiones olei ipsius caloribus proportionales esse recte assumpsimus.

Translation

Table of Quantities and Degrees of Heat

Equal parts of heat	Degrees of heat	Descriptions and Signs of Heats
0	0	Heat of the air in winter when water begins to freeze. This heat is determined accurately by placing the thermometer in crushed snow when it is thawing.
0. 1. 2.		Heats of the air in winter.
2. 3. 4.		Heats of the air in spring and in autumn.
4. 5. 6.		Heats of the air in summer.
6		Heats of the air at midday about the month of July.
12	1	Greatest heat which the thermometer reaches in contact with the human body. This is approximately the same as the heat of a bird hatching its eggs.
$14\frac{3}{11}$	$1\frac{1}{4}$	Heat of a bath near the maximum which one can bear for a considerable time with the hand immersed and constantly moving. This is about the same as that of blood freshly drawn.
17	$1\frac{1}{2}$	Greatest heat of a bath which one can bear for a long time with the hand immersed and remaining immobile.
$20\frac{2}{11}$	$1\frac{3}{4}$	Heat of a bath at which molten wax floating on it begins to solidify and lose its transparency.

361

Equal parts of heat	Degrees of heat	Descriptions and Signs of Heats
24	2	Heat of a bath in which floating wax, on becoming hot, melts, and is kept liquid continuously without boiling.
$28\frac{6}{11}$	$2\frac{1}{4}$	Heat intermediate between that at which wax melts and that at which water boils.
34	$2\frac{1}{2}$	Heat at which water boils vigorously, and a mixture of two parts of lead, three parts of tin and five parts of bismuth on cooling solidifies.
		Water begins to boil at a heat of 33 degrees and on boiling barely reaches more than $34\frac{1}{2}$ degrees. Iron, as it cools to 35 or 36 degrees of heat, ceases to cause any boiling when hot water is dropped upon it; the same if the iron is of 37 degrees of heat and cold water is dropped upon it.
$40\frac{4}{11}$	$2\frac{3}{4}$	Lowest heat at which a mixture of one part of lead, four parts of tin, and five parts of bismuth on becoming hot melts and remains liquid continuously.
48	3	Lowest heat at which a mixture of equal parts of tin and bismuth melts. This mixture on cooling coagulates at a heat of 47 degrees.
57	$3\frac{1}{4}$	Heat at which a mixture of two parts of tin and one of bismuth is melted, as also a mixture of three parts of tin and two parts of lead; but a mixture of five parts of tin and two parts of bismuth on cooling solidifies at this heat. Similarly a mixture of equal parts of lead and bismuth does the same.
68	$3\frac{1}{2}$	Lowest heat at which a mixture of one part of bismuth and eight parts of tin melts. Tin by itself melts at a heat of 72 degrees, and on cooling solidifies at a heat of 70 degrees.
81	$3\frac{3}{4}$	Heat at which bismuth melts, and the same with a mixture of four parts of lead and one of tin. But a mixture of five parts of lead and one of tin when melted solidifies at this heat on cooling.
96	4	Lowest heat at which lead melts. Lead on becoming hot melts at a heat of 96 or 97 degrees, and on cooling solidifies at a heat of 95 degrees.
114	$4\frac{1}{4}$	Heat at which ignited bodies on cooling cease completely to glow in the dark, and conversely by becoming hot begin to glow in the same darkness though with a very faint light which can barely be perceived. At this heat a mixture of equal parts of tin and regulus of Mars[3] melts, and a mixture of seven parts of bismuth and four parts of this same regulus hardens on cooling.
136	$4\frac{1}{2}$	Heat at which bodies ignited glow in the darkness of the night,

362

Equal parts of heat	Degrees of heat	Descriptions and Signs of Heat
		but not at all in twilight. At this heat a mixture of two parts of regulus of Mars and one part of bismuth, as well as a mixture of five parts of regulus of Mars and one of tin, hardens on cooling. Regulus by itself hardens at a heat of 146 degrees and a mixture of two parts of regulus and one of copper solidifies at a heat of $140\frac{1}{2}$ degrees.
161	$4\frac{3}{4}$	Heat at which ignited bodies in twilight, immediately before the rising of the Sun, or after its setting, glow clearly but not at all in the clear light of day, or only very faintly.
192	5	Heat of coals burning in a small kitchen fire made up of bituminous pit coals burning without the use of bellows. The heat of iron glowing intensely in such a fire is the same. The heat of a small kitchen fire, made from wood, is a little greater, i.e. 200 or 210 degrees. And the heat of a large fire is still greater, especially if urged by bellows.

In the first column of this table are the true quantities of heat computed in arithmetical progression,[4] beginning with the heat at which water begins to turn to ice, being as it were the lowest degree of heat or the common boundary between heat and cold, and regarding the external heat of the human body to be 12 degrees. In the second column are the degrees of heat in geometrical progression, in such a way that the second degree is twice as great as the first, the third twice as great as the second, and the fourth twice as great as the third, and making the first degree the external heat of the human body in its normal state. It is clear by this table that the heat of boiling water is almost three times greater than the heat of the human body, and that the heat of melting tin is six times greater, and that of melting lead, eight times greater; the heat of melting regulus of Mars is twelve times greater, and the normal heat of a kitchen fire 16 or 17 times as great as that of the human body.

This table was constructed by the aid of a thermometer and red-hot iron. By means of the thermometer I found the measure of all the heats up to the heat at which tin melts, and by means of the red-hot iron I found the measure of the remainder. For the heat which heated iron gives up to cold bodies in contact with it in a given time, that is, the heat which iron loses in a certain time, is as the whole heat of the iron,[5] and so if equal times of cooling be taken, the degrees of heat will be in geometrical progression and can therefore easily be determined by a table of logarithms.

First therefore I found by means of a thermometer constructed from linseed oil that if the oil, when the thermometer was placed in melting snow, occupied a space of 10,000 parts, the same oil rarefied with one degree of heat, or that of the human body, occupied a space of 10,256; in the heat of water just beginning to boil, a space of

10,705, and in the heat of water boiling vigorously, 10,725; in the heat of molten tin as it begins to cool and take on the consistency of an amalgam, 11,516, and when all the tin has become completely solid, a space of 11,496. Therefore the oil was rarefied and expanded by the heat of the human body in the ratio 40 to 39, in the ratio 15 to 14 by the heat of boiling water, and in the ratio 15 to 13 by the heat of molten tin as it begins to coagulate and solidify on cooling, and in the ratio 23 to 20 by the heat of tin when in the process of cooling it is completely solidified. The rarefaction of the air was, with the same heat, ten times greater than that of oil and the rarefaction of the oil almost fifteen times greater than the rarefaction of spirits of wine. On the basis of these findings, by making the heats of that oil proportional to its rarefaction, and calling the heat of the human body 12 parts, we obtain the heat of water as it begins to boil 33; of water boiling vigorously 34; of tin when either it melts or begins to solidify on cooling and to assume the consistency of amalgam 72, and when on cooling it solidifies and hardens 70 parts.

These things being known, in order that I might investigate the remainder, a sufficiently thick piece of iron was heated till it was red-hot. It was then taken out of the fire with a pair of pincers which were also red-hot. It was then placed in a cold place where the wind blew continuously about it. By putting on it particles of different metals and other fusible bodies I noted the times of cooling until all the particles, having lost their fluidity, became hard, and the heat of the iron was the same as that of the human body.[6]

Then by assuming that the excess of the heats of the iron and of the solidified particles above the heat of the atmosphere found by the thermometer were in geometrical progression when the times are in arithmetical progression, all the degrees of heat were determined. The iron was placed, not in calm air but in a uniformly moving current of air so that the air heated by the iron was always carried away by the wind and that cold air might replace it with a steady motion. For thus equal parts of air were heated in equal times and received a heat proportional to that of the iron. The heats thus found had the same ratio one to another with the heats found by the thermometer, and therefore we were right in assuming the rarefactions of the oil to be proportional to the heats.

NOTES

(1) This paper was read to the Society on 28 May 1701, and entered in the Early Letter Book (N. 1. 62). It was published with slight modifications, and without Newton's name, in the *Philosophical Transactions*, **22** (1701), 824–9, under the title *Scala Graduum Caloris. Caloris Descriptiones & Signa*. It is particularly noteworthy inasmuch as it is one of the very few papers of a chemical nature to be found among Newton's published works, despite the fact that the subject had occupied his attention for many years. See Letter 546. It is also worthy of note that Newton here assumes the constancy of freezing and boiling points, although a distinction is made between the heat (i.e. temperature) at which water begins to boil and that at which it boils vigorously. In the table, Newton altered *deferviendo* to *defervendo* but did not make the corresponding alteration in the rest of the manuscript.

(2) The square brackets seem to have been inserted later on Newton's original copy. They do not appear in the copy in the *Philosophical Transactions*, and they have been ignored in the translation given.

(3) Stellate Regulus of Mars was metallic antimony, the term antimony being used in Newton's day for the ore (antimony sulphide). The 'regulus' or metal was obtained by fusing the ore with small pieces of iron and so reducing the sulphide to the metal which gave star-shaped crystals on cooling, hence the name *stellate regulus*.

(4) For *computatæ* Newton wrote *computum*. The words *in proportione arithmetica* do not appear in Newton's original copy, but they do appear in the copy in the *Philosophical Transactions*. They have been included in the translation given. *In ratione geometrica*: the numbers in the *first* column (not the second) are in geometrical progression, the common ratio being approximately 1·19.

(5) This was later known as Newton's Law of Cooling: the rate of cooling of a body at any instant is proportional to the excess of its temperature over that of its surroundings, i.e. if θ denote this excess of temperature $d\theta/dt = -k\theta$, whence $\theta = Ce^{-kt}$.

(6) A comparison of the temperatures measured on Newton's linseed oil thermometer, and their equivalent centigrade values, is given below:

	Newton's linseed thermometer	Corresponding centigrade value	Modern centigrade value
Freezing water	0 parts	0°	0°
Body temperature	12 parts	35°	37°
Boiling water	34 parts	100°	100°
Lead (20%)-tin (30%)-bismuth (50%) alloy hardens	34 parts	100°	
Lead (31%)-tin (19%)-bismuth (50%) alloy melts			94°
Bismuth melts			
Lead melts	81 parts	238°	271°
Antimony melts	96 parts	282°	327°
Iron hardens		420°	630°
Iron melts	146 parts	431°	
			1533°

As it is unlikely that the metals used by Newton would have been fully refined, the presence of impurities would account for the melting points being lower than those accepted today.

637   WALLIS TO FLAMSTEED

3 JUNE 1701

From the original in the Royal Greenwich Observatory.
For answer see Letter 638

Oxford June 3. 1701

Sir,

I have not heard of you a great while,[1] but do believe that you be well imployed. I would be glad to hear that your Observations are in the Press; that so great a Treasure be not lost; of which we are in great danger in case you should dy before they be printed. Those of Hevelius were in a good forwardness before he dyed;[2] but, I presume, it would have been better that he had lived to see them printed. And I do not know that you have any Amanuensis who doth so throughly understand yours, as to publish them if you were gone. I understand that, in Germany, (as M. Leibnitius signifies in a letter to me) that they are going to erect large Instruments for observing the Earths Annual Parallax; which is, I suppose, in pursuance of what you have done.[3] I wish you would pursue that busyness yet further. And particularly, that you would examine your Observations of *Lucida Lyræ*;[4] which is a great Star, (& may be presumed nearer us than those that seem lesser) and is as near to the Pole of the Zodiack as is the Pole star; and, (being bigger than it, & brighter,) may be fitter for that purpose. I am not likely to live so long as to see your Observations published;[5] but, however, I would not have the publike loose them. I am

Sr

Yours to serve you,

John Wallis

For Mr Flamsteed, at the
   Royall Observatory in
     Greenwich.

[Added in Flamsteed's hand] Recd 24th Junij 19 1701

NOTES

(1) Vol. 33, fo. 55 (from back). With Flamsteed's letter of 10 January 1698/9, correspondence between the two men appears to have ceased. There had been many delays in preparing Flamsteed's observations for the press (see the series of letters which begins 11 December 1704); Wallis was aware of this, hence his anxiety that the result of Flamsteed's labours should be available to the public during his lifetime.

In his reply Flamsteed assured Wallis that in the event of his death there was no danger that the result of the labour of thirty years would be lost. The work, *Historia Cœlestis Britannica*, was not finally published until 1725, by which time Flamsteed had been dead nearly five years. It was seen through the press by his assistants, Sharp and Crosthwait. (A pirated copy under the editorship of Halley appeared in 1712.)

(2) Hevelius died on 28 January 1687. Wallis is referring to Hevelius's *Machinæ Cœlestis Pars Prior* (1673) and *Pars Posterior* (1679).

(3) See Wallis's letters to Flamsteed concerning *Parallax.*

(4) Vega. After Arcturus, this is the brightest star in the northern hemisphere of the sky. Its distance is 26 light years.

(5) Wallis died 28 October 1703. He was in his 86th year when he wrote the above letter.

638 FLAMSTEED TO WALLIS
24 JUNE 1701
From a copy in the Royal Greenwich Observatory.[1]
In reply to Letter 637

To Dr Wallis[2]   *The Observatory June ye* 24 1701
Reverend Sr.

Tis not for want of respect but onely to gain Time to perfect my Catalogue of ye fixid Stars that I have for born to write to you this two Yeares You very well apprehend that my Observations would be of little or no use If printed without it. Last week I received the places of ye fixed Stars of 9 Constellations computed by a Calculator I have hired in Derbyshire.[3] my Domesticks have Compared them wth ye Same done here and I hope they will be Inserted into ye Catalogue before this week be ended after wch I shall have ye places of 2200 fixed Stars determined in it and only 5 Constellations remaining viz Cepheus Cassiopeæ Draco ye Greater and the Lesser bear of wch in Draco there remainis a good part in ye Lesser bear only some few Stars to be Observed.

Whilest these have been carrying on I have determined above 100 places of ye Planet ♂ from my own Observations taken betwixt ye Yeares 1671 & 1701 wth the new places of ye fixed Stars and Compared them very lately wth ye Rudolphin Numbers[4] *whereby Something is discovered in his Motions that the Theorists think nothing of*

I think I told you formerly that I had by me 30 places of ye moon near her Northern Limite and as many about her Southern & above 60 on her Quadratures, and Some more on her oppositions to ye Sun derived from Observations take[n] here betwixt ye Yeares 1689 and 1701 wth ye New places of ye fixed Stars.

Those 200 places of ye Moon[5] I imparted to Mr Newton taken in all places of her Orbit extend no further then from ye year 1689 to 1695 and were determind by ye help of a Small Catalogue of ye fixed Stars on wch (tho they were much more acurate then Tychoes[)] I durst not rely; a many of them are repeated in those above specified and ye rest shall be recalculated when I have Leasure.

Of ♄ and Jupiter I have by me above 100 places computed (wth the New places of ye fixed Stars) from Observations taken wth ye mural Arch betwixt ye Years 1689 and 1701

The Observations themselves Ly all fair transcribed in ye Same Order wherein they are to be printed. and my Amanuensis James Hodgson[6] knowes very well wt Corrections are to be made in any of them and how to find them out when required

He is a Sober Young man about 22 Years of Age. A very good Geometrician and Algebraist Understands ye Series and Fluxions Tho I have not suffered him to spend much time in them because I could not spare him from ye Calculation work he understands ye Latin Tongue indifferently, haveing got since he became my Servant he knows my method and is acquainted wth all my Labors and will easily finish and print them If god should call me hence before I shall have perfected them myself.

But since the Allwise Creator of ye heavens has thus far prospered my Endeavours beyond my hopes or expectations I cannot doubt but he will afford me both Life health and means to finish and publish them myself.

My Youngest Se[r]vant Tho Weston[7] has been Educated wth Learning has a good Talent at drawing and I design to set him to draw ye Mapps of ye Constellations this Summer and perhapps to Engraveing plats for them for those that draw well seldom fail of Ingraving as well

I give you this account of ye present posture of my affairs that you may see that *If I should dye before ye Catalogue be finished there is not ye least danger of loseing either it or any part of my Long and painfull Labours* to perfect wch after ye Catalogue shall be finished.[8] The places of all the planets are to be Calculated from ye measure taken wth ye Sextant betwixt ye year 1676 (when I first sat down) and ye Year 1689 (when I built the Large mural Arch) which being Numerous and much more dificult to manage then those taken wth it will require good and Skillfull Assistances to Calculate them.

As allso ye places of ye planets since observed wth ye meridional Arch for I have only calculated the principle at ye □ and ☍ of ye Sun or at remarkable times as I have given you an hint before

These cannot be Set upon till ye Catalogue is finished but ought then to be done to render my work Compleat

for Tho wee have all Thyco Brahe's Observations by us Yet I find not that Kepler Bulliand[9] Wing[10] Street[11] or any of our Theorists have been at ye paines to Compute the places of ye Planets from any of them but take such of them as they found ready Calculated to their hands

Mr Newton had done nothing in ye Theory of ye Moon If I had only given him ye Observations here made

* 3 yeares pains at spare times whilst I was laying in a stock of observations wth ye murall Arch

I was forced to give him ye p[l]aces Computed by myself and servants from them and repeated carefully *[12] as allso her places computed in like manner and repeated from my own Tables (grounded on ye Horroccian Theory) wth all ye Ellements of ye Calculation whereby he was shewed at Once, *in wt parts or positions of her Orbit in respect of ye Sun ye notable Errors happened (and compareing them wth such as ought to be according to his Theory of Gravitation how they might be taken away*[)]

368

Since the World will have ye Use, and the King and Nation the Honour of ye Work under my hands it ought to be rendered as Compleat and perfect as it Can. I shall publish Tables wth it that will render ye Calculation of ye planets places from them easy and expeditious but our Theorists are Clamorous and will Complain If ye whole paines be not Spared them. I will doe wt Lyes in my power that it may and If I cannot procure ye help requisit for this purpose I shall let ye World know that 'tis not my fault but theirs (who have for reasons I will not mention) misrepresented my Labors

Of wch that you may have a truer and more perfect apprehension you will permitt me to tell you that the Calculations I have been Obliged to make or made by my Servante fill above a dozen hansom 4to Volumes besides wt has been done by my Country Calculator and a Couple of folioes of Collections and Synopses of ye Constellations employed in ye work. Our Theorists know little or nothing of this Yet by Clamoring and Calling for my observations as If they were as easily wrought up as a Set of Theorems and Corollaryes[, they (the Theorists)][13] have given to the World a false Idea of my Labors and prejudiced and hindered me from Obtaining ye help and Assistance I have need of to render my work as Compleat as it ought to be. *If you would perticularly advise Your Colleague Dr Gregory to have a Care of discourseing of things he is not acquainted wth and has only false, imperfect, or prejudcied information of* You would oblige me much: had he pleased to visite ye Observatory when he was lately in Town as Mr Keile[14] did I should have taken it kindly he should have met wth a Civil reception and found (as Mr Keile did) that I would not have remembered wt passed 30 months agone.[15] I wish him health and Success in his Labors.

I am glad to find that tho our Nation takes little notice of my letter concerning ye parallax of ye Pole Star.[16] Yet forreigners are excited by it to build large instruments on purpose to Examin it. My state of health permitts me not any longer to sit up for hours to geather, in ye Night for Observing as formerly Nor I bless God for it have I now ye reason I had I have a large Stock by me and it fully Imploys mine and my servants time to work them up Nevertheless I take Care to Observe *myself* the Planets and ye Eclipses of ye Satellites as formerly in Order to correct their motions as soon as my Servant shall have gotten Skill enough in ye doctrine of Gravitation to Settle the moĉon of ♃ their primary Planet wch I find will creat more trouble then at first I Expected I must threfore Leave ye further Enquiry into ye Parallax of ye Orb to my Young men or Forreigners till I can build larger and better Instruments wch I may not hope for at ye publick Charge I may perhapps in good time make at my owne

Your life and health is ever heartily prayed for

by Sr

Your respectfull & humble Servt

J. FLAMSTEED

Tho Your Letter was
dated June ye 3rd it came
not to my hands till ye 19th
wch is ye reason You have not
this answer sooner

To the Reverend Dr. Wallis at his
hous in Oxford   these

NOTES

(1) Vol. 33, fo. 55, numbered from the end.

(2) Only these words are in Flamsteed's hand. The rest of the letter, including the signature, is in the hand of a copyist.

(3) Possibly William Bossley, or Luke Leigh (see Flamsteed's *Memoranda* (491), note (1), p. 82). Each of these came from Derbyshire.

(4) See Letter 483, note (12), p. 56.

(5) See Flamsteed's *Memorandum* (516), pp. 135–6.

(6) James Hodgson (1672–1735) was Flamsteed's assistant from 1696 to 1702. He acted as co-editor of the *Historia Cœlestis Britannica*, and was elected F.R.S. in 1703. He married Flamsteed's niece, and Flamsteed had a very high opinion of his abilities. He became mathematical master at Christ's Hospital in 1709, in succession to Samuel Newton, and he continued in that post till his death in 1735. See Letter 499, note (5), p. 105.

(7) Thomas Weston came to the Observatory as assistant to Flamsteed early in 1698/9. He was a skilled draughtsman and was responsible for the drawing of Flamsteed's constellation maps. He left the Observatory in May 1706 to become a teacher of mathematics.

(8) See Letter 637, note (1), p. 366.

(9) Bullialdus. See Letter 484, note (13), p. 60.

(10) See Letter 484, note (2), p. 60.

(11) See Letter 474, note (7), p. 32.

(12) A marginal insertion, still in the hand of the copyist.

(13) This seems to be what Flamsteed meant. Actually the copyist has written: 'as easily wrought up as to Sat of Theorems and Corollaryes have given to the world a false Idea of my Labors.'

(14) John Keill (1671–1721), Scottish mathematician; Professor of Astronomy at Oxford.

(15) That is, about the end of 1698.

(16) See Index to the present volume under 'Parallax'.

639   A MANUSCRIPT BY NEWTON

?JULY 1701

From the original in the Royal Mint[1]

Directions about the Triall of the monies of Gold & Silver in the Pix.[2]

About a month before the triall of the monies in the Pix his Majty or his Council appoints the time & place for that triall and names (if he pleases) the Lords of the Council[3] before whom it shall be made & notice thereof is sent to the Lord High Chancellour & Lord High Treasurer of England & from the Treasury to the Officers of the Exchequer & Mint.

The Lord Chancellour presently after this Order is to be waited on by ye Officers of ye Mint to send his Letter to ye Warden & Company of Goldsmiths to return him the names of an able Jury for that Triall. and a paper is to be delivered to his Lordship conteining the form of the triall.

The return of the names being made to his Lordship he sends his Warrant by his Serjeant at Arms or Serjeants Deputy to summon the Jury by their names to meet at the place & time appointed for the triall.

The Chancellour of the Excheqr is also to be attended about a dinner for ye Jury. The place has usually been at the Dogg Tavern[4] in ye Palace yard.

The Lord High Treasurer or Chancellour of the Exchequer sends an order for the Standard Troy weights & the Indented Tryall pieces of his Majts crown gold Monies & standard silver monies kept in the Treasury of the Excheqr to be delivered upon the day of the Triall for the use of the Jury

In the mean time the King's, Warden's, Master's & Comptrollers Clerks severally extract out of all their books the weight & tale of all the monies taken up from time to time since the Triall of the last Pix & the tale of all the monies in the present Pix & when their extracts are all compared & agreed 40 or 50 Copies thereof are written fair to be distributed to the Lords & Jury at the Triall

On the day of the Triall the Officers of the Mint cause the Pixes to be brought before nine in the morning & placed on a Table where the Lords are to sit.

About tenn the King if he pleases ye Ld Chancellour & other Lords of the Council take their places at ye Table aforesaid, the Officers of ye Mint Excheqr Officers & Goldsmiths attending: and the Pixes are opened by the Officers of the Mint either before the Lords sit down (if they direct it) or after they are sat.

The Lord Chancellour calls to his Serjeant at Arms for his Warrant for summoning the Jury & appoints the same to be delivered to the Remembrancer's Deputy who attends there wth the copy of their Oath.

The gold & silver monies in the Pixes being poured on the Table the Lord Chancellour causeth ye said Oath to be administred to ye Jury.

The Standard Troy weights & the Indented Tryall pieces of his Majts Crown Gold & standard silver monies kept in the Treasury of ye Exchequer being delivered for the use of the Jury, the Lord Chancellour gives ye Jury in charge to make triall of ye said Gold & Silver money by ye said standards of his Majts Treasury according to ye rules set down in the Indenture of the Mint & to do it justly wth all possible care & exactness it being a business of a very publick concern. In the absence of ye Ld Chancellour the Chancellour of ye Exchequer administers the Oath & gives the charge

The Charge being given the Ld Chancellour appoints the Jury when & where to attend his Lordship wth their Verdict & then departs wth ye rest of ye Lds unless any of them be appointed to see the Assays. And the Warden Master & Comptroller lock up the monies & with the Jury withdraw into another room where they take their places according to ye order of the Mint for trying the monies.

The Officers of the Mint examin the Indented Triall pieces of the Exchequer by ye like triall pieces of the Mint in the Wardens custody & ye Jury doth ye same by the Triall pieces of ye Company of Goldsmiths. They also inform themselves about the triall & for that end inspect the Indenture of the Mint delivered to them by the Master & Worker & so proceed to ye triall.

This triall is made by weight fire & water after the same manner as is done in the Mint at the Pixing of the Monies before deliverance the Jury being the Kings Assay masters Weighers & Tellers sworn in the room of the Assaymaster Weigher & Teller of ye Mint to try the monies according to ye rules in the Indenture & by consequence in presence of the Warden Master & Comptroller who as Officers of the Mint are to see that this triall as well as those made in the Mint be duly performed & the Warden & Comptroller or their Deputies are to enter of Record how much the monies prove too feeble or too strong & make a true accompt thereof to his Majty.

The Pix being opened the Jury tells & weighs all the money therein (Gold & Silver apart) & reccon how much it makes in the pound weight. And when they have told out of every species of Gold or Silver monies so much as should make a pound or a pound & an half or two pound weight they weigh it & melt it into an Ingot & weigh ye Ingot & ye grains & if the wast exceed not 8 or 10 grains in ye pound weight they assay it, but if the wast be too great they tell out other money & melt that into an Ingot & assay it, & the Assaymaster reports the Assay in ye same manner as is done in the Mint.

Then the Jury draw up & signe their Verdict expressing the weight & Tale of all the monies in ye Pix & how much they make in the pound weight &

that they are agreeable to standard or better or worse according to ye report. And in the mean time the Warden & Comptroller may write down how much the monies are too feeble or too strong in weight & allay for their private satisfaction[5]

At the time appointed by the Lord Chancellour the said Officers of the Mint & Jury attend his Lordship, & the Jury presenting their Verdict to him the Remembrancers Deputy attends to read it & keeps it when tis read, & enters it of record for the Warden & Comptroller who are to make[6]

NOTES

(1) Newton MSS. I, 228. There are several drafts of this in the Royal Mint. See also Memorandum 614.

(2) On 6 August 1701 Newton experienced the first Trial of the Pyx of gold and silver coins made during his mastership and of those made during the last six months of Neale's occupation of that office. The pyx was a box in which sample coins were locked. Periodically these coins were tested for weight and fineness by a Jury of the Company of Goldsmiths named by the Company and summoned by the Lord Chancellor with representative members of the Privy Council and certain legal officials. See also Letter 725 and Craig, *Mint*, pp. 394–407; also J. H. Watson, *Ancient Trial Plates* (1962), pp. 6–9.

(3) Newton uses this term 'Lords of the Council' to mean no more than Privy Councillors.

(4) There were several taverns of that name. There was a 'Dog' in Holywell Street, and another in Ludgate Hill, but it does not seem likely that either of these was the one referred to here.

(5) The last four words have been substituted for 'in order to enter the same of record'.

(6) The account of the procedure stops short at this point.

640 NEWTON TO THE TREASURY
28 SEPTEMBER 1701
From the original in the Public Record Office[1]

To the Rt Honble the Lords Commrs of his
Majties Treasury.

May it please your Lordships

By the late Edicts of the French King for raising the monies in France, the proportion of the value of Gold to that of Silver being altered, I humbly presume to give your Lordships notice thereof.[2] By the last of those Edicts the Lewis d'or passes for fourteen Livres & the Ecus or French crown for three

Livres & sixteen sols. At wch rate the Lewis d'or is worth 16s 7d sterling supposing the Ecus worth 4s 6d as it is recconed in the course of exchange & as I have found it by some Assays. The proportion therefore between gold & silver is now become the same in France as it has been in Holland for some years. For at Amsterdam the Lewis d'or passes for nine Guilders & nine or ten styvers[3] wch in our money amounts to 16s 7d & it has past at this rate for the last five or six years.

At the same rate a Guinea of due weight & allay is worth 1*lib*. 00s. 11d.[4]

In Spain Gold is recconed (in stating Accompts) worth sixteen times its weight of Silver of the same allay, at wch rate a Guinea of due weight & allay is worth 1*lib*. 2s. 1d, but the Spaniards make their payments in gold & will not pay in silver without an abatement. This abatement is not certain but rises & falls accordingly as Spain is supplied wth Gold or Silver from the Indies. Last winter it was about five per cent.

The state of the money in France being unsetled, whether it may afford a sufficient argument for altering the proportion of the values of Gold & Silver monies in England is most humbly submitted to your Lordships great wisdome

Is. NEWTON.

Mint Office
Sept. 28. 1701

NOTES

(1) T. 1/76, no. 36. All this is in Newton's hand.

(2) One of Newton's recurrent problems at this time was the maintaining of a stable ratio between gold and silver coins which were circulating side by side. This becomes clear from the fact that large amounts of foreign currency, notably the French louis d'or, had been circulating freely in this country and had been passing for 17s. 6d. each. Newton had already indicated (Letter 631) that, compared with the guinea at 21s. 6d., the gold value of these coins was no more than 17s. 0½d., or 17s. 1d. Since the discovery in 1546 of rich silver mines in Potosí in Bolivia, silver had flowed into Europe disproportionately to gold. In England the gold-silver ratio was further reduced by premiums on the guinea from 14½ to 1 when it was first coined (1663) to 15½ to 1 to which after the 1696 enhancement it returned in 1699. Silver coins were therefore being melted down and exported abroad in large quantities where as bullion they realized a higher price. See *The History of the Gold Sovereign* (1962), by Sir Geoffrey Duveen and H. G. Stride, pp. xiv, xv, xvi and 22.

(3) Styver, or stiver: a small copper coin of the Netherlands, equal to one-twentieth of a guilder.

(4) For the fluctuations in the value of the guinea see Letter 593, note (1), p. 283.

641 NEWTON TO THE TREASURY
SEPTEMBER 1701
From the original in the Royal Mint[1]

> To the Rt Honble the Lords Commrs of
> his Majties Treasury

May it please your Lordships

Upon your Lordships Reference of the 7th of August last concerning the qualifications of the Petitioners for the Weigher & Teller's place I humbly conceive it my duty to give your Lordships what light I can into the matter by laying before your Lordships faithfully & according to the best of my knowledge or information the qualifications & services of the Petitioners in Mint affairs: in wch respect one of them at least deserves in justice a particular character. Mr Haynes[2] has been in the Mint about fourteen years except two short intermissions & while he acted there had a general reputation amongst us for integrity sobriety good humour & readiness in business. He has a steady hand, writes very fairly, is a very good Accomptant & skilled in all the business of the Mint & in the Recoynage instructed the Officers & Clerks of the five Country Mints & did other great service. For these reasons the Officers of the Mint (Mr Neale, Mr Hall & my self) then recommended him earnestly to the Lords Commrs of his Majties Treasury as a fit person to execute the Office of Comptroller under the two late Comptrollers in that time of great business & when the Comptrollers insisted on Mr Berisford[3] a stranger, Mr Haynes was appointed to take care of the Comptrollers business till Mr Berisford could qualify himself. Since the Recoynage Mr Haynes has been imployed in the Excise Office about two years by Mr Hall above mentioned who is able to give a further character of his abilities & behaviour in both offices. Yet Mr Neale in recompence of his services & of his examining & setling the accompts of the five Country Mints continued him his Clerk with a salary of 100*lib* per annum and duely paid till his death December 99: wch salary being now ceased Mr Haynes has nothing remaining in lieu of a setled business wch as he represents was of more profit & wch at the instance of some of the principal Officers of the Mint he quitted to serve the publick in the Recoynage in hopes of being further considered when there should be an opportunity, so that at present he is a loser of that service. By reason of his abilities I have ever since wished for him back into the Mint, and if he be not now brought back as he has been once already when we could not be without him he may be so engaged in other business that we cannot have him when we may want him.

375

If any of the principal Officers or their Clerks or Deputies should at any time dye or leave the Mint, Mr Haynes is qualified to assist till the place can be supplied anew that ye business of the Mint receive no stop, & upon any extraordinary occasion to help in our Accompts or other business of the Office of Receipt. for we want men of skill. These things I represent humbly conceiving it for the service of the Mint, to encourage & imploy those who by their past services appear best qualified to serve in it.

All which is most humbly submitted &c

Is. NEWTON.

I do humbly certify your Lordships
the truth of this Report
    THO: HALL.

The principal business of the Weigher & Teller is to weigh the Ingots of Gold & Silver from the Importers into the Custody of the Warden Master & Comptroller who enter the weight in their books & by that weight deliver the Ingots to the Melter & after coynage thereof he weighs it from the Monier to the said three Officers & from them to the Importers by the same weight justly distributed according [to] every mans share & proportion that there be no occasion of complaint to the discredit of the Mint & discouragemt of Importers to whom I am answerable for their Bullion & whose satisfaction is my Interest.

Is. NEWTON.

NOTES

(1) Newton MSS. I, 121.

(2) Hopton Haynes (1672–1749), prominent Unitarian. He was employed at the Mint from boyhood. He was appointed Weigher and Teller, 27 November 1701 (Royal Mint Record Books, v, 30), and was promoted Assaymaster, 1723. He wrote *Brief Memoires Relateing to the Silver and Gold Coins of England* (1700), as well as some theological works. See *Gentleman's Magazine* (1750), pp. 93, 524; John Nichols, *Illustrations of the Literary History of the Eighteenth Century* (1822), II, 150; W. Whiston's *Memoirs* (1753), p. 178; *D.N.B.*

(3) The Beresford here mentioned served as Deputy Comptroller between 1708 and 1717.

GOLDEN SQUARE TABERNACLE ACCOUNTS

642 GOLDEN SQUARE TABERNACLE ACCOUNTS

1701

From a draft in the Royal Mint[1]

The Tabernacle Accompts.

The Accts of Warren relating to the Tabernacle neare Golden Square[2]

Purser
Treasurer

The Accts of Warren Agent for the Trustees Governers & Directers of the Charity of his Grace Thomas A. B. of Cant[3]

Received by Mr Warren for the Trustees in the Quarter ending at Christmas 1700.

Lloyd	Feb 14.	For seats in the Tab. as per Books	11.06.00
Lloyd		Mr Lloyd's Benefaction till Mich.	2.10.00
Sr W. Rich	Mar 8	due at Mich.	8.15.00
Warren		due at Mich	4.05.00
			26.16.00
		For Sr Wm Rich's house a Qtrs rent due at	
		For Warrens house a Qters rent due at	
		Mr Lloyd's Benefaction till	
		Arrears to be received	
Seats		For seats in the Tabernacle till Christm	
Interest		For Interest of 500*lib* at 5 per cent due Oct. 1 1700	6.05.00
	Received by Mr Warren for the Trustees in ye Qter ending at Christmas 1700		
Tabernacle		For seats in ye Tabernacle as per Book	11.06.00
Rich	Nov. 7	For Sr Wm Rich's house a Qters rent due at Mich	8.15.00

377

Warren		For Warrens house a Qters rent	
		due at Mich	4.05.00
St Pauls	Dec 10	Interest of 500*lib* till Mich.	
		pp Mr Spencer	6.05.00
Lloyd	Dec 16	Mr Lloyd's Benefaction till Mich.	2.10.00
			33.01.00

Due before Christmas 1700 & not yet
receiued
For seats in ye Tabernacle till Mich
For Sr Wm Rich's house a Qters rent due
at Mich

Received in Qter ending at Christmas 1700

Ballance	Sept 29	By the Ballance of the last Acct	12.10.08[4]
Warren		For Warren's house a Qters rent	
		due at Mich	4.05.00
Rich	Oct 4	For Sr Wm Rich's house a Qters	
		rent due at Mich	8.15.00
Chest	Oct 15	Out of the Chest of ye C Tr.	
		Dir. & Gov. &c	00.00.00
Lloyd	Oct 20	Mr Lloyds Benefaction till	
		Mich 1700	2.10.00
	Nov. 25	For Interest of 500*li* till	
		Mich 1700 pd by Mr Spencer	6.05.00
Tabernacle		For seats in ye Tabernacle as	
		per book	11.06.00
			33.01.00

Due before Christmas 1700 & not yet
received
      For seats in the Tabernacle
      For &c

Jan 10 1600/1 Received of Mr Warren   To Dr Bird
the summ of 20*li* 00*s* 00 by us the   Mr West
Trustees Governors & Directors of   Mr Holloway
ye Charity &c   Mr Bachelor
                                      Mr Swain
                                      Mr Warren as
                                          Master

Paid by Mr Warren in the same Quarter
To the Trustees Governors & Directors

Ground rent	Feb 13. 1700	To Mrs Looker a Qters rent due at Mich. 1700	4. 5. 6
Tax	Octob	To ye Qters Tax of 2s per pound due Nov 15. 1700	0. 9. 6
Scavinger	Nov. 15	To ye Scavinger for ye Qter ending at Mich 1700	0. 1. 6
Chandler	Nov 20	To Mrs Price for two dozen of Candles	0.10. 8
Pewkeeper	Nov 22	To Mrs Cranwell Pewkeeper for ye Qter ending at Mich 1700	1.00.00
Washing	Nov 30	To for washing & surplices	0.06.00
Ringer	Dec 7	To Widdow Bates Ringer ye Qter ending at Mich	0.05.00
Ballance		Remaining in Mr Warrens hands	19.17.10
			26.16.00

Due before Christmas 1700 & not yet paid

To Dr Bird
To Mr West   Feb 17 1600 The Accts above
To Mr Holloway   written were allowed by the Govern-
To Mr Bachelor   ors &c & the summ of   was put
     Mr Swain   into the Chest, and there remained
                     in ye hands of   Warren the summ of
     Mr Warrent as Master

	Paid in the Qter ending at Christmas 1700	
Ground rent	Oct 8 To Mrs Looker a Qters ground rent due at Mich 1700	4.05. 6
Chest	Oct 15 To ye Chest of the Trustees Directors & Governors &c	12.10 8
Tax	Nov 20 To the Qters Tax of 2s per pound due Nov 15 1700	0.09. 6
Smith	Nov 22 To Pierson the Smith	7 7
Scavenger	Nov 23 To the Scavinger for ye Qter ending at Mich 1700	0.01. 6
Chandler	Nov 25 To Mrs Price for two douzen of Candles	0.10. 8

379

Pewkeeper	Dec 1	To Mrs Cranwell Pewkeeper for ye Qter ending at Mich 1700		1.00.00
Coales	Dec 1	To	for half a chaldron of coales	0.13.06
Washing	Dec 4	To	for washing eight surplices	0.06.00
Necessaries	Dec 4	To	for a cieling Brush & nailes	2.06
Ringer	Dec 10	To the Ringer Widdow Bates for ye Qter ending at Mich 1700		0.05.00
Ballance		Remaining in my Mr Warrens hands[5]		25.12.10
				33.01.00

NOTES

(1) Newton MSS. II, 642, in Newton's hand. These appear to be a succession of revised attempts to prepare an account, in which there are numerous additions, deletions and alterations.

(2) The Tabernacle—it was also known as the King (now Kingly) Street Chapel or Archbishop Tenison's Chapel—is now St Thomas's, Regent Street. It was built as a chapel of ease of St James's, Piccadilly. The Tabernacle, a temporary structure of timber, was erected in 1688 and survived until 1702, when the present church was built. Newton was one of the original nine trustees of the chapel, appointed in 1700, and remained in that office for twenty-two years. There are references to 'the Tabernacle neere Golden Square' in Evelyn's *Diary*.

The Editor is obliged to the Reference Librarian and Archivist of the City of Westminster Public Libraries for this information.

(3) Archbishop of Canterbury (Thomas Tenison).

(4) This was inserted after totalling the other entries.

(5) The three words after 'my' have been deleted.

643   NEWTON TO THE TREASURY

LATE 1701

From a draft in the Royal Mint[1]

To the Rt Honble the Lords Commrs of his
Majts Treasury.

May it please your Lordships

In obedience to your Lordships Order of Reference signified to us by Mr Lowndes[2] we have considered the Memorial of the Chancellour of Ireland about erecting a Mint in that kingdom [& also the Report of the late Warden & Master & Worker][3] & ye other Papers accompanying it, & finding upon enquiry into the state of the coyn of Ireland that the forreign coyns wch make a great part of their silver monies are generally in great pieces wch are inconvenient for change in marketings & other small expences & that by the want of smaller silver monies the coyning of greate quantities of Copper monies for change hath been so much encouraged as to be complained of: we are humbly of opinion that this inconvenience may deserve to be remedied by recoyning the said forreign monies or some part thereof into smaller monies of the same weight allay species & impression wth the monies of England, adding only such a mark of distinction as his Majty shall think fit.[4] And we beleive it cheapest & best for Ireland & safest for England to have this coynage dispatcht at once by erecting a Mint in Ireland for some short time (as eighteen months or two years) under the same laws & rules with this in the Tower but wth less salaries & fewer Officers & by lowering the value of the forreign monies to bring them into this Mint, And we are ready to promote such a designe & particularly to supply that Mint wth standard weights, Tryal pieces, Dyes & Coyning Tools & to try their money.

The reasons by which the former Officers of the Mint in some of the Papers referred to our consideration, opposed the coyning by different standards are so much of force against a difference in the Denomination or extrinsick value of the same pieces of money & in the proportion of Gold to silver, that we are of opinion (wth most humble submission) that the agreeing of Ireland wth England therein may deserve the consideration of the Government as a Preliminary to a Mint whenever they shall provide for the charge of erecting one

The Directions in the Warrant of 14 Car. II referred to our consideration being applied & restrained to the present way of coyning by the Mill & Press without a Seigniorage & without fine & ransome upon two penny weight remedy in the single pieces we approve of.

If a standing Mint be desired for coyning from time to time the Bullion of Merchants & others we beleive it may put Ireland to a greater charge then to coyne such Bullion in London & how it may in time affect the trade or governmt of England we do not know. Such a Mint may have been often & much desired but has either not been granted or not suffered to continue. For to use the words of our Predecessors in one of the Papers referred to our consideration, 'It hath been the policy & caution of Kings & Queens of 'England to stock their Realm of Ireland with moneys (both for quantity & 'quality) coyned in their Mint in the Tower of London whereof one part yet 'retains the name of the Irish Mint;[5] and King James (of happy memory) by 'his Indenture of the Mint caused his monies stampt for Ireland to be charged 'with an Harp crowned for distinctions sake whose Reasons and Examples (as 'we submissively conceive) may well admit your Lordships first consideration'

If the Government of Ireland shall think fit to discourage the exportation of English money from thence by setting a value something lower upon the forreign so that when Merchants or others have occasion to export Gold or Silver they may chuse rather to export the forreign monies then the English, there will be but little occasion for a Mint in Ireland hereafter. A difference of one or two per cent in the value of the monies may be sufficient for this purpose & will mend their Exchange without hindring the importation of the monies of Spain or Flanders into Ireland by Trade.

All wch &c.

NOTES

(1) Newton MSS. II, 214 (in Newton's hand). The letter is undated, but it appears to have reference to the subject of an Irish Mint which was discussed at the Council on 6 and 10 June 1701, with the result that the King was requested 'to have an Instruction given to Ld Lt of Ireland to Repo[r]t upon his arrivall in that Kingdome whether the want of smaller species of English mo. be very great and very prejudiciall to the Trade of that Kingdome, what the charge may be for Erecting a Mint to coyn the foreign Silver mo. now current there into the Smaller Species of English mony, and whether there be a sufficient quantity of Forreign mony there for that purpose'.

After the Restoration a patent was granted for the making of copper tokens in Dublin but was revoked soon after. A year later another patent authorizing the minting in Dublin of some silver coins for twenty-one years was granted on payment of a seignorage of one shilling on each pound of coin, but this was at once withdrawn on protests by the London Mint. After the Revolution the Dublin Government pressed for a full local Mint. The Mint Board, with Newton as Master, pointed out the higher cost involved as well as the limitations of a smaller Mint, but, as the above letter shows, he offered a compromise. See Craig, *Newton*, pp. 46–7, and *Mint*, chapter XXI.

(2) Secretary to the Treasury from 1695 to 1724. See Letter 571, note (3), p. 248.

(3) The square brackets are Newton's.

(4) The silver coins struck from the reign of Henry VIII had a harp for the standard emblem (Craig, *Mint*, p. 368).

(5) The new 'upper house' at the east end of the Tower Mint was originally devoted to Irish coinage and retained the name of the Irish Mint.

644   NEWTON TO ?

EARLY 1702[1]

From Brewster, II, 215

Sir,

I wrote lately to Mr Vice-chancellor,[2] that by reason of my present occasions here, I could very ill come down to your University to visit my friends in order to be chosen your burgess. I would have it understood that I do not refuse to serve you, (I would not be so ungrateful to my Alma Mater, to whom I owe my education, nor so disobliging to my friends,) but by reason of my business here I desist from soliciting, and without that, I see no reason to expect being chosen. And now I have served you in this Parliament, other gentlemen may expect their turn in the next. To solicit and miss for want of doing it sufficiently, would be a reflection upon me, and it's better to sit still. And tho' I reckon that all one as to desist absolutely, yet I leave you and the rest of the gentlemen to do with all manner of prudence what you think best for yourselves, and what pleases you shall please

Your most humble and most obedt. servant

[ISAAC NEWTON][3]

NOTES

(1) This letter refers to the election of 1702, and not, as Brewster declares, to the election of 1705. Newton represented the University of Cambridge in the Parliament which sat from 30 December 1701 until its dissolution on 2 July 1702. He did not represent the University, or any other constituency, in the Parliament which sat from 1702 to 1705. His name does not appear in the *Official Return* of 1878, nor in the *London Gazette*, no. 3829. William III died on 8 March 1702. Parliament could continue in office for a further six months, but the above letter, which apparently follows an invitation from the Vice-Chancellor to stand, could hardly come before the election was in view, if not actually ordered. The political atmosphere was strongly Tory, and Newton, who was a Whig, may have felt that he had little hope of success, and this may help to explain his reluctance to stand. See More, p. 519 and Sotheby's *Catalogue of the Newton Papers* (1936), p. 37.

(2) This was probably Thomas Richardson, who was Vice-Chancellor, 1701–2; he was succeeded by Charles Ashton, 1702–3.

(3) The letter bears no address. It was probably sent to some prominent person in Cambridge. The date may have been early 1702, when the imminence of an election was being freely discussed.

645   ROYAL WARRANT TO NEWTON
[? 9 MARCH 1701/2]
From a copy in the Royal Mint[1]

Whereas wee are Inform'd that the Ind[entu]r[e] of the Mint bearing date the 23d day of Decr 1700 made Between our late Dear Bror King William of the one parte And Isaac Newton Esqr Master & Worker of our Mints of the Other parte so by the Decease of our sd Bror become void: And Whereas a new Indr cannot be Suddenly made, Wee have thought fitt for removeing the Obstrucčons

of our sd Mints, wch may be of great prejudice at this time to Our selfe & Our
Good People And for Carrying on so usefull & publick a Service as the Coyning
of Gold & Silver Mony's to signifie Our Will & pleasure & accordingly Wee
do hereby Command and Authorize You, untill a new Indr shall be made for
the well Establishing of our Gold & Silver Mony's by the Mill and Press,
carefully to observe the Rules & Orders Appoynted by the sd Indr of the
23d Decr 1700 aforesaid, in the Assaying, Weighing receiving, rateing Com-
mixing, Melting, Coyning, pixing and paying or Delivering of our own as well
as our Subjects Gold & Silver when Converted into our Current Mony's, And
all our officers of our Mint, & all others Concern'd are required to take notice
of this Our Will & pleasure And for so doeing this shall be your Warr[an]t.

Given at our Court
To Our trusty & well beloved
Isaac Newton Esqr Master &
Worker of our Mint.
Warrant to ye Master & Worker of the Mint
By her[2] Maties Command

NOTES

(1) Newton MSS. I, 416. This is in the hand of a copyist.
On the death of William III, Newton was authorized by this Warrant to continue to
act under his Indenture of 23 December 1700 (see Warrant dated 3 February 1699/1700,
note (1), p. 322) until a fresh Indenture should be concluded with Anne. This was done on
14 January following.
There is a draft of this warrant (Newton MSS. I, 414) with notes for Anne's Coronation
medals on the reverse. These medals were delivered on 22 April 1702.

(2) The copyist has written 'his'.

646   NEWTON TO THE TREASURY

15 APRIL 1702

From a draft in the Royal Mint[1]

To the Rt Honble the Lords Commrs of
her Majties Treasury

May it please your Lordships

The time for making the Coronation Medals being very short[2] I humbly pray
your Lordships speedy Orders concerning them. The silver Medals may be
15dwt better then standard & a pound weight Troy divided into 22 Medals.

384

The Gold may be 1 car 2 gr better & a pound weight Troy divided into 20 Medals. If a pound weight Troy of Gold were divided into eighteen Medals they would take the impression better, for the former gold Medals were too thin. The Gold will cost about 4*lib.* 6*s.* per ounce & the silver about 5*s.* 7*d.* The Coynage Duty in the opinion of the Attorney General being not applicable to this service, other monies will be requisite.

All wch is most humbly submitted &c

Is. NEWTON

Mint Office
15 *Apr.* 1702.

NOTES

(1) Newton MSS. III, 319 (in Newton's hand). There is another copy in the Public Record Office (T. 1/79, no. 40).

(2) Anne acceded on 8 March 1701/2, and was crowned on 23 April following. This, and similar drafts, is probably a summary of notes made by Newton, together with proposals by Edward Northey, the Attorney-General. Designs of various medals submitted to, and approved by Newton, are to be found in the British Museum (Add. MS. 18757).

For receipt from the Treasurer of the Queen's Household, see Letter 649.

647   NEWTON TO GODOLPHIN[1]
15 APRIL 1702
From the original in the Public Record Office[2]

To the Rt Honble Sidney Lord Godolphin
Lord High Treasurer of England.

May it please your Lordship

I lately laid before your Lordship an Accompt of the charge of Medals of Gold & Silver for her Majts Coronation amounting to 2485£. 18*s.* 3½*d*: towards the discharge of wch I have in my hands 1886£. 6*s.* 3½*d* being the remainder of the Ballance of Mr Neales Accompt. Which money was given by Act of Parliament to defray the expence & charges of his late Majts several Mints & to clear the monies due to the several Importers there. Out of this remainder there have been paid severall summs on Accompt of the Country Mints as in the Bill annexed, & other summs are still to be paid on ye same Account all wch exceed the said Remainder &, if your Lordship think fit, being placed on the Coynage Duty, the said Remainder of 1886£. 6*s.* 3½*d* will become clear, excepting such further summs as your Lordship shall think fit to allow out of the same for rebuilding the Press house & paying to Mr Weddell[3] a salary of 60£ per annum to wch the Coynage duty is not applicable. Wherefore I humbly pray your Lordships Warrant to the Auditors of the Imprests for allowing out of the said Remainder such summ as your Lordship shall think fit towards the discharge of the Accompt above mentioned & that the residue of this Accompt may be paid to me out of such other fund as your Lordship in your great wisdom shall think meet.

All wch &c
Mint Office   ISAAC NEWTON.
15th *Apr.* 1702.[4]

NOTES

(1) Sidney, first Earl of Godolphin (1645–1712), of whom it was said that 'no man ever had more friends or fewer enemies'. He was M.P. for various boroughs; created Baron Godolphin of Rialton, 1684; Earl, 1706; a Commissioner of the Treasury at various periods between 1679 and 1701; Lord High Treasurer, 1702–10. He was one of the three commissioners to William of Orange in December 1688. Although he supported the setting up of a Regency, he took the Oath of Allegiance to William and Mary. See *Handbook of British Chronology* (1961), ed. Powicke and Pryde, p. 104; and *D.N.B.*

(2) T. 1/79, no. 56.

(3) Formerly Deputy Warden at Chester; Clerk for Prosecutions (from 1700). It was the salary for the latter service which Newton is questioning. No regular finance was provided for criminal work until 1708. See also Letters 660 and 726, note (2), p. 497.

(4) The date (15 April 1702) cannot be correct since Godolphin did not become Lord High Treasurer until 6 May. See also Letters 648 and 649.

648   ORDER IN COUNCIL TO NEWTON
17 APRIL 1702
From a copy in the Royal Mint[1]

At the Court at St James the 17th day of
Aprill 1702
Present
The Queens Most Excellent Majesty in Councill.

It is this day Ordered by her Majesty in Councill that the Master of her Majestys Mint do cause to be made and prepared the number of Three hundred Medalls of Gold, and the Number of Twelve hundred medalls of

Silver of her Majesty according to the Pattern presented to and approved by her Majesty, The value of each of the Medalls of Gold and Silver to be the same as at the Last Coronation. The said Medalls to be delivered by the Master of the Mint to the Earl of Bradford[2] Treasurer of her Majesty's Household to be distributed by his Lordship at her Coronation

EDWARD SOUTHWELL[3]

Lett the Master & Worker of her Majestie's Mint take care that the Number of Medalls above mentioned be prepared and delivered as by the above written Order of Councill is directed.

Cock pitt Treasury Chambers 21st Aprill 1702.

STE: FOX[4]
H. BOYLE[5]
RICHD HILL[6]
TH. PELHAM[7

NOTES

(1) Royal Mint Record Books, VIII, 16. See also Letter 646.

Apparently the Order in Council was not sent directly to the Master of the Mint but to the Commissioners of the Treasury who then passed it on to the Mint.

(2) Francis Newport (1619–1708), son of Richard Newport, first Baron Newport; Christ Church, Oxford, 1635; M.P. for Shrewsbury in the Short and Long Parliaments. He engaged in Royalist plots, 1655 and 1657. He was created Viscount Newport, 1675, and Earl of Bradford, 1694. See *D.N.B.*

(3) Edward Southwell (1671–1730), statesman, son of Sir Robert Southwell; Merton College, Oxford; M.P. for various boroughs; Secretary of State for Ireland. F.R.S., 1692. See *D.N.B.*

(4) Sir Stephen Fox, a Lord Commissioner of the Treasury. See also Letter 548, note (3), p. 202.

(5) Henry Boyle (d. 1725), politician, nephew of Robert Boyle; M.P. forTamworth,1689–90, Cambridge University, 1692–1705, Westminster, 1705–10; Chancellor of the Exchequer, 1701–8; Lord Treasurer of Ireland, 1704–10; Principal Secretary of State, 1708–10; created Baron Carleton, 1714; Lord President of the Council, 1721–5. See *D.N.B.*

(6) Richard Hill (1655–1727), diplomatist; St John's College, Cambridge; Ambassador at the Hague, Envoy Extraordinary to the Elector of Bavaria; Commissioner of the Treasury, 1699; Hon. D.C.L., Oxford.

(7) Sir Thomas Pelham (1650–1712), fourth baronet and first Baron Pelham of Laughton; M.P. for East Grinstead, 1678, Lewes, 1679–1702, Sussex, 1702–5; he was made a Commissioner of the Treasury, 1689. See *D.N.B.*

649 BRADFORD TO NEWTON
22 APRIL 1702
From a copy in the Royal Mint[1]

April the 22d 1702

Received then of Is. Newton Esqr Master and Worker of her majestie's Mint Three hundred Meddalls of Gold Weighing One hundred Seventy Nine ounces Sixteen penny weights twelve grains, which he was directed to provide for her Majestie's Coronation by order of Councill dated the 17th Instant. I say received by me

BRADFORD.

Received More Twelve hundred Meddalls of Silver Weighing Six hundred thirty four ounces five penny weight provided Likewise for her Majestie's Coronation, I Say Received by me

BRADFORD.

NOTE

(1) Royal Mint Record Books, VIII, 16.

650 MINT TO GODOLPHIN
7 JULY 1702
From the original in the Public Record Office.[1]
In continuation of Letter 640

To the Rt Honble Sidney Lord Godolphin
Ld High Treasurer of England.

May it please your Lordship

According to your Lordships direction we have examined the values of several forreign coyns[2] & endeavoured to inform our selves of the values of Gold in proportion to silver in several nations[3] & considered the ways of preserving the coyn. And by the Accompts we have met with, Gold is higher in England then in France by about 9d or 10d in the Guinea, then in Holland by 11d or 12 pence in the Guinea, then in Germany & Italy by 12d in the Guinea or above. In Spain & Portugal Gold is higher then in England by about 11d in the Guinea. For the great quantity of Silver coming from the

West-Indies[4] has brought down the price of Silver in all Europe in proportion to Gold & principally in Spain where the Bullion first arrives. The low price mends the market and thereby carries silver from Spain into all Europe & from all Europe to the East Indies & China, the Merchant bidding more for it then it goes for among the natives. In Spain the Merchants advance about six per cent or above for silver: At which rate a Guinea is worth about 21s. $3\frac{3}{8}d$ & sometimes less. In England they advance 3d or 4d per ounce, and at the rate of 3d per ounce advance a Guinea is worth but 20s. $6\frac{1}{6}d$.

Gold is therefore at too high a rate in England by about 10d or 12d in the Guinea. And this tending to the decrease of the silver coyn we humbly conceive that one way of preserving this coyn is to lower the price of Gold suppose by taking 6d, 9d or 12d from the price of the Guinea so as that Gold may be of the same value in England as in the neighbouring parts of Europe. France has set us an example for in the last war when the Lewidor was raised there to 14 livres the Ecu was raised only to 72 sols but it is now raised to 76 sols tho the Lewid'or be raised only to 14 livres as before. So that Gold in respect of Silver is lower in France now then in the last war in the proportion of 76 to 72 that is by above $13\frac{1}{2}d$ in the Guinea.

The liberty of melting forreign monies into ingots in private shops & houses for exportation gives opportunity of melting down the money of England for the same purpose. For restraining of wch a law might be usefull against exporting any Ingots of silver melted down in England except in a publick Office to be appointed or erected for that purpose.

The law by barring the exportation of forreign silver after it is coyned prevents the coynage thereof because the Merchant cannot afterwards export it, & tends to discourage the importation of silver into England because the Merchant can make no use of it whilst it stays here in the form of Bullion. The bringing of silver to the market of England & the turning it into money should rather be encouraged as the proper means of encreasing the coyn, silver being more apt to stay with us in the useful form of money then in the useless form of Bullion. If the merchant might export what he coyns, some part of what he coyns would be apt to be laid out here. And this liberty may be allowed him after some such manner as is described in the scheme hereunto annexed.

The licensing the exportation of Bullion whilst the exportation of the money is prohibited makes silver worth more uncoyned then coyned[5] & thereby not only stops the coynage but causes the melting down of the money in private for exportation. For remedying this mischief it may be perhaps better on the contrary to prohibit the exportation of Bullion & license that of money, & whenever the money is in danger to licence the exportation of so much money only as shall from time to time be coyned out of forreign Bullion.

The safety & encrease of the coyn depends principally on the ballance of trade. If the ballance of trade be against us the money will be melted down & exported to pay debts abroad & carry on trade in spight of laws to the contrary, & if the ballance of trade be for us such laws are needless & even hurtfull to trade. If trade can be so ordered that no branch of it be detrimental to the nation the money will be safe. For wch end luxury in forreign commodities should be checkt & the exportation of our own commodities encouraged. If a law were made & well executed against trading with more gold & silver by any Merchant or company of Merchants then in certain proportions to the value of the goods exported, such an Addition to the Act of Navigation[6] might put Merchants upon searching out sufficient ways of vending our commodities abroad & as we humbly conceive, be more effectual for preserving the coyn then the absolute prohibition of the exportation thereof.

As for the alteration of the standard we are humbly of opinion that if the value of the several species to be hereafter coyned be diminished without changing the denomination, it will occasion the melting down & recoyning the species already coyned for the profit that may be made thereby. And if the value be encreased the Merchants & people will value their goods by the old money already coyned in wch they are to be paid, & the new money of greater value (if any shall be coyned) will be pickt out for exportation & the Importer who coyns it will lose the overvalue to the discouragement of the coynage, & in payments made by tale to forreigners the nation will also lose the overvalue.

But if it be proposed to retain the value of the several species or quantity of fine silver therein & only to alter the allay, we are humbly of opinion, that if small money which by continual use weares away fast & is apt to be lost, were coyned of coarse allay as is done in several countries abroad, provided it were well coyned to prevent counterfeiting, such money would weare longer & be less apt to be lost then the small money now in use. By small money we understand Groats, Three pences, Two pences & pence,[7] unless the penny by reason of its smallness should be made of copper.

All wch is most humbly submitted to your Lordships great Wisdome.

J STANLEY[8]
Mint Office   ISC. NEWTON
7 *Jul.* 1702   JN ELLIS.[9]

NOTES

(1) T. 1/80, no. 105. This report is in Newton's hand.
(2) See Letter 593, Letter 631, note (3), p. 353, and Letter 640.
(3) This letter is a further example of Newton's attempt to establish a gold-silver ratio. The

troy pound of 925 silver was minted into 62 shillings from 1601 to 1816. The number of nominally 20s. pieces of the same weight of 916·7 gold was raised, after previous increases, to 44½ in 1663. Owing to the discovery of rich silver mines in Bolivia, the market raised the value of these guineas to 21s. 6d., thus increasing the gold-silver ratio from the planned 13½ to 1 to 14½ to 1. In 1699 the guinea was forced back to 21s. 6d. after a slight inflation which broke into a panic price of 28s. to 30s. for some months. Both before and after the silver recoinage of 1696–8 the English silver circulation was eroded by melting down and exporting of silver coins while the country remained a net importer of gold coins. Newton devoted much work in 1701–2, and later in 1717, to deducing the gold-silver ratios abroad.

(4) That is, Spanish America.

(5) In England at this time, silver coin had to be melted, often twice, in order to disguise it for export. The exportable silver resulting from the remelting of the coin had to be, and was, 1d or 2d dearer than the coin from which it was obtained.

(6) The Navigation Act of 1651 was designed to strike a blow at Dutch supremacy on the high seas. It enacted that goods coming into this country should be carried in English vessels, or vessels belonging to the country supplying them. It remained in force, with certain modifications, for about two centuries.

(7) These four silver coins still survive as Maundy Money.

(8) See Letter 635, note (6), p. 357.

(9) John Ellis, an Under-Secretary of State, was Comptroller from 1701 to 1711.

651 DAVID GREGORY TO NEWTON
30 SEPTEMBER 1702
From the original in the University Library, Cambridge[1]

Oxon. 30 Sept.
1702.

Much Honoured Sir

I remember that some time since, you told me that there are some Propositions in Euclids Data designed for the Resolution of Biquadratick Equations which want the second and fourth terms, but that one of them is corrupted. And the Euclid in Greek & Latin,[2] that is printing by the University, being in a good forwardness, and they now at work on the Data; I give you this trouble to desire of you to acquaint me which Propositions those be, and how (or to what purpose) you think that which is corrupted ought to be restored. Your speedy answer to this, and what else you may think fitt to impart concerning this work, Will be a great honour, and of great advantage, to the Edition.

I hope Sir you have caused call for my Books, that are due to you, at Mr Bennetts[3] bookseller at the Half Moon in St Pauls Church Yard.

I am

Much Honoured Sir

Your most oblidged and most humble Servant

D. GREGORY

For
The Much Honoured
Isaac Newton Esquire
at his House in Germin Street
near St James's Church
Westminster   London

NOTES

(1) Add. 3960, no. 10.

(2) *Euclidis quæ supersunt omnia, ex recensione D. Gregorii* (1703).

(3) Thomas Bennet (*c.* 1665–1706), publisher and London agent for some Oxford books. See *The Note Books of Thomas Bennet and Henry Clements*, ed. Hodgson and Blagden, 1956 (Oxford Bibliographical Society, new series, vol. vi, pp. 3–6).

652   NEWTON TO GODOLPHIN

16 OCTOBER 1702

From the original in the University Library, Basel

To the Rt Honble Sidney Ld Godolphin
    Lord High Treasurer of England

May it please your Lordship
For the good execution of the Office of Master & Worker of her Majties Mint[1] I humbly propose the following security, vizt

> Isaac Newton in 2000 *li*[*b*]
> Tho Hall Esqr—1000
> John Francis Fauquier[2] Gent   1000

The acceptance of wch is most humbly submitted to your Lordships great wisdome

Mint Office   Is NEWTON
16 *Octob.* 1702

392

(1) It was customary for the Master to give £2000 security to the king before the sealing of the Indenture. See Letter 628.

(2) John Francis Fauquiere. See Letter 635, note (4), p. 357.

653 NEWTON TO GODOLPHIN

21 DECEMBER 1702
From the original in the Public Record Office[1]

To the Rt Honble Sidney Ld Godolphin
Lord High Treasurer of England.

May it please your Lordship

There is due to me for Medals for he[r] Majties Coronation the summ of 2485£. 18s. $3\frac{1}{2}d$[2] and for the paymt of this debt I humbly conceive the Civil List to be the proper fund. I am now to make up my Accompts for the year ending this Christmas and that I may be able to ballance this part of my Accompts I humbly [pray] your Lordship that the money above mentioned may be imprest to me before the year expire.

All wch is most humbly submitted to

your Lordships great wisdome

Is. NEWTON

Mint Office
21 *Dec.* 1702.

NOTES

(1) T. 1/83, no. 27. This letter was written by Newton, probably in some haste, judging by the two editorial additions which have been necessary.

(2) See also Letter 647.

654 NEWTON TO ?LOWNDES

Late 1702
From a draft in the Royal Mint[1]

Sr

By the Indentures of the Mint under the broad Seale the Irish Mint[2] belonge to the Offic of the Mint, It was taken out of our possession in the latter part of the reign of King Charles the second, we do not know by what authority.

We[3] have heard that it was in exchange for the Ground on wch the house & shops of the Smith of the Ordnance now stand: but that ground never was restored to us. Upon the recoinage of the hammered money the grounds on wch the Barracks in the Irish Mint are built was redelivered to us & the Barracks were turned into Millrooms. And after the coinage was ended the Millrooms were redelivered to the Office of Ordnance. except two of them wch by the verbal consent of some of the Officers of the Ordnance were kept for the coinage of copper money. They are now filled with Tinn. And if his Mty pleases to give order that the Tinn be removed out of the Mint into Warehouses where we may not be answerable for it, we are ready to deliver those Millrooms to be turned into Barracks. The Letter of K. William for restoring the Barracks to ye Office of Ordnance We have not seen, but beleive that the King was not informed of the rights of the Mint & that the Indenture of the Mint under the broad seale is of greater authority then that Letter.

We desire that no soldiers may be lodged in the Mint between the two gates thereof least it render the custody of the gold & silver unsafe & discourage the Merchants & Goldsmiths from bringing their gold & silver into the Mint to be coined contrary to the intent of the coinage Act.[4]

NOTES

(1) Newton MSS. III, 433. There is another draft, *ibid.* III, 439. On the reverse is a calculation of exchange rates which was *prima facie* linked with Newton's concern with the guinea. The letter was probably written when the Mint held stocks of tin without responsibility for sale.

(2) The Mint was enlarged by new buildings during the reign of Elizabeth I, to be devoted entirely to her recoinage. This upper Mint was afterwards used for coinage for Ireland and retained the name 'Irish Mint' throughout the eighteenth century. See Craig, *Mint*, p. 119, *et seq.*

(3) This sentence has been crossed out.

(4) At the foot of the sheet, Newton has added: 'From the Warr Office.'
'An Act for Encourageing of Coynage.' See Letter 726, note (2), p. 497.

655   NEWTON TO GODOLPHIN

?1702

From a draft in the Royal Mint[1]

May it pl. your Lordship

To enable the Master & Worker to carry on the coynage wthout depending upon the King or Queen's Assaymaster, he is allowed 60 pounds per annum towards the charge of keeping an Assaymaster for himself whenever he shall need one. But now for a long time the Gold & Silver has been received into the Mint by the assays of ye King or Queens Assaymaster & the Mr & Wr allows the 60 *lib* per annum to a Deputy who has power to examin the assays of the said Assaymaster by his own or those of any other Assaymaster as often as shall be thought fit. If your Lordship therefore pleases that the Queens Assaymaster be still trusted in the receipt of Bullion into ye Mint I am also willing to trust him, he being a sworn officer & an indifferent person & I having met with no complaints against him & his Assays being recconed exacter then those at Goldsmiths Hall, & the gold & silver being safe in his hands

But if your Lordship chuses rather that ye ☉ & ☽ should be received into the Mint by the Assays of another Assaymaster to be imployed under me, I am ready to provide one so soon as a new Assay Office with other conveniences can be had & made fit in the Mint.

All wch is most humbly submitted

NOTE

(1) Newton MSS. i, 103 (in Newton's hand).

656   NEWTON TO GODOLPHIN
1702
From a draft in the Royal Mint [1]

To the Rt Honble Sidney Lord Godolphin
Lord High Treasurer of England.

May it please your Lordship

Upon your Lordships Order of Reference of the Petition of Mr Bull,[2] one of the Engravers of the Mint, for the renewal of the Warrant for his salary: we humbly lay before your Lordship that he has represented to us that the salary of fifty pounds per annum wch hath hitherto been allowed him is too little & that he can get much more money by buisiness abroad & must apply himself to other business unless he may be allowed above 60*lib* per annum. And therefore considering that we shall want his service & that the money for salaries & repairs of buildings is limited by Act of Parliament so that what is added to the salaries is taken from the buildings, we are humbly of opinion that his warrant be renewed, and that he be allowed therein a Salary of sixty pounds per annum, and also that for every serviceable master Puncheon which he shall make he be further allowed so much money as to the Warden[3] Master & worker & Comptroller of the Mint shall seem reasonable, not exceeding six pounds a piece for the largest Puncheons nor three pounds a piece for the Puncheons for half Guineas and sixpences. All wch is most humbly submitted to your Lordships great wisdome

NOTES

(1) Newton MSS. i, 163 (in Newton's hand). The letter is undated, but the evidence suggests late 1702.

(2) Samuel Bull was employed by Harris 'for making Dyes' at 20 shillings a week, and from Christmas 1698 was appointed probationer engraver in the Mint at a salary of £50 a year. From 7 April 1705 he became assistant engraver at £80 a year. Coinage punches and medals were both engraved by him. He died in 1726.

(3) Newton wrote 'Warder'.

657   MINT TO GODOLPHIN
JANUARY 1702/3
From a draft in the Royal Mint[1]

To the Rt Honble Lord High Treasurer of England

May it please your Lordship

In obedience to your Lordships Reference of 23 November[2] last we have enquired into the case of Mr Anthony Redhead[3] Master & Worker of the Mint in Norwich and find it truly stated in the annexed paper[4] on which the said Reference is endorsed. And whereas Mr Redhead in discharge of a debt of 2497li[b] 16s 3d owing to Mr Leonard Blofield a Receiver of Land Taxes in Norfolk produces three Receipts for 2500li[b] paid out of that Mint to Mr Blofields Order, We find that these three Receipts are in the form of private Receipts so that they cannot be allowed by an Auditor as Vouchers in Mr Redheads Accompts. If any thing be due upon them it is to be recovered in a Court of Justice by Mr Redhead or by him or them who shall answer Mr Redheads debt to her Majty But upon examining the matter we are humbly of opinion for the reasons specially reported in another annexed paper that the money paid upon those Receipts hath been accounted for & tended upon the printed Mint Tickets & that the Receipts are cancelled by him Mr Redhead insists not peremptorily on them but submits them to your Lordships wisdome.

We humbly conceive therefore that for bringing these matters to an issue the

Accompts of the Country Mints should be passed without regard to these Receipts & Mr Neale set insuper & then Mr Neales Administratix or Security charged with the debt. For Mr Redhead seems poor & unable to make satisfaction & not unfit to be set at liberty so soon as it may be done by the consent of all parties concerned.

We further lay before your Lordship that Mr Redhead at the conclusion of the Mint was indebted to Mr Chaplyn a Receiver of Land Taxes for 2573oz 5dwt of hammered money at 5s per oz the summ of 643£. 6s. 3d & that in Mr Chaplyn's Acct wch is past & declared & a Quietus obtained Mr Redhead is set insuper for this debt at 5s. 8d per oz whereas Mr Neale should have been set insuper & that only at 5s per oz.

All wch is most humbly submitted &c.

NOTES

(1) Newton MSS. II, 481.

(2) This letter has not been found.

(3) Anthony Redhead was Deputy Master of the Norwich Mint. He was put in Ludgate Gaol from September 1699 to August 1701 on Neale's accusation that he had not paid over all the Norwich Mint balance. This was the imprisonment referred to in Letter 661, note (5), p. 402. A customer, Leonard Blofield, was now trying for double payment. See Letter 658, note (2), p. 399.

(4) See Letter 658.

658   NEWTON TO THE TREASURY

15 JANUARY 1702/3

From a copy in the Royal Mint[1]

The Case stated specially between Mr Redhead[2] & Mr Blofield.
By ye Day bookes of the late Mint at Norwich & the Cashbookes of Mr Redhead & his Clerk & by another book Composed & sign'd by the Warden of yt Mint we find that Mr Redhead upon ye breaking up of that Mint was indebted 2497l: 16s: 03d to Mr Blofield for hammer'd mony at 5s per Ounce.
In Discharge of this Debt Mr Redhead produces three Rects for 2500l: paid to Mr Blofields ordr out of that Mint, The one dated Augt 20th 1697 for 500l The other Two dated Augt ye 30th & Sepr 1st 1697 for 1000l each To wch Mr Blofield & his friends answer that whatsoever Sums were payd to him or his ordr out of that Mint upon such private Rects were paid in Course for Silver Imported & intended upon ye next accounting to be brought to acct & set

397

off upon ye printed Mint Ticquetts then payable in Course & accordingly were all of them faithfully brought to acct as he is ready to make Oath, but ye Rects were not alwaies taken up & Cancell'd as they should have been The endorsmt on ye printed Mint Ticquetts implying (as they Concieved) according to ye order & Course of the Mint that all Summs paid untill ye day of the Endorsmt were then Accounted for & sett off upon ye printed Ticketts & thereby all private Rects of such Summes untill that day discharg'd & made voyd, wch made him & his agents less carefull to take them up. They say also that abt Three weeke after the date of the sd Three Rects vizt on 22th Sepr 1697 Mr Blofield accounted wth Mr Redhead and endorsed 14922 *l.* on Two Mint Ticketts including all Summs paid to him untill that time & that in so short a time as Three Weeks So great a Summ as 2500*l* could not be forgott & that in Decr following he accounted again & endorsed 6443*l*: 13*s*: 09*d* on Two other Mint Ticketts including all further Summs untill that day, By wch Accountings & Endorsemts the sd Three Rects (if they be true ones) being looked upon as Discharg'd, they were not mencōned any further by Mr Redhead while that Mint stood nor for a long time after but lay neglected till he thought fitt to produce them, as he did also some other Rects of the same kind wch ye Importrs neglected to take up & Cancell & wch are now allow'd to be voyd. In examining this matter we find therefore that Mr Redhead did pay severll Summs of mony to Mr Blofield & some other Importers upon private Rects without endorsing ye Summs upon ye printed Mint Ticketts untill they Came to a Generall Reckoning upon the Tickett or Ticketts next payable in Course, & that the Importers did sometimes upon such a Reckoning neglect to take up their Rects That Mr Blofield did endorse 14922*l*: on Two Ticketts 22th Sepr 1697 & 6443: 13: 09 more on Two others in Decr following as he alledged & we humbly Concieve those Endorsemts by ye Course of the Mint to be full of all monys paid upon those Ticketts so as to voyd the sd Three Rects unless Mr Redhead can positively prove ye paymt of more mony by 2500*l* upon ye Two first of those Tickets then were Endorsed upon them wch proof is wanting & would inferr ye Crime of undue Preference, The Officers of that Mint for preventing Misreckonings took an Acct every Two or Three dayes & sometimes dayly of all ye mony's new Coyned & pd away & of wt remained in ye Treasury whereby a Misreckoning of 2500*l* might soon have been discovered, Whereas those Three Rects lay neglected till about Michae' 1699 wch was Two yeares after the Endorsmt[,] Mr Redhead representing that he then found also & produced a Rect of 1858*l*. left in yt Mint by Mr Dashwood ano[the]r Importer of publick Hammer'd mony, but by an Affidavitt of Mr Tho: Allen Clerk to the Warden of that Mint made before my Ld Chief Baron Ward ye 3d of July 1701 & by other Circumstances it appeares to us,

That Mr Dashwood did acct for that mony & neglected to take up & Cancell the Rect & this is now acknowledged also by Mr Redhead and abt the same time the sd Mr Redhead produced also Two other Rects of the aforesd Mr Blofield besides the Three above menconed both dated in ye sd Month of Augt 1697 The one for 500*l* The other for 1000*l* but by ye aforesd Affidavitt of Mr Allen Those Summs were accounted for upon ye same 22th Sepr 1697 & Mr Redhead insists no further upon ym seeing therefore that Mr Blofield upon accounting on ye sd 22th Sepr 1697 did neglect to take up & Cancell his Rects then accounted for or some of ym it may be supposed That the Three now produced were of that Number, For the sd Mr Allen in ye Affidavitt above menconed affirmes further that he hath heard & believes that Mr Red-head hath ano[the]r Note of Mr Blofield for 1000*l* & believes that Mr Blofield forgott to take up that Note when he accounted for the mony & sign'd the printed Rect or Tickett which Note wee take to be one of the Three Rects now produced by Mr Redhead, and Mr Redhead affirmes nothing further of the sd Three Rects then yt he found ym amongst his papers sometime after that Mint broke up & believes ym truely to be sign'd by Mr Crowne who was imploy'd by Mr Demee ye Agent of Mr Blofield.

Janry 15 1702.

Is: NEWTON

NOTES

(1) Royal Mint Record Books, VIII, 35–6. See Letter 657.

(2) Early in 1703 Newton had to deal with Mint transactions going back to 1697 concerning Anthony Redhead. Acknowledgements for £2500 of hammered coin paid into the Norwich Mint were produced by Leonard Blofield who demanded payment. Redhead maintained that payment had been made, but he was unable to produce official receipts. After much investigation it appeared reasonably certain that Blofield had been paid at the time but that Redhead had carelessly accepted informal receipts in some cases, and had lost receipts in others. (See Craig, *Newton*, p. 62.) The Treasury called for a further report on 22 March 1702/3 (see Letter 661).

659   AN INVENTORY
15 FEBRUARY 1702/3
From the original in the Public Record Office[1]

An Inventory of the Plate & other things delivered into the Treasury of the Mint by the Honble Her Majts Comrs for Prizes. 15th Feb. 1702/3.

One Draught weighing 83lb.wt.[2] 06oz. 00dwt.

A Benitier or Vessel for Holy-water faced wth Philligram work.

A Plate wth four Cups & four Potts about it & a Salt wth a cover in ye middle.

A Salver three dishes & an old Salt.

A Cup, two Candlesticks, & an old Pan.

32 new Plates, 22 Forks, 39 Spoons & a piece of a Spoon.

A Mermaid guilt wth a stone in ye body

A pair of Snuffers wch seem to be of coarse silver.

Twelve thimbles, six doz: of brass buttons & two copper boxes wth 18 Rings in them

A sand-box & an ink-box of Pewter & some pieces of a broken Antimonial cup.

Draught 2d weighing 7lb.wt. 3oz

An hundred pieces of eight[3]

A Crucifix wth a golden chain

A golden Dolphin with a chain.

A set of Beads, six Fans, & a gold ring wth nine false stones, not weighed.

Draught 3d weighing 77lb.wt. 11oz. 00dwt

Six Chocolate Potts, a water-Pott, two Candlesticks, 25 Silver Plates, two salts, two dishes, an Ewer, a small Tumbler & eight Cakes of silver.

Draught 4th weighing 42lb.wt. 07oz. 10dwt.

Two new Basins & Ewers well made.

Draught 5th weighing 75lb.wt. 02oz. 00dwt.

The bottoms of two silver Lamps with chains to hang them by & a Fane for one of them & a Pot for oyle & two Cakes of Silver

Is. NEWTON

NOTES

(1) T. 1/84, no. 89. This document (in Newton's hand) is an inventory of part of the treasure captured at Vigo Bay. See Warrant of 1702/3 (663), note (2), p. 404, and Letter 665.

(2) See Manuscript 579, note (9), p. 259.

(3) Spanish American dollars. See Letter 593, note (7), p. 283.

660 NEWTON TO ?LOWNDES

18 MARCH 1702/3

From the original in the Public Record Office[1]

Mint Office 18*th March* 1702/3.

Sr

In answer to yours of Mar. 16th, I humbly desire you to acquaint my Lord High Treasurer yt I have in my hands some money imprest formerly to Mr Neale,[2] out of other Funds then the Coynage Duty, out of wch I pay a salary of sixty pounds per annum to Mr Robt Weddell[3] by Warrant for his services in prosecuting Clippers & Coyners, and I am humbly of opinion that out of this money the summe of forty pounds may be paid to Mrs Ann Morris[4] in satisfaction of her husbands like services. I am

Sr

Your most humble Servant

Is. NEWTON.

NOTES

(1) T. 1/85, no. 9. This letter is in Newton's hand and was probably addressed to William Lowndes, the Secretary of the Treasury.

(2) See Letter 549, note (5), p. 204.

(3) Robert Weddell, formerly Newton's deputy at Chester. See Letter 582, note (3), p. 386, and Letter 647.

(4) Widow of Richard Morris, a Mint employé and professional informer.

661 THE TREASURY TO NEWTON AND OTHERS
22 MARCH 1702/3
From a copy in the Royal Mint.[1]
In continuation of Letters 657 and 658

Gentlemen

By Order of my Ld Treasurer I send you ye inclosed papers relating to Mr Redhead[2] late Master & Worker of the Mint at Norwich His Lordship desires you to Considr ye same & Report to his Lordship your opinion what you think fitt to be done therein. I am

Gentlemen

Your Most humble Servt

Treasury Chambers
22 March 1702/3

WM LOWNDES[3]

To Samll Travers Esqr her Mats Surveyr Generall
Edwd Harley[4] Esqr one of ye Audrs of her Mats Imprests
Sr John Stanley Barrt Warden of her Mats Mint
Isaac Newton Esqr Master Worker
John Ellis Esqr Comptroller
& Thos Hall Esqr Chief Clerk of ye Mint.[5]

NOTES

(1) Royal Mint Record Books, VIII, 33.

(2) See Letter 658.

(3) See Letter 571, note (3), p. 248.

(4) Edward Harley (1664–1735), M.P. from 1698 to 1722; an original trustee of the National Land Bank. He was the brother of Robert Harley who later became Earl of Oxford. See Letter 628, note (2), p. 349.

(5) In Sotheby's *Catalogue of the Newton Papers* (1936), there is mention of a 'Draft of a Warrant, with Autograph corrections by Newton, for the release from prison for three months of Anthony Redhead, late Deputy Master and Worker of the Mint at Norwich, that he may better help in the discovery of embezzlements committed at the Norwich Mint, 26 Mar. 1700; and Counsel's opinion that Redhead cannot be released, 5 *Feb*. 1703/4.' See Letters 657 and 658.

662 EXTRACTS FROM MEMORANDA BY DAVID GREGORY

MARCH 1702/3

From the original in the Library of the Royal Society of London[1]

Notata Mathematica.   *Martio* 1702/3

There is a M.S. of Euclid de Levi & Ponderib in St Johns Library, Oxon.

Mr Fatio[2] gave Mr Newtons Corrections of his own book to M. Hugens, and he to Libnitz who has published them in a Dutch book, of no kin to this purpose.

···   ···   ···

The Comet whose Orbit Mr Newton determins may sometime impinge on the Earth. Origen[3] relates the maner of destroying the Worlds by one falling on another

Mr Newtons Lemma V. Lib. 3 pag. 481[4] ought to have added to it, *et quæ proxime accedet ad lineam rectam*[5]

An easy way of determining *quam proxime* the Trajectoria of a Comet is by traceing on a Globe the way of a Comet, & Connecting with great Circles the places of the Sun in the Ecliptick & Comet, at the same points of time and drawing a third great Circle which shall be divided by the first connecting circles in proportion of the times of the motion. You may thereby see the Angles in which lines from the immoveable Sun cut the Comets Trajectory....

Ld Brouncker,[6] Dr Ward[7] & Sr C: Wren laughed at Mr Hook when he said that he suspected Gravity towards the Earth is the cause of the Moon not going on in a streight line. Mr Newton pursued the Motion of the planets first on supposition of ane equal gravity, but afterwards wrote to Mr Hook asking what he suspected the Law of Gravity to be.

··· ··· ···

At the End of the Schol: about the Comet.
The Comett 1680/1 its Orb would be exacter or agree better with observation, if in stead of Parabolick it were Made or assumed Elliptick, so as to make the Comets period about 2000 years or between 1000 years and 3000 years.

··· ··· ···

He thinks I insist too much on the Astronomers opinions about the precession of the Equinoxes. Mr Flamsted's Lunar Tables, give not the Moons place sometimes within 20 minutes of the Heavens.

NOTES

(1) R.S. Greg. MS. fo. 87.

(2) According to W. W. Rouse Ball (*Essay*, p. 125), Newton had at one time seriously entertained the idea of allowing Fatio to undertake the preparation of a second edition of the *Principia*. See Letter 307, note (1) (vol. II, p. 477), and Letter 346, note (3) (vol. III, p. 45).

(3) Origen (surnamed Adamantius) lived about the third century A.D. He was one of the greatest of the early Christian fathers and he made a praiseworthy attempt to reconcile the ancient learning with Christian faith. He was imprisoned during the Decian persecution, and though released on the death of Decius did not long survive him.

(4) *Invenire lineam curvam generis Parabolici, quæ per data quotcunque puncta transibit.* (To find a curved line of the parabolic kind which shall pass through any number of given points.)

(5) 'And which will approximate very closely to a straight line.'

(6) William, Viscount Brouncker of Castle Lyon (1620–84). One of the original members of the Royal Society. After its incorporation he became its first President, an office which he held for fifteen years. In 1662 he became Chancellor to Queen Catherine. See *The Royal Society Its Origin and Founders* (1960), p. 147; and vol. I, note (15), p. 8.

(7) Seth Ward (1617–89), D.D. (1654), an eminent divine and an early fellow of the Royal Society (1663). From 1649 to 1661 he was Savilian Professor of Astronomy. Bishop of Exeter (1662); translated to Salisbury (1667). He wrote many theological treatises and was also a capable mathematician, having received some of his mathematical training from Oughtred. He proposed Newton for election to the Royal Society. See also Letter 29, note (2) (vol. I, p. 74).

663   ROYAL WARRANT TO NEWTON
1702/3
From a copy in the Royal Mint[1]

Anne R

Whereas wee are Informed that a considerable quantity of Gold and Silver has been taken by Our Royall fleet at the Late Expedition at Vigo,[2] Our Will and pleasure is, and Wee do hereby require and Authorise you, to cause to be coyned all Such Gold and Silver as shall be brought into Our Mint, and delivered unto you in the Name of [3] with this Inscription, VIGO, In small letters under Our Effigies, which we Intent as a Marke of distinction from the rest of Our Gold and Silver Moneys, and to continue to posterity the remembrance of that glorious Action, And for so doing this shall be your Warr[an]t. Given at our Court. &c

To our Trusty & well Beloved
Isaac Newton Esqr. Master and
Worker of our Mints

NOTES

(1) Newton MSS. III, 277. This is undated, but the subject suggests early 1703, i.e. shortly after the date of Letter 659.

(2) The fleet was not English, but English and Dutch combined, under the command of Admiral Sir George Rooke. The English Infantry on board were under the command of the Duke of Ormond.

After failing to take Cadiz, an expedition was made against the Spanish treasure galleons returning from South America and their French convoy in Vigo Bay, 12 October 1702. See Trevelyan, *Blenheim*; also a summary account in Clark's *The Later Stuarts*, which stated that the booty captured was worth a million pounds of which £95,000 went to the Mint.

To commemorate this success the Queen ordered some of the spoil to be converted into coin (mainly crowns, half-crowns, shillings and sixpences), with the word VIGO imprinted below her effigy. This practice was not uncommon. The booty captured by Anson, in 1746, in his voyage round the world was similarly distinguished. (See *Commodore Anson's World Voyage*, Appendix E, Boyle Somerville, 1934.)

(3) A gap has been left here.

664   NEWTON TO LOCKE
15 MAY 1703
From the original in the Bodleian Library, Oxford[1]

Sr   *London* 15 *May* 1703

Upon my first receiving your papers I read over those concerning the first Epistle to ye Corinthians, but by so many intermissions that I resolved to go over them again so soon as I could get leasure to do it with more attention. I have now read it over a second time & gone over also your papers on the second Epistle. Some faults wch seemed to be faults of ye Scribe I mended with my pen as I read the papers. Some others I have noted in the inclosed papers.

In your Paraphrase on 1. Cor. VII. 14, you say, The unbeleiving husband is sanctified or made a Christian in his wife.[2] I doubt this interpretation because the unbeleiving husband is not capable of baptism as all Christians are. The Jews looked upon themselves as clean holy or separate to God & other nations as unclean unholy or common & accordingly it was unlawfull for a man that was a Jew to keep company with or come unto one of another nation. Act. x. 28.[3] But when the propagation of ye Gospel made it necessary for the Jews who preached the Gospel to go unto & keep company with the Gentiles, God shewed Peter by a vision[4] in the case of Cornelius that he had cleansed those of other nations so that Peter should not any longer call any man common or unclean, & on that account forbear their company: & thereupon Peter went in unto Cornelius & his companions who were uncircumcised & did eat with them. Act x. 27, 28 & XI. 3. Sanctifying therefore & cleansing signify here, not the making a man a Jew or Christian but the dispensing wth the law whereby the people of God were to avoyd the company of ye rest of ye world as unholy or unclean. And if this sense be applied to St Pauls words they will signify that altho Beleivers are a people holy to God & ought to avoyd the company of unbeleivers as unholy or unclean, yet this law is dispensed with in some cases & particularly in the case of marriage. The beleiving wife must not separate from the unbeleiving husband as unholy or unclean nor the beleiving husband from the unbeleiving wife, for the unbeleiver is sanctified or cleansed by marriage wth the beleiver, the Law of avoyding the company of unbeleivers being in this case dispensed with. I should therefore interpret St Pauls words after the following manner

'For the unbeleiving husband is sanctified or cleansed by the beleiving wife 'so that it is lawful to keep him company & the unbeleiving wife is sanctified 'by the husband, else were the children of such parents to be separated from 'ym & avoyded as unclean, but now by nursing & educating them in your 'families you allow that they are holy.'

This interpretation I propose as easy & suiting well to ye words & designe of St Paul, but submit it wholy to your judgment.

I had thoughts of going to Cambridge this summer & calling at Oates[5] in my way, but am now uncertain of this journey. Present I pray my humble service to Sr Francis Masham & his Lady.[6] I think your paraphrase & commentary on these two Epistles is done with very great care & judgement. I am

Your most humble & obedient servant

Is: NEWTON

For John Lock Esqr at Oates in
    Essex

NOTES

(1) MS. Locke, c. 16, fo. 155. The letter is reproduced in Lord King's *Life of John Locke*, vol. I, p. 420. In this letter Newton criticizes Locke's paraphrase of 1 Corinthians vii. 14. In the autumn of 1702 Newton visited Locke at Oates, who showed him his 'essay upon the Corinthians, with which he seemed very well pleased, but had not time to look it all over'. Locke therefore sent the Essay to Newton just before Christmas for his further perusal, and not hearing anything from him wrote to him again a month or six weeks later. As no answer to this was received, Locke, on 30 April 1703, wrote to his nephew, Peter King, requesting him to wait upon Newton to discover the reason of his long silence, 'You will do well to acquaint him, that you intend to see me at Whitsuntide, and shall be glad to bring a letter to me from him.' (King, *op. cit.*, vol. II, pp. 38–9). Newton's reply, reproduced above, was sent on the eve of Whit Sunday. See King, *op. cit.*, vol. I, pp. 420–3, and Edleston, p. lxx, note (144).

The letter illustrates Newton's profound interest in theology (see also Letters 357, 358, 359, 360, 362, 388, vol. III).

(2) 'For the unbelieving husband is sanctified by the wife, and the unbelieving wife is sanctified by the husband: else were your children unclean; but now are they holy' (A.V.).

(3) 'Ye know how that it is an unlawful thing for a man that is a Jew to keep company, or come unto one of another nation; but God hath shewed me that I should not call any man common or unclean' (A.V.).

(4) Acts x. 11–16.

(5) The residence of the Masham family (see note (6) below), near Bishop's Stortford, in Essex.

(6) The Mashams were close friends of Locke. From 1691 he resided with them permanently. Lady (Damaris) Masham (1658–1708) was a daughter of the Cambridge Platonist, Ralph Cudworth, an eminent divine, sometime Master of Christ's College. See Letter 362, notes (1) and (4), vol. III, p. 148.

665 NEWTON TO ?LOWNDES

16 JUNE 1703

From the original in the Public Record Office[1]

Sr   *Mint Office June* 16. 1703.

Upon stating Accompts yesterday wth the Commrs for Prizes concerning the Gold & Silver taken at Vigo & sent into the Mint, we found the same as follows.

The Accompt of the Silver

	lb.wt[2]	oz	dwt
Silver delivered into ye Mint at several times (besides an Altar piece & Salver not weighed)	4504.	2.	0
Coyned & paid to the Commrs for Prizes 1000£ weighing	321.	2.	13
Paid to Sr Cloudsly Shovel[3] 1000£ in pieces of 8/8 weighing	289.	10.	5
Pieces of 8/8 remaining in the Mint	171.	1.	10
Three Ingots imported & 48 others melted out of old Plate, Cakes, Wedges &c & remaining in the Mint	3283.	6.	0
Plate reserved as valuable for its fashion	415.	4.	4
Wast in the melting arising chiefly from the fuming of the leady Cakes, Virgin Silver & Soder[4]	16.	7.	11
Iron & Brass taken out of the Plate 6 lb.wt. 8oz 6dwt. Allay put in to make the money standard 2oz. 9dwt. The difference	6.	5.	17
	4504.	2.	0

The Accompt of the Gold.

	[lb]	[oz]	[dwt]	[gr]
Gold delivered into the Mint at several times	7.	8.	16.	6
Coyned & paid to ye Commrs for Prizes 267£. 7s. 9d weighing	5.	7.	5.	12
Remaining in the Mint	2.	1.	10.	18
	7.	8.	16.	6

I beleive it may be proper to acquaint my Lord High Treasurer with this Accompt. And whenever his lordship pleases to give further directions about disposing of what remains in the Mint I shall be ready to assist in the execution as far as I can be serviceable. I am

Your most humble & most obedient
Servant

Is. NEWTON.

(1) T. 1/86, no. 41. See also the Inventory, No. 659, p. 399.

(2) Throughout this letter Newton has consistently written lwt.

(3) Sir Cloudesly (Clowdesley) Shovel (1660–1707), Admiral of the Fleet. See Letter 615, note (4), p. 315.

(4) That is, 'solder'.

666   NEWTON TO GODOLPHIN
13 JULY 1703
From the original in the Royal Mint[1]

To the Rt Honble Sidney Ld Godolphin
Ld High Treasurer of England.

May it please your Lordship

In obedience to your Lordships Order of Reference of the 16th of June last past upon the annext Proposal of Mr Abel Slaney[2] for himself & partners for a new Coinage of 700 Tuns of half pence & farthings, We do humbly acquaint your Lordship that We have enquired into all the Coinages of that sort since the year 1672 & do find that in the Reign of K. Charles ye 2d & K. James the 2d & in the beginning of the Reign of the last King & Queen the coinage of half pence & farthings was performed by one or more Commissioners who had money imprested from the Excheqr to buy Copper & Tin, & coined at 20*d* per pound Haverdupois & accounted upon oath to the Government for the produce thereof.

That upon calling in the Tin half pence & farthings by reason of the complaints made against them, there was a Patent granted to the Proposer & others who contracted to change the Tin farthings & half pence, & to enable them to bear that charge they were allowed to coin 700 Tuns at 21 pence per pound weight without being accountable to the Government. Which reason now ceasing we are humbly of opinion that[3] the former method by Commission is most advantageous to the Government.

We do not hear there is any demand of halfpence & farthings at present, & tho there should be a want in some places it seems to proceed from an unequal distribution, for we are informed they are overstockt with them in others, as at ye General Post Office, about Newcastle & at Leicester.

We are further of opinion that the coynage of halfpence & farthings in this Kingdome should be to the intrinsick value,[4] the charges of ye coynage &

Incidents deducted; but if that be not thought advisable at present for fear of stopping the currency of those that are already abroad, We humbly conceive that whenever a new Coinage shall be thought convenient, it should be done in small quantities as her Majesty from time to time shall appoint, to supply the decrease & loss of those already coined without danger of new complaints by overstocking the Nation

All wch is most humbly submitted to your Lordships great Wisdome.

Mint Office   Is. NEWTON.
13 *Jul.* 1703

NOTES

(1) Newton MSS. ii, 301.

(2) Abel Slaney was a member of a private company of licensees. On 17 April 1694 the House of Commons decided on a coinage of English copper of 'the intrinsick value', and a licence was granted to Sir Joseph Hearne, Sir Francis Parry, George Clark, Abel Slaney and Daniel Barton for the coinage of halfpence and farthings to a total of 700 tons spread over seven years from Midsummer, 1694. See Craig, *Mint*, pp. 182 and 417; also C. Wilson Peck, *English Copper, Tin, and Bronze Coins*.

(3) Newton wrote 'the'.

(4) An Order in Council, 28 May 1684, had ordered the coinage of halfpence and farthings in tin, with the two-fold object of increasing the royal revenue and helping the tin industry. This ceased in 1692 in deference to the popular clamour against the difference between 'intrinsick' and face value, and the ease with which the coins could be counterfeited.

667   NEWTON TO GODOLPHIN
30 OCTOBER 1703
From a draft in the Royal Mint[1]

To the Right Honble the Lord High Treasurer[2]
of England.

May it please your Lordship

In obedience to your Lordships Order I have considered what may be requisite for lodging her Majties Tin[3] in the Mint & delivering it out at a certain price & paying the money into the Exchequer, and am humbly of opinion that I can do it wth two Clerks added to my own to enter the number & weight of the Blocks of Tin received & delivered & compute the price keep an account of incidents, & wth so many Porters as shall be necessary upon

409

occasions, one of wch may be constant if continual attendance shall be required; and wth the use of the Master & Workers Offices & Rooms so far as they may be wanted & spared from the coynage, & of the Cranes of the Office of Ordnance, & liberty of carrying the Tin between Tower Wharf & the Mint over the Draw-bridge, an Officer of the Customs being directed to attend the ships there.

Some things are also to be provided as Scales & Weights, Sledges Pulleys & Stamps for numbring the Blocks. And it may be convenient that their weight be stamped on them either in Cornwall or at their Receipt in the Tower.

And since the Tin is to be delivered out at a certain price I am humbly of opinion that the Blocks should be delivered as they come to hand without giving leave to the Merchant to pick & chuse, setting aside only unlawfull Blocks to be reexamined by your Lordships Order if any shall occurr, & remelted or otherwise disposed of. And that the Tin be delivered only for ready money, & the money be paid into the Exchequer as often as it rises to a certain summ to be named by your Lordship, & accounted for annually.

All wch is most humbly submitted to

your Lordships great wisdome

[Is. NEWTON][4]

Mint Office
30 *Octob.* 1703.

NOTES

(1) Newton MSS. III, 476. There is an earlier draft, dated 28 October 1703 (Newton MSS. III, 492).

(2) Lord Godolphin; see Letter 647, note (1), p. 386.

(3) The right to pre-emption of tin was vested in the Duke of Cornwall, and reverted to the Prince of Wales as such on Anne's death. She held it because there was no Duke, and the money or liability was treated as part of her personal estate.

This conversion of her unexercised right into a liability was a revolution—but was confined to Anne's reign. Anne's first contract was for seven years, 1703–10, and was followed by a second for another seven.

(4) This draft is in Newton's hand; the signature has been torn out.

668   NEWTON TO GODOLPHIN
?1703
From a draft in the Royal Mint[1]

To the Rt Honble the Lord High Treasurer
of England.

May it please your Lordship

Upon reviewing the Mint I find that there is room for coyning twenty thousand pounds a week of silver & eighty thousand pounds a week of gold & at the same time lodging above two thousand Tunns of Tinn[2] without bringing any Tin into the Rooms of ye Gold & Silver

All wch &c

Is. NEWTON

NOTES

(1) Newton MSS. III, 560.

(2) See Newton's letter to Godolphin, Letter 667.

669   NEWTON TO ?GODOLPHIN
6 JANUARY 1703/4
From the original in the Public Record Office[1]

Jan 6. 1703/4.

May it please your Lordship

The Warden & Comptroller of the Mint having sometimes exprest a desire that the charge of her Majts Tinn should be in the board of the Mint, with a common Treasurer or Receiver under them, & the Assaymaster having very much desired some imployment about the Tinn & being of the board for things below stairs:[2] I humbly beg leave to lay their desires before your Lordship, & to signify that when your Lordship shall resume the consideration of lodging the Tinn in the Tower I am ready to serve her Majty in such a manner as your Lordship in your great wisdome shall think fit. I am with all submission

My Lord

Your Lordships most humble
and most obedient Servant

Is. NEWTON.

NOTES

(1) T. 1/89, no. 6.

(2) The three officials concerned were Sir John Stanley, John Ellis and Daniel Brattell.

411

670   CHAMBERLAYNE[1] TO NEWTON
2 FEBRUARY 1703/4
From the original in the British Museum[2]

Petty France Westmr
2 Febr: 1703/4

Honor'd Sr

I suppose you have not forgot the Famous Conference appointed to be this day at Gresham College between Mr George the Formosan,[3] the bearer hereof, & le Pere Fontenaye[4] a Jesuite lately come from China. I have engaged Mr George & am to carry him thither this afternoon in my Coach, but without telling him the Reason; I beg therefore the same Caution & Secrecy on your side, and you will much oblige

Honor'd Sr

Your most Humble Servt

JOHN CHAMBERLAYNE

For the Honor'd
Isaac Newton Esqre

NOTES

(1) John Chamberlayne (1666–1723), miscellaneous writer, and student of modern languages. Entered Trinity College, Oxford, but left without taking a degree, and went to Leyden. On his return he held various offices about the Court under Queen Anne and George I. He contributed three papers to the *Philosophical Transactions*, and was elected F.R.S. in 1702. In the Sloane MSS. in the British Museum are a number of letters from Chamberlayne. His most important work was his translation of Brandt's *History of the Reformation in the Low Countries* (1720–3), 4 vols. He also continued *The Present State of England*, which had been begun by his father, Edward Chamberlayne. See *Biographia Britannica* (I, 1282).

(2) Sloane MSS. 4063, fo. 234. A copy of this letter is in the Birch MSS. (4292, fo. 121) in the British Museum. It differs slightly from that here reproduced. It is headed: 'John Chamberlayne Esq., to Isaac Newton Esq., President of the Royal Society.'

(3) George Psalmanazar (1679–1763), described in *D.N.B.* as a literary impostor. He was an able scholar and had a particular gift for modern languages. At the age of sixteen he set out for Rome, and by means of a forged passport, in which he described himself as a native of Japan converted to Christianity, he wandered over Europe, earning his livelihood by the roadside as a mendicant. His Jesuit preceptors had instructed him in the history and the geography of China; this knowledge, which he supplemented by a study of Bernardus Varenius's *Descriptio Regni Japoniæ* (Amsterdam, 1649), enabled him to pose as an authority on the little-known island of Formosa. But he overreached himself when he declared that Formosa was

part of the empire of Japan (instead of China). Father Fontenai, a Jesuit missionary who had returned to London from China, exposed Psalmanazar's claims. As a result, Psalmanazar was invited to meet his critics at a public meeting at the Royal Society on 2 February 1703/4. This was the meeting referred to in the above letter; it does not appear that Newton attended the meeting, an account of which Psalmanazar gave in the preface to a work which he published the following year. This work is described in the title-page as a *Description de l'Ile Formosa en Asie. Du Gouvernement, des Loix des Mœurs & de la Religion des habitans: Dressée sur les Mémoires du Sieur George Psalmanaazar. Native de cette Ile. Avec une ample & exacte Relation de ses Voiages dans plusieurs endroits de l'Europe, de la persécution qu'il a soufferte, de la part des Jesuits d'Avignon, & des raisons qui l'ont porté à abjurer le Paganisme, & à embrasser la Religion Chrétienne Reformée. Par le Sieur N.F.D.B.R.* (Amsterdam, 1705). According to his own version, Psalmanazar successfully rebutted the charges made by Father Fontenai. Sir Hans Sloane invited the disputants to dine with him; among the guests was the Earl of Pembroke (see Letter 695), who became one of Psalmanazar's most generous patrons. See *D.N.B.*, also *Gentleman's Magazine* (1765), p. 78, where he is described as 'an ingenious man, and a good scholar, but is thought by some to be a counterfeit, and a *Jesuit* under the character of a Japonee'.

(4) Pierre-Claude Fontenai (1683–1742) entered the Jesuit novitiate in 1698. He became Professor of Humanities, and later Rector of the College of Orléans. He continued the *Histoire de l'Eglise gallicane*, which had been begun by his predecessor, P. Longueval. In 1740 he fell a victim to an attack of paralysis which enforced complete rest. He retired to La Flèche where he died, after two years of suffering, at the age of 59. See *Biographie Universelle*, **15**, 215. See also Letter 678.

671   A REPORT BY NEWTON
c. 16 FEBRUARY 1703/4
From the original in the Public Record Office[1]

The substance of the Report given in by Mr Newton concerning Mr Whites proposal about pasting a Mill-mark upon Paper & Parchment

The mark is obscure, being not well visible unless by holding the Paper between the eye & the light. It is also unornamental, appearing upon the Paper like a patch. And for these reasons it will scarce please the people.

It may be counterfeited several ways & by the Makers of paper it may be counterfeited more exactly then the stamps can be by Gravers.

It does not appear to me that the dammage sustained by counterfeiting the stamps hath equaled the charge wch the government would be at in making & pasting on the Mill-mark

Is. NEWTON

NOTE

(1) T. 1/89, no. 68 (in Newton's hand). This is the substance of a report which Newton submitted to the Treasury on 15 May 1704 (Letter 673).

672   *ADVERTISEMENT* TO *OPTICKS*[1]

1 APRIL 1704

From the first edition in the Royal Society

Part of the ensuing Discourse about Light was written at the desire of some Gentlemen of the *Royal Society*, in the Year 1675.[2] and then sent to their Secretary, and read at their Meetings, and the rest was added about Twelve Years after to complete the Theory; except the Third Book, and the last Proposition of the Second, which were since put together out of scattered Papers. To avoid being engaged in Disputes[3] about these Matters, I have hitherto delayed the Printing, and should still have delayed it,[4] had not the importunity of Friends prevailed upon me. If any other Papers writ on this Subject are got out of my Hands they are imperfect, and were perhaps written before I had tried all the Experiments here set down, and fully satisfied my self about the Laws of Refractions and Composition of Colours. I have here published what I think proper to come abroad, wishing that it may not be Translated into another Language without my Consent.

The Crowns of Colours, which sometimes appear about the Sun and Moon, I have endeavoured to give an Account of; but for want of sufficient Observations leave that Matter to be further examined. The Subject of the Third Book I have also left imperfect, not having tried all the Experiments which I intended when I was about these Matters, nor repeated some of those which I did try, until I had satisfied my self about all their Circumstances. To communicate what I have tried, and leave the rest to others for further Enquiry, is all my Design in publishing these Papers.

In a Letter written to Mr *Leibnitz* in the Year 1676,[5] and published by Dr *Wallis*, I mentioned a Method by which I had found some general Theorems about squaring Curvilinear Figures, or comparing them with the Conic Sections, or other the simplest Figures with which they may be compared. And some Years ago I lent out a Manuscript containing such Theorems, and having since met with some Things copied out of it, I have on this Occasion made it publick, prefixing to it an *Introduction* and subjoining a *Scholium* concerning that Method. And I have joined with it another small Tract concerning the Curvilinear Figures[6] of the Second Kind, which was also written many Years ago, and made known to some Friends, who have solicited the making it publick.

I. N.

NOTES

(1) The title is: *Opticks: or, a Treatise of the Reflexions, Refractions, Inflexions and Colours of Light*. The *Advertisement* is undated, but in the fourth edition the date is given as 1 April 1704.

(2) In 1675, on 7 December, Newton sent to Oldenburg: 'An Hypothesis explaining the Properties of Light discoursed of in my severall Papers' (see Letter 146, vol. I, pp. 362–92). His first communication on this subject appeared in the *Phil. Trans.* **6** (1671/2), 3075–87; it bore the title: 'A Letter of Mr Isaac Newton, Professor of the Mathematicks in the University of Cambridge; containing his New Theory about Light and Colors.' It is referred to in Birch (III, 9) under the date 8 February 1671/2.

'Five letters to Mr Oldenburg were read:

...   .

'Of Mr Isaac Newton from Cambridge, 6 February, 1671/2, concerning his discovery of 'the nature of light, refractions and colours; importing, that light is not a similar, but a 'heterogeneous body, consisting of different rays, which had essentially different refractions, 'abstracted from bodies through which they pass; and that colours are produced from such 'and such rays, whereof some, in their own nature, are disposed to produce red, others green, 'others blue, others purple, &c. and that whiteness is nothing but a mixture of all sorts of 'colours, or that it is produced by all sorts of colours blended together.

'It was ordered, that the author be solemnly thanked, in the name of the Society, for this 'very ingenious discourse, and be made acquainted that the Society think very much of it, if 'he consent to have it forthwith published.'

(3) For the nature of the 'Disputes about these Matters' with Hall (Linus), see Letters 173, 200, vol. II; with Lucas, Letters 161, 185, 196, 200, 215, 220, 221.

(4) Hooke had died the previous year. See 'Robert Hooke, F.R.S.' by E. N. da C. Andrade, in *The Royal Society Its Origins and Founders* (1960), p. 143: 'There had previously been trouble, carefully fostered by mischief-makers...between Newton and Hooke on optical questions.'

(5) This is the 'Epistola Posterior' (see Letter 188, vol. II, pp. 110–61). In subsequent editions of the *Opticks* this was omitted.

(6) With the *Opticks* were published: *Enumeratio Linearum Tertii Ordinis* (Enumeration of lines of the third order) and *Tractatus de Quadratura Curvarum* (Treatise on the Quadrature of Curves).

673 NEWTON TO GODOLPHIN

15 MAY 1704

From the original in the Royal Mint[1]

To the Rt Honble the Lord High Treasurer
of England

May it please your Lordship

According to your Lordships Order, Mr White has been with me several times to make out his Proposal for preventing the counterfeiting of stampt Paper, but without convincing me of its fitness to be used

For whereas he proposes to past or glue a Mill-mark upon the Paper, this Mark thus becomes a faint one, not well to be seen unless by holding the paper between the eye & the light; and it is also unornamental, appearing on the paper like a patch. And for these reasons it will not please the people.

It may be counterfeited several ways, & by the Papermakers it may be counterfeited more exactly then the stamps can be by Gravers

And it doth not appear to me that the dammage susteined by counterfeiting the stamps, equals the charge wch the Government would be at in making & glueing on the Mill-mark.

However finding that Mr White is not satisfied in this Report I am willing if your Lordship thinks fit that the matter should be reexamined by whom your Lordship pleases.

All which is most humbly submitted to your Lordships great wisdome

May 15*th*   Is. NEWTON
1704.

NOTE

(1) Newton MSS. III, 447. This letter, in Newton's hand, is based on Newton's report (671), p. 413. A tax had recently been imposed on paper or parchment for deeds, etc. The Mint engravers made dies for impressing paper that had paid duty. White proposed to affix a duty mark on the paper to exclude the possibility of forgery. See Craig, *Newton*, p. 67.

674   MINT TO GODOLPHIN
23 AUGUST 1704
From the original in the Royal Mint[1]

To the Right Honble the Lord High Treasurer of England.

May it please your Lordship.

In obedience to Your Lordship's order of Reference of August 8th & 10th wherein Wee are directed to consider the Qualifications of Coll Parsons,[2] Mr Croker[3] Mr Ross and Mr Fowler to succeed Mr Harris[4] in the place of Graver of the Mint, Wee humbly lay before your Lordship that the Master of the Mint upon considering what inconveniency the Mint lately suffered and may again suffer by taking in Cutters of Seales into the Chief Graver's place; did upon his succeeding Mr Neale putt a Clause into the Indenture then made, between the Crown and himself, whereby the Graver's salary of 325£ per Annum upon the next voydance of the place should cease in order to anew Establishment. For the Roetiers[5] brought up no new Gravers under them, and Mr Harris who succeeded them being a Cutter of Seales and not skilled in that sort of Graving which is proper for the money, Imployed Mr Croker to do all that Work for an Allowance of 175£ per Annum and retained to himself the remaining 150£ per Annum, and Mr Croker was not bred up in the Service of the Mint, but now by long Practice works very well, and Wee are humbly of opinion that he is the fittest person to be made first Graver of the Mint. [Colonell Parsons does not grave himself, but only employs good Workmen and has a good Fancy and Judgement in those matters; neither does Mr Fowler grave but only designs, and Mr Rose desires to succe[e]d Mr Harris only in his place of Graver of Seales to her Majy, and by a Seale which he has graved for the Dutchy of Lancaster he seems to be a good Workman for Seales and sufficiently qualifyed for that place. *But Mr Le Clerk*[6] *is a nimble & skilful Graver very fit to be received into the Mint.*]

It is humbly proposed therefore for the advantage and Security of the Coyn against counterfeiting that there should be a sett of Gravers constantly brought up in the mint, who having once attained to perfection may keep their Art amongst themselves and propogate it to Probationers or Apprentices and be succeeded by them, For which end Wee humbly propose that Mr Croker be now made the first Graver of the Mint with a Salary of 200£ per Annum for maintaining himself and a Servant to file and polish the Dyes and turn the Press, and be allowed the use of that part of the Graver's House in which he now lives, and that Mr Samuell Bull[7] who is now a Probationer and has a Salary of 60£ per Annum [with some other allowances worth about 10 or 20£ per Annum] be made the second Graver of the Mint with a Salary of 80£ per Annum if your Lordship shall approve thereof, and be allowed the use of the other part of the Graver's House, and that *Mr Le Clerc* be taken into the third place and be allowed *also a salary of 80£ per Annum &* two Rooms to lodge and Work in over the great Press Room and over Mr Croker's Shop. *And that the salaries of Mr Croker & Mr Bull commence from the death of Mr Harris, but by reason of the charges wch the Mint has been at by the dammages done by the great winds last Autumn,*[8] *we are humbly of opinion that for saving money to defray those charges the salary of Mr Le Clerc do not commence till next Midsummer.*

All which is most humbly submitted to your Lordship's great Wisdom

Mint Office. the 23th August 1704.

J STANLEY
IS. NEWTON
JN ELLIS

NOTES

(1) Newton MSS. I, 170. This is in another hand, except for amendments by Newton, which are shown in italics; it is signed by the members of the Mint Board. There are also four other drafts of this letter.

(2) Probably William Parsons (1658–1725), chronologer; entered Christ Church, Oxford, 1676; lieutenant-colonel, 1687; published *Chronological Tables of Europe.*

(3) Memorandum 630, note (3), p. 351.

(4) Harris had died recently, on 3 August 1704. See also note (3) above.

(5) See Letter 568, note (2), p. 241.

(6) 'The probationer's post was perversely conferred on Gabriel le Clerk, who seems to have been valued as a channel of communication between English ministers and the rising sun of Hanover. Employed in Zell and other German mints, he appeared at times in London, sought leave for family affairs, drew his pay and did nothing in return till he was got rid of in 1709' (Craig, *Mint*, p. 202).
The square brackets used here, and elsewhere in this letter, are in the original.

(7) See also Letter 656, note (2), p. 396.

(8) G. M. Trevelyan wrote of 'brick-built London that had suffered so severely in the Great Storm of 1703' (*Ramillies and the Union with Scotland*, 1932, p. 200).

675   MINT TO GODOLPHIN
13 SEPTEMBER 1704
From the original in the Royal Mint[1]

To the Rt Honble the Ld High Treasurer of England

May it please your Lordship

In obedience to your Lordships Order of Reference of Mr Le Clerks[2] Petition to succeed Mr Harris[3] in the place of Graver to her Majts Mint, we have considered the same, and are humbly of opinion that Mr Le Clerk is a skilfull & expeditious Graver fit to be received into her Majts Mint in the third place, that is, next after the two Gravers Mr Croker[4] & Mr Bull[5] who were imployed under Mr Harris, & that he be upon the same foot wth those two Gravers wth a salary of eighty pounds per annum & the Lodgings appointed the third Graver in our former Report. But by reason of the charges wch the Mint has been at by the dammages done by the great winds last Autumn, we are humbly of opinion that for saving money to defray those charges his salary do not commence till next Midsummer.[6]

All wch is most humbly submitted to your Lordships great wisdom

Is. Newton
Jn. Ellis[7]

Mint Office. 13 *Sept.*
1704.

NOTES

(1) Newton MSS. 1, 167.

(2) See Letter 674, note (6), p. 418.

(3) Letter 630, and Letter 674, note (4), p. 418.

(4) See Letter 630, notes (2) and (3), p. 351.

(5) Samuel Bull was appointed under-graver when Croker became graver. See Letters 656 and 674.

(6) See Craig, *Newton*, pp. 54–5.

(7) Letter 650, note (9), p. 391.

676   NEWTON TO GODOLPHIN

12 OCTOBER 1704

From a copy in the Royal Mint[1]

To the Rt Honble the Lord High Treasurer of England

May it please your Lordship

A Question being moved about a Clause in the Gravers Patent I humbly beg leave to lay the matter before your Lordship. All persons haveing a liberty to make Medals unless restrained by the Government the Gravers of the Mint have by a clause in their Patent been allowed and all others prohibited to make Medals wth the Effigies of ye King or Queen.[2] And this Place of Medal maker to the Crown has been sometimes encouraged by a large salary out of the Civil List & sometimes granted to strangers and is no part of the constitution of the Mint For by the standing constitution of the Mint the Moneyers coyn whatever the Government wants whether Money Medals or Healing pieces, the metal weight allay & form of the money & medals being first appointed by the King or Queen by the advice of the Council, and the Graver only makes the Stamps. This I take to be the proper way of coyning such Medals as the Government approves of & I am humbly of opinion that no other Medals should be coyned by the Mint. The Gravers privilege of making other Medals for their private advantage is an encouragement to them to improve themselves and to be content with less salaries. If it be continued they may be obliged to set their names or the first letters thereof on their own Medals to distinguish them from Medals made by the Mint or be otherwise limited as your Lordship shall think fit, if abrogated they may want some other encouragement to improve themselves & may expect to be paid for the Dyes &

419

27-2

Puncheons they make for Medals, Whether it shall be continued & in what manner or be abrogated is most humbly submitted to your Lordships great wisdom

Mint Office Oct. 12   [Is. NEWTON]
    1704.

NOTES

(1) Newton MSS. I, 172. Only the date is in Newton's hand; the rest has been written by a copyist.

(2) To supplement their salaries, as well as to give them an opportunity of developing their skill, the engravers were at liberty to sell replicas in any metal of official medals and to design, strike with mint presses and mint labour, and sell medals of their own. The private medal work of Croker had raised a minor political storm, because in catering for the public taste he seemed to the Tories to glorify Marlborough unduly (see Craig, *Mint*, p. 203).

677   AN ESTIMATE BY FLAMSTEED

8 NOVEMBER 1704

From a copy in private possession

An Estimate of the Number of Folio Pages, that the *Historia Britannica Cœlestis*, may contain when Printed.[1]

The First Part will contain some Observations of Mr *William Gascoigne*[2] (the first Inventor of the Way of measuring Angles in a Telescope, by the Help of Screws, and the first that applied Telescopical Sights to Astronomical Instruments) taken at *Middleton*, near *Leeds*, in *Yorkshire*, betwixt the Years 1638, and 1643, excerpted from his Letters to Mr *Crabtree*,[3] with some of Mr *Crabtree's* Observations of the same Years: As also, Observations of the Sun and Moon's Diameters, Configurations, and Elongations of *Jupiter*'s Satellites from him; small Distances of Fix'd Stars, with Appulses of the Moon and Planets to them, observ'd with a Telescope and Micrometer at *Derby*, betwixt the Years 1670, and 1675; with the larger intermutual Distances of Fix'd Stars, of the Planets from them; Eclipses of the Sun, Moon, and *Jupiter*'s Satellites; Spots on the Sun; Comets and Refractions; taken with a large *Sextant*, a voluble *Quadrant*, and the above-mention'd Instruments, betwixt the Years 1675, and 1689, at Her Majesty's Observatory, rank'd under proper Heads, in about,   *Pages* 500

The Second Part will contain the Meridional Distances from the *Vertex* and *Transits* over the Meridian; of the Fix'd Stars and Planets, observ'd with the

420

large *Mural Arch*, betwixt the Years 1689, and 1704; with other Observations of the Eclipses of the Luminaries, Satellites, Variations of the Compass, &c. Interspers'd according as they happen'd in order of Time, in about,

Pages 630

The Third Part will contain, 1. *Ptolemy's* Catalogue of 1026 Fix'd Stars, in *Greek*, with a new *Latin* Version of it, and *Vlug-Beig's*[(4)] Places, annex'd on the *Latin* Page with the Corrections, in about,
  P. 46
 2. Small Catalogues of the *Arabs*, in about
  P. 4
 3. *Tycho-Brahe's*,[(5)] of about 780 Fix'd Stars in a proper Order, in
  P. 22
 4. The *Landtgrave* of *Hesse's*,[(6)] 386, in
  P. 6
 5. *Hevelius's*, of ---- 1534, in a proper Order, in
  P. 32
 6. A small Catalogue of *J.F.* to the Year 1686, entring, and a *French* one to the same Year ended, in
  P. 3
 7. Mr *Halley's* Catalogue of the Southern Fix'd Stars, to the Year 1677; with the *French* small Catalogues of them, in
  P. 7
 8. The *British* Catalogue,[(7)] containing the right Ascensions, Distances from the Pole; Longitudes and Latitudes of near 3000 Fix'd Stars, with the *Variations* of their right Ascensions and Polar Distances, whilst they change their Longitudes one Degree; rendring them easily Useful to our Sailors, and Perpetual, in *Latin* and *English*; with Tables of the Sun's Places and Declinations for the Four first Years of this Century, and Variations to make them serve for the Whole, for the Use of our Sailors, in about,
  P. 100

The *Liber Prolegomenos*, or *Preface*, which will be large, can't be estimated at present, nor how many Pages the Places of the Planets deduced from these Observations by the Help of this Catalogue, will contain. Allow

  P. 100

Pages in all, about 1450

The New Figures of the Constellations, or the Ancient Ones restor'd, (those in *Bayer*,[(8)] and on our Globes being false, and different from all the Catalogues in all Languages) in about Sixty Copper Plates, each near Two Foot broad, and Twenty Inches deep, with a *Preface*.

The first Part may be put into the Press immediately: The Second may follow it as soon as the First is wrought off; the Charts of the Constellations may be put into the Hands of the Designers, drawn, and engrav'd; the necessary Calculations finish'd; and the Third Part (if God spare me Life and Health, and necessary Assistance be afforded me) may in the mean time be fitted for the Press; most of the Papers which it is to contain, lying ready for Transcription.

In drawing up this Account, I have purposely omitted all the Observations of Meridional *Zenith* Distances, taken with the *Sextant*, in the Year 1676, 77, 78, 79, and 80.

As also all the like Distances, taken with a large, but slight *Mural Arch*, betwixt the Years 1683, and 86, by reason I esteem'd them not sufficiently certain.

And moreover, not a few Solar Observations taken with the strong New *Mural Arch*, when those which I design'd to take the Night following of the Stars or Planets, were not observ'd, by reason of Clouds or Accidents.

Those I give are no Closet Speculations, but the actual Labours of Thirty Years, continued by Day and Night, with frequent Injuries to my Health, and great Expences in Instruments, and Assistance, at my own proper Charge.

The Charge of Drawing and Engraving about Sixty Copper Plates of the Constellations, and Printing such a *Work*, as never before passed under the Press in *England*, is like to be Great, and therefore not fit for a private Under-taker.[9]

It was begun at the Command, and by the Encouragement of King *Charles* the Second, and has been continued in the Reigns of his Successors, for the Service of the Publick, and *Improvement of Navigation*, purposely that our Sailors might want none of the Helps necessary for making true Charts of our Earth, or finding the Places of their Ships at Sea.

And now if Her Majesty, and His Royal Highness,[10] think fit to bestow it on the Publick, by contributing and affording such Helps as are wanting to its Publication; Ingenious Men of all Nations, especially our Sailors, will own Them as potent and happy in Arts as Arms, and celebrate Their *Memories* with Applause, so long as Ships sail on the Seas, or Ingenious Men contemplate the Heavens on Land.

The Observatory,
  Nov. 8. 1704

JOHN FLAMSTEED, M.R.[11]

NOTES

(1) Even before he had received the warrant of his appointment as King's Astronomer (4 March 1674/5), Flamsteed had entertained the idea of presenting to the world the result of his labours. But there were many difficulties, not the least of which was the inadequacy of his salary of £100 per annum. Towards the end of the century, many learned men began to express impatience at the delay in the publication of his observations, and this fact induced Flamsteed to draw up an estimate showing the extent of the work he hoped to publish. This estimate, here reproduced, was read at a meeting of the Royal Society on 15 November and received the wholehearted commendation of the fellows. A fortnight later, Prince George, the Queen's Consort, was elected a Fellow of the Society, and as soon as he learned of Flamsteed's

difficulties he generously undertook to bear the expense of publication. See Letter 680, particularly notes (4) and (5), p. 430.

(2) William Gascoigne, or Gascoyne (1620–44), an inventive astronomer. He devised a micrometer which is described in the *Phil. Trans.* **2** (1667), 457. His observations were interrupted by the outbreak of the Civil War in 1642, in which he took a prominent part. He was killed at the Battle of Marston Moor in 1644.

(3) William Crabtree, an English astronomer, born near Manchester early in the seventeenth century. He was a friend of Horrocks, with whom he carried on a long correspondence upon astronomical matters and with whom he observed a transit of Venus in 1639.
   Extracts from the observations of Gascoigne and of Crabtree form part of the first volume of Flamsteed's *Historia Cælestis Britannica.*

(4) Ulugh Begh (1394–1449) of Samarkand, grandson of the oriental conqueror Tamerlane. He revived the study of astronomy in Turkestan, where he built an observatory about 1420. He also published tables of the planets and a catalogue of stars comprising those of Ptolemy as well as new observations of his own, which Flamsteed made use of in his *Historia.*

(5) Danish astronomer. See Letter 709, note (3), p. 476.

(6) Probably Charles, Landgrave of Hesse-Cassel, son of William IV. His chief astronomer was Christopher Rothmann, a Copernican, who maintained an extensive correspondence with Tycho Brahe.

(7) The *British Catalogue,* in its final form, was published in the last volume of the *Historia Cælestis Britannica.* It contained 3000 stars. The introductory matter included Ptolemy's catalogue of 1026 stars, a catalogue of the Arabs, the observations of Tycho Brahe, Kepler and Hevelius together with the observations made by Halley of the stars in the southern hemisphere.

(8) Johann Bayer (1572–1625). See *Allgemeine Deutsche Biographie* where he is described as *Rechtsanwalt* (lawyer). He was also a competent astronomer, and his famous work, *Uranometria* (1603), was a complete celestial atlas. See Hutton, *Mathematical Dictionary,* I, 219. Bayer was the first to adopt the plan of designating the stars by the name of the constellation, prefixed by the letters of the Greek alphabet, usually assigned in the order of magnitude. See H. Spencer Jones, *General Astronomy,* p. 286.

(9) At this time the word 'undertaker' was sometimes used as synonymous with 'publisher'. See *O.E.D.* But Flamsteed here wishes to make it clear that the charge would be too great for a private individual.

(10) Prince George of Denmark; he married the future Queene Anne in 1683. He is described by Trevelyan (*Blenheim,* p. 178) as 'a kindly, negligible mortal'. See *D.N.B.*
   At this time he was Lord High Admiral, and this accounts for his interest in Flamsteed's *Historia.* He died in October 1708, two years before the printing of the *Historia* was begun.

(11) *Mathematicus Regius.*

678 FLAMSTEED TO POUND

15 NOVEMBER 1704

From the original in the Royal Greenwich Observatory[1]

The Observatory No: 15. 1704

Kind Sr

I have yours of June 7 last from Condore by Capt Monke who is now as I am told comeing up ye River. I hope by him I may receave ye two small potts of Tea & ye Fans you present my Wife. I have also your two former letters by Capt Roberts & Capt Hurle....

Since I wrote to you last Mr Newton has publisht his treatise of Light & Colours (he calls it *Opticks*)[2] wherein he has given us a many curious experiments whereby he proves fully that the raies of light are differently refrangible. I was delighted with it till I came to pag 72. 73 where I found he committed a great fault which runs t[h]rough the deductions in the following parte of his book.[3] for* in ye *Phil. Transactions* Num: 80 pag 3079 he assumes *that ye Object glass of a Telescope cannot collect all ye raies which come from one point of an object so as to make them convene at its focus in less room yn in a Circular space whose Diameter is ye 50th part of ye Diameter of ye Aperture* on ye Object glass. This is a Naturall consequence of his Theory. but now in his *Opticks* pag 72 he contracts this distance without takeing any Notice of the exorbitancy of the former, for sayes he Very Warily. *The sensible image of a lucid point is scarce broader yn ye 250th parte oj ye diameter of ye aperture of a good object glass of a telescope; and therefore in a Telescope whose aperture is 2 inches & length 20 or 30 feet.*[4] it (yt is every visible lucid point & consequently every fixed star) may be 5 or 6 seconds (diameter) *and scarce above*, which is absolutely false: for of the Vast Number of fixt stars yt may be seen with a telescope of betwixt 20 & 30 foot with a 2 inch aperture, 4 parts in 5 have not one second diameter, if I had said 19 parts in 20 I beleive I had not erred, Had Mr N. held to his assertion in ye *P. Trans*: each point of light or star must have been 30 seconds Diameter. but he was to help a freind and his affections mislead him. I orderd my former servant Mr Hodgson[5] (who Now teaches Mathematicks in London) to discourse of this publickly. he did it & thereupon Mr N. came down to see me I shewed him ye error of his assertion & an easy experiment whereby it might be proved at our earth yt ye raies of light spreading had no such monstrous effects as he imagined. he gave me the heareing quietly & made me no answer. The book Makes no Noyse in Town as the *principia* did (which I hear he is prepareing againe for the press with Necessary corrections), for tho the experiments are very good & sincerely related, his deductions are a many of them Mistakes of which he lays ye

* in ye Miscellanea pag 103.

ground in the forementioned *Ph. Tr.* pag 3083 Prop 8 & 9[6] where he affirmes 1st That *whiteness is a compound of all sorts of Colours mixt* & 2dly yt Whiteness *is ye usuall colour of light* he means native light yt of ye Sun or Stars for he calls ye colour of yt representation of the Sun which is made by ye collection of his raies into ye focus of a deep convex (or burning) glass *Whiteness* which he may call *candor* but tis far from *albedo*[7] for being mixed *with ye prismaticall blue* it makes *green* with ye *Red Orange* whereas proper *white* or *paint* mixed with *blue* makes onely paler blue, & a paler red with red. & therefore Native light is rather yellow or of ye forme of yellow.

Now Mr N. concludeing that a Mixture of all colours makes Whiteness, argues hence that ye mixt raies of a lucid point collected in ye distant base of a telescope must forme a lucid object there 5 or 6 seconds in diameter which experience proves they doe not whereby tis intimated that onely the Raies of pure or Native light serve to paint ye object in ye distinct base, the colourd raies serveing onely to compass it with a ring of Colours, visible when viewd through ye Edges of a very convex Eye glass, but scarce seen when viewd on ye pole or Axis of it as I doubt not but your own experience either has convinct you or easily may.

I send you this book & with it another of tracts collected from ye *Philosophicall transactions* in which you will find what Mr N. published concerneing light and colours in ye *Ph: Trans.* You heard no doubt of Mr Lowthorps[8] designe to excerpe & abridge ym under proper heads. this seemes done to spoyle his but I am apt to thinke will render his more acceptable, for he pretends to be more correct yn ye *Transactions* are when I find but one error Correct in ye Miscellanies & many new made yt were not in ye Originall as you will see in ye Theory of ye Moon.

With it you will also receave a book of a young man lately arrived here from ye Island of *Formosa.* he had a meeting with Pere Fontenay[9] at ye R Society where ye Jesuit declared him (wth ye usuall confidence of yt Order) to be an imposter: the Main Argumt was because he affirmed formosa to be under ye Jurisdiction of ye Japonese whereas he (the Jesuit) affirmed it under the Emperor of China. pray enquire under which it is & informe us as also what more you can learn of yt country for we are here in the dark about it. The Author studies now at Oxford. is very pregnant understands latin greek High Dutch French & Italian your thoughts of his book will be very acceptable. This book is presented you by *Dr Gastrell a Canon of Christ Church* who has Married Dr Mapletofts[10] daughter & is thereby become my Neighbour

Our R Society produces nothing worth imparteing to you nor I fear is like to doe till its present constitution be altered: Mr Newton at this time is president. Mr Thirnhaus[11] has contrived an extraordinary burning glass.

The French have bought it at a great price & sent over an account of its performances. Mr N. has contrived one yt consists of 6 concave glass speculums placed about a 7th each near a foot diameter. I have not seen it but am told it performes well & melts Gold readily in ye ☉, but the Gold I am told fumes much, as all other mettalls doe, before it fluxes. I am apt to thinke a burning glass Made to performe by Refractions would performe better, but wee are fond here of Theorys & reflections, tho our success in either is not much to be boasted of:

I am glad to hear you have receaved ye 3 foot quadrant safe; tho injured wth rust & canker in ye passage: that fault I hope you will soon & easily remedy, I would entreat you to get it fixed on a wall as neare Meridionall as you can, & to observe with it ye Meridionall Zenith distances of ye southern fixed stars. which (without derogateing from Mr Halleys previous labors) I must tell you are still wanteing his paines, will be a good guide to you, but 1st he left a many fixed stars unobserved & even some intire constellations. 2. he had no good assistant. 3d with such an one as had no relish of his worke it was almost impossible for him to avoyde faults in writeing downe what he observed & I fear a many are committed not so much through his hast as his assistants negligence & Indifferences for his business. but 4th he calculated their places from his observations I fear but once himselfe & had no skillfull calculator to repeate them, the Calculations in his Cases were long perplext & troublesome so yt it was almost impossible for him to avoyd faults & what he has committed are on yt account excusable. I have heard him say he had a mind to repeate his Calculations. he took no Zenith Meridionall distances. when you have these taken it will be easy examineing his observed places. the right Assentions will be readily given as in my Method. You may make charts of ye Constellations by them & ye transits & when you come home persons will be found yt will readily assist you in any further Calculations to be made. The Work will bring you a great reputation & that will draw a recompense after it if the company should not afford you one which certeinly they will doe since nothing can better or More secure their Navigation in ye Southern Seas then your true Catalogue of ye Southern fixed Stars which all our Ingenious Saylours complaine they want. There are a many impudent fellows mixt amongst ym that boldly hazard & lose their ships for want of skill. these want nothing or are ashamed to owne their ignorance. & therefore rashly censure all such works as useless or rally them sharply for which little witt will serve their turnes, these I hope you have learnt to despise. & will not suffer the more ingenious and skillfull to suffer for their raillerys & impudence

I have translated Ptolemys Catalogue anew out of the original greek into latin. because I found the old translators understood him not for want of skill:

drawn new figures of the Constellations conformable to his descriptions, with which all the printed Catalogues whether french Italian Spanish or latin nearely agree. save that they make some stars to be in the backs of the human formes whereas Ptolemy allwayes represents them with their faces to us & in such postures yt the backs cannot be seen. So yt your Globes yt represent their backs to us must be conceaved transparent & ye eye at the center viewing the faces of them. which causes me to thinke Ptolemy did not designe his figures for ye Globe but his Astrolabe which they very well fit. Bayers mapps which our Globemakers Copy are all false. Hevelius's are onely fit for Globes & no body can learne ye Constellations by them because the stars yt are on ye right hand in his figures are to the left in the Catalogues besides he makes his signes succeed from the left hand to the right just contrary to what they doe in the heavens. My chartes avoyd all these incoherencys & inconveniences they are each neare 2 foot broad & 20 inches deep. I have got halfe a dozen of them curiously drawn by an Able Master the Prince[12] has seen them. & the Providence of God that has conducted my Worke hitherto seemes now designing to perfect it for I am told he will be at the Charge of the Impression. that all the Glory and honor of it may be Gods. the advantage to ye publick. I have drawne up an estimate[13] of it & find to my surprise it will not be conteind in less yn 1500 pages. the observations alone will take more yn 1100 of them without the Prefaces which cannot be estimated. The rest will scarce conteine all the Catalogues extant & mine with them, besides which there will be near 100 pages of Tables not accounted into ye estimate

I give you this full account of my Worke purposely to excite & encorage you to set upon & goe through with Yours vigorously and cheerfully. I shall prepare Chartes for you against your returne that as soon as you have determined the stars places they may be inscribed & the larg volume of the heavens layd open to our Navigators and such Ingenious persons as take pleasure to contemplate ye Workes of God in ye heavens. And I doubt not but your labors will be as kindly received by all knowing men as those of dear Sr

Your sincere freind & servant

JOHN FLAMSTEED MR

NOTES

(1) Vol. 36.
James Pound (1669–1724), astronomer; M.A. (Gloucester Hall, Oxford), 1694; M.B., 1697. Having taken orders, he entered the service of the East India Company and went out to Madras in 1699 as chaplain to the merchants of Fort St George, whence he proceeded to the British settlement on the islands of Poulo Condore. On 3 March 1705 the native troops mutinied, and Pound was one of eleven who escaped to Malacca and ultimately reached Batavia. His collections and papers were destroyed. On Flamsteed's death he was presented with his living

as Rector of Burstow in Surrey. He was elected F.R.S. in 1699. He mounted Huygens's 123-foot object glass, lent to him by the Royal Society in 1717, in Wanstead Park, Essex, on the maypole removed from the Strand and procured for the purpose by Newton.

Newton employed, in the third edition of *Principia* (pp. 390, 392 of Sir W. Thomson's reprint, 1871), Pound's micrometrical measures of Jupiter's disc, of Saturn's disc and ring, and of the elongations of their satellites. He also supplied data for correcting the places of the comet of 1680. He trained James Bradley, his nephew, in astronomy, and many of their observations were made together.

See Bradley MS. no. 24 (Bodleian Library), and *D.N.B.*

(2) The *Advertisement* to *Opticks* (672) is reproduced on p. 414.

(3) See *Phil. Trans.* **6** (1671/2), 3079.

(4) In *Opticks* (1704), p. 72, Newton wrote: 'The sensible Image of a lucid point is therefore scarce broader than a Circle, whose Diameter is the 250th Part of the diameter of the aperture of the Object glass of a good Telescope, or not much broader, if you except a faint and dark misty Light round about it, which a Spectator will scarce regard. And therefore in a Telescope whose aperture is four Inches, and length an hundred Feet, it exceeds not 2″ 45″ [a slip for 45‴], or 3″. And in a Telescope whose Aperture is two Inches, and length 20 or 30 Feet, it may be 5″ or 6″, and scarce above. And this Answers well to Experience: For some Astronomers have found the Diameters of the fixt Stars, in Telescopes of between twenty and sixty Feet in length, to be about 4″ or 5″ or at most 6″ in Diameter.'

(5) James Hodgson, who left Flamsteed's service to become mathematical master at Christ's Hospital. See Letter 499, note (5), p. 104.

(6) Newton wrote: '8. Hence therefore it comes to pass, that *Whiteness* is the usual colour of *Light*; for, Light is a confused aggregate of Rays indued with all sorts of Colors, as they are promiscuously darted from the various parts of luminous bodies. And of such a confused aggregate, as I said, is generated Whiteness' (*Phil. Trans.* **80** (1671/2), 3083).

(7) Albedo is defined as the fraction of the total amount of sunlight incident upon the Moon which is reflected from it in all directions (see H. Spencer Jones, *General Astronomy*, 3rd edition, p. 133).

(8) See Letter 626, note (1), p. 332.

(9) Psalmanazar and Fontenai. See Letter 670, notes (3) and (4), p. 412 and p. 413.

(10) John Mapletoft (1631–1721), physician; Scholar and Fellow of Trinity College, Cambridge; F.R.S. 1676. A close friend of John Locke until about 1680. He was Gresham Professor of Physic from 1675 to 1679; he then abandoned medicine and gave himself up to the study of Divinity, becoming D.D. in 1689. See Ward: *Lives of the Gresham Professors*, pp. 274–9.

Francis Gastrell (1662–1725), Christ Church, Oxford. M.A. 1687, D.D. 1700. He was appointed Queen's Chaplain in 1711; three years later he became Bishop of Chester. He published *Christian Institutes* (1707). He was a friend of Mapletoft, whose daughter he married.

(11) Ehrenfried Walther von Tschirnhaus (1651–1708). He wrote extensively on various branches of mathematics. See vol. I, p. 355, note (5): also *G.M.V.*, pp. 315, 382. See also vols. II and III for his correspondence with Collins and others.

(12) Prince George of Denmark; see Letter 677, note (10), p. 423.

(13) See Flamsteed's Estimate (677).

679   NEWTON TO ?THE LORD CHAMBERLAIN
24 NOVEMBER 1704
From a draft in the Royal Mint[1]

My Lord

Mint Office 24 *Nov.* 1704.

Some designes for Medals[2] having been communicated to your Lordship by others, I humbly beg leave to present the enclosed.[3] Her Majesties effigies may be on one side with the usual inscription, & this Designe on the other; & instead of Britannia on a globe the Queen may be placed in a chair.

And if for saving her Majty & your Lordship the trouble of approving Medals your Lordship shall think fit that the Gravers be empowered to make such Medals as I shall approve of under my hand in writing, I am ready to undertake this trust, or otherwise to act in such manner as your Lordship in your great wisdom shall think fit, being

My Lord

Your Lordships most humble

and most obedient Servant

Is. NEWTON

NOTES

(1) Newton MSS. III, 288 (in Newton's hand). There is no address on the letter; it was perhaps sent to the Lord Chamberlain.

(2) The medal was to commemorate the institution of Queen Anne's Bounty. From the description of the medal, which was struck, Newton's suggestion was adopted: *Obv.*, bust of queen to the left, laureate and draped; *Rev.*, Anne seated on throne presenting charter to kneeling bishops (W. J. Hocking, *Royal Mint Museum Catalogue*, 1910, vol. ii, p. 228).

The 'first-fruits and tenths', originally paid into the Papal Exchequer, had, after the Reformation, been annexed by the Crown and became a source of privy income. This inflicted considerable hardship on the poorer clergy, and to remedy this, Anne instituted her famous bounty. She not only remitted all arrears of first-fruits and tenths to poor clergy who were in debt on that account, but she also made over the fund for the increase in stipends. The announcement of Queen Anne's Bounty was made on her birthday, 6 February 1704, and an Act of Parliament was passed that year to give it effect.

Gilbert Burnet, Bishop of Salisbury, had repeatedly urged William III to divert first-fruits and tenths to the relief of poorer clergy but his financial straits prevented him from acceding to this request. See Trevelyan, *Blenheim*, pp. 47–8.

In 1947 Queen Anne's Bounty was transferred to the Church Commissioners.

(3) The enclosure has not been found. But there are several designs for medals, approved by Newton, in the British Museum (Add. MS. 18757). Newton showed a scholarly interest in medal design.

680   GEORGE CLARK[1] TO NEWTON
11 DECEMBER 1704[2]
From a copy in the Royal Greenwich Observatory[3]

Sr.

The Prince[4] has perused the Estimate[5] of the Intended Historia Cœlestis Britanmia [*sic*] which you presented him his Royall highness is perswaded of Mr Flamsteeds fitness for A work of this nature and being unwilling that the Observations designed for ye benifit of navigation and Encouraged So well in ye begining should want any Necessary Assistance to bring them to perfection he has been pleased to Command me to desire your Self Mr Roberts[6] Sr Christopher Wren Dr Alburthnet [*sic*][7] and others of your Society as you think proper and will share the trouble wth you to inspect Mr Flamsteeds Papers and consider it is fitt for the press and when his Royall Highness knows your Opinions[8] you may be sure he will do any thing that may conduce to the making ym of use to ye Publick

I am

                                        Sr Your most humble Servant

                                                          GEO. CLARK

To Mr Newton President
   of ye Royall Society

NOTES

(1) George Clark, or Clarke (1661–1736), Brasenose College, Oxford. Was successively Member of Parliament for East Looe (1705), Winchelsea (1708) and Launceston (1711); D.C.L. (1708); Warden of the Cinque Ports, and one of the Lords Commissioners for the Admiralty (1710–14). He was private secretary to Prince George (see Letter 677, note (10), p. 423) from 1702 to 1705. See *Bulletin of the Institute of Historical Research*, v, 179.

(2) The letter is undated, but we learn from the Referees' letter to the Prince (Letter 685) and Newton's summary of the correspondence relating to the publication of the *Historia Cœlestis Britannica* (Number 749, p. 529) that it was sent on this date. Newton did not receive it until 17 December (see Letter 681).

(3) Vol. 33, fo. 48. The address is in Newton's hand.

(4) See Letter 677, note (10), p. 423.

(5) As early as 1703 it was known that Flamsteed's catalogue was ready for the press. On the invitation of the Royal Society, Flamsteed submitted 'an estimate of ye Number of pages yt my books of observations and catalogues when printed might be comprehended in & by

J. Hodgson imparted it to ye R S at one of their meetings It was agreed yt it was fit to be recommended to ye prince who was chosen into ye R.S. No: 30 following' (R.G.O., vol. 35, fo. 47 *et seq.*). This is the estimate reproduced on pp. 420–2. On 7 December Newton communicated this information to the Prince, who lost no time in replying, through his secretary, George Clark, with the above letter. This letter was read to the Society on 20 December and the President (Newton) 'was desired to return the most humble Thanks & Acknowledgements of the Society to the Prince, for so great a Favor, in such a manner as he shall think fitt' (Journal Book of the Royal Society of London, 20 December 1704).

(6) Francis Robartes (1650–1718), politician and musician; son of Sir John Robartes, first Earl of Radnor; sat in Parliament from 1673 until his death. Elected F.R.S. (1673). He wrote many works on the theory of sound.

(7) Dr John Arbuthnott (1667–1735), physician and wit; M.D. (St Andrews), 1696; settled in London, where he taught mathematics. Elected F.R.S. in 1704; physician to Queen Anne; F.R.C.P., 1710; Censor, 1723; Harveian Orator, 1727. He was acquainted with many literary men of his day, including Swift and Pope. His principal works are political satires; he wrote *History of John Bull* (1712).

(8) Newton's report was sent on 23 January 1704/5 (Letter 685).

681   NEWTON TO FLAMSTEED
18 DECEMBER 1704]
From the original in the Royal Greenwich Observatory[1]

Mr Flamsteed

I received last night a letter from the Prince wherein his Highness expresses that he is unwilling that your Observations designed for the benefit of Navigation & encouraged so well in the beginning should want any necessary assistance to bring them to perfection & therefore desires me, Mr Roberts, Sr Chr. Wrenn & some others of your friends to inspect your papers & consider what is fit for the Press & when his Highness knows our Opinions he is ready to do any thing that may conduce to the making your Observations of Use to the publick. This is the substance of the Letter wrote by ye Princes Secretary by his Highnesses order.[2] And to morrow Mr Roberts, Sr Chr. Wrenn and the rest of the Gentlemen to whom his royal Highness has referred the inspection of your papers, are to dine with me in order to consider of this matter & speak wth you about it. And therefore I desire the favour of your company at dinner with them, & if you please to come in the morning & bring your papers with you or such parts or specimens of them as may be sufficient you will oblige me & the rest of your friends to whom the inspection of them is referred & promote the dispatch of this affair.[3] If you bring the papers themselves you expedite your buisines, & you may rest assured that they shall not goe out of your hands.

I am   Your very loving Friend

& humble servant

Jermin Street.   Isaac Newton.
18 *Decem.* 1704

NOTES

(1) Vol. 35, fo. 37.   (2) Letter 680.

(3) Flamsteed attended the meeting, bringing with him some of the documents, but as the Referees had not sufficient time to examine these, a further meeting was arranged for 27 December. As a result of this, and of subsequent talks between Flamsteed and Newton, the Referees drew up a report (Letter 685) which was submitted to the Prince on 23 January 1704/5, and received his approbation.

682   NEWTON TO FLAMSTEED
26 DECEMBER 1704
From the original in the Royal Greenwich Observatory.[1]
For answer see Letters 683 and 684.

Sr

I thank you for ye information you give me about the charges for printing. I am sorry your servant is ill, but if you do not bring your papers[2] there will be nothing done. For the buisiness of the meeting is to view your papers according to ye Princes order & give him an account of them. We had but little time at my house to view those you brought & did not meet at the Coffee house for that purpose, but appointed to morrow morning at ten a clock that we might have time to view them & come to a conclusion. And I hope you will not disappoint your friends.

I am

Your humble servant

Is. Newton

London:
26 *Decem*
1704

For the Rnd Mr John Flamsteed
at the Observatory at
Greenwich

NOTES

(1) Vol. 35, fo. 41.   (2) See Letter 683, note (4), p. 434.

683   FLAMSTEED TO NEWTON
2 JANUARY 1704/5
From a copy in the Royal Greenwich Observatory.[1]
In reply to Letter 682

To Isaack Newton Esqr
at his house in Jermin
Street London.[2]

The Observatory Jan. 2, 1704/5

Sr

Yesterday I sent You according to Your desire & my Promise (by Mr Hodgson) my Old Charts of Orion, Ophiuchus, Aquarius & Pisces these have all ye circles of Longitude & Parallells of Latitude inscrib'd in ym as You will find ym in ye New one of Orion, wch I think is Compleately finishd yt by it you may judge of ye Rest wch I have design'd of ye same bigness, I have added ye New Constellations of Aquarius & Ophiuchus whereby You will see ye Charts cannot conveniently be made less, but I have not Caus'd ye Circles of Longitude & Latitude to be inscrib'd in those be cause I Esteem'd it Needless, The Chart of Orion alone serving sufficiently to show how they will Fill ye Copper Plate wn they are inscrib'd. Wth these I have sent You my Greek Ptolemy & my Latin Version Concerning wch I must informe You yt because I differ From ye Common Translations I have thought it necessary to keep as close to ye Greek as I handsomely could, & therefore you must carry ye *Quæ est* that begins ye head or second Line of Every Constellation to ye beginning of every Line Following it, And further yt if You think it not advisable to Print Ptolemy's Greek text, it will be best to put those Words, in whose interpretation I differ From others in ye Margin, For if they be inserted wth a Parenthesis in ye Text, they will make ye Line too Long.

I have also sent you a Bayer,[3] yt by compareing his Descriptions & Figures wth ye Greek & Mine, You may see how (by takeing ye Meaning of Greek Words From Lexicons) he makes all those Stars to lye on ye left sides leg of Aries &c wch ye Originall & all ye Printed Catalogues (except his Own) Put on ye Right, & Vice Versa; & moreover Often thrusts such Greek Words into his Text as are No Where to be found in Ptolemy. I would have sent You a Scale to ye Charts but that I have never a Cleane one by me; if ye Ill weather Should hinder me from coming to London to Morrow[4] as I Intend (God continuing my health) I shall cause a New one to be drawn an wn a Good Day affords me an Opportunity You shall be attended wth it by Sr.

Your oblig'd & humble Servt

JOHN FLAMSTEED MR

I wish you many happy years.

NOTES

(1) Vol. 33, fo. 51.

(2) Only this is in Flamsteed's hand. The letter is in the hand of a copyist, probably Weston (see Letter 638, note (7), p. 370).

(3) See Letter 677, note (8), p. 423.

(4) Flamsteed visited Newton on 3 January following (R.G.O., vol. 35, fo. 47). On this occasion Newton said he desired copies of some of Flamsteed's papers. On the same day Newton reported that he 'had waited on the Prince with the Thanks of the Society: and that the Publication of Mr Flamsteeds Papers is carrying on as fast as may be' (Journal Book of the Royal Society of London, 3 January 1704/5).

684 FLAMSTEED TO NEWTON

5 JANUARY 1704/5

From a draft in the Royal Greenwich Observatory.[1]
In further reply to Letter 682

The Observatory Jan 5. 1704/5

Sr

Herewith my Servant will deliver you the papers you desire if his fit prevents him not & force him to send them by ye peny post; if he reaches your house pray send me by him yt Volume of Petronius[2] yt has ye fragmt of Eudoxus[3] in it: perhaps it may be of use to me I shall take especiall care of it & return it safe when ever you please. I have hastned these yt the want of them may be no hindrance to your makeing your report to ye Prince or concludeing about the Impression of the Observations & Catalogue

Num 1. is a page of ye Observations of ye Stars' intermutuall distances taken with ye Sextant these are such as you conclude cannot be printed double on one page of paper

Numbr 2. (ye second page on ye same sheet) is copied from ye book of lunar Observations taken with ye sextant the 5 last lines of this ought to stand a line or two lower to make room for tytles to stand over ye heads of ye Columns of Numbers I was absent when my Servt copied it but shall take care the like error be not committed when it shall be copied for ye Press.

Numb 3. is ye first page of ye Observations of ye year 1699 & make a part of ye Second Volume, I chose it because it appears by the lines I had wrote at ye entrance of ye year, that I had determind 6 years agone to transcribe them all thus as you lately hinted you thought it would be best & have caused them to be so copied ever since

434

Numb 4. is a part of a double page of my Catalogue I have caused my Servant to transcribe onely a few lines of it, they will be sufficient for a Specimen, the rest of the Page may safely be conceaved filld up in ye same Manner I pray ye Good God who has blest my labors hitherto with success to bless your endeavors for their publication send you many happy yeares & am with all due respect Sr

Your obliged humble Servant

JOHN FLAMSTEED M R

N B the report here Mentioned was made signed by Mr Newton Mr Roberts & the other gentlemen concerned before Mr Newton went to Cambridg[4] & I saw it not till his return Feb: 26 at London see it ye last page but one before this[5]

☿ Feb. 28. I was at London met wth Mr Newton accidently at Garways[6] talkt wth him about ye printing & an honorable recompense for my paines & 2000 in *lib*. expense: had been with Sr C. W. ye monday before. 26.

March ye 3d at London met wth Mr Newton stayd at his house till ½ past 3 in expectation had a letter from him at my return directing to meet him at ye Castle Tavern[7] in Pater Noster Row on Monday ye 5. Expense 3*s*. 00*d*:

NOTES

(1) Vol. 35, fo. 51. This is a draft of a letter which Flamsteed sent on 6 January, for at the head he has written: 'Ye 3 waited on Mr Newton who told me he had not yet made his report to the Prince desired copies of some papers which I sent him Jan: 6 with ye following letter by T. Weston.' See Letter 638, note (7), p. 370.

The three persons referred to were Wren, Robartes, and David Gregory. The report was sent on 23 January 1704/5. See Letter 685.

(2) Petronius, Gaius (first century A.D.), companion of the Emperor Nero, who regarded him as chief arbiter of the imperial pleasures (*Elegantiæ Arbiter*); and author of the novel *Satyricon*, a medley of prose and verse, giving a description of Roman life.

(3) Eudoxus of Cnidos (fourth century B.C.), an important mathematician and astronomer. There is no connexion between Petronius and Eudoxus. What Flamsteed here refers to may well be a composite volume, or a volume of Petronius with a fragment of Eudoxus loosely inserted in it.

(4) Newton frequently visited Cambridge about this time, probably in connexion with the forthcoming election. It was whilst on one of these visits that he was knighted (16 April 1705).

(5) This appears to refer to Letter 685.

(6) See Letter 594, note (4), p. 285.

(7) A tavern in Paternoster Row, not so well known as Garraways. It was burned down in 1770, but it appears to have been closed as a tavern before that date.

685   THE REFEREES TO PRINCE GEORGE
23 JANUARY 1704/5
From a copy in the Royal Greenwich Observatory[1]

May it please your Royall Highness[2]

According to your highnes order signified to us by Mr Clarks Letter[3] of Decembr: 11th Last we have inspected Mr flamsteds Papprs & are humbly of opinion that *all* the observations wch he proposes to be printed in ye first & second parts of ye works are proper to come Abroad together with his tow Catalogues of ye fixed stars in Latin all wch are Redy for ye prese & with prefaces will take up about 1200 pages[4] in folio when printed wch agrees with what he has Represented: and the Expenses of printing 400 Copies According to ye estimate will be as follows

		p	s	d
for 283 Rea[m]s of Demy paper of $16\frac{1}{2}$ inches by 22 att the Rate of 20s a Ream		283	00	00
for Composing & press work of 300 Sheets att ye Rate of 20s a Sheett		300	00	00
for the Charges of an Emanuensis to Copy & Correct the pr[e]sse & to Compare and examin ye papers. . .		100	00	00
	Totall	683	00	00

it may be Also very proper to print the places of ye moon and planets & comets derived from ye observations: of this six hundred are Alredy Computed & fourteen hundred Remain to be Computed: and the Charges of tow Calculators[5] to finish them & of ye paper press work and printing will be about 180p: So that the whole Charge will be about 863p

This set of observations we Repute the fullest & Complatest that has ever yet Been made: and as it tends to the perfiction of Astronomy & Navigation: so if it should be Lost, the Los would Be irreparable & we have no prospect that a work so expensive will ever see the light unless your Highness will please to be at the charge of publishing it. We are

May it please your Royal Highness

Your Highnesses most humble

and most obedient Servants

Signed by Mr Roberts Sr Chr. Wren. Mr Newton. &c

London. 23 Jan.
1704/5.

436

NOTES

(1) Vol. 35, fo. 33. There is a draft of this, in Newton's hand, in the University Library, Cambridge (Add. 4006, fo. 7).

(2) This, as well as the names of the signatories at the foot of the page, is in Flamsteed's hand. The remainder of the letter is in the hand of a copyist except for the final lines beginning 'ever see the light' and ending 'most obedient Servants' which were written by Newton.

(3) Letter 680.

(4) Flamsteed, in his estimate of the number of pages (Number 677) says: 'Pages in all, about 1450.' This includes all the three parts, together with the *Liber Prolegomenos*, or *Preface*. The figure here quoted by the Referees ('about 1200 pages') suggests that they only recommend to be printed what is ready for the press and this did not include 'the Third Part [which] (if God spare me Life and Health, and necessary Assistance be afforded me) may in the mean time be fitted for the Press'.

(5) The copyist has written 'Caludators'.

686   NEWTON TO FLAMSTEED
2 MARCH 1704/5
From the original in the Royal Greenwich Observatory[1]

London. 2 Mar. 1705 [Friday].

Mr Flamsteed

The Gentlemen to whom his Royal Highness has referred the care of printing your Observations, have agreed to meet on munday morning at eleven a clock at the Castel Tavern[2] in Pater Noster Row to set forward the printing thereof & I desire earnestly that you would be pleased to meet us there at the time appointed that we may agree wth you about an Emanuensis & Calculators[3] & what else you have to propose for dispatching the work. I am

Your humble servant

ISAAC NEWTON

John Flamsteed
MR at the Greenwich Observatory

NOTES

(1) Vol. 35, fo. 45.

(2) See Letter 684, note (7), p. 435.

(3) See Flamsteed's *Diary*, 5 March 1704/5 (Number 687).

687   EXTRACTS FROM FLAMSTEED'S DIARY
RELATIVE TO THE PRINTING OF
THE *HISTORIA CŒLESTIS BRITANNICA*

MARCH–APRIL 1704/5

From the original in the Royal Greenwich Observatory[1]

☽ Martij 5 1704/5. Met ye gentlemen at ye taverne saw ye specimens printed[2] discourst of Calculators Mr Churchill[3] not there
Expense   5s. 04d
March 12 1704/5 ☽ J. Weston[4] began to copy ye observations & distances & Mr Witty[5] afternoon Mr Gascoignes[6] observations God give us success.
March 19 ☽ I had mislayd Mr Newtons first letter to me on the occasion of printing my workes I have orderd it & some of Mr Newtons to be hereunder copied to preserve them, the rest shall be ordred God willing as they arrive if they be proper to be preserved
    I was in Surry from July 16 to August ye 22.
1704/5 March 24 ♄ Went to London. Ye specimens not done
            28 ☿ againe there. Specimens done[,] corrected part[:] wholly next morning[.] Sent them back to be printed
Apl 3 ♂ at London. waited on Sr C W and Mr Roberts.

NOTES

(1) Vol. 33, fos. 54–9.

(2) On the following day (6 March) Flamsteed wrote to Sharp: 'Yesterday I was at London with Sir C. Wren, Mr Roberts, Mr Newton, and some other gentlemen, to view some specimens of a printer's page of my book of Observations: they were ill done. I am causing them to be copied again in order to have them printed by another, if the gentlemen think fit' (Baily, p. 235). Flamsteed had not been consulted when the Referees decided to appoint Churchill.

(3) Awnsham Churchill (died 1728), a leading bookseller and publisher. He was connected with the family of Churchills of Colliton, Dorset. He worked alone from 1681 to 1690 when he was joined by his brother, John, at the sign of the Black Swan in Paternoster Row. They were the leading Whig publishers and they appear to have done a great deal of government printing. They were regarded as competent, for quality and finance, for publishing Flamsteed's *Historia*.

(4) See Letter 638, note (7), p. 370.

(5) John Witty became Flamsteed's assistant on 30 March 1705, and he remained with him for at least three years, residing at the Observatory, Greenwich. He is described as 'an expert calculator'. He helped with the publication of the *Historia*.

(6) See Letter 677, note (2), p. 423.

688 HALIFAX TO NEWTON
17 MARCH 1704/5
From the original in King's College Library, Cambridge[1]

Sir,

I send you the Addresse of the House of Lords to wch the Queen made so favourable an answer, that the enemy[2] are quite enraged, the Paragraph in her speech against the Tackers[3] provokes them still more then this: and whatever the Ministers may think they will never forgive them for either; I beleive they begin to think so, and will take measures to make other friends. I was in hopes by this Post to have sent you an account of several alterations that would have pleased you, but they are not yet made, thô you may expect to hear of them in a very little time, among other expectations Wee have, Wee do depend upon a good Bp, Dr Wake[4] is likely to be the man, Wee are sure Sr William Dawes[5] will not. I think this will have great influence in the Place where you are and therefore I think you may mention it among your friends as a thing very probable, thô it be not actualy settled. He is to hold St James's in comendum[6] and Dr Younger[7] will be Dean of Exeter. Mr Godolphin[8] will go down to Cambridge next weeke and if the Queen goes to Newmarket, and from thence to Cambridge[9] she will give you great Assistance the Torys say she makes that tour on purpose to turn Mr Ansley[10] out. He is so affraid of being thrown out that Ld Gower[11] has promised to bring him in at Preston, which they should know at Cambridge. If you have any commands for Me here I desire you would send them to one who shall be very ready to obey them I am

Your most Humble and most Obedient servt

HALIFAX

17 March [1704/5][12].

NOTES

(1) Keynes MS. 102.

(2) The address was presented to the Queen, and answered by her on 14 March. The 'enemy' was the Tory party. Halifax and Newton, it should be noted, were both Whigs.

(3) Tacking, in parliamentary language, was the practice employed by the House of Commons for attaching to a money bill a measure for some other purpose in order to force it through the House of Lords. In 1704 the tackers tried to attach clauses against Occasional Conformity to a supply bill. See G. M. Trevelyan, *England under Queen Anne*, II, 14–16.

(4) Dr Wake (1657–1735), Archbishop of Canterbury; M.A., Christ Church, Oxford; D.D., 1689. He was appointed Bishop of Lincoln (in which diocese Cambridge was situated), 1705, and was translated to Canterbury, 1716. He published many theological treatises. He bequeathed his library to Christ Church, Oxford. See *D.N.B.*

439

(5) Sir William Dawes (1671–1724), second baronet; fellow of St John's College, Oxford; Master of St Catharine's Hall, Cambridge, 1696; D.D., 1696; Chaplain in ordinary to William III, 1696; Bishop of Chester, 1708; Archbishop of York, 1713. A writer of religious poems and treatises.

(6) When the incumbent of a parish became a bishop he was sometimes permitted to retain his living for a period along with his new preferment; he was then said to hold the living *in commendam*, or as a *commendam*. In the present case, Wake retained the living for a year.

(7) Dr John Younger was chaplain to Queen Anne before her marriage. See Joseph Foster, *Alumni Oxonienses*.

(8) Godolphin, Hon. Francis (1678–1766), son of Sidney, the first earl (see Letter 647, note (1), p. 386). Educated at Eton and King's College, Cambridge; M.A., 1705. Represented various constituencies in Parliament from 1702 to 1712. Viscount Rialton, 1706. Second Earl of Godolphin, 1712. He held important offices in the government and in the royal household. He was one of the Lords Justices, 1723, 1725 and 1727; Lord Privy Seal, 1735–40.

(9) Queen Anne went to Newmarket on 12 April. Four days later she went to Cambridge, where she was received by Dr Bentley, the Master of Trinity. On this occasion she conferred the honour of knighthood upon John Ellis (Ellys), the Vice-Chancellor, on James Montague, and on Newton. See Number 692, also Cooper, *Annals*, IV, 70–2, and Trevelyan, *op. cit.*, p. 28.

(10) The member for Preston was Francis Annesley, grandson of first Viscount Valentia. Arthur Annesley was re-elected Member for Cambridge University in May 1705, and again in 1708. He was the son of James, Earl of Anglesey, and succeeded as fifth Earl in 1710. He is described as a Tory. It would seem very likely that Arthur Annesley had been promised Preston if he failed to win his seat at Cambridge and that Francis Annesley having been elected for Preston on 15 May 1705, only two days prior to the date of Arthur's return, would have been willing to relinquish his seat in favour of the latter. For this information, the Editor is indebted to D. Menhennet of the Research Department of the House of Commons.

(11) Sir John Gower (Leveson-Gower), first Baron Gower (1675–1709). Represented Newcastle-under-Lyme in Parliament; created Baron Gower of Stittenham, 1703; Privy Councillor, 1703; Chancellor of the Duchy of Lancaster, 1704–6.

(12) The letter clearly relates to the election of 1705. Under the Triennial Act, the Parliament which had met on 20 August 1702 was approaching its end; it was, in fact, dissolved on 5 April 1705. Newton, despite the diffidence he had shown on a previous occasion (Letter 644), consented to stand. See Letter 689, note (1), p. 441.

689   NEWTON TO ?

MARCH 1704/5

From a draft in the Library of New College, Oxford[1]

I understand that Mr Patrick[2] is putting in to be your Representative in the next Parliament, & beleive that Mr Godolphin[3] my Lord High Treasurer's son will also stand. I do not intend to oppose either of them they being my friends, but being moved by some friends of very good note to write for my self, I beg the favour of you & the rest of my friends to reserve a vote for me till I either write to you again or make you a visit, wch will be in a short time, & you will thereby very much oblige

Your most humble

and most obedient Servant

Is. NEWTON.

NOTES

(1) MS. 361/2. The letter is without date or address. It appears to have been intended for some prominent person in Cambridge, either the Head, or the Fellow, of a college. The date is surmised. Parliament was dissolved on 5 April 1705, and the letter may have been written about the time of one of Newton's visits to Cambridge in connexion with the forthcoming election which took place on 17 May. It was clearly written prior to Halifax's letter of 5 May to Newton (Letter 693).

Newton had represented the University in the Convention Parliament of 1689 and in the short-lived parliament of 1701 (see Letter 644). He was persuaded to stand for the 1705 election and he had reason to hope for success because the popularity of the Tory party was on the wane. The Queen's Ministry looked to the moderate Whigs for support for the prosecution of the war, rather than to the high Tories whose cry, 'The Church in danger', seems to have swayed the electorate (composed for the most part of country parsons). Flamsteed did not rate Newton's chances very highly, for a fortnight before the election he wrote to Sharp: 'Mr Newton is knighted; stands for parliament man at Cambridge; and is going down thither this day or to-morrow, in order to his election. 'Tis something doubtful he will succeed or no by reason he put in too late' (Edleston, p. lxxiv). For the result of the election see Letter 693, note (4), p. 445.

(2) It is not known who Mr Patrick was. Possibly Simon Patrick, son of Simon Patrick (1625–1707) who was successively Bishop of Chichester and Ely.

(3) See Letter 688, note (8), p. 440.

690   FLAMSTEED TO NEWTON

5 APRIL 1705

From a copy in the Royal Greenwich Observatory[1]

Sr   *The Observatory Aprill* 5 1705

Since You went hence I have got almost 30 Sheets of my Observations copied for ye Press. I shall cause ye figures belonging to ym to be transcrib'd upon a folded Sheet in order if you like that Method best to have ym engrav'd, but I desire to discourse wth you again about ym before you fully resolve.

441

I have got specimens done by another Printer[2] as that ought to be, wch shall be left at your House to be ready for You at your return, wch I fear will be delayd by ye Queen's & Prince's coming to Cambridge in Easter Week.[3] I am going into Surry on Saturday next but hope, God sparing me health to be back here ye Wednesday following in ye mean time ye Copies go on and ye Calculation Work is preparing, that there may be no Stop on this account.

I have got some further information concerning ye booksellers & Printer's Practices: I find ye latter dare not disoblige ye former, & yt ye Paper Stationers are so in wth ym that I can't now learn ye Prices of Paper from ym wch before I had to do wth a Printer, was no difficulty. I have taken another Course to be inform'd, & therefore desire You not to proceed to any Agreemts wth ye bookseller, till I have waited on You wth Mr Roberts, Sr Ch. Wren &c yt we may take ye best way to save ye Princes bounty, & make it reach as far as we can; for it will be a terrible Reflection on us, if we suffer a bookseller to devour that as his gaines, wch ye Prince design'd to employ for ye Honor of ye Nation & ye Queen: Good success in your Affairs, health & a happy return[4] is heartily wisht by   Sr

Your Obliged & humble servt

JOHN FLAMSTEED M R

This not sent: he
return'd too soon[5]

NOTES

(1) Vol. 33, fo. 60. Flamsteed was in London on 21 April and met Newton.

(2) Flamsteed had delivered specimens of his work to another printer named Barber. He had informed Newton of this when they met on 7 March.

(3) See Letter 688, note (9), p. 440.

(4) This was written on the day of the dissolution of Parliament.

(5) Only this is in Flamsteed's hand. The rest of the letter is in the hand of a copyist.

691   MINT TO GODOLPHIN

5 APRIL 1705

From the original in the Public Record Office[1]

To the Rt: Honble ye Lord high Treasurer of England

May it please your Lordship.

In Obedience to your Lordships order of Reference of the 15th of March last, upon the Memoriall of Mr Willm Shepard Mr N: Shepard, and Mr Geo: Freeman, wherein they desire to have a patent for coyning forty or fifty Tunns per Annum of Copper halfpence or Farthings, for the space of Eight or Ten years, of equall weight and fineness with those now Curr[en]t as also your Lordships Order of Referrence of the 22d of March last upon the Peticõn of the Provost and Monyers of the Mint, wherein they pray for a License to coyn thirty Tunns of Copper mony a year of the same value, under the direction of the Officers of the Mint, and under such restrictions as your Lordship shall think fitt, to relieve them under the great Straits they are in for want of employment in the Mint, they being allow'd no salarys[2]

We humbly take leave to remind your Lordship of a Report we made the 1st of July last upon a Peticõn of Mr Abell Slaney[3] to coin half pence and farthings, wherein we Acquainted Your Lordship, that the coinages of Copper money in the Reign of King Charles the 2d King James ye 2d and in the begining of the Reign of the late King & Queen had been carried on at the charge of the Crown, under the care and direction of the Principall Officers of the Mint, who kept due Entrys thereof in Books, and the whole Profitt and Advantage was Accounted upon Oath to the Crown, which we then thought the best & safest Method to be followed

We have not heard any reasons since to Alter our opinions, and therefore humbly propose to your Lordship that whenever her Majesty shall think fitt to order Copper mony to be coined, that it may be performed as near as may be According to the Method of the Mint, that either a Genll Importer or the Master worker of the Mint may have money Imprest to him to buy copper to be coined into halfpence and farthings, under the same cheques that are Observed for the coinage of Gold and Silver, and to be coined by small quantitys, Sufficient only to answer the demands of persons, without danger of complaints by overstocking the Nation. out of the clear profitts whereof, her Majesty may have it in her power either to relieve the wants of the monyers of the Mint, many of whom we must inform your Lordship are in a Starving condition, or to gratify any other persons

> All which is humbly Submitted to your
> Lordships great wisdome.

Mint Office   J STANLEY
Apr. 5. 1705   IS. NEWTON
                                                                          JN ELLIS.

NOTES

(1) T. 1/94, no. 8. This letter is written by a clerk and signed by the Warden, Master, and Comptroller.

(2) The coinage of copper had been in the hands of licensees who employed Mint personnel for making dies and stamping blanks which were supplied by the licensees. After the licence

granted in 1694 ran out in 1701, many petitions for the granting of new licences were submitted, but Newton maintained that the coinage of copper should be in the hands of the Mint for Exchequer account; this would remove any possibility that profit motives might lead to inferior work on the coin. Moreover, he argued, the country was amply supplied with copper coins. Consequently no halfpence or farthings were coined for the next sixteen years. By 1713 copper coins had become so scarce as to cause inconvenience. Newton admitted this inconvenience, but delayed the supply for a further five years. The coinage of copper was authorized by Royal Warrant of 13 September 1717, and the issue began on 15 February of the year following. See Craig, *Newton*, Chapter IX.

(3) See Letter 666, note (2), p. 409.

692   NEWTON'S KNIGHTHOOD
16 APRIL 1705
From the *London Gazette*[1]

Cambridge, April 16th.

Her Majesty went up into the Regent-House; where, as is usual upon so great and extraordinary a Solemnity, Degrees in the several Faculties were by Her Majesty's Special Grace conferred upon Persons of High Nobility and Distinguishing Merit....From the Schools Her Majesty went to Trinity-College, the Master whereof, Dr Bentley, received Her Majesty likewise with a very dutiful Speech; and Her Majesty was pleased to Confer the Honour of Knighthood upon John Ellis Esq; Doctor in Physick, and Vice-Chancellor of the University, James Montague Esq; Council for the University, and Isaac Newton Esq; formerly Mathematick Professor, & Fellow of that College.

NOTE

(1) The ceremony is described in the *London Gazette*, number 4116, 'Thursday 19th April to Monday 23rd April'. The Queen was received by the Chancellor, His Grace the Duke of Somerset, after which 'Mr Ayloff, the publick Orator, made Her a Speech full of Loyalty and Obedience in the name of the University'. The Queen then proceeded to the Regent House, where two persons are mentioned as receiving the degree of Doctor in Divinity, and twenty-one as receiving the degree of Doctor in Law. There is no specific reference to Doctors in any other faculty. Following this the Queen went to Trinity College, and there she knighted Newton, as well as the other two persons mentioned. It was unusual for the honour of knighthood to be conferred in recognition of scientific advancement. See article by E. N. da C. Andrade in the *Newton Tercentenary Celebrations* (1946), p. 16: 'Newton was a national figure: when in 1705 Queen Anne knighted him she awarded an honour never before, I believe, awarded for services to science. For, strangely enough, it was not as Master of the Mint that he was knighted: Conduitt tells us that the Queen, "the Minerva of her Age", thought it a happiness to have lived at the same time as, and to have known, so great a man.'

W. Ayloff, LL.D., of Trinity College, was the successful candidate for the office of Public Orator in 1696. He had the support of Newton.

HALIFAX TO NEWTON 5 MAY 1705

693 HALIFAX TO NEWTON
5 MAY 1705
From the original in King's College Library, Cambridge[1]

Sr.

I have writ to My Ld Manchester[2] to engage Mr Gale[3] for Mr Godolphin, but I am affraid his Letter will not come time enough. There can be no doubt of Ld Manchesters sentiments in this affair, Mr Gale may be sure He will oblige Him, and all his friends by appearing for Mr Godolphin, and He can do you no good any other ways. I am sorry you mention nothing of the Election,[4] it does not look well but I hope you still keep your Resolution of not being disturbed at the event, since there has been no fault of your's in the Managemt, and then there is no great matter in it: I could tell you some storys where the Conduct of the Court has been the same, but complaining is to no purpose and now the Die is cast, and upon the whole Wee shall have a good Parliamt. I am

Your most Humble and

most Obedient Servt

5. *May.* 1705   HALIFAX

To   Sr Isaac Newton at
   Trinity College in
      Cambridge

NOTES

(1) Keynes MS. 102B.

(2) Charles Montague (1660–1722), first Duke and third Earl of Manchester. (Halifax and Manchester were cousins.) He was Ambassador extraordinary at Venice (1697–8), Paris (1699–1701), and again at Venice (1707–8). He joined William of Orange's northern supporters in November 1688 and served under him in Ireland. See *D.N.B.*

(3) Roger Gale (1672–1744), son of Dr Thomas Gale, Dean of York; he was considered one of the most learned men of the day. He was the first Vice-President of the Society of Antiquaries: Fellow (1717) and Treasurer (1728–36) of the Royal Society. He was M.P. for Northallerton from 1705 to 1713. According to W. W. Bean, *Parliamentary Representation of the Six Northern Counties*, Hull (1890), p. 951, he 'got himself returned by bribery and indirect practices'. In 1715 he was appointed a Commissioner of Excise, a post he retained for twenty years.

(4) The result of the election was:

Hon. Arthur Annesley	182
Hon. Dixie Windsor	170
Hon. Francis Godolphin	162
Sir Isaac Newton	117

(See Cooper, *Annals*, IV, 72–3, and Edleston, p. lxxiv, note (153).)

445

694   NEWTON TO FLAMSTEED
8 JUNE 1705
From the original in the Royal Greenwich Observatory[1]

London Jermin Street
June 8th 1705[2]

Sr

The Gentlemen to whome the Prince has referred your matters are to meet
at my house on munday next[3] at twelve a clock & Mr Churchill[4] being
returned to London will be one of the company. Pray do me the favour to
meet us at the time appointed & dine with me that we may set the Press a
going as soon as possible. I am

Your most humble servant

ISAAC NEWTON.

For the Rnd Mr John Flamsteed
Her Majtys Astronomer at the
Observatory in
          Greenwich

NOTES

(1) Vol. 35, fo. 51.

(2) Although no correspondence appears to have passed between the two men for two
months, we learn from Flamsteed's Diary (R.G.O. vol. 33, fo. 54 *et seq.*) that they met several
times during the months of May and June, and discussed the progress of the work. Newton
appears to have been growing impatient at the delays: Flamsteed on his part repeatedly
complained, 'not a word of any recompense for 30 years' pains, and extraordinary expense'
(Baily, p. 220).

(3) 11 June.

(4) See Number 687, note (3), p. 438.

695   NEWTON TO SLOANE
14 SEPTEMBER 1705
From the original in the British Museum[1]

Jermin street Sept 14 [1705]

Dr

I beg the favour of you to get Mr Hawksbee[2] to bring his Air-pump to my
house & then I can get some philosophical persons to see his Expts who will
otherwise be difficultly got together. But first know when my Lord P.[3] can be

at leasure & let me know the time, & let Mr Hawksbee bring his Pump that evening by a Porter & I will give him two guineas for his pains. I am

Your humble servant

Is. NEWTON

For Dr Sloane.

NOTES

(1) Sloane MSS. 4060, fos. 72–3.

(2) Francis Hauksbee (died 1713). Elected F.R.S. 1705. In 1706 he designed the first electrical machine, which consisted of a glass globe rotated by hand against a rubber of coarse woollen cloth. In 1707 he was frequently called upon by the Society to prepare experiments and for this service he was paid, though he does not appear to have borne the title of Curator (*Record of the Royal Society of London*, p. 30). He also suggested important improvements in the design of air pumps, one of which bears his name; for a description of this machine, see article by E. N. da C. Andrade: 'The Early History of the Vacuum Pump', *Endeavour*, 16 (1957), p. 35. Hauksbee contributed many important papers to the *Philosophical Transactions*. See also 'A Lecture with Experiments on Various Subjects giving an Account of several surprising *Phenomena*, touching Light and Electricity', by E. N. da C. Andrade (*Journal of Scientific Instruments*, vol. IV (1926–7), p. 130).

(3) Thomas Herbert, eighth Earl of Pembroke and Montgomery (1656–1733). A close friend of Locke, who first made Newton's acquaintance at the weekly meetings given by Lord Pembroke (see Rouse Ball, *Essay*, p. 116). Elected F.R.S. in 1685; First Lord of the Admiralty, 1690; Lord Privy Seal, 1692; Lord Lieutenant of Ireland, 1707; Lord High Admiral, 1702 and 1708. He was elected President of the Royal Society in 1689, and he held that office for one year. See *Record of the Royal Society of London*, p. 335.

696   NEWTON TO FLAMSTEED

17 SEPTEMBER 1705

From the original in the Royal Greenwich Observatory[1]

You have now been I think above a fortnight in town & no step yet made towards putting your papers into ye Press.[2] And now Mr Churchill is going out of town for a for[t]night. But however he has left matters with his brother[3] till his return so that your papers may go into the Press as soon as you please. If you stick at any thing pray give Sr Chr. Wren & me a meeting as soon as you can conveniently, that what you stick at may be removed. I am

Your humble servant

Is. NEWTON

Jermin St in St Jameses
Sept. 17. 1705

NOTES

(1) Vol. 35, fo. 49.

(2) Flamsteed had written again and again explaining why he had not complied with Newton's requests. ''Tis very hard', he wrote on 29 August 1705, ''tis extremely unjust that all Imaginable care should be taken to Secure a certain profit to a bookseller and his partners out of my pains and none taken to secure me the reimbursement of my large Expences & in carrying on my work above 30 years. 'Tis a great dishonour to the Queen his Royal Highness and the Nation that no reward is proposed for so long difficult and Laborious a Work, and that the small one I might justly expect is cast upon those that have no part in my Labor and Expence, nor hazarded their health nor felt my severe pains of ye stone and other distempers caused by my night Watches and day Studys. My instruments are my own my Assistance has been hired at my own Charge. *Impius hæc tam culta novalia** My Copy is my own I am ready to deliver it to the Press on just and reasonable conditions it concernes others to make them just and honourable but I am weake. they must make hast and doe those things which may make me easy If they intend to preserve my Labors Aug 29 1705 J.F.' (R.G.O. vol. 35, fo. 55).

* *Impius hæc tam culta novalia miles habebit?* (Is a lawless soldier to possess the lands I have tilled so well?). This pertinent quotation is from Virgil, *Eclogues*, i, line 70.

(3) John Churchill, brother of Awnsham Churchill. See Letter 687, note (3), p. 438.

697   NEWTON TO SLOANE
17 SEPTEMBER 1705
From the original in the British Museum[1]

Jermin street. Munday
Sept 17th 1705

Sr

My Ld Halifax, the A.B. of Dublin[2] & Mr Roberts are out of town & therefore I desire that Mr Hawksbee's shewing his Experimt[3] here, may be put off for a while. I am

Sr

Your most humble servant
Is. NEWTON.

NOTES

(1) Sloane MSS. 4040, fo. 68.

(2) William King (1650–1729). He became Archbishop of Dublin in 1703.

(3) Hauksbee's experiments with the air pump were carried out before the Society the following month, and are described in *Phil. Trans.* **24** (1705), 2129. See Letter 695.

698   NEWTON TO SLOANE
18 SEPTEMBER 1705
From the original in the British Museum[1]

Tuesday night [18 *September* 1705]

Sr

My Ld Pembrook[2] has appointed Thursday a little before six in ye afternoon to introduce us to the Prince. And therefore I beg ye favour that you would be in ye Anti-chamer on ye Princes side about a quarter before six, where you will meet me & others of ye Society. I am

Your most humble servant

Is. NEWTON

Dr Sloan

NOTES

(1) Sloane MSS. 4060, fo. 76. See also J. Nichols, *Illustrations of the Literary History of the Eighteenth Century*, IV, 59.

(2) See Letter 695, note (3), p. 447.

699   FLAMSTEED TO NEWTON
25 OCTOBER 1705
From the original in the University Library, Cambridge[1]

The Observatory. Octob: 25 ♃ 1705

Honored Sr

I have sent you by the bearer my servant the Ptolemy I promised you desireing that by him you will please to return my old one togeather with ye mapps of Orion Ophiuchus & Serpentarius, the two last of want to have ye lines of Longitude & parallels of latitude inserted.[2]

The distance of this place from London was a great impediment in ye building of it, & will prove no less to the publication of my work here done if you doe not cordially & freindly assist me, especially at our setting out. My frequent ill health permitts me not to come to towne to consult you or give you an account at leasure of such things as cannot be discourst of at a meeting where other things are on ye table. You seemed not to like my begining ye *Historia Cœlestis* with ye Observations here made & placeing those made before at Derby in ye Liber Prolegomenos or Preface. I shall give you my reasons for so doeing which I beleive may satisfie you that I have not designed amiss.

The Derby measures of ye Suns & Moons diameters & small distances of
fixed stars I take to be very exact. for I had good leasure to examine my New
Measurers very often & carefully. In those no accurate limitation of ye times
of the Observations was needfull. it was sufficient to know ye hour & height
of the ⊙ or ☽ to a single degree which is commonly noted. I had no Movemt
to shew the times & therefore I made use of a Wood Quadrant with a paper
limbe divided as exactly as I could with My owne hand the first I used was
8 or 10 inches Radius the last about 20 with these applyed to the side of ye
tube wherewith I observed I took ye height of ye Stars by which ye moon
passed when I observed her appulses to any but by reason of the smallness &
coursness of the instruments I used & yt I had no assistance but of an ignorant
meniall servant I could not be sure of the times nearer then to a single Minute
of time I am persuaded I never erred too

The Instruments I used to take ye height of ye ⊙ & Stars with at the Tower
& in Greenwich before I sat down at ye Observatory had their imperfections
as well as those I used at Derby and therefore I esteem ye times got by them
something doubtfull tho not as much as in my Derby Observations.

But soone after I entred ye Observatory I fitted [o]ut a Quadrant of my own
of 3 foot Radius with a brass limbe yt being carefully used performed tolerably
well. & Sr Jonas Moor borrowed Mr Hooks Quadrant[3] of the R Society which
being very well made I divided its limbe very exactly hung it conveniently &
made use of it for 3 yeares togeather or More till upon ye decease of Sr Jonas,[4]
Mr Hook got it remanded[.] I restored it wth thanks & ever since it lies useless
as it was before in the Repository

By this Instrument I could be sure of the heights taken to a minute of a
degree & consequently of my times to ye 4th & 6th parte of a minute of time:
before I parted with it I made a larger of 50 inches Radius at my own charge
which was stronger & more convenient for my purpose & consequently
performd better yn Mr Hooks & I was still more certeine of the times derived
from ye heights taken with it

You see my Derby Observations are not so accurate as to the times as those
taken at the Observatory tho they be equally good as to the Measures I cannot
therefore pretend to set them equaly with those made here in which both are
good & certeine.

Lucius Barettus or Albert Curtius[5] a judicious Jesuit that Publisht Tychos
observations begins the *Historia Cœlestis* with the yeare 1582 where they were
accurate & certeine putting some less accurate into his liber prolegomenos in
their proper place & year. which is sufficient Warrant for my doeing ye same
upon a greater occasion.

I have sent the two first Sheets of My Observations beginning wth those here made to Mr Hudson with Orders to alter the title & if you approve it to let it be as at the head of the following page. I love not long nor over particular titles & therefore shall make all I have occasion for as breife & comprehensive as I can. I doubt not but that Good providence that has blest my labors hitherto will assist me in publishing of them & afford me health & all things needfull for that end. If he spares it mee next Week I intend to Wait on you at your own house & there to pay you the respects of Sr

Your very humble Servant

JOHN FLAMSTEED MR

Historiæ Cœlestis
Pars Prior
Fixarum in Constellationibus Eclipticæ incumbentibus
Distantias ab invicem
Et a Circumjacentibus insignioribus
Sextante captas
a
Johanne Flamsteedio M R
complectens
Anno MDCLXXVI
this serves that part with which I begin[:] the Generall one may be:
Historiæ Cœlestis
Pars Prior
Fixarum ab Invicem Distantias
Et Planetarum a fixis
Sextante Captas
Jo: Fla: M R
complectens.

NOTES

(1) Add. 4006, fos. 14–15.

(2) See Letter 683.

(3) For a description of Hooke's astronomical instruments see the Wilkins Lecture, *Robert Hooke*, by E. N. da C. Andrade (*Proceedings of the Royal Society of London*, B, 137, 1950, p. 166).

(4) Sir Jonas Moore died in 1679. See Letter 488, note (2), p. 73.

(5) Albert Curtz, born in Munich and died there in 1671. He taught mathematics and philosophy at Ingolstadt. He wrote *Historia Cœlestis ex libris Commentariis Manuscriptis Observationum vicennalium viri generosi Tychonis Brahi Dani* (1666) and *Historia Cœlestis Observationum ab anno 1582 ad annum 1601* (1667). The latter is listed under the name Lucii Barretti, an anagram of Alberti Curtii.

700   AGREEMENT BETWEEN THE REFEREES
AND FLAMSTEED[(1)]

10 NOVEMBER 1705
From the copy in the Royal Greenwich Observatory

Whereas it is agreed by Sr. Isaack Newton knt President of ye Royall Society. Sr Christopher Wren knt Surveyour Generall of Her Maties Works, the Honorable Francis Roberts Esq, &c. in order to Printing all the Astronomicall Observations of John Flamsteed her Maties Professor of Astr[on]omy at her Observatory in Greenwich Parke in ye County of Kent under ye Title of *Historia Cœlestis*, at the sole charge of his Royall Highness ye Prince of Denmarke. And that Mr * [Churchill] shall be the Undertaker for printing ye same. It is covenanted and agreed betwixt the said Sr Isaack Newton Francis Roberts Esq, Sr Christopher Wrenn * [Dr Gregory and Dr Arbuthnott] and John Flamsteed to and with the said Undertaker for printing of ye said *Historia Cœlestis*

1. That neither the said Undertaker nor his heyres executors or assignes shall have or clayme any Right title Interest or property in ye Originall Copy or Copys or in any of the printed Copy's but that the same shall after the printing be & belong to ye said John Flamsteed his executors Administrators &c.

2 That the said Undertaker shall print no more yn * [400] Copys of the said *Historia Cœlestis*, & that he shall print them all on the same paper with ye specimen hereto annexed at ye Rate of * [34 shillings per] Sheet. without changing the paper. that he shall cause them to be fair printed & use all possible expedition in printing of them. and to prevent errors he shall not suffer any Sheet to be printed off till ye second proof has been e[x]amined & corrected at ye press. yt he shall take care yt errors be avoyded as much as possibly he can. & shall send all the proofs to be examined & corrected to ye Observatory at his own Charge.

3 That in order to prevent any more copys then the * [400] agreed upon from being printed. he the said John Flamsteed his Agents and Servants shall have access to ye press at all times and it shall [be] lawfull & permitted to him or them by his order to stand by the Press to see the said Number of * [400] Copys wrought of & as soone as that just Number shall be wrought of to breake the press, without any delay let hinderance or Molestation from him the said Undertaker his printers & his or their Agents or Servants

4. And it is further ordered & Agreed wth ye consent and approbation of ye said Sr I. Newton. Mr Roberts & Sr Ch: Wren &c betwixt the said Undertaker * [Mr Churchill] aforesaid & the abov[e]said Mr Jo: Flamsteed:

PLATE V. PORTRAIT OF JOHN FLAMSTEED (ARTIST UNKNOWN)

that in order to secure to the said Mr Flamsteed the fraists [?fruits] of his labors that through Gods blessing have been continued above 30 yeares wth great Industry & expense out of his own proper income. Hee the said * [Mr Churchill] the Undertaker for printing of them. shall & will at any time & at all times on the reasonable demand of him the said John Flamsteed by himselfe his Servants or Agents made *deliver*(2) or Cause to be delivered where he shall appoint, all the printed Sheets of the said *Historia Cœlestis* cleane & undefaced. to be by him (God spareing him life) or his executors presented to his Royall Highness. Whose life health & happiness May the God of Heaven long preserve to the joy of her Matie & all her Subjects.

I request further that the Undertaker Mr * [Churchill] May give 2000 *lib.* bond security for the Performance of these Articles

aftter which I shall imediately put the first Volume which is fair Copyd. into the hands of such persons as they the said Mr Roberts Sr Is. Newton & Sr C Wren shall require to be put into ye press:

NOTES

(1) This is a draft in Flamsteed's hand of the proposals for printing the *Historia Cœlestis* which were drawn up by Flamsteed and the Referees on the date mentioned. Its main provisions are contained in the final agreement which was drawn up and signed a week later. In the copy here transcribed (vol. 35, fos. 67–8) Flamsteed has left a number of gaps. These are indicated by asterisks and the missing words and phrases, supplied by comparison with the final agreement and with entries in Flamsteed's Diary, are inserted in editorial brackets.

Flamsteed had already discussed many of the above proposals with the Referees as is clear from the following entries in his Diary:

'June 11. 1705. Dined with Sr I. Newton: agreed wth Mr Churchill at 34*s* per sheet: Mr Roberts and Dr Gregory there: I dissented.

'June 13. At Garways. saw Sr I. Newton; gave him a note that the Undertaker was to have no interest in the Copy, nor any printed copies: that he might give £1000 security to print no more than the 400 agreed on: that he should not change the paper: that care ought to be taken that the printed Copies may be put into such hands that my executors may come by them without trouble in case of my mortality, or Mr Newtons or other account.

'October 12 Met Mr Roberts, Sr C Wren and Sr I Newton at Sr C Wrens office: showed my paper of articles: 'twas laid by: Sr Isaack Newton would like nothing I proposed, though he could not say it was unreasonable: drew up another paper; appointed another meeting on the 18th following.

'October 22 when we met: read over all the Articles very carefully: I did not assent to many of them: much talk, little done.'

(2) This word was underlined, probably by Flamsteed.

453

701   NEWTON TO FLAMSTEED
14 NOVEMBER 1705
From the original in the Royal Greenwich Observatory[1]

Jermin street Nov. 14. 1705.

Mr Flamsteed   [*Wednesday*]

On Saturday next about twelve a clock the Referees meet at my house to finish the agreement & signe the Articles about printing your book & I shall be glad to have your company here at the same time & that you will be pleased to dine wth me. I am

Your humble servant

ISAAC NEWTON.

[Below this, Flamsteed has written]

I was there & signed the Articles but covenanted that the Catalogue of the fixed stars was mentiond (to make a part of the first Volume) it should not be printed but with the last.[2] Dr Arbuthnot was there wth Mr Roberts & Mr Churchill but neither Sr Ch: Wren nor Dr Gregory:

NOTES

(1) Vol. 35, fo. 57.

(2) At this meeting Flamsteed insisted that the catalogue of fixed stars should appear in the second volume. Newton wanted it in the first, but Flamsteed was adamant. This was the only concession he obtained. See Number 702, note (3), p. 459.

702   ARTICLES OF AGREEMENT BETWEEN THE REFEREES AND FLAMSTEED[1]
17 NOVEMBER 1705
From a copy in private possession

ARTICLES of Agreement Tripartite concluded and made the Seventeenth Day of November in the fourth Year of the Reigne of Our Soveraigne Lady Ann by the Grace of God of England Scotland France and Ireland Queen Defender of the faith Between Sr Isaac Newton Knt President of the Royall Society, the Honourable Francis Robarts Esqr, Sr Christopher Wren Knt David Gregory Dr of Physick and John Arbuthnot Dr of Phisick of the first part, the Reverend John Flamsteed her Majesties Astronomer at the Observatory in Greenwich of the second part and Awnsham Churchill of the Parish of st Gregories London Bookseller of the third part as follows (that is to say

WHEREAS his Royall Highness the Prince out of his great generosity and propension to encourage Arts and Sciences has been pleased to offer to defray the charges of Printing the Astronomicall Observations made by the said John Flamsteed at the said Observatory and comprized in a Book entituled Historia Cœlestis and to refer to the care and and management of the Impression thereof to the said Sir Isaac Newton Francis Robarts Sr Christopher Wren Dr David Gregory and Dr John Arbuthnot And whereas the said Referrees have treated with the said John Flamsteed and Awnsham Churchill for printing the

1   same It is hereby Covenanted and agreed between the said parties 1 That the sd book shall be printed in two Volumes the first to contain the *Catalogues of the fixed Starrs* the Observations of the fixed Starrs Planets and Cometts made by the great Sextant Voluble Quadarant Telescope and Micrometer at Darby and Greenwich from the year One thousand six hundred and seventy to the year 1689 inc[l]usively and the true places of the Planetts and Cometts completed from those Observations together with a generall preface and tables for getting the places of the Planetts and Starrs from Observations

2   The second to containe the Observations of the fixed Starrs Planets and Comets made by the great Meridionall Arch Telescope and Micrometer in and after the Year one thousand six hundred and eighty nine until the finishing of the Impression and the true places of the *Planets and Comets computed from them.*

3   Item the said John Flamsteed doth thereby for himselfe his heirs Executrs and Administratrs covenant and agree to and with the said Referrees their Executrs and Administratrs that he the said John Flamsteed shall with all convenient speed prepare and deliver in to the said Referrees or their Order fair and correct Copies of his *Catalogue of the fixed Stars* containing the Longitudes Latitudes right Ascentions Declinations Magnitudes Motions of about three thousand Stars and also fair and correct Copies of the said Tables and Observations of the fixed Stars planets and Comets to be printed in the said two Volumes with fair and correct schemes of a size of a printed page of the said book containing the figures of Eclipses and other Telescopicall Observations to be graved in Copper plates.

4   And further that the said John Flamsteed shall with all convenient dispatch compute or cause to be computed the places of the Planets which remaine to be computed and deliver in to the said Referrees or their Order fair and correct Copies of their true places in Longitude and Latitude computed from all the Observations and well corrected to be printed in the said two Volumes at the end of the Observations.

5   And further also that he the said John Flamsteed his Exrs Admrs and Assignes shall suffer the said Referrees and their Order or any of them to collate the sd MS copies and Schemes and also the printed Sheets with all or any part of the Originall papers in his custody from whence the sd MS Copies and Schemes were taken and with the first minutes from whence those papers were drawn up and for that end shall at the request of the said Referrees send the said papers and Minutes or any of them to the Order of the said Referrees the person to whome they are sent giving a Receipt of the same *with a promise in the said receipt to return* the same to the sd John Flamsteed in a *reasonable time*.

6   Item the said Awnsham Churchill doth hereby for himselfe his heirs Exrs and Admrs covenant and agree to and with the said Referrees their Exrs and Admrs that he the said Awnsham Churchill his Exrs Admrs and Assignes shall print or cause to be printed four hundred Copies well corrected and only four hundred Copies of the said book upon the same paper and with the same Letter with the paper and Letter in the Specimen hereunto annexed.

7   And further that he the said Awnsham Churchill his Exrs Admrs and Assignes shall at his own proper costs & charges send the corrected proof of every Sheet to the place appointed or to be appointed by ye sd Referrees in Greenwich or in London to be there further corrected & compared wth ye Originall and allowed by ye said John Flamsteed or his Corrector before the same be printed off.

8   And the said John Flamsteed doth hereby covenant and agree to and with the said Referrees that the sd John Flamsteed or his Corrector shall speedily and diligently collate correct and allow the same so that no stop be put to the press.

9   Item the said Awnsham Churchill doth hereby for himselfe his heirs Exrs and Administrators further covenant and agree to and with the said Referrees their Exrs and Administrators That all the printed Copies of every Sheet within three Months after the Sheet shall be printed off shall at the charge of the said Awnsham Churchill be sent by the Printer to the order of the said Referrees to be kept for his Royall Highness till the whole be printed off excepting the two last Copies of every Sheet or two Copies last printed off which at the charge of the said Awnsham Churchill shall be sent by the said printer the one to the sd John Flamsteed or his Corrector the other to the Order of the said Referrees to be examined and collated with the last proof and with the MS. Copy and originall papers and Minutes of the said John Flamsteed And that every Sheet in which any error shall be found which ought to be corrected and is not the error of the M.S. Copy or corrected proof shall be reprinted at the sole cost and charges of him the said Awnsham Churchill both for paper and printing.

10   And further That the said printer upon demand shall make Oath that all the Copies printed off have been sent and delivered according to Order.

456

11 And further that he the said John Flamsteed his corrector or either of them may at all times have access to the press and stand by it while any Sheet is printing off and after the four hundred Copies with some Supernumerary Copies (not exceeding twelve) allowed to make good the faulty Sheets shall be printed off may break the press without delay lett or molestation.

12 And further that the said Awnsham Churchill his Exrs Admrs and Assignes shall print at least twenty Sheets per Month abateing a Sheet for every holyday provided he be supplyed with sufficient Copy and sufficient dispatch be made in the correcting.

13 Item the said John Flamsteed doth hereby covenant and agree to and with the said Referrees That he the said John Flamsteed shall assist a Graver with his directions for graveing the Schemes above mentioned on Copper Plates and examine the Plates and correct their faults so that the Schemes may be exact.

14 Item the said Awnsham Churchill doth further agree and covenant to and with the said Referrees That he will roll off or cause to be rolled off four hundred Schemes and not more from every plate upon four hundred Sheets of the same paper with that of the book in such a manner that they may be readily and conveniently laid open to the Reader and then deliver up the said plates and printed Schemes to the Order of the said Referrees.

15 Item the said Referrees do hereby Covenant and agree to and with the said Awnsham Churchill his Exrs and Admrs that they the said Referrees or the Major part of them shall within two Months after the said book shall be in the press signe an order for the said Awnsham Churchill to receive of the Treasurer of his Royall Highness the Summe of two hundred and fifty pounds advanced in part of Payment for the paper and printing of the said book And after the Impression of the said book shall be finished the said Referrees or the Major part shall signe a further Order for the said Awnsham Churchill to receive the remainder of the Money which after the rate of thirty four shillings per Sheet shall then be due to him the said Awnsham Churchill for the paper and printing the whole book.
Item the said Awnsham Churchill doth further covenant and agree to and with the said Referrees that he the said Awnsham Churchill shall not claime any right or title to the said M.S. Copy or printed Copy thereof.

16 Item the said Referrees do hereby covenant and agree to and with the said John Flamsteed That they the said Referrees or the Major part of them shall also signe an Order for the said John Flamsteed to receive of the said Treasurer the summe of two hundred and fifty pounds for his charges in employing assistants and Servants to Calculate Observations Copy papers and Schemes and correct the press the one halfe thereof to be paid so soon as all the M.S. Copies of the first Volume shall be delivered to the said Referrees in order to

457

be collated as aforesaid and the other half thereof to be paid so soon as the Second Volume of the said book shall be printed off Provided the said John Flamsteed shall well and truly observe perform fullfill and keep a[ssi]gnd Singular the Articles Covenants and Agreements above mentioned specified and declared which on his part ought to be observed performed fullfilled and kept.[2]

Sealed and delivered by the within
named Sr Isaac Newton   Fra: Robarts
Jo: Arbuthnot   John Flamsteed &
A. Churchill in presence of
       Francis Child
       John Moors
Sealed and delivered by the within
named
Sr Chr: Wren & D: Gregory (the papers
being first stamped) in the presence of us
       Fra: Child
       John Moors

Isaac Newton
Fra Robarts
Chr: Wren
D. Gregory
Jo: Arbuthnot
John Flamsteed
A Churchill

[Flamsteed's marginal notes. See note (1), last paragraph.]

1. This Article impossible for the observations to compleate it were not all gotten till five years after.[3]

2. These he [Newton] has had in his hands 7 years and I know not what is become of them.

3. About half ye Catalogue was delivered into S[ir] I N hands, before we began to print, at his own request, sealed up,[4] which he promised should not be opened till I had compleated it, but contrary to his promise he broke it open gave it with wt I had added to Mr Halley, who printed it, haveing corrupted it in a multitude of places and altered all ye Names to make it agree with his own false Maps.

5. This was done, but one of the M.S. of the first Notes is still detained by him, contrary to this Article, after haveing kept it nine years to no purpose.[5]

6. How many Copies he printed, or what is become of them I know not.

7. Two sheets were printed off, without my revising, false and uncorrect

8. This was done.

12. He never printed above six Sheets a month

13. No graver was ever sent to me about yt buisiness neither do I know yt any Schemes were ingraved.

15. I never consented to give thirty four shillings per Sheet.

16. I never Recd but 125 pounds although I had disbursed 173 pounds for Calculators and yt not till 2 years after Sr I.N had stoped ye press and forced me to dismiss my Calculators for want of timely paymt of this allowance.[6]

He has had this Volume seven years in his hands and not printed it. I desire it may be returned[7]

NOTES

(1) According to Baily (p. 81), four drafts of this agreement have been found, all of which are unsigned. Although they differ slightly from each other, they nevertheless agree in the following main particulars:

(*a*) 400 copies and no more were to be printed off.

(*b*) Flamsteed was to supply the copy promptly and he was to be entirely responsible for correcting the proofs.

(*c*) Flamsteed was to have access to the press at all times, and to break it when the stipulated number (400 copies) had been printed.

(*d*) The publisher was to print at the rate of twenty sheets per month.

(*e*) The publisher was to have no further interest in the work once it was completed.

(*f*) The Referees were responsible for the disposal of the Prince's grant.

The copy here reproduced is in the hand of a copyist. The numbers on the left were inserted by Flamsteed. In the right-hand margin are Flamsteed's comments; these are in the same hand as the copy.

(2) Flamsteed was far from satisfied with the provisions of the above agreement. A few days later (20 November) he wrote to Sharp: 'Sir Isaac Newton has at last forced me to enter into articles for printing my works with a bookseller, very disadvantageously to myself' (Cudworth, p. 89).

Flamsteed was a difficult man to do business with, nevertheless a perusal of the agreement, and of Flamsteed's marginal notes, show clearly that he had reason to be dissatisfied. In the first place he had not been consulted as to the constitution of the Board of Referees, most of whom, it will be noted, were well disposed towards Newton, and who would naturally follow his lead. Nor was he consulted as to the choice of the printer. Finally, although some provision was made for the payment of his assistants, not only was this inadequate, there was no mention of any recompense for his own labours. Nevertheless, Flamsteed was compelled to accept the terms of the agreement since otherwise he saw no hope of his work ever being published.

(3) The place where the catalogue of fixed stars was to appear was one of the main points of difference between Newton and Flamsteed. Despite Newton's opposition, Flamsteed insisted that it should not appear in the first volume. 'They [my observations] will be of no use without a correct Catalogue of the fixed stars and that could not be made till I had a Competent Number of Observations to ground it on' (Letter 495).

(4) The catalogue, in an uncompleted form, was deposited in a sealed packet with Newton at his own request in March 1705/6. There appears to be little doubt that it was later opened with Flamsteed's knowledge and consent. 'We met on March 20, 1707/8', Flamsteed wrote some time later, '& then Sr Isaack had opened the Catalogue & desired me to insert the Magnitudes of ye stars to their places for they had not allwayes been inserted in it.' Whether it was then resealed, and later opened without Flamsteed's knowledge, must remain a matter for conjecture. Flamsteed, in letters to Sharp, repeatedly accuses Newton of a gross breach of

faith; he declared that he was forced to trust in the hands of Newton an imperfect copy of the catalogue which he very treacherously broke open though it was at his own desire sealed up and so delivered into his hands (see Baily, p. 298). Although Brewster (II, 228) vigorously defends Newton's action, it is difficult to regard Newton as completely blameless throughout this regrettable affair (see Baily, pp. 727–8).

(5) See Flamsteed's letter to Wren, 19 July 1708 (Letter 747), in which he categorically repudiated the charge that he had been in any way responsible for the repeated delays; on the contrary, he maintained that he had done all he could to hasten and expedite the work.

(6) On 26 March 1708 (Letter 738) the Referees authorized Newton to pay Flamsteed £125, and the receipt of this sum is acknowledged by Hodgson at the foot of the letter. It would appear from Flamsteed's comment that this was merely towards paying the £175 which he had already paid his assistants. See also Letter 711.

There is an account in the University Library, Cambridge, Add. 4006, fo. 23 (Number 755), of Newton's expenses in connexion with the printing of Flamsteed's work.

(7) Some years later (1716) Flamsteed wrote a formal letter to Newton requesting the return of the catalogue and the 175 sheets of observations. Newton paid no heed to this, or to similar requests. A previous request had been sent on 30 June 1715.

703   LE NEVE[1] TO NEWTON
24 NOVEMBER 1705
From the original in the Babson Institute Library

Sr.

The office on thursday last Received your desires[2] very kindly & ordered the proofs to be examined by some proper persons who are to report their opinions to the next chapter which will be on the first Thursday in december when I question not but it wilbe settled for you & which in the meantime I will see shalbe examined as being Sr

Your most humble

Servant

24 *Nov*: 1705

PETER LE: NEVE.

Norroy.

If[3] you please to send the Rentroll to me to the Heralds office on Munday next for I shall not be at the excheqer that day

To
Sr Isaac Newton Kt.
In Jermyn Street
   by St James Church
      Middx

NOTES

(1) Peter Le Neve (1661–1729), Norfolk antiquary. He became the first President of the Society of Antiquaries in 1707, and he remained in that office until 1724. He was elected F.R.S. in 1711. He was Rouge Croix Pursuivant from 1690 to 1704, when he was promoted Richmond Herald; shortly afterwards, he became Norroy King of Arms. He left many interesting manuscripts in the College of Heralds.

(2) No sooner had Newton been knighted (16 April 1705) than he proceeded to establish his status. Before he could be granted his coat of arms, however, he had to submit his pedigree. According to his own sworn testimony, both he and Sir John Newton (see Letter 583, note (1), p. 265) were descended from John Newton of Westby in Lincolnshire. Isaac's great-great-grandfather, John Newton, and Sir John's great-grandfather, William Newton, were brothers.

The following extract, in the hand of Peter Le Neve, is taken from the records in the College of Heralds, by courtesy of the Somerset Herald:

'Sr John Newton of Thorpe in the Parish of Hatter in the County of Lincoln, aged about 53 Years certifieth that this Deponent hath heard his Father Sr John Newton Baro[ne]t aged about Seventy and 3 years at the time of his Death, Speak of Isaac Newton Master of Arts and Fellow of Trinity Colledge in Cambridge, now Sr Isaac Newton Knight, Master and Worker of his Maties Mint in the Tower of London and of St James Parish in the County of Middlesex as of his Relation and Kinsman. He finds one John Newton of Westby in the said County mentioned to have had 4 Sons vizt John Newton Thomas Newton Richard Newton and one William Newton of Gunnerby in the said County which said William Newton was his great grandfather and believe the said Sir Isaac Newton to be descended from the said John Newton son to John Newton of Westby the common ancestor.' This was signed on 22 November 1705. Newton's pedigree, together with a certificate by Newton himself, testifying to the accuracy of it, is deposited in the Library of the College of Heralds (vol. D. 14). See also Turnor's *Grantham*, p. 169.

For some interesting—and contradictory—correspondence concerning the relationship between the two men, see *Notes and Queries*, 3rd Series, I (1862), pp. 158 and 190.

(3) This sentence was added (as a postscript) in the margin.

704   FLAMSTEED TO SHARP[1] (Extracts)

12 DECEMBER 1705–12 JULY 1707

From the originals in the possession of F. S. E. Bardsley-Powell, Esq.

The Observatory December 12. 1705

...   .

I have put 100 sheets of my first volume conteining all the Observations made here betwixt Sept 1676. & Sept 1689 with the figures for them into Sr I. Newtons hands last Saturday so yt now there can be no pretence of obstruction or delay on my parte. if he formes any New. he must beare the blame I will not. I have 70 sheets of ye 2d volume in good forwardness. You shall be acquainted with our proceedings upon every proper occasion. but I will deale plainely wth you if he deales fairly & does really promote ye press I shall attribute it wholly to ye over[r]uling hand of Gods providence for when I consider his temper & behaviour tis more yn I expect if he does I have work for your New Tables & can gratifie you for your paines. . . .

The Observatory January ye 17 ♃ 1705/6

...   .

I have no more to adde but that now my Manuscripts have been 7 weeks in ye Referees hands I am told that the undertaker has not yet agreed wth a printer: which makes me expect slow proceedings. . . .

The Observatory Feb: 2. 1705/6

...   .

I hope in a fortnights time to have an intire sheet from ye Presse. & that afterwards wee shall goe on vigorously tho I fear not so fast as some people imagine I am prepareing ye 2d volume. neare 100 sheets are transcribed by T Weston.[2] who is this day out of his time, & 20 of them are ready filled & fitted up by my owne hand under which all of them must passe & each will cost me more yn an hours labor to insert the correct zenith distances from those transcribed from the Instrument. . . . I shall send you the first printed sheet as soone as I conveniently can after it comes to my hands & an account of our proceedings as God spares me health & opportunitys happen. . . .

The Observatory ♄ *March ye* 2. 1706/5

...   .

You desire to hear how my works goe on. I shall give you a breif account.
In November last I delivered in 100 sheets of Copy. About a moneth agone Sr Is: desired to see my first Notes & I sent the first book of them. this day sevennight Mr Hodson[3] told me he had been wth Sr Is: & had seene 4 or 5 folio pages of differences he had noted betwixt ye originall & copyes I visited him last Monday & desired to see them he told me Dr Gregory had collected them. the Dr soon came, when we sat down to examine them Sr Is. told me he did beleive them to be errors but desired that himselfe & the Dr might be informed of my ways of observeing. they were proper judges in the meantime I ordered James[3] to come to me for I have resolved not to talk with them without good Witness he came in good time we got to work & found a great many differences but all of ye Drs Makeing[4] he had formed a table for turning ye revolves & parts into degrees minutes and seconds.[5] & supposing the thrids of the screw every where equall wonderd that his æquipollent degrees minutes and seconds[5] agreed not with mine. I told them I wonderd

he should adventure to make this table. wiped out[6] his emendations from my Margin. engaged to give them an account of the other differences dined wth them & returnd home next day I caused my own larg table to be copyed & the day following sent it them with the rest of my first Notes to Sept 1689. to be compared & now expect to hear by Wednesday next what will be done. I told them I had been at great expense in this work & expected a recompense but I fear Sr Is: had rather stop it then give himselfe any further trouble for he finds I doe not court him, & his temper wants to be cried up & flattered.

I have allways hated all such low practices. but carried [with] that care that I have not afforded him any opportunity to recede, he thruste himselfe into ye business purposely to be revenged of me because I found the fault both of his *Opticks* & Corrections of My Lunar Numbers & would not suffer him to recommend my works privately to [the] Prince when he desired it about two yeares agone.[7] however I take no notice of this but carry as if I thought he onely wanted better information & take care to oblige him with enough of it. I have got neare 100 sheets of ye 2d volume copied all ye Right ascensions & distances of the Planets from ye Pole are gone throw by Mr Witty:[8] the moone perfectly the superior planets as far as 1699 so yt I hope in a Week or twos time to have them finisht. As soon as wee are throw these I shall cause them to be copied & let them be seen & if he then goes back all the world [must] acknowledge that I have done what I ought & all the blame will be at his door.

...   .

Sr Is: carrys himselfe very cunningly. I deal plainly & sincerely wth him & doubt not but God will let me see a good effect of it....

Burstow Aug 12. 1706 ☽ *Mane.*

...   .

...All things are made as difficult to me as can be, easy to others.... Mr Witty[8] I have dismist & he is now a Chaplaine & Companion to a young gentleman in Hampshire on better terms than I could afford him....

The Observatory Sept 14 ♄ 1706

...I thank you for your offer of assistance. & should make use of it if I could doe it to your advantage. but Sr Isaack is ye same man he ever was I have not receavd a farthing from him (who has drawn the Princes mony into his hands & forced himselfe into my business.) though I have beene at a large expense upon it. I must be patient. No one ever yet servd his country honestly & honorably but he was ungratefully used for his paines

Mr Pound[9] is returned from India. has lost his books papers & Instruments at Pulo Condore is now in Holland returns for England in a short time....

The Observatory Octob 12. ♄ : 1706

...Some shortsighted people put all the obstacles they can in our way upon suggestions & pretences.... S. I. N takes particular care yt I shall not receave a farthing for all my expenses nor what Mr Witty & Weston or Calculators have cost me these are great discouragements. but to be born with till God sees fit to raise us better freinds....

The Observatory Dec ☽ 9. 1706

...plainely it has cost me more yn 150 *lib* out of my pocket to forward this edition. S. I. N plays all the tricks he can to keep me from receaveing one peny towards the reinbursemt of this expense conceales what he has receavd from the Prince, tho at the same time he owns that he has received Monys for a *useless* undertake[r] & has paid him for ye paper (which, by the by, is far from being what it ought) but continues pretexts to delay pa[y]ing the monys due to me. out of which I designed to have made you a further acknowledgmt. I am patient at present but shew the account to every body I handsomly can to make him ashamed of his false behaviour for the Truth is he designed by what I can collect *absolutely to hinder the publication of this work* & I had no other way to prevent him but by consenting to Conditions altogeather unreasonable....

The Observatory Jan: 20. 1706/7 ☽æ

...   .

My business of the press goes on very awkwardly yet I have hopes I may with Gods blessing find some way to quicken it ere long what successe I have in my Endeavors I shall informe you

...   .

The Observatory July ♄ 12. 1707

...   .

J Hodgson now takes care of ye press I scarce know how it goes on but I helpt him to correct ye sheet Hhhh on Monday last I am not at all concernd at S. I. Ns false dealeing tis what I expected. he injures not me but himselfe & the Work & loses his reputation by it....

NOTES

(1) Abraham Sharp (1653–1742) was born at Little Horton, near Bradford, Yorkshire. He went to Bradford Grammar School and then moved to Liverpool where he opened a day school. Whilst at Liverpool he made contact with Flamsteed, and in 1684/5 he went to Greenwich as Flamsteed's assistant. Here he worked on the Catalogue of British Stars and constructed a 140° mural arc with which many of the observations were made. Following a breakdown in health, in 1691 he moved to Portsmouth. In 1694 he returned to Little Horton

where he remained for the rest of his life. He was in constant correspondence with many of his contemporaries, including, in addition to Flamsteed, Halley and Thoresby. In 1695 Halley offered to help Sharp to obtain the Mathematical Mastership at Christ's Hospital, but he declined to proceed with his application.

While at Little Horton, he devoted much time to constructing mathematical instruments, telescopes, sundials, micrometers, quadrants and sextants; he also made his own tools. His skill as an instrument maker was of the highest order, and he is generally regarded as the first to divide astronomical instruments with any pretensions to accuracy. See *The Life and Correspondence of Abraham Sharp*, ed. by William Cudworth, 1889.

In the prolegomena to his *Historia Cœlestis Britannica* (1725), vol. III, p. 108, Flamsteed has this to say of Sharp: 'Mense Maio 1688 *J. Stafford* meus Amanuensis diem obiit supremum, & Augusto proximè sequenti *Abrahamum Sharp* illius vice conduxi, qui *Mechanices* perquam expertus, pariter ac *Mathesios* peritus, *Limbi* Marginem *Cochleis* firmavit, Gradus exsculpsit, Indicem adaptavit, omnesque ac singulas ejus partes tam affabrè fabricavit, ut omnibus Artificibus expertis, qui opus illud conspexerunt, admirationi fuerit.' (In May 1688 J. Stafford, my amanuensis, departed this life, and in the August next following I employed in his place, Abraham Sharp, a man much experienced in mechanics and equally skilled in mathematics. He strengthened the rim of the limb with screws, scribed the degrees upon it, affixed an index, and made all and each of its parts so skilfully that it was a source of admiration to every experienced craftsman who beheld it.) See Robert Grant, *History of Physical Astronomy* (1852), p. 469 n., and Taylor, *Mathematical Practitioners*, p. 265.

(2) See Letter 638, note (7), p. 370.

(3) See Letter 638, note (6), p. 370.

(4) For other causes of Flamsteed's suspicion of Gregory see Letter 597, note (3), p. 288.

(5) Flamsteed wrote 'degs ' and ''s' and 'deg 's & ''s'.

(6) Flamsteed wrote 'wipt out'.

(7) In an earlier letter to Sharp, sent from the Observatory on 4 May 1704, Flamsteed wrote: 'My discourse about the faults of Mr Newton's *Optics* and corrections of my lunar numbers brought the subtle gentleman down here the 12th past. I thanked him for his book. He said then he hoped I approved it. I told him loudly "no", for it gave all the fixed stars' bodies of 5 or 6 seconds diameter, whereas four parts in five of them were not 1 second broad. This point would not bear discussion; he dropped it, and told me he came now to see what forwardness I was in. The Books of Observation were shown him, my Catalogue with Tycho's and Hevelius's, as also the chart of the fixed stars. He seemed pleased, and offered to recommend them privately to the Prince, but was told he must do it publicly, as he could for some good reasons, which not being able to answer, he was silent. Plainly, his design was to get the honour of all my pains to himself, as he had done formerly, and to leave me to answer for such faults as should be committed through his management; but having known him formerly, and his sole regard to his own interests, I was careful to give him no encouragement to expect I should give him anything gratis, as I had done formerly. I showed him also my new lunar numbers fitted to his corrections, and how much they erred, at which he seemed surprised. and said "it could not be", but when he found that the errors of the tables were in observations made in 1675, 1676, and 1677, he laid hold on the time, and confessed he had not looked so

far back, whereas if his deductions from the laws of gravitation were just, they would agree equally in all times' (Cudworth, p. 80).

For Flamsteed's criticism of the *Opticks* see Letter 678.

(8) See Letter 687, note (5), p. 438.

(9) See Letter 678, note (1), p. 427.

705   A MANUSCRIPT BY NEWTON
1705
From the original in the Royal Mint[1]

The 1600 Tuns of Tin bought annually in Cornwal amounts in Merchants weight to $1714\frac{2}{7}$ Tunns yearly & in all the seven years to 12000 Tunns. Between Apr 6th 1704 & Sept 12th 1705 there has been sold by the Pewterers 600 Tons, by the Officers of the Mint 488 Tons & by Mr Drummond[2] about 1000 blocks or 144 Tonns, in all 1232 Tons wch is after the rate of 860 Tons per annum. From the 12000 Tons deduct the 488 Tons & the 144 Tons already sold & there will remain to be sold 11368 Tons. Deduct further 860 Tons per annum for the $5\frac{1}{4}$ years to come & at the end of the term of the bargain there will be sold of 5148 Tons & remain in the Queens hands 6853 Tons. Which will require 8 yrs more to sell it all supposing a cessation of digging Tin in Cornwall all that time & without such a cessation it may require a much longer time.

The Queen pays annualy for Tin 112000£ Salaries 3000 carriage by sea 2000 incident charges about 1200£ in all 118200£ in all the seven years 827400 besides interest. Her Majty has receved for 632 Tons already sold 48032£ & will receive further for 860 Tons 65360£ annualy & in all the $5\frac{1}{4}$ years to come 343140. Deduct the 65360£ from the 118200£ & the 343140 from the 827400 & the Q. will run the debt 52840 yearly & at the end of the time of the bargain be in debt 484260 besides interest wch amounts to about 72000£ more, so that ye whole debt will be about 556000£.

The price of Tin in Cornwal has sometimes fallen to 50s & 48s and even 45s pr [cwt][3] merchants weight: & the great glut of Tin when the time of ye bargain expires may make it fall down to 45s pr [cwt] in London or lower. The 6853 Tons of Tin at 45s pr [cwt] will bring in 308385£ wch deducted from the said debt of 556000£ leaves the Queen 257615£ a loser by the bargain besides the interest of the debt while the tin is selling off; wch interest if the Tin be eight years selling off or above will amount to 194787£ or above so the Queen will lose at least 452402 by the bargain.

By the like computation I find that if for ye future Tin should be sould after the rate of 1000 Tons pr annum, there would remain in ye Queens hands at the end of the term 6118 Tons & the Queen would be then in debt 380368£ besides interest of money taken up to carry on the service wch interest would amount to about 58320£ & make the whole debt 438688£. And the 6118 Tons sold at 45s[4] pr Ton would bring in 275310£ wch deducted from the debt would leave the Queen 63378£ a loser besides the interest of the debt while the tin is selling off, wch may amount to above 108372£, & so make ye Queen above 271750£ a loser by the bargain.[5]

NOTES

(1) Newton MSS. III, 520.

(2) Probably the agent referred to in Letter 712.

(3) Newton used the sign ⊕.

(4) This is written '45s', but should read '45£'.

(5) The outcome was much better than Newton had forecast. Sales rose during the further part of the contract to bring the average for the whole period to 1560 tons avoirdupois per annum.

Newton altered these figures, sometimes more than once, and they are not always consistent.

706   FURTHER EXTRACTS FROM FLAMSTEED'S DIARY

23 FEBRUARY 1705/6–15 APRIL 1707

From the original in the Royal Greenwich Observatory[1]

1706/5 Feb 23 ♄ J Hudson[2] acquainted Me that Sr I N had shewd him 3 or 4 pages of errata in my Manus[cript] Copy of the observation.

1705/6 ☽ 25 I went to Sr Is. who told me first that he & Dr Gregory desired to be informed better concerneing my way of observeing very civilly. & shewed me the papers. J Hudson was not yet come thither. Dr Gregory came soon after me we set to examine them I found none Materiall. The Dr had made a Table for turneing ye Revolves of the screw into degrees minutes & seconds by considering my partes equipollent & compareing them wth ye Revolves & Cents but supposeing them more equally then they were had erred in it. & made a great number of faults where there were really none Some slight mistakes in my copy were found but none that could be of any consequence. I dined wth Sr Is: ye Dr & JH[2] came home yt night, caused my own Table to be copied out next day & with the 2d book of observations & as much of the 3d as reached Sept 10. 1689 left it wth J H on ye 26 ☿ to be delivered to Sr Is. that the Dr might continue his examination. without danger of ye Like mistakes hereafter.

♀ Martij 1. Examined their pretended faults & answered them togeather & wrote a paper yt shews what care I have taken to prevent error & that probably none committed in the measures taken with the Sextant can corrupt ye Catalogue to be made use of hereafter.

Martij 4 ☽: Wrote a letter to J H directing him to call for ye Articles. take care of my MSS. & told him I would put the imperfect Copy of the Catalogue into ye Presdt hands of ye R.S Sr Christopher Wrens Mr Roberts & his to be printed (or a More compleat Copy in its Room) as soon as ye two first Volumes of Observations were printed of provided I might receave monys to pay my Calculators & Amanuensis at the same time that I delivered [it] & that it might be kept sealed up[3] till all ye observations were printed I shewed him also a correction of ye errors found in ye Copy by Dr Gr. & Sr Isaack & that they were of No mom[en]t.

6 ☿ James[2] was here againe offerd to send up ye Catalogue by him but he could not take it by reason of ye ill weather.

*8 ♀ Was at Sr C. Wrens left ye Catalogue with him for James, to be seald up & delivered to Sr IN when 10 Sheets are printed & ye 125 *lib* paid, complaind of Sr Is s dilatoriness. he seemd much concerned

15 ♀ Met Dr Bently[4] at Garways[5] Sr Isaack was there. we discourst first about Dr Plumes[6] Astr. professorship ye Dr would have had my hand to a paper for ye election of Mr Cotes[7] to be professor I refused till I saw him. he told me Mr Whiston[8] & Mr Cotes should wait on me next weeke then we began to discourse of my press bu[s]iness Sr Is. told me he heard I had left the Catalogue in London I told him I had left it in Sr C Wrens house to be delivered to James. he seemed netled as If I would not trust him wth it but on my telling him that James was to seale it up & leave it in his hands he recalled his threates & told me then he would take it into his keeping & receave 800 *lib* of ye princes mony but not a word of paying me for my Amanuensis & Calculators.[9] James was come to Greenwich wth Bp Burnets sonn[10] I wrote a letter to order him to seale up ye Catalogue & leave it wth Sr Is: which he tells me he did & a meeting is appointed on Saturday next ye 23d instant God give us good success:[11]

March 23 ♄ went wth Sr C Wren to Sr Is: Ns met Dr Arbuthnot[12] Dr Gregory & Mr Churchill[13] there. they agreed to draw ye Princes monys. Sr Isaack askt me if things went not now to my Content I returned yt it was strang[e] that I should be so little taken notice of who was the person mainly concernd at which he seemed chagrin. before we parted he askt me what ye first copy cost me transcribeing. I answerd I could not tell for yt was Not the Whole work of my Amanuensis but he was to prepare & Copy work for Mr Sharp[14] & Witty[15] besides.

March 26 Dr Gregory came to see me: I shewed him Mr Wittys work Mr Sharps paper & some other things. he came *tanquam explorator*: enquired what I had done in order to frameing Statutes for Dr Plumes professorship. I said I had thought of something particularly that no one should enjoy it above 10 or 15 yeares. & for his employmt. he urged me to talk more on this head I forbore.

1706 Apr 4 at London. hear yt all ye mistaken errors are quitted & yt ye first Sheets will goe to ye presse this Week

♃ 18 Mr Hudson here told me if I would goe up Sr Is: would goe to the Princes Treasurer with me. urged me much I went

♀ 19 *mane* Sr Is: was very grave told me yt ye Prince haveing subscribed a great summ to ye loan to ye Emperors ye whole mony could not be receaved. yt he had taken up monys for Mr Churchill would say nothing when I asked if he had not taken up also to pay Me for my Calculators. but that he must give bond to Mr Churchill. I told him he had my Catalogue & papers in his hands he answered sleightingly yt ye Catalogue was imperfect which he knew when he receaved it sealed up & was contented wth it. I desired my MSSs back to correct ye faults of ye press. he told me we must goe on slowly at first quicker after yt in a few weeks he would return my MSS. Dr Greg is at Oxford I suppose will not return till after Trin terme: he must be paid for ye needless collations. & they cannot be finished till his return all this unsincere practise I must bear so long as God thinkes fit. May his goodness deliver me speedily May 16 ye first Sheet is wrought of: was promised a 2d but it was not ready. 23 at London no 2d Sheet ready yet promised on Thursday but not ready

Copy of my letter to Mr A & J Churchill

The Observatory

May ye 24. 1706 ♀ *mane*

Sr by your Covenant wth Sr Is: Newton you are to print 5 Sheets per week[16] of ye Historia Cœlestis, tis some weeks since ye first Sheet was wrought of, & tho I have been severall times in London & sent often to Mr Mathews for ye proof of the second I have not yet received it I wish for your own interest as well as mine you would enquire into ye cause of this neglect, & informe me of it yt care may be taken to prevent ye like for ye future & that you may make good your agreemt whereby you will oblige Sr

Your humble Servt

JOHN FLAMSTEED M R:

Tis 3 weeks since ye first Sheet was wrought of: if I am truely informed J F: This letter was not sent till ☽ May 27

Maij 28 ♂ was at london corrected ye 2d Sheet gave orders to have all ye notes put in Italick & to keep a space betwixt the constellations of about an inch always for ye future

Junij 3 ☽ ye proofs of the 2d Sheets are not come down tho promisd on ♄ day night gave the following note to J Hodgson

Observatory June ye 3d 1706
Mr Hodgson.[17]

Pray take 2 proof Sheets of every one printed, marke the faults in both exactly alike as you possibly can return one to the Printer to be keept for his Vindication, & keep the other by you for Vindication of your selfe & bring it down hither to be preserved by you both to vindicate your self & Sr Yours

JOHN FLAMSTEED MR.

The Observatory June ye 7 1706 ♃[18]

Sr I wrote to you May 24 last to complaine of the slowness of the press on ye 28 I corrected ye 2d Sheet which I am told is wrought of tho there are none of it sent me. This morneing I have corrected ye 3d proof. To morrow God spareing me health I intend to call upon you before noon to take care wth you both that the printer[19] may performe his business according to your agreemt & that the first proofs may have fewer faults & be better Corrected by him. I am Sr Yours

JOHN FLAMSTEED MR

June ye 7 I was with Mr J Churchill: sent for Mr Mathews ye printer but he came not because ye Sheet was not finished on Sunday he sent his Man Jackson with ye 2 & 3d Sheet B & C & a first proof of the 4th D which was returnd by Mr Hodgson on Monday Morneing

♂ 11 June recev'd ye 4th Sheet D to correct ye 2d time returnd it by ye coach, was at ye gate for horsback. found it sterk[20] lockt, it began to rane continued violent I returnd home sent up ye correct Sheets by ye coachman at 8½ ☿[21] *mane* with a letter to J. Hodgson:

11 corrected ye 5th Sheet E

☿ 19 was at London corrected ye 6th F Met Monsr Fatio. & on ye next day ♃ sent him Sr I Werdens account of ye Aur[ora] Bor[ealis] Nov 29 1681 of Mr Sharp March 24 last & ye eclipse at Bern May 1 1706

♃ 27. corrected ye 8th Sh[eet] wth proof of ye 9th brought by Isaack returned by Mr Hudson. 28 *mane*

470

July 1 1st proof of K

3 at london 1st of L 2 of K passed: Lent 6 *lib*[22] expenses 3*s*. 00*d*.

17 ☿ at London waited on Sr Is: about printing 100 or 150 more copys. represented yt I thought it needless. contrary to our agreemt &c hee seemd to assent & that we should goe on on ye old foot I suggested that it was probable Mr C had caused more to be printed yn he ought by 200 that if any besides my selfe had copys to sell I should not make any thing of mine & he agreed yt no body but I ought to have copys to sell & that as I desired the plates should be put into my hands that I might cause ym to be engraved & drawn of: promised to pay me 100 *lib* & I to send J H to him to inform him about ye princes Tresurer. promised to wait on him next week. expense 2 nights 10*s*:

good frid

1707 ♀ Apr. 11 was at London met the Referres at ye Castle Tavern in pater noster Row carried Mr Witty & Mr Weston there shewed Receipts Sr Isaack Newton was perverse: yet promised to come downe to Greenwich tho J Hudson says it was concluded otherways. Mr J Churchill there wth Mr Mathews. this meeting hindred my Jorney to Burstow. &c

♂ April 15. Sr I N. came down with Dr Gregory viewed ye 2d Volume & Catalogue declared they would stop ye press & pay me nothing till they had both in their hands. I dined with ym afterwards we parted quietly I can not say very freindly they had seene ye bill of my disbursemts ye friday before

NOTES

(1) Vol. 33, fos. 220–3. So far as is known, these extracts from Flamsteed's Diary have never been previously published *in extenso*. They help to an understanding of Flamsteed's impatience at the slow progress which was being made in regard to the publication of the *Historia*. It will be noted that fifteen months had elapsed since Flamsteed had submitted the Estimate which is reproduced on p. 420. The financial backing of Prince George had been obtained, and it might reasonably have been assumed that the preparation of the work would now proceed harmoniously. These extracts show that this was far from being the case. Flamsteed repeatedly complained of the lack of financial assistance afforded him.

(2) Flamsteed's assistant, James Hodgson (see Letter 638, note (6), p. 370, and elsewhere). Throughout this letter he is variously referred to as Hudson, or J.H., or simply as James.

(3) The sealed packet, containing the Catalogue in an incomplete form, had been deposited with Newton by Flamsteed as a guarantee that he would supply a more perfect copy when the occasion demanded. See Number 702, note (4), p. 459.

(4) See Letter 517, note (7), p. 139.

(5) See Letter 694, note (4), p. 285.

(6) Thomas Plume (1630–1704); admitted Christ's College, Cambridge, 1649. M.A., D.D. (1673); was appointed Vicar of Greenwich in 1658 by the Lord Protector (Richard Cromwell); Archdeacon of Rochester, 1679–1704. He left a considerable sum for various

purposes, including the erection of an Observatory and the establishment of a Professorship of Astronomy and Experimental Philosophy at Cambridge (the Plumian Professorship). See *D.N.B.* and Cooper, *Annals*, iv, 69.

(7) Roger Cotes, an English mathematician whose early death probably deprived him of the distinction of being one of Newton's most brilliant contemporaries. Educated St Paul's School, London; Trinity College, Cambridge (Fellow, 1705). Appointed Plumian Professor of Astronomy in 1706. Elected F.R.S. 1711. Ordained priest, 1713; in the same year he brought out a second edition of Newton's *Principia*. Wrote *Opera Miscellanea* (1722), *Hydrostatical and Pneumatical Lectures* (1738).

(8) William Whiston (1667–1752), a noted divine who became Chaplain to John Moor, Bishop of Norwich. He succeeded Newton as Lucasian Professor in 1705. Flamsteed, in a letter to Sharp (31 March 1702) wrote: 'He [Newton] has lost his professorship at Cambridge & put Mr Whiston into it tis sayd by some Malitious people that Augustus left a Tiberius to succeed him purposely to render his own fame the more illustrious.' The point of the allusion is this. On the occasion of the funeral of Augustus, the critics of his régime are represented by Tacitus (*Annals*, i, 10) as saying that Augustus had adopted Tiberius (a son of his wife Livia Drusilla by a former marriage) as his successor, not out of any affection, or regard to the interests of the State, but because, having studied Tiberius's savage and arrogant temper, he wished to glorify himself, posthumously, by contrast (*comparatione deterrima*).

Whiston was banned from the University and from his professorship on account of his *Essay of the Apostolic Constitution*, which expounded Arian doctrines. He wrote many theological and mathematical treatises, and he published Newton's Cambridge lectures, under the title *Arithmetica Universalis sive de Compositione et Resolutione Arithmetica Liber*.

(9) Item 16 of the Agreement, dated 17 November 1705, stipulated that 'the said Referees or the Major part of them shall also signe an Order for the said John Flamsteed to receive of the said Treasurer the summe of two hundred and fifty pounds for his charges in employing assistants and Servants to Calculate Observations Copy papers and Schemes and correct the press'. Flamsteed's comment was: 'I never Recd but 125 pounds although I had disbursed 173 pounds for Calculators' (see Letters 702, 711 and 738).

(10) William Burnet (b. 1688), eldest son of Gilbert Burnet, Bishop of Salisbury (see Letter 600, note (12), p. 296).

(11) By the side of the letter, Flamsteed has inserted a list of expenses for coach hire, etc., from 25 February 1705/6 to 22 May, amounting to £1. 14*s.* 5*d.*

(12) John Arbuthnott. See Letter 680, note (7), p. 431.

(13) Awnsham Churchill. See Number 687, note 3, p. 438.

(14) Abraham Sharp. See Letter 704, note (1), p. 646.

(15) John Witty. See Letter 687, note (5), p. 438.

(16) Item 12 of the Agreement of 17 November 1705 stipulated that 'the said Awnsham Churchill...shall print at least twenty Sheets per Month'. See p. 457.

(17) Here is a further marginal list of expenses from 28 May to 11 June, amounting to 18*s.* 10*d.*

(18) This should be either 'June ye 6 1706 ♃' [Thursday], or 'June ye 7 1706 ♀'.

(19) J. Matthews (see above, 24 May 1706). He printed the 1712 edition of the *Historia*.

(20) 'Sterk', a dialectical form of 'stark', O.E. *stearc* = rigid, stiff. (Joseph Wright, *English Dialect Dictionary* (1898), vol. 5.)

(21) That is, half past eight on Wednesday morning, 12 June.

(22) The word following '6 *lib*' is indecipherable.

707   FLAMSTEED TO NEWTON[1]
26 FEBRUARY 1705/6
From the copy in the Royal Greenwich Observatory[2]

The Observatory Feb: ♂ 26 1705/6

Sr

I returne you my thanks for your enterteinmt yesterday[.] had you acquainted me with your designe before I had saved both you Dr Gregory & My self the trouble I gave you for I designed allways to compare the printed sheets with the first notes & could have done it 6 moneths agone if you had not urged me for my Copy: but it happens well. you now see that there was not one Materiall fault committed for I depend not on ye Arches got by the Revolves of the screw but where I find these conformable to those given by ye diagonalls, or where several distances give the same place of a ∗ or planet

You will ask the reason why the Diagonalls were not inscribed at ye first, tis an history too long to be told you in a letter I shall give it in its proper place in my Works

I have got me one of my young people[3] to copy my table for turneing ye Revolves into Arches I hope twill be transcribed by to morrow & sent you with ye books of observations I had no copy of it by me otherways you had received both this Morneing togeather.

I have examined those few notes I brought from you but want the Manuscript for one of them. I wish the Dr to goe on orderly hereafter to compare ye Minits & Copy togeather & writes his notes as he has hitherto done one under another but with a larger margin for my emendations & then I shall easily give them. the more strict he is, the more he will oblige me

I forgot through my earnestness to go throw wth Dr Gregorys Notes to speak to you about the Copy of the Articles which you have not yet given me and about Dr Plumes[4] professor of Astronomy which Dr Bently[5] has determined without ever so much as letting me know yt hee was about such a business & I fear directly contrary to ye Archdeacons designe wherewith I am apt to

473

thinke none of ye Trustees in Cambridge were so well acquainted as I am. I had not known of it but by an accident I have wrote about it to Mr Whiston[6] who tells me the thing is done as to the nomination of a Professor & past remedy. I am sorry for it because the first election will be a president[7] for ye future, and fear a very ill one

NOTES

(1) There is no address of the intended recipient of this letter but the text clearly indicates that it was meant for Newton.

(2) There are two copies of this letter. The one reproduced here, in Flamsteed's hand, is in the Royal Greenwich Observatory (vol. 33, fo. 66). The other, in the hand of a copyist, is in the possession of the Earl of Macclesfield (Shirburn Castle Library, 101.H.2). They differ only in minor details (spelling, punctuation, etc.).

(3) Until 1696 the assistants at the Observatory were supplied by the Board of Ordnance, who paid them £26 per annum. After that date, Flamsteed had to make his own arrangements regarding assistants. The 'young people' employed by Flamsteed about this time were:
Thomas Weston, 2 February 1698/9–12 May 1705
John Witty, 30 March 1705–1708
Abraham Ryley, 1705–1708
Isaac Woolferman, 10 January 1705/6–5 March 1708/9
Joseph Crosthwait, 18 September 1708 to the death of Flamsteed (31 December 1719).

(4) See Extracts from Flamsteed's Diary, Number 706, note (6), p. 471.

(5) Richard Bentley (1662–1742). See Letter 517, note (7), p. 139, and vol. III, p. 156, note (1).

(6) See Number 706, note (8), p. 472.

(7) This spelling was commonly used in the seventeenth century for 'precedent'.

708   RECEIPT BY NEWTON
28 FEBRUARY 1705/6
From the original in the University Library, Cambridge[1]

Received this 28th of February 1705/6 of Mr John Flamsteed by the hands of Mr James Hodgson two Minute Books[2] of Astronomical Observations, to be collated with a Copy of Observations taken from them & then returned

Is. NEWTON

NOTES

(1) Add. 4006, fo. 27.

(2) See Statement by Newton, Number 755.

709   ARBUTHNOTT[1] TO NEWTON
30 JULY 1706
From the original in the University Library, Cambridge

Windsor July 30
1706

Sir

His Royal Highness ordered his Secretary[2] to write about the observations of Tycho Brahe[3] if ther was any thing remaining that was not yet published: I have sent you, by his Royal Highnesses order, a copie of the Answer which you may communicate to those concerned and to Mr Hally. The prince likewise orderd me to tell you that he will use his interest to procure the said observations and will publish what shall be thought fitt for publick use, so I would have you to consult how to proceed in the matter: it is likely by these observations having been sent into France that they contain at least some things not published for Mr Romer[4] who sent them could not but know what the importance of them was. my opinion is that we should draw up a letter to Mr Romer giving him an account of the substance of Mr Flamsteds observations[5] that we are now publishing which will be obliging, and at the same time desire the favour of him that he would give us an abstract of what those eight Volumes of observations contain, or perhaps it may be allwayes worth the while to have those eight Volumes wrote by Tychos own hand in our Custody and no body indeed would refuse so fair a proffer, and besides I find the prince is mighty desirous to have them I submitt all to your better judgement Dr Gregory and Mr Hally are both in town I have not wrote to either when you have talk'd of the Matter together please to lett me hear from you, and I shall obey your commands

I am with all respect

Sir
Your Most humble Servant
Jo: ARBUTHNOTT

direct for me at my lodgings in Windsor
Castle the coaches & post come every day.
I know you will not lett Flamsted know
that you consulted Mr Hally

NOTES

(1) Add. 4006.

(2) The Prince's secretary at this time was John George Hugk, a Dane. Clark (see Letter 680, note (1), p. 430) left the Prince's service on 25 October 1705.

(3) In 1576 Frederick II, King of Denmark, bestowed upon Tycho Brahe (1546–1601) the Island of Hveen, with funds to equip and maintain an observatory there. From this Observatory—Uraniborg, or Castle of the Heavens—Tycho made many observations. He was succeeded by Kepler in 1602. See J. L. E. Dreyer, *Tycho Brahe* (1890).

(4) Olaus Römer (1644–1710), the noted Danish astronomer. See vol II, p. 299, note (13).

(5) This letter was sent early in the following year (see Letters 714, 715).

710 NEWTON TO SLOANE
?1706
From the original in the British Museum[1]

Sr   Thursday night.

Lady Betty Gayer[2] being engaged for tomorrow, & at liberty on munday or tuesday, I beg the favour we may wait on you on either of those days at three a clock, & that you will let us know which of those two days you can be most at leasure.

Your most humble Servant

Is. NEWTON

For Dr Sloane

NOTES

(1) Sloane MSS. 4060, fo. 71. As is the case with a number of letters in this collection, Newton has omitted the date.

(2) Lady Elizabeth Annesley, daughter of James, the second Earl of Anglesey, married Robert Gayer of Stoke Poges, grandson of John Gayer, Lord Mayor of London, 1647. She was the sister of Arthur Annesley, who defeated Newton in the Election of 1705 for Cambridge University (see Letter 693). According to Burke's *Peerage*, Lady Betty died in 1725. See A. E. Gayer, *Memoirs of the Family of Gayer* (1870).

711 FLAMSTEED TO NEWTON
14 SEPTEMBER 1706
From the original in the Royal Greenwich Observatory[1]

The Observatory ♄ *Sept* 14 1706

Sr

I have Consulted Tychos[2] *Mechanica*[3] where he says that at that time when he wrote it he was 50 Yeares of Age compleate and that his Volumes contained the accurate observations of 21 yeares which shews they commenced in the Year 1575. Tycho was borne in 1546: Decembris 13.22h.17′[4]

But the Observations of the *Historia Cœlestis* begin no sooner then the Year 1582 so that by this Account there are 7 Yeares observations wanting in the very beginning

Besides all the Observations of the year 1593 wch were not to be found in Germany. In the same place he sayes he had observed seven Comets whereas in the *Historia Cœlestis* there are no observations that I can find either of that of the year 1582 or 1590, of which he gives an Account in his Epistles. The first part of his *Progymnasmata*[5] gives his Tables for Calculateing the ⊙s & ☽s places with his Observations of the new Starr of 1572[6] & deductions from them[:] the 2d part is Concerning the Comet of 1577.[7] so that we have the Observations of but 3 of his 7 Comets & of these only such as he thought fit to employ. this makes me think that his Observations of the Comets made a book by themselves and that probably it is still to be found in Denmarke with the 7 or 8 Yeares Observations that are missing.

Whatever his R. Highness determines concerning the rest of Tychoos works it may be much for his honour to bestow these on ye World with ye Errata of the German Edition if he can procure ye Originalls, as I doubt not but he easily may to be sent into England.

After this Account it will be needless to send you the *Progymnas[m]ata* but if you have a desire to see them please to Intimate it by a Note and I will send them as you shall direct.

Allow me to mind you that by ye Articles I was to have a note signed by the Referrees[8] for the paymt of 125 *lib*:[9] to me as soon as ten Sheets were printed That Number was printed off before I went into ye Country I have dismist[10] my Amanuensis & Calculators because they lay me in a greater Summe & I could not promise them recompenses suitable to their Work. I would desire a Meeting of ye Referre's to sign ye Order that I might have what you agreed to readily paid me, and we may take care together to prevent the Press from making delayes upon false pretences. where you will oblige Sr

Your very humble Servant

JOHN FLAMSTEED M R

To Sr Is: Newton at his
house in Jerman Street
    St Jamess[11]

NOTES

(1) This letter (vol. 33, fo. 67) is in the hand of a copyist.

(2) See Letter 709, note (3), p. 476.

(3) *Astronomiæ instauratæ Mechanica* (1596). This contains accurate observations over a period of twenty-one years beginning from the year 1576. It actually appeared in 1598 at Wandsbek. The work contains a description of Tycho's instruments and an account of his discoveries.

(4) This sentence was added by Flamsteed.

(5) *Astronomiæ Instauratæ Progymnasmata*, 2 vols. (1602–3). The chief catalogues of the fixed stars in use at this time were those of Tycho Brahe and Kepler. In the *Progymnasmata* the former gave the latitudes and longitudes of 777 stars; these formed the bulk of the material which his successor, Kepler, inserted in the Rudolphine Tables. See vol. III, p. 165, note (4).

(6) On 11 November 1572 Tycho discovered a new star in the constellation of Cassiopeia. The account was published in his *De Stella Nova* (1573).

(7) First observed on 13 November 1577. It is the most conspicuous of the seven comets observed in his time. See J. L. E. Dreyer, *Tycho Brahe* (1890), ch. vii.

(8) This word is in Flamsteed's hand.

(9) See Article 16 of the agreement between the Royal Society and Flamsteed, 17 November 1705. The authorization for the payment of this sum was issued on 26 March 1708 (see Number 738).

(10) See Flamsteed's letter to Wren (Letter 747).

(11) This address is in Flamsteed's hand.

712   MINT TO GODOLPHIN
23 SEPTEMBER 1706
From the original in the Public Record Office[1]

To the Rt Honble the Lord High Treasurer
of England.

May it please your Lordship

In obedience to your Lordships Orders of Reference dated May 31st & July 17th 1706 upon the annexed proposals of Mr Holt & Mr Williams to raise the price of Tin,[2] We humbly take leave to acquaint your Lordship that the same Gentlemen made Proposals of the like nature last year for raising the price of Tin, wch were referred to us by your Lordship. Upon wch we desired a meeting with some of the principal Merchants of the City trading to the East Indies, who were then of opinion that great quantity of Tin might be had in the East Indies, & if the price of her Majesties Tin were raised here, it would encourage the importation of Tin from thence, & thereby hinder the consumption of English Tin in Europe. Which opinion of the Merchants We then thought proper to report to your Lordship.

Since your Lordship's second reference of Mr Holts proposal, he brought to us a Dutch Merchant, who informed us that he had been a long time at Siam, where the only East-India Tin Mines are, & assured us that those Mines could not yield more Tin good & bad together, one yeare with another, then fifty or sixty Tunns per annum; that the Indians had occasion for more Tin then their Mines supplied them with, that the price of Tin in the Indies was about sixty Gilders[3] per [cwt], wch is dearer then here, & therefore it was never imported by the Dutch but for want of other commodities for ballast, & that it was scarce used in Europe for any thing else then glazing earthen ware, being bought up chiefly by the workmen of Liege for that purpose.

Upon this information we thought it proper to enquire more fully into a matter of that consequence, & therefore delayed our report to your Lordship till we could be well informed of the character of the Dutch Mercht, & writ to Mr Drummond[4] in Holland to enquire into the nature of the East India Tin, & what quantity their mines might produce.

We therefore now humbly acquaint your Lordship, that we find the Dutch Merchant has such a character both here & in Holland, that we think his credit no ways to be relied on, & that we have advice from Mr Drummond, that the best East India Tin is finer then the English, & will beat into a finer leaf without cracking & draw into threds, like silver & gold, & that the mines in India are larger then ours. & if we rise up the English Tin above 45 Gilders per [cwt] (wch used formerly for many years to be under 38) the Dutch will soon supply not only themselves but also other places from India, they importing 80 Tunns the last year, & having imported formerly above 300 Tunns in one year, & that a high price will also encourage the Germans to work their mines faster. Which account of Mr Drummond agreeing so well wth the opinion of the Merchants here wch we formerly reported to your Lordship we are humbly of opinion that the price of English Tin cannot be raised without diminishing the consumption, & hurting the trade

All wch is most humbly submitted

to your Lordships great wisdome

Mint Office the 23d Septembr   J STANLEY
1706.   IS. NEWTON
                                                        JN ELLIS.

NOTES

(1) This letter (T. 1/99, no. 97) is in Newton's hand.

(2) Unsold stocks of tin were piling up in the Tower (see Letters 667 and 705) when John Williams pressed the Government to raise the sale price of tin from £76 to £100 or £120 per ton on the grounds that England had a monopoly of the metal. Newton opposed this. He maintained that he could not sell all the tin he had purchased. As sales had fallen notably in 1705 he urged a curtailment of purchases. See G. R. Lewis, *The Stannaries* (1924).

(3) Guilder: originally a silver coin current from the fourteenth to the seventeenth century in the Netherlands, where it was also known as a *gulden*.

(4) Drummond of Amsterdam was one of the agents commissioned to negotiate sales of tin on the Continent. See Craig, *Mint*, p. 208.

713   REFEREES TO PRINCE GEORGE'S TREASURER

28 NOVEMBER 1706

From a draft in the Royal Mint[1]

November 28, 1706

Sr

We desire you to pay to Mr John Flamsteed her Maties Astronomer upon account the summ of fifty pounds towards defraying the charges of preparing his Astronomical papers for the publick & correcting the Press, & the same summ shall be allowed in your account to his Royal Highness the Prince.

NOTE

(1) Newton MSS. I, 183. This is in Newton's hand. The Prince had his own treasurer (Letter 738.)

714   REFEREES TO RÖMER[1]

(?) JANUARY 1706/7

From the original in Trinity College Library, Cambridge[2]

Viro   Roemer, Franciscus Roberts,

Chr. Wren, Is. Newton, David Gregory, & Jo. Arbothnot Salutem.

Vir celeberrime

That his R.H. the Prince of Denmark out of his great inclination to promote sciences has been pleased to beare the charges of printing Mr Flamsteeds Astronomical observations for ye last 30 years wch will take up two Volumes in folio & in all probability would otherwise have been lost. These Observations are of the fixt Stars, Planets, Comets, Sun & Moon, With the calculated places of almost 3000 fixt stars & of the five superior[3] Planets & the Comets & many of the Moon. And we reccon it will be the completest set of Observations extant, and that his Catalogue of ye fixt stars will be the fullest & exactest. There are about 60 sheets printed and we expect that the first Volume will be printed of by Michaelmas[4] & when the book is printed off we hope you will please to let a coppy of it have room in your library.

His Highness has been pleased to referr the care of this matter to us who write this letter, & hearing that Tychos Observations were left in the K. of Denmarks Library written in Tycho's own hand, he is desirous that those Observations or as many of them as may be of use in Astronomy & are not yet printed or not correctly printed should be printed here & come abroad with Mr Flamsteed's. And therefore we desire the favour of you to let us know what books of Observations Tycho has left in MS & what are their contents & how many years they reach & in what method they are written and what your judgment is about printing them or any part of them. Your great reputation & particular skill in Astronomy has prompted us to write to you about this matter. Farewell.[5]

NOTES

(1) Olaus Römer (1644–1710). See vol. II, p. 299, note (13).

(2) U.L.C. Add. 4006, fo. 3. It is in Newton's hand throughout, and is evidently a draft of a letter drawn up by the Referees, following the suggestion of Arbuthnott (see Letter 709), for transmission to Römer. The letter was put into Latin by David Gregory, who before sending it to Römer submitted it to Newton, who made certain alterations (see Letter 715, note (1), p. 484). It appears that Gregory, having noted Newton's amendments, translated the letter back into English and retained the copy which is now in the Library of the Royal Society of London (R.S. Greg. MS. fo. 20), since it contains two additions not found in the letter here transcribed. In the first place, Gregory has added the date: 'London 31 January 1706/7' at the head of the letter. Secondly, after the words 'about this matter' there follows the sentence (in Gregory's hand): 'We beg you will transmitt yours to the Princes Secretary Mr Hook directed for Sir Isaac Newton President of the Royal Society', and above this after the words 'any part of them' Gregory has placed a mark I, and this refers to a passage on the reverse of the letter which reads: 'Wee are assured of a very precise account of them from you if the books are in your power: But if they are not at present so; yet we desire such as your memory will furnish you.'

'The Princes Secretary Mr Hook' can only refer to Hugk (see Letter 709, note (2), p. 475). Hooke died in 1703, and at no time did he act as Secretary to the Prince.

(3) The word 'superior' is crossed out, both in the original and in Gregory's draft.

(4) Printing had begun in May 1706, but it proceeded with discouraging slowness. The death of Prince George in October 1708 brought about a complete cessation which lasted for some years. In the spring of 1711 Queen Anne ordered a resumption of the work. The following year the first volume appeared. The preface, written by Halley, contained some very ungenerous reflexions upon Flamsteed. See Robert Grant, *History of Physical Astronomy* (1852), p. 475.

(5) At the foot of the letter is a list of books, still in Newton's hand, apparently taken from a bookseller's catalogue. It includes '*Basnagij: Annales Politico Eccles.* 3 vols. Rotterd. 1706', and '*Jul. Pollux: Onomasticum. g. l. Com. var.* 2 vol. Amst. 1706'. This helps to fix the date of the letter.

715   REFEREES TO RÖMER[1]
From the original in the University Library, Cambridge[2]

Celeberrimo Viro D. Olao Roemer &c.

Franciscus Robarts, Christophorus Wrenn, Isaacus Newton, David Gregory, Joannes Arbothnot Salutem.

Vir Celeberrime

Cum Clarissimus Flamstedius noster per triginta jam annos proxime elapsos suas de Fixis, Planetis, Cometis, Sole, & Luna Observationes summâ curâ atque solertia in Observatorio Regio Grenovicensi excoluerit, insuper et Fixarum fere 3000, quinque Planetarum, Cometarumque et Lunæ Loca calculando designavit [calculo ad Eclipticam reduxerit,] ne opus tam elaboratum penitus intercideret, suo sumptu curavit Serenissimus apud nos Daniæ Princeps qui cum Regiæ suæ Celsitudinis favore Literas omnes dignetur, tum singulari studio Matheseos scientiam et ipse amplectitur, et ubivis gentium promovere amat.[3] Jussit itaque pro Regia Sua Munificientia ut hæ Flamstedij lucubrationes prælo mandarentur, nobisque qui hanc ad Te Epistolam damus, id curæ commisit. Implebunt vero Tabulæ hæ Astronomicæ bina in Folio Volumina, quorum prius (sexaginta [quinquaginta] enim Schedæ dudum impressæ sunt) ante proximum Sti Michaelis Festum absolvere, [hyemem proximam elapsam] Tibique V.C. illius exemplum in Bibliotheca vestra collocandum commendare speramus. Interim Serenissimus Princeps, cum audierit Tychonis Observationes Autographas fuisse Bibliotheca Serenissimi Regis Daniæ repositas, obnixe rogat [cupit] ne tantus thesaurus ulterius delitescat, sed uti illæ, aut saltem quotquot illarum Astronomiæ ornandæ [perficiendæ] inservire possunt, et vel nondum omnino, vel minus accurate editæ sunt, tandem una cum Flamstedianis prodeant. Hanc igitur Nos a Te Gratiam petimus, ut ex Te certiores simus, quæ qualesque fuerint Hæ M.S. Tychonis Observationes, ad quot annos pertineant, quâ methodo conscriptæ sint, quem præcipue scopum spectent, [vel quemnam usum jam habeant] et quid Tu de iis vel omnibus vel aliquibus juris publici faciendis [quotquot usus esse possint in lucem edendis] existimes. Si penes Te jam fuerint, clariorem de iis notitiam ab humanitate tua facile nobis pollicemur: Sin Tibi jam ad manus non sint, quantum de iis memoriæ tenes, nobiscum amice communicabis. Ut hæc ad Te liberius scribamus impulit celebris Tua humanitas et Eruditio, et insignis præsertim in rebus Astronomicis Scientia.[3] Quicquid

482

hisce literis responsi dederis, si ita Tibi visum est, ad Dominum Hook[4] Serenissimi Principis Secretarium, Isaaco Neutono Equiti, et Regalis Societatis Præsidi [Honorabili Viro Francisco Robarts, Armiger] trandendum mittas.

Vale.

Translation

To the illustrious Mr Olaus Roemer

greetings from Francis Robartes, Christopher Wren, Isaac Newton, David Gregory, and John Arbothnot.

Illustrious Sir,

Since our distinguished fellow countryman Flamsteed, over a period of thirty years now recently completed, has with the greatest care and skill pursued his observations of the fixed stars, the planets, the comets, the sun, and the moon at the Royal Greenwich Observatory, and since moreover he has by calculation mapped out the places of nearly 3000 fixed stars, five planets, the comets, and the moon, [by calculation had reduced them to the Ecliptic], His Serene Highness the Prince of Denmark dwelling here among us has taken steps, at his own expense, to prevent the utter loss of a work so carefully elaborated—His Serene Highness, I say, who while regarding all literature as worthy of his royal favour, yet embraces with special and personal interest the science of mathematics[3] and is accustomed to advance its progress everywhere in the world. Therefore with characteristic royal generosity the Prince has ordered that these studies of Flamsteed's should be printed for publication and has entrusted this commission to us the writers of this letter. These astronomical tables will indeed fill two folio volumes, of which the first (since sixty [fifty] sheets have already been printed) we hope to complete before Michaelmas [before next winter has passed], and we hope to send to you, Noble Sir, a copy of it for inclusion in your library. Meanwhile His Serene Highness the Prince, having learned that the observations of Tycho, written in his own hand, had been deposited in the library of His Majesty the King of Denmark, earnestly requests [desires] that so great a treasure should no longer remain hidden, but that these observations, or at least as many as could serve for the adornment [perfecting] of astronomy, and which are either not yet published at all, or are inaccurately published, should be published together with those of Flamsteed. We seek this favour of you, therefore, that you would let us know what, and of what kind were these manuscript observations of Tycho, to how many years they relate, by what method they were written, what purpose they have mainly in view [what use they now serve] and what are your views about making either all of them, or some of them, public property [or what number of benefits can exist in publishing them]. If they shall prove to be now in your possession, we readily promise ourselves clearer knowledge of them in virtue of your gracious kindness; but if they are not now accessible to you, will you be so good as to share with us whatever memory you have of them. Your well known kindness and erudition, and particularly your unique knowledge of astronomy have induced us to write to you somewhat freely. Whatever reply you give to this letter, if it seems good to you, please direct it to Mr Hook,[4] Secretary to His Royal Highness the Prince, to be transmitted to Isaac Newton, Knight, President of the Royal Society [the honourable Francis Robartes, Esquire].

Farewell.

NOTES

(1) This is the letter referred to in Letter 714. It is in David Gregory's hand. The original letter which was sent to Römer is lost. Newton crossed out certain words and phrases and substituted others above the words so crossed out: in the copy here reproduced such alterations are shown in the Latin, and in the translation, in square brackets, following the words deleted.

(2) Add. 4006, fo. 5. The letter was probably sent towards the end of 1706 or early in 1707. The date 1706 appears in the top corner in an unknown hand; this date is confirmed by reference to Letter 714, note (5), p. 481.

(3) The reference to Prince George's particular skill in science was much exaggerated. He, Tycho and Römer were fellow-countrymen.

(4) See Letter 714, note (2), p. 481.

716   NEWTON TO FLAMSTEED
9 APRIL 1707
From the original in the Royal Greenwich Observatory[1]

Jermin street. Apr 9. 1707.

Sr

The Referees meet on Fryday next[2] at four a clock in the afternoon in Pater-noster Row at the next Tavern to Mr Churchil the Bookseller. You will hear of them at Mr Churchils. I desire you would not fail to meet them because after the Queen returns to Windsor, they will scarce have an opportunity of meeting any more before next winter. And yt all things may be now setled[3] & ajdusted, I desire that Mr Witty & your Emanuensis may be there & that you will bring your Bill & the three or four folio leaves of MS copy which you had from the Printer. I am

Your humble Servant

Is. NEWTON.

NOTES

(1) Vol. 35, p. 75.

(2) Good Friday, 11 April.

(3) According to Flamsteed's Diary (Number 706) this was far from being the case. 'Sr Isaack Newton was perverse', wrote Flamsteed. Gregory summarizes the conclusions reached at this meeting in Manuscript 718, p. 487.

717   MINT TO GODOLPHIN

12 APRIL 1707

From the original in the Public Record Office[1]

To the Rt Honble the Lord High Treasurer of England.

May it please your Lordship

In obedience to your Lordships Order of the 3d. instant we have considered the annexed Inventory therewith sent us of Tools & other necessaries proposed to be provided in London for the use of her Majties Mint in Scotland, & have here set down the prizes of such of them as seem to us proper & necessary to be furnished for the said Mint, supposing they are not already in that Mint & cannot be had cheaper from other places then from London.

For the Office of Receipt.

	li	s
One Pyle of standard weights	2.	10
One set of Bell-weights for gold	2.	00
One set of Bell-weights for silver	4.	10
Two sets of penny weights, coyn-weights & grains	0.	09
One pair of gold-scales for weighing 15 lb wt of gold wch we call a journey[2] of gold	4.	00
If a pair of scales be wanting for weighing 60 lb wt of silver, wch we call a journey of silver,[2] they will cost	10.	00
Two books of Vellum, suppose of 80 leaves each, at 18d a skin or leaf will cost 12 li. They are not absolutely necessary. We use books of thick demy paper	12.	00

li   s   d
35. 09. 00

For the Assay Office.

A pair of Assay-ballances with a Lanthorn	5.	00.	00
Assay-weights	2.	00.	00
Two Copell molds		10.	00
Two pair of iron molds or Calms to cast two sorts of Assay bullets		10.	00
Five dozen Mufflers at 18d a Muffler	4.	10.	00
Fine bone ashes one bushel		12.	00
Charcoal an hundred bushels at 8d a bushel	3.	06.	08

16. 08. 08

485

For the Melter.

Brass patterns for making molds to cast the barrs of gold & silver at 18d per lb wt, vizt 6 Guinea barrs, 3 half Guinea barrs, 2 Crown barrs, 4 half crown barrs, 9 shilling bars & 2 sixpenny barrs	2. 06	
A sweep mill	6. 00	
Fifty pounds of Quicksilver at 5s a pound	12. 10	*li* *s* *d*
Thirty bushels of bone-ashes for tests at 4s a bushel	6. 00	26. 19. 00
Half a dozen of frail brushes for cleaning the barrs of gold & silver	0. 03	

For the Moneyers.

Two dozen of sizing scales wth stands & counter-poises, at 11s a piece	13. 04	
Twenty four dozen of sizing flots (if by this name they mean Rasps for sizing the blanks) at 1s per lb wt	40. 00	
Five hundred weight of Alume at 18 or 20s per hundred	05. 00	62. 06. 00
Two hundred weight of red wine-stone or Tartar[(3)]	3. 00	
One dozen of soft brushes	0. 06	
One dozen of small scratch brushes for cleaning the Dyes, ten for the Moneyers & two for the Graver	0. 16	

For the Graver Engineer & Smith.

Blistered steel 500 lb wt at $4\frac{1}{2}$d per lb wt	10. 10	
A table Vice & Files of several sorts, vizt Rubbers at 1s a pound wtt, smooth Files & bastard smooth files at 3s a file & other files at 18d a file	16. 00	28. 10. 0
Two Oyle-stones 20s. Two Grind-stones 20s	2. 00	
Total		169. 12. 08

Cast Rollers are not to be bought. The man who makes them keeps the secret to himself & only lends the Rollers to the moneyers at 10s a day. Hammered Rollers cost 6 *li* a pair, but are not so serviceable. What is meant by a sizing Mill we are not certain. If it be the Mill for drawing the barrs of gold & silver to a just thickness, such a Mill with three hammered Rollers in the late recoinage cost 177 *pounds*. The Moneyers have some of those Mills remaining in their hands wch they can afford cheaper. German steel is scarce to be met with in shops fit for their use. Our Smith sends into Germany for it. It costs from 8d to 12d or 14d per lb wt according to the goodness & scarcity.

Its chief use is for making the Dyes & Puncheons, & we conceive it best to have the money of both nations made from the same puncheons. For the variety of impressions makes it more difficult to know good money & more easy to adulterate it. What is meant in the Inventory by large scratches half wier, we do not understand. Besides the things above mentioned there should be two indented Trial pieces of crown gold & two of standard silver one for making the money of due allay the other for examining it before delivery. They will cost the value of the gold & silver conteined in them, & are to be made by a Jury of Goldsmiths with four other Trial pieces for England as we mentioned in our late Report of ye 24th March last.

All wch is most humbly submitted to

your Lordships great Wisdome.

Mint Office. Apr. 12
1707

J STANLEY
Is. NEWTON
JN ELLIS.

NOTES

(1) T. 1/101, no. 91. This is in Newton's hand. A briefer copy is in the Royal Mint (Newton MSS. III, 70).

(2) See Letter 555, note (6), p. 214.

(3) 'Cream of tartar' or 'Argol' (potassium hydrogen tartrate) which is deposited during the fermentation of grape-juice.

718   MANUSCRIPT BY DAVID GREGORY

15 APRIL 1707

From the original in the Library of the Royal Society of London[1]

A Paper offered to Mr Flamsteed to be signed ♂ 15 Aprile 1707, at Greenwich by Sr Isaac Newton & Dr Gregory sent by the Referees

1. Mr Flamsteed promises to deliver to the Referees 174 Sheets of the 2d volume containing his observations down to the end of the year 1705.
2. Then he promises to give up a Copy of his own book that contains the Catalogue of the Fix't Starrs to correct & supply the Catalogue of the fixt starrs delivered some time since to Sr Isaac Newton, or to collate & perfect the said Catalogue in Sr Isaac Newtons hand by his own said Catalogue supplied from Observations already made; and to let the Referees examine the said Collation in what Sheets they shall think fitt.

487

3. The 174 Sheets of the 2d volume he promises positively to deliver to Sr Isaac Newton by the feast of Whitsunday[2] next in this year 1707.

4. He undertakes to deliver to Sr Isaac Newton the Corrected & Supplyed Coppy of the Catalogue of Fix'd starrs as above, & to doe the other things in the 2d Article by the   [3] day of the   [3] month in the year   [3].

5. How soon these articles are fullfilled, & all the forsaid Sheets delivered then (and not until then) shall Mr Flamsteed receive the Money mentioned in the triparte Contract, betwixt the Referees, Mr Flamsteed & Mr Churchill;[4] and Mr Flamsteeds part of the said Contract shall be fully acquitted & discharged: Deducting notwithstanding from the said money so to be pay'd to Mr Flamsteed, what is absolutely necessary to put those papers into the condition named & described in the forsaid contract. And in this deduction Mr Flamsteed may be assured of all allowances & favour from the Referees, who (notwithstanding this) are positively resolved not to print any further after this day, nor to give Mr Flamsteed any money until these conditions are fullfilled.

At Greenwich. 15 Aprile 1707

Mr Flamsteed would not sign the forsaid Papers, but solemnly promises on the word of an honest man, that he will goe about fitting up his papers, for being delievered as soon as ever he can. And thinks he will deliver the 174 sheets of the 2d volume by Whit Sunday next, and the Catalogue suppyed & finished in Septr or Octr or at farthest in November in this year 1707; but undertakes nothing positively.

NOTES

(1) R.S. Greg. MS. fo. 19. This appears to be a summary, drawn up by Gregory, of what transpired at the meeting held on 11 April. See Letter 716, note (2), p. 484.

(2) 1 June 1707.

(3) Gregory has left gaps in these places.

(4) See Item 16 of the Agreement of 17 November 1705, p. 457, and Flamsteed's protest that he 'never Recd but 125 pounds although I had disbursed 173 pounds for Calculators'.

719   NEWTON TO SIR JOHN NEWTON[1]
APRIL 1707
From Edleston, p. 307

Sir John

I was very much surprized at the notic of Mr Cook's[2] death brought me this morning by the bearer who being an undertaker came to me to desire that

I would speak to you that he might be employed in furnishing things for ye funeral. He having married a near kinswoman of mine I could not refuse troubling you with this letter in his behalf beleeving that he will do it well if you are not otherwise provided. I had an opinion that my Cousin was not in danger tho weak, wch makes my concern greater for the loss. I am

Your affectionate Kinsman

and most humble servant

Is. NEWTON

Jermyn Street, Apr. 1707
For Sr John Newton, Baront
at his house in Soho Square.

NOTES

(1) Sir John Newton (died 1734), of Barrs Court, Gloucestershire, and later of Thorpe, Lincolnshire, became the third baronet in 1699. According to Newton himself, both he and the recipient of the above letter were descended from John Newton of Westby, in Lincolnshire. Isaac's great-great-grandfather, John Newton, and Sir John's great-grandfather, William Newton, were brothers. See Letters 583 and 703, note (2) p. 461.

(2) Edward Coke (died 1707), of Holkham, great-great-grandson of Chief Justice Sir Edward Coke (1552–1634), who married Cary, sister of the above. Her son, Thomas, was created Earl of Leicester in 1744.

720   MINT TO GODOLPHIN

26 MAY 1707

From the draft in the Royal Mint.[1]
In continuation of Letter 717

To the Rt Honble the Earl of Godolphin
Lord High Treasurer.

May it please your Lordship

The things directed by your Lordship to be provided by us for her Majts Mint in Scotland are put on board to be sent thither, excepting the weights & scales[2] wch are already made, but the weights are not yet sized for want of authority to make & mark the standard weights.

We therefore humbly desire her Majts Order may be directed to the Warden of this Mint[3] to make two exact Piles of Standard weights & to examin & mark them with the new mark before the Officers of this Mint & to deliver one of the Piles to the General of the Mint of Scotland by Bill indented according to the directions of the Indenture of this Mint.

489

And we further lay before your Lordship that for setting on foot the coynage in Scotland it will be convenient that ye Lord Chancellour of great Britain[4] do summon a jury of Goldsmiths to make six new indented Trial pieces of crown gold & as many of standard silver to be delivered as has been usual: two of wch are to be sent into Scotland as we mentioned to your Lordship in a former memorial.

All wch is most humbly submitted to your Lordships great wisdome.

Mint Office
26 *May*. 1707.

J. STANLEY
Is. NEWTON
JN ELLIS.

Memorial of the Officers of
the Mint concerning the
Weights & Trial Pieces for
the Mint in Scotland

NOTES

(1) Newton MSS. III, 97.

(2) The weights were to be made by the London Mint under the nominal supervision of its Warden.

(3) That is, the Warden of the London Mint, Sir John Stanley.

(4) William Cowper, Lord Keeper of the Privy Seal; created Chancellor 4 May 1707. At this time he was Baron Cowper; later he became Earl Cowper. He was the first Lord Chancellor of Great Britain, the Act of Union having come into force on 1 May 1707.

721   A MANUSCRIPT BY FLAMSTEED[1]
27 MAY 1707
From the original in the Royal Greenwich Observatory[2]

Maij 27 ♂ 1707. mane

I have plowed sowed reaped brought in my Corne, with my own hired Servants & purchased Utensills. Sr I N. haveing been furnished from My Stores would have me thrash it all out my selfe & charitably bestow it on ye publick that he may have ye prayse of haveing procured it. I am very desirous to supply the Publick wth my Stores if it will but afford me what I have layd out in tillage and harvesting in Utensills & help. & afford me hands to work

PLATE VI. THE EDINBURGH MINT ('THE CUNZIE HOUSE')

it up since the labor is both too hard & much for me. for an adequate recompense I doe not expect but I must stand upon a reasonable one since God has blest my labors with large fruites & not to doe it were not to acknowledge his goodness; & my Countries Ingratitude would be attributed, by Sr I N himselfe, to my *Stupidity*[3]

...   .

Laudibus digna faciat, qui laudes velit duraturas.[4]

NOTES

(1) No correspondence appears to have passed between Newton and Flamsteed from 9 April 1707 until the Referees sent Flamsteed a copy of the resolution on 13 July 1708 (Letter 746).

(2) Vol. 48.

(3) Underlined by Flamsteed.

(4) 'He who would wish his praises to endure, let him do works worthy of praise.'

722   MINT TO GODOLPHIN
31 MAY 1707
From a copy in the Royal Mint[1]

> To the Rt Honble Sidney Earl of Godolphin
> Lord High Treasurer

May it please your Lordship

Attending my Ld Chancellour of Scotland[2] on thursday last about the Mint at Edinborough, his Lordship proposed to us to lay before your Lordship a Memorial concerning some things to be considered by the Committee of Council wch is to sit to day upon the affairs of Scotland.

His Lordship is considering whether the Pix[3] of that Mint may not be still tried before her Majties Council in Scotland as formerly. And if so, We are humbly of opinion that that in conformity to the Act of Union[4] there may be made seven indented Trial pieces of Crown gold & seven of standard silver; two of each metal to be kept in the two Treasuries to try the Pixes of the two Mints; two of each for the Wardens of the two Mints to try the moneys before delivery, & to decide questions between the Master & Importer about the fineness of the bullion; two of each for the Masters of the two Mints to make the moneys by, & one of each for the Wardens & Company of Goldsmiths to try their plate & manufactures of gold & silver. It will be also convenient that a Pile of standard Troy weights be made for the Treasury in Scotland by the Deputy Chamberlains of the Exchequer, if it be not already done. But if both Pixes be tried by the original standard weights at Westminster, six Trial pieces will be sufficient

The Indenture of the Mint directs that two Piles of standard Troy weights be made by the Warden of the Mint & one of them delivered to the General of the Mint in Scotland. They may be printed like the weights sent to the Corporations & in the absence of the General delivered to your Lordships order.

For setting the coynage speedily on foot in that Mint, her Maty may please to issue out her Warrant to the General & other Officers with a copy of the Indenture of the Mint in the Tower annexed to it, authorising & requiring them to act under those Rules there set down wch relate to their several Offices, & particularly that the Master coyn all the moneys of the weight & fineness & within the remedies[5] there set down & take care that the several pieces be not lighter then their counterpoises, & that the Warden & Counter-warden survey & Cheque the proceedings of all the other Officers & Ministers & see that the moneys be well & duly coyned in all respects & that an account be kept of the Dyes & Puncheons.

Another Warrant may be directed to the Master of that Mint authorising him to command the Graver to make Puncheons & Dyes for coyning the money of the same form with the money coyned in the Tower of London, & to use such master Puncheons as shall be sent him from the Tower untill the Puncheons made by him self shall be approved by the Officers of that Mint, & to set the letter E (the first letter of the name of the City of Edinborough) or such other mark as shall be appointed under her Maties effigies, as in the specimens of Dyes wch will be sent him from the Tower. that the money of the two Mints may be thereby distinguished.

An Instrument may be drawn up at any time hereafter, either in the form of an Indenture between her Maty & the Master of that Mint, or in the form of a Warrant, prescribing the duty of every Officer in particular.

In the late recoinage of the hammered moneys in England, one tenth part of the silver was coyned into sixpences & four tenths into shillings; & the same proportion may be prescribed in one of the Warrants above mentioned, if it be thought fit.

The Gravers tell us that Puncheons & two pair of Dyes for shillings with the new arms, will be ready to be sent to Edinborough within a fortnight. And while they are coyning shillings in that Mint, there may be Puncheons made for coyning other money. For it may be convenient to send them the first Puncheons from hence that the money of both Mints may be exactly alike.

And if more Puncheons be at any time desired we are ready to furnish them, the same being paid for out of moneys belonging to that Mint

All which is most humbly submitted
to your Lordships great wisdome

Mint Office. May 31.
1707.

Is. NEWTON
JN ELLIS

Memorial of the Officers of
the Mint concerning the
Mint at Edinborough.[6]

NOTES

(1) Newton MSS. III, 190. This is in Newton's hand.

(2) James Ogilvy (1664–1730), fourth Earl of Findlater, and first Earl of Seafield. F.R.S., 1698. Appointed Lord Chancellor of Scotland, November 1702, and again in March 1705; he remained in office, after the Union, until 25 May 1708. Scottish representative peer from 1707; Privy Councillor of England, 1707; Chief Baron of the Court of Exchequer of Scotland, 1707. He was appointed Commissioner for the Union with England in 1706. He succeeded to the title of Earl of Findlater in 1711.

(3) For an account of the Trial of the Pyx, see Memorandum 639, p. 371.

(4) The kingdoms of England and Scotland were united by the Act of Union which came into force on 1 May 1707. Article xvi of the Act stated: 'That from and after the Union the Coin shall be of the same Standard and Value throughout the United Kingdom as now in England, and a Mint shall be continued in Scotland under the same Rules as the Mint in England and the present Officers of the Mint continued subject to such Regulations and Alterations as Her Majesty Her Heirs or Successors or the Parliament of Great Britain shall think fit' (*Statutes of the Realm*, 6° Annæ c. 11. p. 570). The Act guaranteed continuance of the Edinburgh Mint and prescribed that coins should be of English standard throughout the United Kingdom. James VI (James I of England), whilst leaving undisturbed the money value of twelve Scots units to one English, had ordered perfect conformity of the two coinages. Scots silver had since fallen below English in weight and fineness, besides developing new denominations. Edinburgh Mint had now to remint its previous issues as well as Continental silver there circulating, into coins identical with those of the London Mint, except for a Mint mark 'E'. In this, and Letters 728, 729 and 732, Newton insisted on identity of administration and processes.

(5) Remedy; permissible tolerance in weight and fineness. See Letter 736, note (3), p. 512.

(6) Endorsement on the outside of the folded sheet.

723   NEWTON TO THE TREASURY
5 JUNE 1707
From a draft in the Royal Mint[1]

For recoining the moneys of Scotland in her Majties Mint at Edinborough
An Order of her Maty or Council is humbly desired, that ye Warden of her
Majties Mint in the Tower do forthwith make or cause to be made two Piles
of Troy weight of great Britain, in the most exact & perfect manner that by
his endeavours can be done, & that the greater & smaller weights of the said
two Piles be made & framed proportionable thereunto, & that they be
examined & printed with a Rose & Thistle standing upon one common stalk
& crowned wth one common crown & the date stamped upon them, in the
presence of the Officers of her Majties Mint in the Tower, & that one of the
said Piles be then delivered by the said Warden to the Order of the Lord High
Treasurer to be carried to her Majties Mint at Edinborough & there to
remain with her Majties Officers of the said Mint, to the end that other weights
of Troy may be there made conformable unto them, & the other of the said
Piles to remain with the Warden or Wardens for the time being within the
said Tower of London.

NOTE

(1) Newton MSS. III, 64. The date is that of the copy in the Royal Mint Record Books,
VIII, 144.

724   NEWTON TO GODOLPHIN
24 JUNE 1707
From the original in the Public Record Office[1]

To the Rt Honble the Earl of Godolphin
Lord High Treasurer.

May it please your Lordship

According to your Lordships verbal Order to lay before your Lordship
Memorials of what may be requisite for setting on foot the coynage in the
Mint in Scotland[2] with expedition, I humbly represent that upon being
informed that there is but one Clerk in that Mint for rating & standarding,
that for want of more Clerks errors are sometimes committed & the silver not
rightly standarded, that their assays & rating & standarding & way of Book-

keeping differ from our's & must be set right, & that none of their chief Officers have yet aquainted themselves with our practise: I have spoke with Dr Gregory Professor of Astronomy at Oxford (as I acquainted your Lordship) & with one of the Clerks of the Mint in the Tower, about going to Edinburgh to instruct their Officers & Clerk & assist them in their business till Michaelmas next.[3] And if your Lordship approves thereof, I humbly propose that for setling this matter, a suitable recompense for that service may be appointed, the consideration thereof (if your Lordship pleases) being first referred to such persons as your Lordship shall think fit.

And whereas Dyes & Puncheons are to be sent from hence till such time as the work of the Graver of that Mint shall be approved, I further humbly propose that the prizes of these things may be also referred & setled.

Mint Office.
24 *June.* 1707

All wch is most humbly submitted to your Lordships great wisdome

Is. NEWTON

NOTES

(1) T. 1/102, no. 57. This letter is in Newton's hand, as is a draft in the Royal Mint (Newton MSS. III, 181).

(2) Under the Act of Union (1707), a common silver currency for Scotland and England was to be established. See Letter 722, note (4), p. 493.

(3) On 12 July 1707 a 'Queens Warrand' was issued 'appoynting David Gregory Esq. To direct the Officers of the Mint in Scotland In the Methods of the English Mint' (R. W. Cochran-Patrick, *Records of the Coinage of Scotland*, Edinburgh, 1876, vol. II, p. 305).

725 NEWTON TO THE GOLDSMITHS' COMPANY
JUNE 1707
From a draft in the Royal Mint[1]

After my hearty Commendations Whereas the Lords of her Maties most honble Privy Councell have appointed the [28th][2] day of July next at 9 of the Clock in the morning to have a Triall taken of her Maties Pix[3] at the Court Room of the Dutchy of Lancaster within the receipt of her Maties Exchequer at Westm[inste]r. These are to will & require you forthwith upon the receipt hereof to nominate & set down the names of a competent number of sufficient able Freemen of your Company skillfull to judge of & present the Defaults thereof (if any should be found) to be of the Jury to attend their Lordships at the said day & place, & to certify unto me a schedule of their names of whom

495

you shall make choise, that they may have due summons to attend the said service accordingly. Whereof requiring you not to fail I bid you heartily farewell

Your loving friend

June 1707

To my very loving friends the Wardens of the Mystery of Goldsmiths of the City of London.

NOTES

(1) This draft (Newton MSS. I, 235), in Newton's hand, is unsigned.

(2) The day of the month was left blank, but in another draft (Newton MSS. I, 237) it was given as '28th'.

(3) For a description of the Trial of the Pyx, see Memorandum 639, pp. 371–3.

726   MINT TO GODOLPHIN
JULY 1707
From the original in the Royal Mint[1]

To the Rt Honble the Lord high Treasurer

May it please your Lordship

We humbly take leave to represent to your Lordship that the house in the Mint belongn to the Survey[o]r of the Meltings is so decayd that it cannot be repair'd

The Mony Allow'd by Act of Parliamt for repairs and Buildings in the Mint is not Sufficient to keep all the Houses in the Mint in good repair, most of which are very old, and to build new ones

That Your Lordship being Impowr'd by an Act of Parliamt anno 1705 to dispose of 500 *lib* a Year[2] for Salaries and other Service's of the Mint out of wch Your Lordship has been pleas'd to allow 400 *lib* a Year to the Monyers for their Maintenance, We humbly pray Your Lordship will please to grant us an Ordr for the remaining 100 *lib* a Year for two Years to Commence from Christm[a]s last to enable Us to build a new house for the Surveyr of the Meltings.

All which is humbly Submitted to Your Lordships great Wisdome

J STANLEY

Mint Office   IS. NEWTON.
July 1707   JN ELLIS.

496

(1) Newton MSS. II, 497.

(2) Mint expenditure on certain standing services, of which buildings were a minor part, was limited to £3000 a year. See *Statutes of the Realm*, ' 18° and 19° Car. II, Cap. 5: AN ACT for encourageing of Coynage'. After stating 'Gold and Silver to be coined gratis' the Act continues: 'It is hereby further enacted That noe moneyes levyable and payable by this Act shall be applyed or converted to any use or uses whatsoever other then to the defraying the charge [and expence] of the Mint or Mints and of the assaying melting downe Waste and Coynage of Gold and Silver and the encouragement of bringing in Gold and Silver into the said Mint or Mints there to be coyned into the current Coynes of this Kingdome.…

'And it is hereby further enacted That there shall not be issued out of the Exchequer of the said moneyes in any one yeare for the Fees and Salaryes of the Officers of the Mint or Mints and towards the provideing maintaining and repairing of the Houses Offices and Buildings and other necessaries for assaying melting downe and coyning above the summe of Three thousand pounds of Sterlin money, And the overplus of the said moneyes soe kepte or to be kepte as aforesaid shall be imployed for and towards the expence waste and charge of assaying melting downe and coynage and buying in of Gold and Silver to coyne and not otherwise'. Afterwards an Act was passed 'to impower the Lord High Treasurer or Commissioners of the Treasury to issue out of the Monies arising by the Coynage Duty any Sum not exceeding Five hundred Pounds over and above the sum of Three thousand Pounds yearly for the Uses of the Mint' (*Statutes of the Realm*, 4° and 5° Annæ c. 9.10). This was in order to raise the salaries of the clerks and to give the Moneyers £400 a year because their piece-work earnings had diminished.

The Mint was always in difficulties over buildings. The clerks' £100 was so regularly diverted to building repairs that they did not get their increase until 1722.

727 DAVID GREGORY TO NEWTON
12 AUGUST 1707
From the original in the Royal Mint[1]

Edenborough. 12 August. 1707.

Sir

I arrived here the first instant, & Mr Morgan with me; and the Monyers on the 8th. Since then wee have been as bussy as possible in putting all things in order toward the recoinage, which the Government here urges extremly. Notwithstanding all the things they sent to London for from this; upon a more strict enquyry by the Monyers, the things in the enclosed list[2] are still wanting. I have sent a Copy of it to Sir David Nairn,[3] and it will come regularly to you: but I thought it my duty to send it also to you immediatly. I must beg all dispatch, after it comes before you.

I find the Government here urges the Coyning Crowns & Half-Crowns first; both because greater dispatch can be made, & because more than one half of the moneys of the Equivalent[4] that were sent down, is Sixpences. Therefore, Sir, the Puncheons for the Crowns & Half-crowns, and as many Dyes, made from them as is possible, will be much wanted. Great difficultys arise dayly about the way of receiving in the Coin here, and melting it down into Ingotts which seemes necessary before the Mint doe medle with it. Though this be not my bussines, yet you will much oblidge me to tell Me how this was managed in the great recoinage of England, and Where the Mint began first to be concerned in this matter.

If you send letters directed for me either to the Scotch Secretarys Office[5] in the Privy Garden, or to the common Post; they will come right to hand.

> I am in all duty
>
> Sir
>
> Your most humble and most
>
> oblidged servant
>
> D GREGORY.

The enclosed is from
the Master of the
Mint here.

NOTES

(1) Newton MSS. III, 110.

(2) *Ibid.* III, 111.

(3) David Nairn (d. 1734), Secretary and Accountant to the Scottish Commissioners for the Treaty of Union. He was appointed Secretary Depute (Under-Secretary) for Scotland, 1703; knighted, 1704; Under-Keeper of the Signet, 1708.

(4) The *Equivalent* was a lump sum payable to Scotland at the Union nominally as compensation for its becoming liable for a share in the English National Debt, though actually to ensure a smooth passage of the Act. The Edinburgh recoinage was one charge upon it.

(5) This was in London.

728 DAVID GREGORY TO NEWTON
9 OCTOBER 1707
From the original in the Royal Mint[1]

Edenborough. 9 Octr. 1707.

Sir

I received yours of the 4th instant,[2] and have given directions accordingly. That same day I wrote a long letter to you, to which I wait an answer.

498

I told you I would give you an accurate account of the next melting, which was on tuesday last; & is this. I examined the Pot paper, and attended the whole melting and Essay[3] my self. The Pot contained about 470 Lib wt Troy. I have sent you enclosed pieces of the first Mold or Flask, the Midle Mold, & the last Mold, as you see them marked. Two Flasks are skrew'd together here, as with you. The Pot was just 35 minutes in being laded out, and I am satisfyed it cannot be done in shorter time here. You see that though the first Mold be Standard by the Assay masters report, yet the last is near 2 dwt better. I must again beg your directions what to doe in this case; for this is a great uncertainty.

I have also, as you directed me, sent you a piece of silver which our Assay Master finds to be precisely standard. This is that which is sealed. I must pray you to cause all those four pieces to be assayed in the Tower and give me an account of the tryal.

I saw our Assay master make assay of the standard tryal piece, at the same time that he made the assay of the pot & piece I speak of.

After a strict search into all the boxes that are come, We can find no Puncheons for small Armes & letters for shillings & sixpences, which therefore have not, in my opinion been sent; And there is only one sieve come to hand. All the other things are right. I am

Sir

Your most humble and
most oblidged servant

D GREGORY

For
Sir Isaac Newton
at his house in
Jermin Street near St James's Church
    Westminster

NOTES

(1) Newton MSS. i, 190.

(2) This has not been found.

(3) That is, assay.

729   NEWTON TO GODOLPHIN
14 NOVEMBER 1707
From a draft in the Royal Mint[1]

> To the Rt Honble the Earl of Godolphin
> Ld H. Treasurer of great Britain.

May it please your Lordship

In obedience to your Lordships Order of Reference of ye 12t Instant upon the Memorial hereunto annexed, we have considered the said Memorial & are humbly of opinion that it is reasonable & agreable to the Indenture of her Majties Mints & at present necessary for the Master of her Majties[2] Mint at Edinburgh to allay the silver molten with Scotch coal[2] in such manner & proportion as by experience is found most effectual & exact for making the moneys standard without error. For an error in fineness otherwise then by accident makes the moneys undeliverable. And[3] since by experiments made in that Mint the silver in melting with Scotch coal & lading out refines about three half penny weight in the pound weight Troy as is alledged in the sd memorial & the moneys coyned by the ancient method of that mint have passed the trialls of the Pix & in the last trial proved standard full & there is no time at present for making experiments to bring this matter to an exacter regulation, We are humbly of opinion that the Officers of the said Mint may[4] be still allowed to use their ancient method of allaying the molten silver to make it standard untill the present recoinage of the moneys in Scotland shall be finished. Ye Indenture of ye Mint prescribing no particular method for melting provided it be standard[4]

> All wch is humbly submitted to
> your Lordships great wisdom.

Mint Office 14 *Novemb.*
1707.

NOTES

(1) Newton MSS. 1, 187. This draft is in Newton's hand, except as indicated in note (4) below.

(2) The metallurgical practice in Edinburgh differed from that in England inasmuch as the melting furnace was fired with coal, and not with charcoal, which was unobtainable in Scotland. The coal gave a fiercer heat, and it took twice as long as in England to ladle the contents of the crucibles into the moulds. Meanwhile the crucible stood on the furnace, slowly losing more copper than silver, owing to oxidation, in the enduring heat. The Scots had been in the habit of adding some grains of copper when the crucible was half empty.

Newton appears to have been unsure of the results of the Scots practice, but he reluctantly accepted it, at the same time insisting that the native or foreign silver coins circulating in Scotland should be converted in Edinburgh into English currency. To ensure uniformity, Newton proposed that the Edinburgh Mint must either send men to be trained at the London Mint, or they must accept overseers from London. This led to the appointment of David Gregory as general supervisor of the new coinage. Gregory sent reports on 12 August (Letter 727) and 9 October (Letter 728), and after he returned to London on 21 November (Letter 733).

See Craig, *Newton*, chapter VII, and especially pp. 71–2.

(3) Several lines, following 'said Memorial &' down to this point, are deleted.

(4) The word 'may' and the last sentence have been inserted by another hand.

730   TILY[1] TO NEWTON
24 NOVEMBER 1707
From the original in the University Library, Cambridge[2]

Sr

I take the Liberty once more to begg the honour of seeing you wn you come this way. my broaken Constitution deprives me of ye advantage of attending you. & I promise my selfe, your Compassion will pardon the frequent troubles I have given you. who am most Sincerely

Sr

Your obliged &

faithfull humble

Servt

JOS: TILY

Whitehall
24th. No. 1707

NOTES

(1) Sir Joseph Tily (Tyly) was appointed Director of John Briscoe's Land Bank on 9 October 1695, along with Thomas Neale. See Horsefield, *B.M.E.*, pp. 180 and 271, and Letter 629, note (3), p. 350.

(2) Add. 3965.10, fo. 113.

731 NEWTON TO DAVID GREGORY
1707
From a draft in the Royal Mint[1]

The Tools you wrote for are preparing with all dispatch but will scarce be ready before the end of next week. Since you wrote about the manner of calling in & melting down the old hammered money in England & delivering it into the Mint to be coyned & Mr Haynes[2] as I understand has written to you largely about it: because that method was troublesome expensive & liable to abuses & great abuses were actually committed in it, I will take the liberty to set the method wch seems to me the best & leave it to your judgment. You may compare it wth other proposals & chuse the best. I propose therefore

1   That every mans money brought in to be recoyned be not only told but also weighed & the tale weight & price paid for it in English money be entered in a Book in this manner.

		Tale		Weight			Price		
		Liv.	sous	Lib.	oz	dwt	lb	s	d
Imported by									
John Anderson	Sept. 27	127.	07	3.	4.	7	9.	12.	6
Thomas Pitcairn	Sept. 28	245.	02	6.	5.	13	18.	4.	4

2   That the money thus weighed be put into baggs, & where the parcels are small two or more parcels may be put into one bagg & the weight of the money in every bagg be enterd on a label of paper tied to the bagg & the bagg tyed up & locked up in a Treasury under two or more keys till they shall be delivered to the melter. The number of the baggs may be also written on ye labels

3   A Melting house to be provided with two fire holes for two iron melting potts, & a Smiths pair of Bellows to blow the fires in the two holes by meanes of a leaden pipe coming from the snout of the Bellows to ye two holes. The holes must have iron covers to keep in the fire for heating the potts speedily. [& the Potts may be big enough to contain 40, 50, 60 or 75 pound weight of silver in each Pot. [Which being melted must be poured out into an Ingot mould, & the Ingots must be marked in continual order with the numbers I. II. III. IIII. V. VI. VII. &c.][3] A melter may melt 10 12 or 15 potts a day....

4.   That the old silver money be delivered [out of the Treasury] to ye Melter by exact weight of 15, 20 or 25 pounds Troy in a parcel[,] the Melter giving receipts for the money & yt the parcels be put into ye numbered baggs.... Treasury room adjoyning to the melting house.... is at work & see that the money baggs be not opened....

the melting potts with the money in them about three baggs into each pot, & that no....
into the Melting house in their absence & kind be brought into the Melting house besides the silver in....
& all utensils of iron.
And that the silver being run & cast into Ingots the Ingots be marked in continual numbers I. II. III. IIII. V. VI. VII. VIII. VIIII. X, XI &c and locked up in the treasury of the Melting house under the said three keys till delivery

5. That the Officer appointed by the Lds Commrs of the Treary to deliver the old money by weight to ye Melter, do receive the Ingots back from the Melter by weight and send them to the Mint in the order of their numbers

6 That ye Melter deliver back the same weight in Ingots wch he receives in old moneys abating about 18 grains in the pound weight Troy for wast & to have the sweep towards making up his account, & to bring in his supply in the standard silver & be paid for coales & potts & servants wages & his own pains after a certain rate per pound weight Troy....

7 That the Ingots be delivered into the Mint to be weighed assayed & standarded coyned & delivered back by weight in new moneys according to the course of the Mint & the new moneys at the delivery told into 100 lb baggs...entered in books

NOTES

(1) Newton MSS. III, 160. This may well be a draft of Newton's answer to Letter 727. It must have been written prior to 21 November 1707, when Gregory left Edinburgh.

The paper on which this draft was written is very discoloured; those parts of the text which are rendered indecipherable are indicated by marks of omisson.

(2) See Letter 641, note (2), p. 376.

(3) The contents of the square brackets (which are Newton's, here as elsewhere in this manuscript) have been crossed out.

732   DAVID GREGORY TO GODOLPHIN
13 DECEMBER 1707
From the original in the Public Record Office[1]

An Account
of the new Regulation of her Majesty's Mint at Edinburgh Humbly laid before the Right Honble the Earl of Godolphin
Lord high Treasurer of Great Brittain.
In Obedience to her Majestys Commands in a Warrant dated, 12th July 1707, Doctor Gregory took Journey to Scotland, 21, July, and arrived at Edinburgh

the 31st, And upon Shewing her Matys Warrant to the General,[2] and other Officers of the Mint there, they prepared for the Recoynage in the Methods of the Mint in the Tower.

And when the Moneyers arriv'd at Edinburgh, The Doctor and they Surveyed all the Offices, Tools &c. belonging to the Mint, and what were wanting were soon Supplied from London.

After the Officers of the Mint & Doctor Gregory had compared the former Constitution of the Mint of Scotland with that of the Mint in the Tower, and with her Majestys Instructions concerning the Recoynage; They added three new Clerks (during the Recoynage) to the Queens Clerk who was there before; to witt One for the Master, One for the Warden, & one for the Counter Warden. All these four Clerks were well instructed in the Methods of Rateing and Standarding, and the Formes of Book-keeping used in the Mint of the Tower, by the Clerk sent by her Majesty for this purpose.

The Officers of the Mint & the Doctor considering that by bringing the Mint of Scotland to the same methods with that of the Tower,[3] more Officers became necessary, and being unwilling to increase Charges, aggreed that the Warden & Counter Warden should by turns do the Office of Surveyor of the Meltings; and that the Counter Warden should officiat also as Weigher & Teller, and that his Clerk should be Clerk to those Offices. It was also agreed That the Queens Clerk should be Clerk of the Papers and Irons.

The great difficulty was in the Melting, it being made there with Pit-coal.[4] In this Dr Gregory made several experiments according to directions which he received from Sr Isaac Newton from time to time. And at last such Rules of Allaying[5] were found out & aggreed on, as by experience were found to make the Silver of Standard fineness. And the Essay Master[6] having no Clerk, was allowed an assistant during this coynage.

Matters being thus adjusted, Dr Gregory continued in Edinburgh untill he saw the Methods of the Mint in the Tower well understood, and exactly practised by all concerned, and the Recoynage advanced so, that they coyned Six Thousand Pounds a week; and then at his request, represented by Sr Isaac Newton to your Lordship, And that there was no further Occasion for his staying there, Your Lordship being pleased to Dismiss him, as he was inform'd by Sr Isaac Newton in his of 15th November, he parted from Edinburgh on the 21st of November, after having taken an exact Acco[mp]t of the State & Condition of the Mint at that time, and left directions for encreasing the Coynage.

London : 13 December.   DAVID GREGORY.
          1707.

NOTES

(1) This report (T. 1/103, no. 94) was sent by Gregory shortly after he had left the Edinburgh Mint (21 November 1707). It is not in his hand, although it is signed by him.

(2) That is the Earl of Lauderdale (see Letter 745, note (4), p. 523). The chief officer of the Edinburgh Mint was styled General. The other officers were:

Master, George Allardes,
Deputy Master, Patrick Scott,
Warden, William Drummond,
Collector of the Bullion, Daniel Stewart,
Queen's Clerk, Robert Millar.

(3) For Newton's efforts to secure uniformity of processes and administration at the Edinburgh Mint, see Letters 720, 722, and 729.

(4) See Letter 729, note (2), p. 500.

(5) That is, for maintaining the correct proportion of copper to silver. See Letter 729, note (2), p. 500.

(6) That is, Assay Master.

733   DAVID GREGORY TO NEWTON
28 DECEMBER 1707
From the original in the Library of the Royal Society of London[1]

London. 28 December. 1707.

By looking over the Regiam Majestatem & the old Statutes & Acts of Parliament of Scotland, These Things appear of a nearer Aggreement betwixt Scotland & England than of latter years has been beleeved or truly been.

When the Shyres were empowred to send Commissioners to Parliament, They are at the same time impowred to chuse a Speaker for themselves; & the King declaires that he will by his special Write & Summons call the Peers: In a word that whole Act in K. Ja: Is time shews a design to have two houses as they had in England.

The Silver Coin of Scotland is ordained & Statute to be of the same Standart fineness with that of England.

The Officers of the Mint of Scotland were the same with the Officers of the Mint of England to wit the Custos, the first Officer, & the Monetarius the next.

The Moneys of Scotland were the same denomination & value (and in all probability the Weights too) with those of England. For the Libra (Monete Argentee) was Statute to be divided into 29 Shillings.

The Calendar & beginning of the Year were the same in Scotland until the year 1514 (or thereabout) to wit the 25 day of March; and the number of years since the Incarnation the same; to wit one less than the common Æra.

NOTE

(1) R.S. Greg. MS. fo. 25. David Gregory went to Edinburgh Mint as Newton's agent on 21 July 1707 (see Letter 732).

734   HENRY NEWTON[1] TO NEWTON
DECEMBER 1707
From the original in King's College Library, Cambridge[2]

Florence: Dec: 1707. N.S:

Sr,

I am much obleigd to ye learned Signr Lancisio,[3] ye present Popes Physitian, at Rome, for the having given mee, on ye occasion of presenting his Discourse de Mortibus Subitaneis to ye Royall Society, an opportunity at ye same time of paying my respects to that illustrious Body, & their President; whose Merritts are equally acknowledgd, & their Fame spread, for reall & usefull Knowledge, where-ever Learning prevailes; or Common Life is capable of receiving any Advantage, or Ornament. And Rome is but gratefull for the fav[ou]rs shee has received on this head from Brittain, when upon my sending thither some time since, ye Books of Opticks, one of their Learned men in his answer thence, was pleasd thus to expresse his own thoughts thereon, & those of ye vertuosi there:

x ad[4] cætera modò tuæ gratissimæ Epistolæ descendens, occurrit primo ingentes tibi reddere gratias ob missionem pretiosi Operis Optices Isaaci Neutoni, quod Angliâ usque advehi fecisti, ut me, amicosque omnes beares. Hucusque manibus mei Galliani[5] teritur, qui eas optimè callet disciplinas, transiturus exinde ad Fontaninum,[5] Vignolium,[5] Gravinam,[5] Garofalum[5]. cæterosque hujus Urbis exquisitioris Literaturæ Magnates, et quanta mea gloria, cui primo, munere tuo, contigit thesaurum novum per Urbem publicare, foetum heroici ingenii Equitis Aurati Neutoni, tam celebris per Europam. Nec error displicet Bibliopolæ, qui hoc opus, vice *Principiorum Philosophiæ Mathematicorum* transmisit, ut enim mihi refert dictus Gallianus, theses tam novas, tam peregrinas propositiones considerationesque de Luce, & coloribus[6] in eo reperit, ut ne quidem Ipse Matheseos facilè princeps, novæque Philosophiæ Creator attigerit Renatus de Cartes.[7] Gratias iterum Tibi Domino meo humanissimo, ingentes indefinitas, æternas. x x Romæ xv Kal. Jan: 1706.

Nor is it worth ye while to mention what lead the way to it in mine. xx Jam[8] quoque credo acceperis Isaaci Newtoni Equitis, quocum nihil mihi commune, præter Nomen, et quod cum aliis omnibus habeo commune, famæ ipsius cultum ac Reverentiam; vice *Principiorum Philosophiæ Mathematicorum*, ex errore quidem Bibliopolæ, de Optice Libros; maximi ingenii foetum novissimum, atque nunc primum in Italia visum. Ad calcem etiam operis Leges atque sua Philosophiæ Principia[9]: Dii Boni, quam pauca, quam simplicia, quam cum ipsâ Naturâ Rerum Convenientia! Genuæ pr. Non. Dec: 1706.[10] And I should bee glad of more occasions to receive such returns for presents, so usefull & so acceptable to ye Publick; as any thing of the same hand would ever bee from,

Sr,

Your most humble faithfull Servt

HEN: NEWTON

Father Grandi of Pisa, who is not one of ye least readers or admirers of ye same worke, presents likewise his humblest respects to you

Sr Isaack Newton

NOTES

(1) Henry Newton (1651–1715), M.A., St Mary's Hall, Oxford, D.C.L., Merton College, Oxford 1678; Judge-Advocate to the Admiralty, 1694; Envoy-extraordinary to Florence from 1704 to 1709; Judge of the High Court of the Admiralty, 1714; Knighted, 1715. He was Chancellor of the Diocese of London from 1685 to his death. He does not seem to have been related to Newton, for his name does not appear in the pedigree of Newton published in Turnor's *Grantham* (pp. 168–9). Moreover the writer of this letter says that the only thing that he and Isaac Newton had in common was the name. He published some Latin letters, verses and speeches in 1710. See *D.N.B.* and Foster's *Alumni Oxonienses*.

(2) Keynes MS. 99 F.

(3) Giovanni Maria Lancisi (1654–1720), Professor of Anatomy at Rome. He was Physician in Ordinary to Popes Clement XI and Innocent XII. F.R.S., 1706.

(4) 'Coming now to the remaining points in your most welcome letter, it occurs to me first of all to pay you my great debt of gratitude for your sending me that precious work, the *Opticks* of Isaac Newton, a work which you have had brought all the way from England in order to enrich me and all my friends. So far it is being handled only by my friend Gallianus[5] who has excellent skill in these forms of learning; but from him it will pass to Fontaninus,[5] Vignolius,[5] Gravina,[5] Garofalus,[5] and to all the other experts in the higher literature who belong to this city, involving unspeakable glory for myself, who (thanks to your gift) have been privileged to be the first to publish in our city this new treasure, the offspring of the super-human intellect of the Golden Knight Newton, whose fame extends throughout Europe. Nor

507

have I any regret at the Bookseller's error, who despatched this work instead of the *Principia*; for, as the above-named Gallianus reports to me, he finds in it such novel theses, and such almost revolutionary propositions and considerations about Light and Colours,[6] as not even René Descartes[7] himself, the outstanding leader in mathematics, and originator of the new philosophy, attained to. Again, I express to you, my courteous friend, my thanks, which are great, boundless, and everlasting. Rome, 18 December 1705.'

This is an extract of a letter of 'one of their learned men' to the unknown Henry Newton in which the gift of the *Opticks* is acknowledged in such laudatory terms that he feels that he should send the extract to his distinguished namesake.

(5) It is difficult to identify the persons named here. Fontaninus may be Giusto Fontanini, a Professor of Eloquence at Rome at this time; Garofalus may be Biagio Garofalo (1677–1762), a priest in the service of Pope Clement XI. It has not been possible to identify the other persons mentioned.

(6) In which the composite nature of white light is established.

(7) Descartes's observations on the nature of light are contained in his *La Dioptrique* which was published as an Appendix to his *Discours de la Méthode* in 1637. The 'nova philosophia' is a reference to the proposition *Cogito ergo sum*, which Descartes expounded in the *Discours* and in his *Principia Philosophiæ* (1644).

(8) 'By this time, too, I trust, you will have received Sir Isaac Newton's (no, I have nothing in common with him apart from the name, and apart from something that I have in common with everyone else, namely, a respect and reverence for his fame), books on Optics sent to you by the bookseller's error instead of the *Principia*: This treatise on Optics is the latest product of that great intellect and is now for the first time seen in Italy. There follow at the end of the book the laws and his own principles of philosophy. Good Heavens! how few in number, how uncomplicated, how utterly in keeping with Nature are these principles! Genoa, 4 December 1706.'

(9) The 'laws and his own principles' are to be found in the penultimate and final Queries of the *Opticks*.

(10) 'Pridie Nonas Decembres', i.e. the day before the Nones of December, or 4 December.

735   MEMORANDUM BY NEWTON
1707
From the original in the Royal Mint[1]

King James the first, to signify that he would unite the two kingdoms of England & Scotland, stiled himself Magnæ Britanniæ Rex, & on the Reverse of his broad pieces[2] & XX*s* pieces of gold, put this motto: Faciam eos in gentem unam, I will make them one nation Ezek. 37.22. In reference to this inscription & thereby to signify that her Majty hath finished a great & difficult work, an undertaking of an hundred years standing, I propose the following Medal.

On the first side her Maties effigies with the inscription ANNA.D.G. MAGNÆ. BRITANNIÆ. F. ET. H. REGINA.[3] On the second Her Maty in royal apparel in the posture of Britannia sitting on a globe with a speare in her right hand & a shield standing by her, to represent both her self & her mystical body Britannia. The sheild to be charged with the new Arms of great Britain. In her left hand a Rose & Thistle upon one stalk. The Rose towards her right hand. In the prospect below, two rivers (Tamesis & Boderia) unite into one common stream. Over her head two hands, to signify that this union is the work of heaven, come out of the clouds holding a single Crown to crown her. The motto is, FECI. EOS. IN. GENTEM. UNAM.[4] And in the Exergue,[5] I. MAII. MDCCVII.

In this designe the union is represented by the single crown in two hands, by the Rose & Thistle upon one stalk, by the new Arms of great Britain upon the sheild, & by the two Rivers Thames & Forth uniting; for rivers were anciently the emblemes of Kingdomes. By the Motto the Union is referred to the Queen as the minister of heaven in this work. And altho this Motto may at first seem flat, yet being compared with that on the gold coynes of K. James I, & with the Prophesy of Ezekiel to wch it alludes, it will appear very significant, comprehensive, grave, lively, pious & majestick, & perhaps the most apposite of any that can be thought of. A poetical Motto is not so grave for such an occasion.

Two women hand in hand imply only a federal union, or only such an union as is represented by the Motto on the money of K. Charles I, Florent concordia regna.[6] England & Scotland, after union should be remembered no more, & therefore in the Medal they should be only glanced at & not made too conspicuous. However for variety, I have caused two draughts on the next pages; but prefer that above. The draughts were made in hast, & when the designe in general is resolved upon, the Graver will be more exact.

I.N.

NOTES

(1) Newton MSS. iii, 303 (in Newton's hand). It is undated, but relates to the Union with Scotland Medal, 1707.

(2) See Letter 593, note (11), p. 283.

(3) Anne, by the Grace of God, Queen of Great Britain, France and Ireland.

(4) 'I have made them one nation.' The two nations were united on 1 May 1707.

(5) The Exergue was the small space in the lower part of the coin (or medal) in which was inscribed the date. It is usually separated from the main device by a line.

(6) 'It is the harmonious nations that flourish.'

736   PATRICK SCOTT[1] TO NEWTON
31 JANUARY 1707/8
From the original in the Royal Mint[2]

Sir,

I have the honour of yours of the 22d wherin you are pleased to give me a Solution to a Scruple I proposed to Allardes my Principall with relation to the Remedies,[3] And by which (if I rightly understand it) It would Seem that ther is but 2 pennies of Remedy allowed on both weight and fineness when the Remedie falls to be on both Sides: I mean One penny to ye weight and another to the fineness, So that at Journal[4] one penny B: or W:[5] fine and $\frac{1}{2}$ B: or W: weight in ye same degree Ought not to pass but be remelted. Being a half penny above ye two. And Therfor if the pot Assay a penny B: or W: It ought to be remelted; And also if the money at delivery be a penny B: or W: weight, it ought not to pass but be remelted on ye Moneyers charges.

This I must acknowledge did not appear so clear to my weak Capacity by the Indenture: for I did think that if the melter gave out his bars within 2 pennys B: or W: of ye fineness, He might plead upon ye Remedies that they could not be refixed And the Moneyers did affirm that their delivery is good if not above ye 2 pennies B: or W: in ye weight and that they wer in use so to do. So that we wer in great doubt here how to determine the mater as to the remelting, If any of ye Journal upon Assay should hapen'd Not to be past by reason of ye Moneys being beyond ye Remedies in ye Case above. But yet I think where the Remedies fall to be on One Side that is 2 penny B: or W: in fineness, and the weight standart: (et per contra). The money might be past by the Indenture: Tho at the same I do acknowledge the Remedies ought not to be wrought upon.

I have communicat your Letter to the Wardens and given a Copy to the Moneyers: The difference has never Yet hapen'd, And I hope now all concerned shall take more care it do not.

Our Pots use Sometimes to be remelted for coming up too fine, And that being accidental I understand should be on the Queens charges, But I do not know how to state it.

I proposed another Question to Allardes, If I was obliged to take from ye Moneyers amongst their Scisell[6] Any Brokeage[7] or spoilt peeces? By Brokeage I mean ye peeces spoilt at ye Press In Setting ye Dyes, Or where the Impression is faulty and not good. And by Spoilt peeces I mean those that hapen to be too light drawn at ye miln, And may be cut out, but go no further than ye Cutter, or it may be spoilt in ye Justing or Seizing, And So thrown in amongst ye Scisell. I have hitherto received of both, but a great quantity of ye Latter And as I desire to do fair things and Justice to all: So I would have ye Like done to me, And would not willingly be imposed upon, Especially wher I bear Charge and trust for and represent another, which with an Insinuation I had of my not being obliged to it, made me propose the doubt. And wherin I must beg the favour to be also Solved, Our method here formerly was of ye Light Crowns to make fourty penies, Of the fourty penies twenties & of ye twentys tens &c

Sir I had presumed to trouble you with my Severall difficulties befor this had not my Principal been on the place; Your true friendship to him and Generous Character makes me the more bold to use this liberty and to expect You will forgive me.

Since the Moneyers are obliged to do the money by the Miln and Press It occurs to me also as doubtfull whether they pay the Edging.

Now as to the State of our Coinage That you may know a little of it; By our acts of Parliament (which we reckon here to be yet in force) The Master is allowed 20 pound Scots money Upon the Coinage of ye Scots stone weight of Silver. Twenty pound Scots money is £1: 13*sh*. 4*d* English money.[8] The Scots Stone weight Consists of 16 Scots pound weight, And the pound weight of 16 Scots ounces, which you know: So that the allowance on ye English pd weight or 12 ounces troy by our acts Should be 19 pence $\frac{1}{12}$ and a fraction, And in ye 100 pd weight 7*lib*: 19*sh*: 1$\frac{3}{4}$ and a fraction.

By our acts of Parliament Also all tools and reparations &[c] are to be allowed. And the allowance for the Stone Weight Coinage is ordered to be advanced by the Thesaury As the Bullion is brought in, Other wise the Master [is] at Liberty to stop ye Coinage.

We are in course to receive weekly from ye Bank 9000 pd Sterline Money Weighting about 4600 Eng. pd Troy: So that according to ye acts of Parliament The Master ought to have Imprest to him weekly a Sume proportionall to the Allowance on ye stone or pd weight for its coinage.

We have already Received of foraign money 113000£ sterl. weighing Standart about 31312 pd Troy: And we have Delivered 41600 weight

Ther is yet 19000£ sterl. of foraign Coin to come in from ye Bank wherof is 16000 in Dollars which will be about 12 Worse on the Assay overhead And Consequently Occasion a great refinage, And for which I see no Remedy Since we cannot wait the Inbringing of the other Money: And when the milnd money is brought in therafter (Tho reckoned Standart) Yet by the remeltings may hapen to come up Better And So put me in a difficulty after of providing Course money: All which are Considerations in my humble Opinion for a good stock to be imprest to ye Master for carieing on ye work and making up his great waist

And all he has gott yet imprest to him is 1800£ sterl. Of which I cannot reckon under 500£ for Tools reparations and other Necessaries provided here.

I have already refined Gross weight 1713 pd being 1814 stand And have 150 more not yet assayed.

I Suppose Likewayes the Master amongst his other charges is to be allowed the penny he payes the Moneyers over the Eight pence per pd Weight. 8 pence being their due by ye Indenture, And the other penny payed at discretion of ye officers But to be allowed by the Queen to the Master in his accompts.

All this I humbly Submit to you, And beging again pardon for ye trouble And that you'll also please excuse wherin I may not be so clear or distinct I am

Sir

Your most obedient humble Servant

Edinbr 31 *Janry*   PAT: SCOTT
    1708

NOTES

(1) Patrick Scott was deputy to George Allardes, the Master of the Edinburgh Mint.

(2) Newton MSS. III, 170.

(3) The *remedy* was the allowance made at the trial of the Pyx for the fallibility of workmanship, because according to the Mint Indenture, 'the said monies may not continually be made in all things according to the right standard, but peradventure in default of the said Master and Worker, it shall be found sometimes too strong or too feeble, or too little in weight, in fineness or in both'.

(4) See Letter 555, note (6), p. 214.

(5) B: or W:, i.e. better or worse (than the standard).

(6) See Letter 555, note (3), p. 214.

(7) Brokeage (brockage): a coin imperfectly struck, applied especially to a coin struck on one side only. This was usually caused by a coin jamming in the die, thus interfering with the impression of succeeding coins (C. C. Chamberlain, *Guide to Numismatics*, 1960, p. 17).

(8) See Letter 722, note (4), p. 493.

737 AN AGREEMENT BETWEEN NEWTON AND FLAMSTEED RELATIVE TO THE PRINTING OF THE *HISTORIA CŒLESTIS BRITANNICA*

20 MARCH 1707/8

From a copy in the Royal Greenwich Observatory[1]

London March 20. 1707/8

It is agreed between Sir Isaac Newton and Mr John Flamsteed[2]

1st that the 2d Volume of the Astronomical Observations with the figures of the first Volume shall be presently delivered into Sir Isaac Newtons Hands

2d. That the Catalogue of the fixt Starrs here present shall likewise be delivered into Sir Isaac Newtons Hands

3d. That the Catalogue of the fixed Stars now in Sr Isaac Newtons hands shall be delivered to Mr Flamsteed in order to have the Magnitudes Inserted & to be returned with the Magnitudes after Sixteen days.

4. That upon the Redelivery of that Catalogue, Sr I.N. shall pay to Mr Flamsteed one Hundred & twenty five pounds on the Princes Account[3]

5. That upon the delivery of the Catalogue of the fixt Stars as far as it can be Compleated at this Time Mr Flamsteed shall have ye rest of ye Money Stipulated betwixt him and the Referrees he Undertaking to Correct the Press and appointing Correctors who live in Town that the Work may not be Retarded Memorandum that at the Same time ye 2d Volume of Observations (with the figures mentioned here) was delivered into Sr Issac Newtons Hands together with a Corrected Copy of ye Eclipticall Constellations & all the Southern of ye Catalogue but that I Covenanted yt ye sd 2d Volume should be Returned to Me to be again Revised & delivered to ye Press as the Printers should work it off, and ye Copy of the Eclipticall Constellations returned me as soon as I should return the Copy now in Sr Isa Newtons Hands with the Magnitudes Inserted

JOHN FLAMSTEED MR

[Added in Flamsteed's hand]

There were present at ye Meeting at ye Castle Tavern in Pater Noster Row March 20 1707/8 Mr Roberts Sr Is. Newton Dr Arbuthnot Dr Gregory Mr Churchill Mr Ja. Hodgson myselfe & Isaack Wollferman
Rest due to me as appeares at ye foot of the account pag 62 foregoing

48. 13. 00

NOTES

(1) Vol. 33, fo. 68. See also Letters 700 and 702.

(2) After protracted delays the first volume had been printed off in December 1707, and the sheets were left with Churchill until it was decided whether the catalogue of the fixed stars should be included in it. Flamsteed insisted that it should be printed at the end of the second volume. Apparently he had his way, for the Referees did not include it in the first volume. See More, p. 543.

(3) See Letters 738 and 739 and Hodgson's receipt at the foot of the former.

738   REFEREES TO NEWTON
26 MARCH 1708
From the original in the University Library, Cambridge[1]

26 *March.* 1708

Sr

You having in your hands the summ of one hundred & twenty & five pounds received by you from the Treasurer of his Royal Highness the Prince towards bearing the charges of printing Mr John Flamsteed's Astronomical papers. We do hereby authorize you to pay unto the said Mr John Flamsteed or his Order the said summ of one hundred & twenty & five pounds so soon as he shall deliver or send back to you his first Catalogue of ye fixt starrs wth the columns of their names & places & order according to Ptolomy & Tycho, filled up after the manner of his last catalogue

To Sr Isaac Newton Kt

F ROBARTS
CH. WREN
JO ARBUTHNOTT
D GREGORY

[Below this is the receipt signed by Hodgson][2]

Received this 12 day of April 1708 of Sr Isaac Newton Kt by the Order & for the use of Mr John Flamsteed, the above mentioned summ of one hundred & twenty and five pounds by me
         JA. HODGSON

lib.	*s.*	*d*
125	00	00

NOTES

(1) Add. 4006, fo. 25.

(2) See Letter 739.

739   FLAMSTEED TO NEWTON
10 APRIL 1708
From the original in the University Library, Cambridge[1]

April 10*th.* 1708

Sr

Pray please to pay the One hundred & five & twenty pounds[2] allowed by his Royall Highness Prince George of Denmark in part towards the payment of Calculators & other disbursmts in order to prepare the Historia Britannica Cœlestis for the press, to Mr James Hodgson & his receipt shall be ye discharge for so much paid to Sr

Your most humble Servant

JOHN FLAMSTEED M R

To Sr Isaack Newton

NOTES

(1) Add. 4006, fos. 16–17.

(2) The Referees had authorized the payment of this sum on 26 March 1708. In accordance with Flamsteed's request it was paid over to James Hodgson a fortnight later and the receipt was acknowledged by him on 12 April.

740   NEWTON TO GODOLPHIN
14 APRIL 1708
From a draft in the Royal Mint[1]

To the Rt Honble the Earl of Godolphin
Lord High Treasurer of great Britain.

May it please your Lordship

We humbly beg leave to lay before your Lordship that Mr Daniel Stuart the Collector of the Bullion for her Majties Mint at Edinburgh is newly dead (as we hear by the last Post) & that in our humble opinion the place of Collector of the Bullion being irregular should cease & the said Bullion be henceforth paid by the Under Collectors into the hands of the Cashkeeper of North Britain & kept apart in the Exchequer in a proper Chest under the key of the said Cashkeeper & also, if it be thought fit, under the key of the General of her Majties said Mint (as is directed by the Scotch Act of Parliament whereby this Duty is granted to the Mint,) & that it be issued out thence from time to time by Warrants to the General & Master of the said Mint & kept apart in the Treasury of the said Mint under the keys of the General the Master & the Wardens for defraying the charges of coynage & repairs & paying of Salaries, & be accounted for annually by the Master, so that the two Mints in respect of their cash may be under the same Rules in conformity to the Act of Union & the Indenture of her Maties Mints.

And[2] we are further humbly of opinion that the Executor or Executors of the said Mr Stuart[3] be directed forthwith to pay into the hands of the General & Master of the said Mint such a summ of money as your Lordship shall think fit, suppose the summ of 2500 *lib* or 5000 *lib*, to be kept in the Treasury of that Mint under the keys of the General the Master & Wardens, that the service of that Mint receive no stop for want of moneys, there being as we are very credibly informed, a far greater summ in the hands of the Executor of the said Mr Stuart.

All which is most humbly submitted

to your Lordships great Wisdome.

Is. NEWTON

Mint Office
14*th April*
1708.

NOTES

(1) Newton MSS. III, 46.

(2) A line has been drawn diagonally through the final paragraph.

(3) Stuart's executors would not pay over the £3000 which had evidently been deposited in his own name. Consequently, the salaries of the officers of the Edinburgh Mint could not be paid. The money was not recovered until 1710 (Newton MSS. III, 166). Altogether there are fifteen documents in the Mint and the Public Record Office relating to this case. See also Letter 745.

741   NEWTON TO GODOLPHIN
APRIL 1708
From a draft in the Royal Mint[1]

To the Rt Honble the Earl of Godolphin
Lord High Treasurer of great Britain

May it please your Lordship

Since our late Report presented to your Lordship concerning the melting of Tin-Oar in Reverbarating Furnaces in Cornwal by the Patentees, the Deputy Assaymaster of Tin hath cutt off Assay pieces from several Blocks of Tin in

Cornwall some of wch Blocks were melted in the Furnaces & others in the Blowing houses, & has assayed the pieces in the Tower & found both sorts much of the same goodness. The Agent of the Patentees sent us other Assay pieces cut off from Blocks melted in the Furnaces wch were assayed also in the Tower by the Deputy Assay Master & found of the same goodness with the former. All these Assays being taken from west country Tin, where very little grain Tin[2] is produced, none of them proved to be grain Tin: but the Patentees affirm that since they began to melt the several sorts of Tin apart, that is in the three last coynages not yet sent to the Tower, a good quantity of their Tin breaks grain.

Some affirm that above one tenth part of all the Tin in Cornwal used heretofore to prove grain Tin, & think that the Furnaces of the Patentees by keeping the Tin too long in fusion evaporates the best parts of it & spoiles the grain Tin, & damages the rest. And if this proves true we are humbly of opinion that the Patent is injurious to the Tin affair. But others represent that not above one twentith part of all the Tin in Cornwall used heretofore to prove grain Tin, & that the Tin is refined from the drossy parts by being kept in fusion & thereby becomes better then before. And the Patentees tell us that their Agent in Cornwall hath by their Order kept a parcel of Tin in fusion in the heat of their furnaces some hours together & found it finer & better Tin after the fusion then before; whereas in their usual way of melting Tin out of the Oar, it continues in fusion but about a quarter of an hour.

We having no Tin Oar nor Blowing houses & Melting houses to decide this important Question, do therefore humbly propose that your Lordship would please to give Order to the Agents in Cornwal, that this matter be there examined by taking out of the same heap of the black Oar of grain Tin, after it has been washed in the usual manner & made fit to be melted & is well mixed together, two equal parcels, & causing the one to be melted in a Blowing house, the other in a Furnace of the Patentees, & run into Blocks fit for coynage in the usual manner, and that the Deputy Assaymaster & such other sufficient Witnesses as the Agents in Cornwall shall think fit to appoint do see that the whole proceeding in both the melting house & the blowing house be done in the usual manner, & that the Blocks produced be assayed in their presence by the Deputy Assaymaster, & the experiment of melting & assaying be repeated once or oftener if need be, for the greater certainty, & the whole matter reported by the said Agents to your Lordship.

We also humbly propose that a parcel of good grain Tin be kept in fusion, in the presence of the same Witnesses, four or six hours, in that heat of a furnace wch is used in melting the Oar, & that at the end of every two hours the molten Tin be well stirred to mixt it eavenly & two assays be taken out of

the furnace one immediately before the stirring & another presently after it &
yt these assays be compared by the Deputy Assaymaster with one another &
with an assay-piece cut off from the Tin before fusion, to see whether & how
much the Tin grows better or worse by the fusion & stirring; & a distinct
Report thereof together wth an account of the dross found in the bottom of the
furnace by keeping the Tin in fusion some time after the last assay is taken out
be sent by the Agents to your Lordship. We also desire that the Agents will
please to inform themselves, as well as they can, of the quantity of grain Tin
which used heretofore to be made annually in Cornwall in proportion to the
other sorts of Tin, & of the quantity, or number of blocks, of grain Tin made
the last coynage by the Patentees, & give your Lordship an account thereof.

All which is most humbly submitted to Your Lordships great Wisdome

Mint Office,
   April 1708.

NOTES

(1) Newton MSS. III, 546. This draft, in Newton's hand, is unsigned.

(2) A solid block of refined tin, near its melting point, is very brittle and is readily broken
up, on hammering, into small pieces, known as 'grain tin'. Also, by stirring molten tin of
good quality, small grains are formed as it cools and solidifies.

742   BENTLEY TO NEWTON
10 JUNE 1708
From the original in the Library of Trinity College, Cambridge[1]

TRIN. COLL., June 10, 1708.

Dear Sir,

By this I hope you have made some progress towards finishing your great
work,[2] wch is now expected here with great impatience, & the prospect of it
has already lower'd ye price of ye former Edition above half of what it once
was. I have here sent you a specimen of ye first sheet, of wch I printed about a
Quire: so yt the whole will not be wrought off, before it have your approbation
I bought this week a hundred Ream of this Paper you see; it being impossible
to have got so good in a year or two (for it comes from Genua) if I had not
taken this opportunity with my friend Sir Theodore Jansen,[3] ye great Paper
merchant of Britain. I hope you will like it & ye Letter too. wch upon trials
we found here to be more suitable to ye volume than a greater, & more
pleasant to ye Eye. I have sent you like wise ye proof sheet, yt you may see,

what changes of pointing, putting letters Capital, &c I have made, as I hope, much to ye better. This Proof sheet was printed from your former Edition, adjusted by your own corrections and additions. The alterations afterwards are mine: which will shew & justify themselves, if you compare nicely the proof sheet with ye finishd one. The old one was without a running Title upon each page, wch is deformd. Ye Sections only made wth Def. I. Def. II. which are now made full & in Capitals DEFINITIO. I. &c. Pray look on *Hugenius de Oscillatione,*[4] wch is a book very masterly printed, & you'l see that is done like this. Compare any period of ye Old & New; & you'l discern in ye later by ye chang of points and Capitals a clearness and emphasis, yt the other has not: as all yt have seen this specimen acknowledg. Our English compositors are ignorant & print Latin Books as they are used to do English ones; if they are not set right by one used to observe the beauties of ye best printing abroad. In a few places I have taken ye liberty to chang some words, either for ye sake of ye Latin, or ye thought it self as yt in p. 4. *motrices, acceleratrices* et *absolutas.* I placed so because you explain them afterwards in yt order. But all these alterations are submitted to your better judgment; nothing being to be wrought off finally without your approbation. I hope you to see in about a forthnight, & by yt time you will have examind this Proof, & thought of what's to come next. My wife has brought me a son lately, who, I thank God is a brave healthfull child. I am,

Yours,

RI. BENTLEY.

Note, yt ye Print will look much better, when a book is bound & beaten

NOTES

(1) MS. R. 4. 47, fo. 19.

(2) The first edition of the *Principia* had become extremely rare within a few years of its publication, and Newton was constantly being urged to prepare a second edition. It is clear from the previous correspondence with Flamsteed that he was seriously entertaining the idea of bringing out a second edition as early as 1694. At a meeting of the Royal Society on 31 October 1694, 'A lre from Mr Leibnits to Mr Bridges was produced and read, wherin he recommends to the Society to use their endeavours to induce Mr Newton to publish his farther thoughts and emprovements on the subject of his late book, *Principia Philosophiæ Mathematica,* and his other Physicall and Mathematicall discoverys, least by his death they should happen to be lost' (Journal Book of the Royal Society of London, p. 171).

Meanwhile, Newton's time was fully occupied with his duties at the Mint, and on 4 July 1700, six months after his appointment as Master, Sloane wrote to Leibniz: 'The Royal Society have laboured to gett his [Newton's] Theory of the Moon, Book of Colours &c: printed but his excessive modesty has hitherto hinder'd him' (Early Letters of the Royal Society of London,

S. 2, fo. 14). At one time he seems to have considered the idea of entrusting the work to Fatio, but he abandoned this notion when he learned that Fatio's 'explanation' of gravity conflicted with his own. From the letter here reproduced it seems that he was willing to allow Bentley to see the work through the press. Bentley had even made a start, but shortly afterwards he abandoned the work, possibly because he found the mathematical work too involved. On his recommendation, Newton agreed to allow Cotes to undertake the task. See Edleston, p. xv.

(3) Sir Theodore Janssen (1658?–1748). Among other business interests, he was director of the South Sea Company. He was knighted by William III and made a baronet by Anne. See *D.N.B.*

(4) *Horologium Oscillatorium sive de Motu Pendulorum.* The work appeared in Paris in 1673.

743 MINT TO GODOLPHIN
25 JUNE 1708
From the original in the Royal Mint[1]

> To the Rt Honble the Earl of Godolphin
> Lord High Treasurer of great Britain.

May it please your Lordship

In obedience to your Lordships Order of Reference of May ye 5t upon the Memorial of the Rt Honble the Earl of Darby[2] for coyning copper money in her Maties Mint to be current only in the Isle of Mann,
We have considered the same, & are humbly of opinion that no other money then her Majties should be coined in her Maties Mint: but leave may be given to the Gravers or Moneyers of her Majties Mint, if it be thought fit, to make any quantity of Medalls for his Lordship of such sizes as his Lordship shall desire, the form & inscriptions being first approved by your Lordship, or by the Officers of her Majties Mint by virtue of her Majties Warrant already granted them concerning Medalls

> All wch is most humbly submitted
> to your Lordships great Wisdome

Mint Office   C: PEYTON[3]
June 25. 1708   IS. NEWTON
                                                                JN ELLIS

NOTES

(1) Newton MSS. II, 456 (in Newton's hand).

(2) The Isle of Man was an independent territory governed by the Stanleys from 1405. The one referred to above, the second James, was the tenth Earl of Derby from 1702 to 1736. In

1708 he requested Godolphin that the Mint should make a few hundred pounds of halfpence, distinguished by the Island arms on the reverse. The Mint recommended that the order be rejected for the reason stated in the letter. The Island found another supplier whose halfpence were not struck but cast. In 1723 and 1758 further supplies were obtained privately. The Isle of Man was purchased by the Crown in 1765, and in 1786 the Mint supplied pence and half-pence with the effigy of George II for obverse and the Island's arms for reverse. See Craig, *Mint*, p. 379.

(3) Craven Peyton, M.P., succeeded Sir John Stanley as Warden, and held that office from May 1708 until December 1714.

744   NEWTON AND GREGORY TO GODOLPHIN
30 JUNE 1708
4 AUGUST 1708
From the original in the Public Record Office[1]

To the Rt Honble the Lord High Treasurer
of great Britain.

May it please your Lordship

In obedience to your Lordships Order of Reference of the 23th Instant upon the annexed Memorial of the Commissioners Mr Rutherford, Mr Brown & Mr Bruce, in wch they desire a summ at present in part of the reward for executing their Commission, & refer the residue thereof to your Lordships consideration after the business shall be over: we have considered the same & their business being to receive the old money of Scotland from the Bank & see the same melted into Ingots & to deliver the Ingots to be coined & keep an account of the whole & certify the deficiency (all wch require attendance & fidelity;) we humbly propose that they may receive at present, One hundred pounds a piece for themselves & sixty pounds for their Clerk & servants, in all 360 pounds to be paid out of the same fund out of wch Mr Allardes[2] is paid for melting the money into Ingots in their presence.

All which is most humbly submitted to your Lordships great wisdom

London   Is. NEWTON
30 *Jun.* 1708.   DAVID GREGORY.

By a Letter from Mr Allardes dated 21th July 1708, I am informed that the Privy Council of Scotland, when my Ld High Treasurers Order was laid before them by the Commrs of the Treasury there did immediately make an Act for paying to the said Mr Allardes a penny per pound weight Troy for melting the money into Ingots & ordeined the Commrs of the Equivalent[3] to pay the same out of the Equivalent money conform to the Article of Union, & that accordinly the said Commissioners have given obedience & paid Mr Allardes some few days ago for what is already melted. The old money is the peoples untill it be melted into Ingots & the Ingots be delivered by the Agents of the people by weight & assay to the Master of the Mint to be coined, & therefore all the charges of this melting are to be born by the people & by consequence out of the Equivalent. For the Act of Union appoints that the losses wch the people may sustein by the recoinage of their money be born out of the Equivalent in the first place.

Aug. 4th 1708.   ISAAC NEWTON.

NOTES

(1) T. 1/108, no. 41.

(2) George Allardes was Master of the Edinburgh Mint. The name is variously spelt: Allardyce, Allardise, etc.

(3) See Letter 727, note (4), p. 498.

745   WILLIAM DRUMMOND[1] TO NEWTON
12 JULY 1708
From the original in the Royal Mint[2]

Much Honoured

Upon the want of money for carraying on the bussines of the Mint here, Occasioned by the death of Mr Daniell Stuart[3] Late Collector of the Bullion money in whose executors hands the Same now Lyes, And the other circumstances of this Mint, The Earl of Lauderdale[4] as Generall has thought fitt to send a Memoriall to my Lord Treasurer this night. Copy wherof I send you inclosed, Allardise the Comptroller and I have drawn up a Memoriall which the Generall also agreed to And Allardise goeing out of town Signed it and it was to have been signed by the Generall, the Comptroller and myself. but my Lord thought fitt to alter somewhat of it and Allardise not being present. My Lord has only signed the Memoriall sent up, The Copy of that Memoriall I also give you the trouble of, which will serve for a memorandum of what we propose to have done, The whole I presume being to be Recommended by my Lord Treasurer to your Self and the other officers of the Mint in the tower, Likewise I send ane account of the present officers belonging to this Mint and their Sallaries and of what Servants have bein added since the Union and of what Surcharge the former officers have taken upon them with respect to the new Establishment in conformity with your Mint. The Scots Act of parliament 1686 Establishing this Mint related to in this Memoriall Sir David Nairn will furnish you with it. Dr Gregorys report to my Lord Treasurer you have also a Copy of it.[5] The Generall desired me to send you this Letter from him, The other inclosed is from the Secretary of the Colledge of Physitians All the officers of the Mint depend verry much upon you in this matter, we doe want to have our Selvis put upon some establishment. for the truth is the Union has disconcerted our foundation intirely, Sir David Nairn is to deliver My Lord Treasurer the Generalls Letter and Memoriall, I have sent Sir David a Copy of it also, He will be verry ready to doe all he can to Serve us, I must intreat the continuance of your favour on this occasion which will put a Lasting obligation upon all the officers of the Mint and in particular upon

Much Honoured

Your most obedient and

most oblidged humble servant

W. DRUMMOND

Edinbrugh July 12t 1708

To

Sir Isaac Newton

NOTES

(1) William Drummond, Warden of the Edinburgh Mint.

(2) Newton MSS. iii, 168.

(3) See Letter 740, note (3), p. 516.

(4) John Maitland, fifth Earl of Lauderdale (1650–1712), was a representative peer of Scotland. As General of the Edinburgh Mint he was the senior officer of that establishment. See Letter 732, note (2), p. 505.

(5) See Letter 732, p. 503.

746   RESOLUTION OF THE REFEREES FOR FLAMSTEED
13 JULY 1708
From a copy in Newton's hand in the University Library, Cambridge[1]

At a meeting of the Gentlemen to whom his Royal Highness the Prince hath referred the care of printing Mr Flamsteeds Astronomical Papers.

It was agreed that the Press shall go on without further delay & that if Mr Flamsteed do not take care that the Press be well corrected & go on wth dispatch, another Corrector be imployed.[2]

White Hall
13th July 1708.

FR ROBARTES
CHR. WREN
IS. NEWTON
D. GREGORY
FRAN. ASTON[3]

[On the reverse, still in Newton's hand:]

A Copy of this Order was sent to Mr
Flamsteed by the penny Post   July 14th 1708

NOTES

(1) Add. 4006, fo. 18. There is a copy, in the hand of an amanuensis, in the possession of the Earl of Macclesfield; at the foot of this, in Newton's hand, are the words: 'Vera Copia Is: Newton.' The copy differs but slightly from that here reproduced.

(2) It is not easy to find an excuse for the tone of this resolution. Flamsteed had repeatedly insisted upon his right to correct the proofs, and at no time had he shown any unwillingness to do so. In the Agreement between him and Newton (Number 737) he undertook 'to correct the Press and [appoint] Correctors who live in Town that the Work may not be Retarded'. According to More (p. 543), the letter here reproduced was sent to conceal the real cause of the delay, for which, in Flamsteed's view, Newton was responsible. Writing to Sharp, 22 November 1708, he said: 'Finding nothing in Sir I. Newton but contrived delays to hinder my work from going on, I however resolved to proceed as I could' (Baily, p. 269). For Flamsteed's reaction to this resolution see his letter to Wren sent a week later (Letter 747).

(3) Francis Aston (1645–1715); educated Trinity College, Cambridge. Elected F.R.S. in 1678, he became Joint Secretary of the Society, with Hooke, in 1681, and continued to hold that office with others till 1685. He was a member of the Committee which drew up the *Commercium Epistolicum* in 1712 (see vol. I, note (1), p. 11; also *Notes and Records of the Royal Society of London*, 3, 1941, pp. 88–92). He was an intimate friend of Newton, which may account for the inclusion of his name in the list of Referees. His name does not appear in Clark's letter (Letter 680) amongst those who were to inspect Flamsteed's papers.

The omission of Arbuthnott's name is significant. Flamsteed seems to have had a very high regard for him, and it is by no means unlikely that Arbuthnott was not willing to be associated with a resolution which clearly caused considerable pain to its recipient.

747   FLAMSTEED TO WREN
19 JULY 1708
From the original in the University Library, Cambridge[1]

The Observatory July $\rangle\!\!\text{æ}$ 19 – 1708

Sr

The Copy of the Agreemt made by the Gentlemen Referrees[2] on Tuesday last, reflecting upon me as if by my dilatoriness I had obstructed the progress of the Press, I find my self obliged, that I may clear my self of so unjust an insinuation, with Your Leave, to Acquaint you

That tho I had got 50 Sheets of the first Volume ready Copied for the Press on May ye 2d 1705, yett upon severall pretences, the printing was Obstructed till it was May 1706 before the first Sheet was printed off

That tho by the Agreemt, the Undertaker[3] was to print off five Sheets a Week,[4] yett it was from May 1706 to October 1707 before we could gett 100 Sheets, comprehending the Observations of the first Volume, wrought of, that is near 75 Weeks, So that taking altogether, the Printer dispatch not a Sheet & halfe per Week.

Tho I did all I could to hasten and expedite the Work, as will appear by the Copies of my Letters to Mr Churchill[5] Mr Mathews & Mr Hodgson that I have by me: I offerd to discharge the expense of the Pennypost Letters that brought the Proofs; if the post brought them in the Evening, I returned them next Morning, if in the Morning, they were sent back that evening after, without fayle; except once on May ye 1. 1706 when the great Eclipse of the Sun hapning, company hindred me from correcting & returning that proof till the Morning following & no longer.

The greatest dispatch, was made, both this Year 1706 & the following 1707, in Autumn, when I was Absent in Surrey, yett that was less than the 5 Sheets a Week and then the Work was allways *worst done*.

At my Return after ye last Years Harvest, I found a whole Sheet had been Omitted by the Printer, who had either lost or mislayd it: I copied it imediately from my Manuscripts, & sent it to him, with directions to print it, & reprint ye next. I caused also Sr Isaac Newton to be acquainted with it, & informed both Sr Isaac & ye Printer that I had about halfe a dozen Sheets more, comprehending the Planetts places derived from the Observations made with the Sextant, contain'd in this Volume, to be added to it; but this was not taken notice of, The 6 Sheets were not call'd for, and the Press has stood still ever since

March ye 12 last I received a Letter from Mr Roberts, Your Self & Sr I

Newton desiring me to meet them in London on ye 20th & bring with me what Papers I had ready for the Press. I Attended them with the 2d Volume containing ye Observations made betwixt Sept. 1689 and 1705 compleat in about 175 Sheets of Paper, I exhibited also at the same time the aforementioned six Sheets, that were to be added to the first Volume. desired that that dropt Sheet might be printed & the next following reprinted, or at least the two first Pages of it, which I thought had been Accorded. The 2d Volume by Agreemt was put into the Referrees Hands. I desired the Press, after the first Volume was compleat, might go on with it. At this meeting also I had 125 *lib*[6] ordered to be paid me in part of above 170 *lib* it had Cost me in repayeing & entertaining three Calculators & Copiers whom I had dismist for want of it at Midsummer 1706. Sr Isaac Newton required that I should insert the Magnitudes of the fixed Stars into a Copy of so much of the Catalogue, as I had gone through with, that I had deposited in his hands; which was done for him, & part of a 3d more perfect Copy left in his hands as a gage for returning it.

At this Meeting the Undertaker Urged to have a Corrector appointed in London; this I lookt upon only as a Contrivance to throw the delays of the Press, caused partly by his own and his Printers Neglect, upon me: & therefore having Answered it then, as I have done in this Paper, to the Satisfaction as I thought of the Referrees present, I took no further Notice of it

Since You now know that the Printer has had the dropt Sheet in his hands full Nine Months, that he may have six Sheets more whenever the Referrees please, that they have also 175 Sheets of ye second Volume in their hands, that I never delayed Correcting & returning the proof Sheets as usually, I hope you are satisfied that I have not been guilty of any dilatoriness or Neglect, & that you will not suffer me to be supposed or insinuated to have been guilty of any

But if Sr Isaac Newton insists upon proceeding to print the Catalogue imediately before the 2d Volume. I cannot at present consent to it; for since the Press has Stopt, I have set my Self to compleat it, & having gotten two payr of hands to help me,[7] have perfected some Constellations, that were not compleat before: I have begun the most difficult, & am going into the Country, as I use allways to do at this time of the Year, to look after my Occasions; there I hope to perfect a good part of what remains, and the whole in a few Months after my Return: Now You will say your Self, were it your own Case, tis not fitt to sett to printing the Catalogue before it be as compleat as I can render it at present, I must say farther that t'is altogether improper to print it before the Observations of the 2d Volume, because t'is almost wholly derived from them. The Observations of the Planetts in this[8] are much more Numerous then in the first and I will add much Exacter, & if any One be of another Opinion, for want of experience I shall bring such Proofs of it, as no Equall & Candid Person shall ever reject.

As for Correcting the Press I am altogether unwilling that the last Sheet shall be printed off in the remaining Volume, till I have seen them my Self, but the Catalogue is of that Importance that I shall never consent that any Page of it should be printed off till I have fully corrected it and received from the Press a Proof without faults.[9]

I am not only willing but desirous that the Press should proceed to finish the first Volume of Observations, I have spoke to Mr Hodgson to take Care of correcting the 2d Proofs, and with him I shall leave the six sheets to be Added; which when they are wrought off, Sr Isaac Newton has 175 Sheets of the 2d Volume in his hands, that the Press may proceed with whilest I am Compleating the Catalogue, So there need be no stop on my Account as there never was, nor hereafter shall be, God spareing me Life & Health & prospering, as I firmly believe he will, my Sincere Endeavours.

I[10] am with all due respect Sr for all your favors

Your gratefull & obliged humble Servt

I thinke to send a copy of this   JOHN FLAMSTEED MR:
letter to Mr Roberts & doubt
not but you will imparte the
contents of it to Sr Isaack Newton:[11]

NOTES

(1) Add. 4006, fos. 20–1. This letter was written by a copyist except where indicated in notes (8) and (10). There are copies, one in Flamsteed's hand in the Royal Greenwich Observatory (vol. 35, fos. 85–8), and another in the Shirburn Castle Library (101.H.2) in an unknown hand.

(2) For the names of the Referees see Letter 746.

(3) That is, Churchill.

(4) Article 12 of the Agreement (Number 702) provided that: 'Awnsham Churchill...shall print at least twenty Sheets per Month abateing a Sheet for every holyday provided he be supplyed with sufficient Copy and sufficient dispatch be made in the correcting.'

(5) On 7 June 1706 Flamsteed wrote to Churchill complaining of the slowness of the press (Baily, p. 225).

(6) See Number 737, para. 4.

(7) Probably John Witty and Abraham Ryly.

(8) The two preceding words were added by Flamsteed.

(9) Article 7 of the Agreement (Number 702) provided for this.

(10) From here onwards, including the postscript, the letter is in Flamsteed's hand.

(11) On the reverse, Flamsteed has written:

> 'To Sr Christopher Wren
> Master Surveyour of
> her Maties Workes at
> his house in Scotland Yard
> near Whitehall London
> present.'

748   JOHN DRUMMOND[1] TO NEWTON
20 JULY 1708
From the original in King's College Library, Cambridge[2]

Sr

Mr Drummond Warden of the Mint delivered in your name to the College of physicians the latin edition of your Opticks.[3] They ordered me their Secretary to return you their heartie thanks for so valuable a present, and to tell you that they are very proud to be taken notice of by Sr Isaac Newton, whose noble and divine intentions have contributed so much to the advancement of natural philosophie, by which medicine and all other usefull sciences have been so mightly improven to the general good and advantage of mankind. That you may enjoy a long life, and perfect health for the further advancement of learning, and benefite of mankind is the hearty wish of our Society, and in particular of

Sr

Your most humble Servant

J. DRUMMOND

Edinburgh July 20
      1708

To
The much Honoured
Sr Isaac Newton

NOTES

(1) John Drummond was Secretary to the Royal College of Physicians of Edinburgh from December 1706 to December 1708.

(2) Keynes MSS. 99B.

(3) At Newton's request a Latin version of the *Opticks* was prepared by Samuel Clarke (for which he received £500 from Newton); it was published in 1706. The English version had already appeared in 1704 (see Number 672, p. 414).

749   NOTE BY NEWTON REGARDING THE
          PRINTING OF THE
HISTORIA CŒLESTIS BRITANNICA[1]

Late 1708

From the original in private possession

Decemb. 11. 1704 the Prince by his Secretary Mr Clark ordered five Gentlemen[2] of the Royal Society to inspect Mr Flamsteeds papers & consider what was fit for the press. And the said Gentlemen having reported that his Observations & his Catalogue of the fixt stars were proper to come abroad his royal Highness was pleased to give directions for defraying the charge & in [the] autumn following the said Gentlemen came to an agreement wth Mr Flamsteed about printing the same. His Observations made with a sextant of seven-foot Radius with his Catalogue of the fixt stars were to compose one book, & those made wth a meridional arch of seven foot Radius, were to compose another, and Mr Flams[t]eed was to correct the press.[3] But the Catalogue of the fixt starrs being delivered imperfect the Press stopt for a time, & after the Princes death Dr Halley[4] examined it & added to it 500 starrs by computing their places from Mr Flamsteeds observations, & reduced the Observations in the second Book into the same order wth those in the first, & took care of the impression.

NOTES

(1) 101. H. 2. This is a summary of the correspondence relating to the publication of Flamsteed's *Historia*. It is undated, but the reference in the final paragraph to the Prince's death, which occurred on 28 October 1708, places it subsequent to that date.

(2) The five persons named in Clark's letter to Newton (Letter 680) were Newton, Arbuthnott, Robartes, Wren and D. Gregory.

(3) According to Robert Grant (*History of Physical Astronomy*, 1852, p. 475) this pledge was not honoured. The catalogue of stars which was eventually published was no other than the admittedly imperfect copy which Flamsteed had deposited with the Referees upon the condition that it should not be published, but merely retained as a pledge for the subsequent delivery of a more perfect copy. It seems that the Committee charged with the task of seeing the work through the press, having grown impatient at the repeated delays, which they attributed to Flamsteed's dilatoriness, resolved to proceed with the printing of the catalogue already in their hands. Grant (*loc. cit.*) characterizes it as 'a wanton display of harshness on the part of the committee in whose hands the manuscripts of Flamsteed were deposited, to have resorted to their publication without beforehand giving him due notice of their intention'. In a footnote Grant adds that it is right to state that this charge rests solely upon the authority of Flamsteed.

(4) Queen Anne having ordered that the publication of Flamsteed's observations was to be resumed, they were published in 1712, in one large folio volume, under the editorship of Halley.

750   ALLARDES[1] TO NEWTON
12 FEBRUARY 1708/9
From the original in the Royal Mint[2]

Edinburgh 12 *Febry* 1709

Honoured Sir

The former kindnesses and Acts of Justice You have done me, makes me presume in the close of our Recoinage to give You the trouble of the inclosed Memorial: As to which I must beg Your advice and direction in every particular, for I depend upon it it will be Just, And I must acknowledge I am but a Stranger to what may be proper for me to do on this occasion.

I also beg leave to mind You again of Robert Miller[3] Her Maties Clerk in this Mint who was appointed by the Government to oversee the weight and take the Specie from the Bank, with a promise of a suteable Reward, He has made application to the Lord High Treasurer by a Petition, whereof I also inclose a Copy for Your information, I doubt not you'l find his desire Just: And therefore I must intreat Your good Offices in his behalf. He is a very honest Man, and his Sallary is but small; and if there were any incouragement to expect some addition to it, I should be glad to be directed in the manner to get it done. I do assure You Sir, I as heartily recommend his concern to Your consideration as it were my own. I hope You'l pardon my freedom, and I would reckon my Self very happy to have any opportunity to testifying gratitude to You and shall still be

Honrd Sr

Your most Faithfull and obedient

Servant.

GEO. ALLARDES

NOTES

(1) George Allardes, Master of the Edinburgh Mint, signed this letter, which is in another hand. See Letter 744, note (2), p. 522.

(2) Newton MSS. III, 35

(3) Robert Miller. The three Commissioners for receiving and melting down the old coin were allowed a clerk for whom a salary of £60 per annum was allowed in June 1708. Miller was appointed to this post, and petitioned repeatedly, with the support of the Edinburgh Mint officials, for his £60. After two nterviews with Lord Seafield, who proposed £50, Newton agreed to £60 (see Letter 756\

751   ALLARDES TO NEWTON
12 FEBRUARY 1708/9
From the original in the Royal Mint[1]

Memoriall from George Allardes Master of Her Majties Mint att Edinburgh

to the honorable Sir Isaac Newton

The Recoinage in Scotland being now brought to an End, & the Moneyers sent down from Her Majties Mint in the Tower being desirous of Leave to return home; Mr Allardes craves liberty to ask Sir Isaac Newtons advice what is proper for him to do in the mater, Since two of them Mr Hopper and Mr Sutton wer sent down since the first Commission upon ye death of Seabrock, And that the Queen and Lord high Thesaurers orders to ye other two Mr Collard and Mr Halley was to stay in Scotland while they had the Thesaurers Leave to return. As to which Mr Allardes is very willing they be recalled whenever the Lord High Thesaurer thinks proper, Since it may be a Charge to keep them longer here And what further bullion may be brought in for coinage may be done by the Old Servants here.

Mr Allardes likewise represents that the fond for the Coinage in Scotland will not only be found very far deficient to Satisfie the Charges of the Recoinage Tho it wer all brought in, But that it will even be a Considerable time befor what is lying out and due of that fond can be recovered from the Collectors and others liable for it, which is a loss to Mr Allardes He being obliged to make up the Wast on his own Credit, And it occasions him also to be deficient to the Moneyers In the payment of ye Nine pence per pound weight according to their aggreement: Wherfor he begs also advice What he shall do for Satisfying the Moneyers, And to know if the Moneyers may obtain payment at London from the Thesaurer of what Ballance may be due to them Mr Allardes Certifieing the Ballance under his hand.

As to the Moneyers Expences in coming to Scotland and their return home, And the allowance of 3sh to Each per diem when they wer not supplied with work Conform to their Commission These being reserved in ye Commission to be payed by the Lord High Thesaurer in such method as he should appoint, Mr Allardes humbly thinks he cannot interpose in that mater further than to give a Declaration of the time When the Work began and when ended and of the Extent of the Coinage in that time And how long the Work was stopd the time of the former Assay Masters death, What else may be propper for him to do in this He will walk by Sir Isaacs advice

He prays also to be advised in what method he shall apply for the triall of ye Pix And to know If the Queen accompts to him both for the Pix and Assay peeces Since ther is a Considerable loss upon the Assay For which end he has preserved all the Assay peeces unmelted till he know If they be to be accompted for, As Some make him beleeve the Queen uses to do, by taking to her self the Pix & Assay peeces and allowing the Value to ye Master

He prayes likewise to be directed in what maner he is to apply for having allowance in his accompts of ye 9th penny paid ye Moneyrs per Lib weight It being a penny more than is allowed by the Indenture of the Tower,

Also what allowance he shall crave or state in his accompts for the Miln horses and their maintainance and Servants And for furnishing Rollers Upholding the files, cutters and tumblers or the Seising or Justing peeces, fire for the Nealing and blenching offices and for the Moneyers room and Candles, Allom and other necessaries to the Moneyers. If on these accounts he ought not to be allowed So much upon the pound Weight of the Coinage. Without being obliged to give in particulars which may be impracticable

He desires further to know In what method to state the Refining And what allowance to crave per Lib weight, The Silver being returned of different fineness, Sometimes at 15, Sometimes at 13 Sometimes only 10 or 8 pr.[2]

He begs also to be informed if on the Recoinage in England Any gratification was given by the Goverment to the Officers Clerks and Servants for their extraordinary trouble, And if ther be any Encouragement to apply for the like, In what method it shall be Done Ther being no allowance for maintainance of melters or Labourers in this Mint otherways than when they are supplied with Work And ther being one James Sheelds a Melter here and very well Skilld in all the parts of the Coinage having wrought in the Mint these many yeares bygone whom the Officers here would willingly encourage It is earnestly also desired to know If he may not be Setled as a Servant in this Mint under the Name of Purveyer or Such like with £15 Sallary And In what method that may be Effectuated

Memoriall for

George Allardes.[3]

NOTES

(1) Newton MSS. III, 177. It is in the hand of an amanuensis. This is the Memorial referred to in Letter 750.

(2) pr = pryme (24 prymes = 1 grain).

10 pr would correspond to a fineness of 833 parts per 1000; as there were 16 oz. to the Scots lb., the 15 oz. would give a fineness of 937 parts.

In Scotland the weights used in the Mint were:

16 pounds = 1 stone
16 ounces = 1 pound (lb.)
24 deniers = 1 ounce (oz.)
24 graines = 1 denier (d.)
24 prymes = 1 graine (g.)
24 seconds = 1 pryme (pr.)

Therefore whether the fineness was given in 'prymes' or 'deniers' would make no difference. In the following extract from the Hopetoun Papers we can understand some of the difficulties experienced at Edinburgh in determining fineness:

'...if any Ingott be for Instance of 12 deniers weight and 11 deniers fynnes and so of 1 denier allay (because the highest denominatione in the fynnes of silver is only 12 deniers as said is) Then everie denier of fynnes in that ingote will ansuer precislie to a denier of weight. That is, als many deniers as the samme is denominat to bee, of fynnes or of allay, so many deniers of weight of vtter fyne silver, or of the basser metall (with the which it is allayed) respective, does it conteine; as in the former instance 11 denieris of weight of vtter fynne silver and 1 denieris of copper or vther allay; And so fourth what ever be the weight of any Ingote in and proportionablie to the denominatione of each twelve deniers. Bot in the denominatione of ounces in respect everie ounce as said is conteines 24 that is tuyce 12 deniers, Then in everie ounce of your ingote, 1 deniere of fynnes will ansuer to 2 deniers of weight. And be consequence in a pound weight (which conteines 16 oz as said is) everie denier of fynnesse will ansuer to sexteine tymes two deniers, that is 1oz 8drs and in everie stonne also to 16lb 8oz...' (see R. W. Cochran-Patrick, *Records of the Coinage of Scotland*, Edinburgh, 1876, vol. I, p. lxvi).

(3) Endorsement on reverse of letter.

752   RÉMOND DE MONMORT[1] TO NEWTON
16 FEBRUARY 1708/9
From the original in King's College Library, Cambridge[2]

A Paris ce 16 *Fevrier* 1709

Je me donne l'honneur Monsieur de vous envoyer un livre dont le tittre est *essai d'analyse sur les jeux de hazard*. Je m'estimerois infiniment heureux Monsieur si cet ouvrage pouvoit meriter votre approbation. Je vous l'envoye comme un gage de la veneration que j'ay pour vous et comme une tribut qui est du au plus grand esprit de l'Europe. Monsieur l'abbé Bignon dont la rang et le rare merite vous sont apparement connus veut bien se charger de vous faite tenir ma lettre et un exemplaire de ce petit ouvrage. Si vous voulies Monsieur m'en accuser la reception et m'honorer de vos amis sur les defauts que vous y trouverez vous obligeries Monsieur le plus ardent de nos admirateurs et l'homme du monde qui respecte le plus vos admirables talens.

Remond de Monmort

Ma demeure est a Paris rue et pres
Ste Croix de la Bretonnerine

Oserois je Monsieur me faire un mente aupres de vous d'avoir fait imprimer icy il y a 2 ans votre traitté *de quadraturis* pour en distribuer une centaine aux Scavants de ce pays qui n'en pouvoient faire venir d'Angleterre. On m'a appris de puis quelques jours la mort de Mr Gregory.[3] J'avois eu l'honneur de le voir a Oxford. il me fit voir une espece de commentaire sur l'incomparable ouvrage *philosophiæ naturalis principia Mathematica.* il me faisoit l'honneur de m'ecrire quelques fois et m'apprenoit l'etat des sciences en Angleterre. On m'avoit dit qu'il faisoit sous vos yeux une seconde edition de ce scavant ouvrage.[4] Voudries monsieur m'apprendre des nouvelles d'une chose qui m'interesse extremement.

NOTES

(1) Pierre Rémond de Montmort (Monmort), (1678–1719); mathematician, pupil of Malebranche; Canon of Notre Dame, Paris. He travelled abroad, visiting England, and was elected F.R.S. in 1716. In the same year he became a member of the Académie des Sciences. He wrote on probability; his *Essai d'Analyse sur les Jeux de Hazards* (1708) enjoyed a wide popularity. See *Eloges des Académiciens de l'Académie Royale des Sciences* (Fontenelle, II, p. 47).

(2) Keynes MS. 147.

(3) David Gregory died on 10 October 1708.

(4) The second edition did not appear until 1713, but the possibility of bringing out a new edition was being seriously considered by Newton as early as 1694. See Letter 742, note (2), p. 519.

753   NEWTON TO ALLARDES

c. MARCH 1708/9

From a draft in the Royal Mint.[1]
In reply to Letters 750 and 751

Sr

I received sometime since a letter from you with an inclosed paper containing several Quæres relating to the buisines of your Mint. I have been slow to return an answer for fear that some of those things may be referred to the officers of our Mint with whom I find it sometimes difficult to agree & there-

534

fore what I now write to you is to be looked upon as coming not from an Officer of ye Mint but from a private friend.[2]

I am of opinion that the Officers of your Mint should sollicit the Barons of the Excheqr to get the Bullion in the hands of ye Executors of Mr Stewart[3] & the under collectors to be paid into ye Mint with all convenient speed & that as it comes in you should pay it in due course & order discharging those debts first which were first owing And when you have cleared all debts & accounts to any certain time (suppose to Christmas or Lady [Day] was a twelvemonth or Midsummer last or to any later time,) you may then make up & pass your accounts to that time if you think fit, & afterwards you may pass your accounts annually according to the course of the Mint in the Tower taking care to clear all the accounts of every year before you make oath to them & pass them, because by passing them you are discharged of all accounts till that time.

As for the Moneyers, I am of opinion that the nine pence per pound weight should be paid them in course as money comes into your hands & that in your accounts the penny per pound weight above the $8d$ be set down in your accounts, & if the Auditor or Barons of the Exchequer scruple it, you may petition my Lord Treasurer. What else is due to the Moneyers on account of your coynage they must petition my Lord Treasurer to be allowed. I beleive my Lord will scarce pay any thing to them at London, because what is due to them can scarce come into any other account then yours.

The Pix & Assay pieces you are to have & place ym in your account as so much money received, & to bring in a bill of the loss of Assays. The Bill must be examined allowed & signed by the Warden & Counter-Warden.

You should also according to the best of your memory & knowledge bring in one or more Bills of charges for the Miln horses & their maintenance & servants, & for furnishing Rollers, upholding the files cutters & tumblers, & fire candles & allome & other necessaries for the moneyers. And this bill also should be examined & signed by the Warden & Counter Warden. And so should all the Bills of Carpenters Brick layers Masons Pairers Glasiers Smiths Plaisterers &c

I pay to ye Smith of ye Mint a farthing per pound weight of all the silver moneys coyned, for his making the Dyes, & whether the like allowance should be made by you towards defraying the incident charges of ye coynage I must leave to consideration.

NOTES

(1) Newton MSS. III, 44 (in Newton's hand).

(2) Decisions on matters of importance rested with the Mint Board, in which the Warden, Craven Peyton, took precedence. See Letter 743.

(3) See Letter 740.

754 NEWTON TO ?ALLARDES

? APRIL/MAY 1709

From a draft in the Royal Mint[1]

I have been long indebted to you for your Letters & was in good hopes that that the Question you wrote to me about would have been decided without me. But understanding that it is still depending, I here send you my thoughts about it.

I imploy a Melter to melt all the gold & silver coyned & allow him thirteen pence per pound weight Troy for melting the gold: whereof I reccon at least 3*d* for potts & fire, & the other 10*d* for wast & charges of making up the sweep. Whence the wast doth not amount to five grains in the pound weight of gold. And the wast in silver cannot be much more.

Because I do not make up the sweep my self I cannot speak of this matter by my own experience. But consulting my Melter about it, he told me that the wast in melting was about 6 grains per pound weight of silver, but in refining it was about double to that in melting. And afterwards he told me that in one parcel he had found the wast in melting amount to 14 grains per pound weight. But this I suspect was by some accident, or falshood in his servants[2] For the Goldsmiths reccon the wast so little that they have perswaded the Crown to make no allowance for it in making the money in our Mint, whereas in your Mint the Master is allowed to put[3] twelve grains of Copper into every pound weight of silver when the silver is molten & they are pouring it off into the moulds, & this is done to make amends for the wast of the copper wch fumes away in the melting.

In the year 1707 when the money current in Scotland was to be recoined, we wrote to the Officers of your Mint that we were not allowed to put any copper into the pot for making recompence for the wast of the copper wch fumed away in the melting, & that they were to conform themselves to ye practise of our mint. But they replied that by the flaming coales wch they used in melting, a greater wast was caused then in our Mint, so that unless they were still allowed to put the 12 grains of copper into the pot, they could not coin the money standard. Whereupon this allowance was connived at.[4]

How much the copper fumes away in your meltings by the flaming of the coals I do not know: but I reccon that when the allowance of 12 grains per pound weight was instituted, it was deemed a sufficient recompence for the wast made by fuming away: whereas in our Mint as I said we have not allowance made for recompencing that wast. And thence I gather that the wast in your Mint after making up the sweep ought to be less by some grains then the wast in our Mint: & that the wast upon the whole coinage (if the sweep be well made up) must be under 234 pounds weight & may be so little as not to exceed 125 pounds weight.

For whilst 12 grains of copper are added to every pound weight of silver in every melting & a pound weight of silver makes but about half a pound weight of money: there are about 24 grains of copper added to every pound weight of money coined. And this addition diminishes the whole wast, & should make it 24 grains per pound weight less in your mint then in ours, supposing the wast by the fuming away of the metal in both mints* And the wast by the flaming coals in your Mint should be 24 grains per pound weight more then in ours to make the whole wast which remains after making up the sweep, equal in both Mints. I am

Your very humble servant

Is. NEWTON.

* I mean that ye 24 grains
are more then enough to make
good all your wast.

NOTES

(1) Newton MSS. I, 181. This draft is in Newton's hand, but he has not dated it.

(2) The five preceding words are deleted.

(3) Here Newton had written 'a half penny', which he deleted.

(4) For the practice in Scotland see Letter 729, note (2), p. 500.

755 STATEMENT BY NEWTON

MAY OR JUNE 1709

From the original in the University Library, Cambridge[1]

AN ACCOUNT OF THE EXPENSE OF PRINTING MR JOHN FLAMSTEEDS OBSERVATIONS

BY ORDER OF HIS ROYALL HIGHNESS THE PRINCE

CHARGE

lib.

Received of the
Treasurer of his
Royall Highness

375. 0. 0

DISCHARGE

lib   *s*   *d*

Paid to Awnsham Churchil
Bookseller (17 Apr. 1706)
the summ of 250 pounds,
whereof 166*lib.* 12*s.* 00*d*
hath been expended by him
in printing 98 sheets by 400,
fifty shillings he desires
to be allowed him for
charges in providing new
stamps wch will be of no
further use to him if the
second volume of Mr Flamsteeds
Observations be not printed, &
thirty pounds is to be paid
to Mr Machin for examining
Mr Flamsteeds copy of his
Minute-books, & correcting
his calculations & 50 *lib*
18*s* remains in my hands
towards the charge of
finishing the first volume

250 00 00

lib

Paid to Mr John Flamsteed
26 March 1708
in part for his charge &
trouble in preparing papers
for ye first & second Volume
of his Observations &
correcting the Press

125 00 00

Total   375 00 00

This Account was laid before the Princes Administrators, after the Order for paying 30 *lib* to Mr Machin & before the payment thereof, that is, after the 7th of May & [before] ye 8th of July 1709 [for] the first Volume except the Cataloge of fixt starrs & the Cutts to ye Observations.

And May 7th 1709 the Referees ordered that these scemes should be forthwith graved. The charge of wch & rolling off &c being deducted from the 50*lib* 18*s* wch remained now in my hands, [there is] left in my hands   25*lib* 3*s* the ballance of my Acct wth the Admistrators at my last recconing with them.

NOTE

(1) Add. 4006, fos. 23–4. Newton autograph.

756   NEWTON TO SEAFIELD[1]

1 JUNE 1709[2]

From a draft in the Royal Mint.[3]

For answer see Letter 757

My Lord

Soon after your Lordship left London the Officers of this Mint made report to my Lord Treasurer upon the petition of the Queens Clerk[4] for allowing him 60*lb* for his pains in assisting the three Commrs in receiving the old money to be recoined. And finding a reference written upon the Moneyers petition (wch I was not aware of when I saw your Lordship last,[)] We have made a report upon that Petition also; wch Report amounted only to this: that the summs of money for wch they petitioned, were due to them but not yet paid by reason that Mr Allardes had not yet received moneys from ye Government sufficient for this purpose. And that of the 9*d* per pound weight wch was due to them for coinage, eight pence was due from Mr Allardes out of his allowance, & one penny was properly to be allowed by her Maty as is done in the Mint in ye Tower of London. I do not know whether these Reports have been yet laid before my Lord Treasurer, because during this vacation his Lordship is seldome in London. I am

NOTES

(1) See Letter 722, note (2), p. 493.

(2) The date is given in the Royal Mint Record Books, VIII, 156.

(3) Newton MSS. III, 164 (in Newton's hand).

(4) Robert Miller. See Letter 750, note (3), p. 530.

757   SEAFIELD[1] TO NEWTON
28 JUNE 1709
From the original in the Royal Mint.[2]
In reply to Letter 756

Edinburgh June 28th 1709

Sir

Mr Allardice[3] was in the Countrey when I came to this place which is the reason that he has not acknowledged the favour of your letter to him sooner, he is very sensible of the Justice and assistance he has receaved from you, he will in all things submitt to your opinion, and hopes that you will continue to give him your advice, Mr Scott, who is his Depute doth promise to send you a state of Mr Allardice acco[mp]tt very speedily and he thinks you will almost understand it by what is contained in the inclosed letter In the mean time the Barons of the Excheqr are stating the acco[mp]tts with the Collectors of the Bullion moneys and I beleeve that Fund will be very quickly cleared. The Lords of the Privy Council of Scotland, appointed Robert Millar[4] to be assisting to the Commissioners appointed for certifying the deficiencies upon the recoynages, and the Officers of the Mint tell me that he was at great pains in that matter; I doe presume to transmitt his petition[5] and Entreat you may lay it before My Lord Tresaurer or recommend it to his Lordship and I doe think he deserves fifty pounds for his pains, he is very glade to find that his affair which was formerly under my Lord Tresaurers consideration is referred to you and the other Officers of the Tower, he is a good Officer and a very honest man but is poor and this small sum will be a very great relieff to his Family

I am with great respect

Sir

Your most Humble and obedient

Servant

SEAFIELD

Sr Isaac Newtone.

NOTES

(1) Lord Seafield, Lord Chancellor of Scotland. See Letter 722, note (2), p. 493.

(2) Newton MSS. III, 123. The letter, signed by Seafield, is in another hand.

(3) George Allardes, Master of the Edinburgh Mint.

(4) See Letter 750, note (3), p. 530.

(5) Newton MSS. III, 120.

758   ALLARDES TO NEWTON
9 AUGUST 1709
From the original in the Royal Mint[1]

Allardes. August the 9th 1709.

Honoured Sir

I thought to have troubled You before this with a double of My Accompts for the Recoinage; but of late I have been valitudinary, my old indisposition of vomiting blood having recurred upon me: but, blessed be God, it has for the time abated, tho' it has brought me very low: I am now making them up, and shall transmit them to You as soon as possible; for I resolve to apply for My payment with all diligence against our next Excheqr terms, having got no more money Imprest to me than what I wrote You last.

And now understanding That on my present indisposition severals have been looking upon my Post: I by this take occasion to let You know that I have been speaking with the E: of Seafield, and others of my Friends if it might not be secured to My son[2] in case of My decease: He is now in the 17th year of his age, and we think his minority may be no scruple, since he may be oblig'd to act by a sufficient Depute, and it may be thought hard to give it to any other so long as I or my Heirs may be unpaid: His Lordship has promised me all the Interest he can make, and as I have had Your favour and Friendship on all former occasions, I must also beg leave to solicite the same in this; and I have no doubt Your Interest and concurrence with his Lordships will get it effectuate. I renew my hearty thanks for Your former kindnesses. I depend upon Your good Officis in this new demand, and praying the Lord to reward You for all, and expecting Your favourable answer I am

Honoured Sir

Your most obliged humble Servant

GEO. ALLARDES

NOTES

(1) Newton MSS. III, 36. The letter is written by an amanuensis, but signed by Allardes.

(2) The application for the succession on behalf of the Master's son was not successful.

759   LAUDERDALE[1] TO NEWTON
10 AUGUST 1709
From the original in the Royal Mint[2]

Sir

The officers and other servants of her Majestie's Mint Are straitned for want of their sallaries and fies especially those who have no other mean of subsistance but the fruits of their Labour; Wherefore I ask Liberty to give you this trouble to know your opinion what is the best way to get this remeded (I am unwilling to give My Ld High Thesaurer of Great Britain trouble upon every occasion) But We having the happiness to be of one Society wth you (suppose at a distance) and that you know the methods of payments made to the officers of their sallaries of her Maties mint in ye Tower, And I humbly Conceive that the same method will be followed here, for by a Late act of ye parliamt of G. Britain for Incourageing of ye Coynage in the mints in Scotland and for prosecuteing offences Concerning the Coyn in England, it appears to me that both her Maties Mints in South and North Britain are to be on one foot; Wherefore Sir it will be a favour Done to her majesties servants here in the Mint, that you would give your self ye Trouble to Cause your Clark Inquire about this affair, & that he may Transmitt to the Master of the Mint here, what is proper to be Done thereanent.

Sir Your former Civilitys are the Occasion of this trouble From

Sir

Your most humble Servant

LAUDERDALE

Haltoun the 10*th August*
1709

NOTES

(1) See Letter 745, note (4), p. 523.

(2) Newton MSS. III, 55.

760   NEWTON TO FRANCIS BURMAN[1]
2 JUNE 1702[2]
From *Itineris Anglicani Acta Diurna*

Sr.

The Gentleman who brings you this, is on[e] of the chaplains to the Dutch Embassadors. I beg the favour that by the leave of the R. Society you would introduce him to see one of their meetings. He has heard Monsr. Volders *Lectures* & has a curiosity about *Mathematicæ* et *Philosophicæ* things. If he brings a friend with him, I beg the favour that you will treat him with respect. I am

Your humble servant

Is. NEUWTON.

Jerounstreet,
June 2. 1702.

NOTES

(1) This, and the two following letters, had not become available when the printing was completed and could not be placed chronologically.

(2) The above letter is in the *Itineris Anglicani Acta Diurna*, of Francis Burman the younger, published in an edition by A. Capadose in Amsterdam in 1828. It describes his visit to England in 1702, and includes the text of a note given by Newton to introduce him to the Royal Society.

For this information, the Editor is indebted to Mr P. A. Hoare, Senior Assistant Librarian of the London Library, who discovered this letter whilst this volume of the *Correspondence* was in the Press.

761   NEWTON TO THE GOVERNOR OF THE CASTLE OF CHESTER
16 APRIL 1698
From the original in the Chester City Record Office[1]

Sr

One William Cook mentioned in the Information of wch the inclosed is an attested Copy, having fled into Ireland to avoyd Justice, hath since been apprehended at Dublin & is at present bailed there, but Orders are sending to ye Lords Justices of Ireland to send him Prisoner into England in order to his being tried for counterfeiting the current coyn. I presume that in a short time he may be sent prisoner to Chester & when he comes I desire yt you'l please to commit him upon the Information of wch the inclosed is an attested Copy, & give me notice thereof that I may order a Habeas Corpus[2] for his removal to Newgate London. I hope you'l excuse this trouble, being it is for the publick service. Pray send me (upon his commitment) the name of the Keeper of your Prison to whom ye Habeas Corpus is to be directed. I shewed the bearer Mr Peers[3] the Original Information & he can satisfy you upon oath that he had the inclosed copy from me, & ye Warden & Controller of your Mint[4] can satisfy you that this Letter is my hand. Direct your Letter to me Warden of the Mints at my house neare St James's Church in Jermyn street Westminster. I am

Your humble servant

London.
April 16. 1698.

Is. NEWTON

NOTES

(1) ML/4/547. The letter is presumably addressed to the civil head or the Governor of the Castle of Chester, the natural port of arrival.

(2) Writs of Habeas Corpus were used by the Crown to transfer accused from one gaol to another. A less formal method was used by Newton in Letters 554.

(3) Peers was possibly the John Peers, a clock maker and associate of criminals, whom Newton employed to investigate from within the plans of Chaloner and Holloway for counterfeiting at Egham (see Memorandum 581).

(4) 'The Warden & Controller of your Mint' were the corresponding deputies in the Chester Mint, Robert Weddell and Edmond Halley. This Mint was in operation from 1696 to 1698 for the requirements of the Great Recoinage.

762 NEWTON TO THE GOVERNOR OF THE CASTLE OF CHESTER

23 NOVEMBER 1699

From the original in the Chester City Record Office[1]

Mint Office in ye Tower of
London Novem 23. 1699.

Sr

Mr Secretary Vernon communicated to me ye copies of the Depositions you sent him concerning Mr Horton & commanded me to answer your letter. I have acquainted Sr Joseph Jekil Chief Justice of Chester with the matter & care will be taken to send down his Majts Commission of Oyer & Terminer[2] directed to proper persons for his Triall the next Assizes. In the meane time tis hoped he will be kept safe. I understand he is committed only upon suspicion of High Treason, if that committment be not thought strong enough I beleive you may commit him absolutely for High Treason by vertue of the late Act of Parliamt wch makes it High Treason to make or mend or begin or proceed to make or mend any of ye coyning Tools mentioned in that Act one of wch is a Press for coyning, or knowingly to buy or sell hide or conceal or without lawfull authority or sufficient excuse for that purpose knowingly to have in his her or their houses custody or possession any of those coyning tools.[3] I beleive it will be thought proper to try him upon this Act & if so the evidence will be of better credit because there is no conviction money to tempt them. If there be any thing wherein I can serve you in this matter, you may command

Your most humble Servt

Is. NEWTON

NOTES

(1) ML/4/561.

(2) A Commission of Oyer and Terminer ('to hear and settle') was the royal authority for a judge or judges to hold an Assize.

(3) Newton claimed (see Memorandum 581) to have inspired the Act of 1697 which made provision or possession of the necessary tools high treason, as counterfeiting the King's coin had been for centuries. The consequent penalty was that the convicted person was dragged on a sledge to the place of execution, but was not drawn or quartered.

TABLE OF ASTRONOMICAL SYMBOLS

Signs of the Zodiac

Aries	♈	Libra	♎
Taurus	♉	Scorpio	♏
Gemini	♊	Sagittarius	♐
Cancer	♋	Capricornus	♑
Leo	♌	Aquarius	♒
Virgo	♍	Pisces	♓

Solar System

Sun	☉	Earth	⊕
Moon	☽ or ☾	Mars	♂
Mercury	☿	Jupiter	♃
Venus	♀	Saturn	♄

Planetary Orbits

First point of Aries	♈	Opposition	☍
Conjunction	☌	Ascending node	☊
Quadrature	□	Descending node	☋

Days of the Week

Sunday	☉	Thursday	♃
Monday	☽	Friday	♀
Tuesday	♂	Saturday	♄ or ♄
Wednesday	☿		

The symbols given above were also used to indicate metals, as shown below:

Gold	☉	Iron	♂
Silver	☽	Tin	♃
Mercury	☿	Lead	♄ or ♄
Copper	♀		

546

GLOSSARY OF ASTRONOMICAL
TERMS

Aberration: the name given to the apparent displacement of a star from its true position arising from the fact that the speed of light bears a finite ratio to the Earth's orbital speed, the amount of the aberration depending upon the value of this ratio.

Annual equation of the Moon's mean motion: see p. 5, note (6).

Anomaly: P is the position of a planet in the elliptic orbit whose focus is S, and whose apse line is ACA', A being the position of perihelion. SR revolves about S along the ellipse at a uniform rate $2\pi/T$, where T is the period of rotation. If n is the planet's

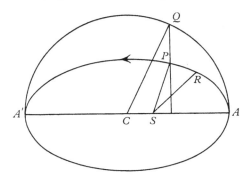

mean angular velocity, then $n = 2\pi/T$, and the *mean* anomaly at a time t measured from perihelion is the angle ASR, or nt. The *true* anomaly is the angle ASP. If Q is the point on the auxiliary circle corresponding to P, the position of the planet on the ellipse, then ACQ is the *eccentric* anomaly. The mean anomaly is related to the eccentric anomaly by the relation $M = nt = u - e \sin u$, where u is the eccentric anomaly. See H. Godfrey, *Treatise of Astronomy*, p. 148, and W. M. Smart, *Spherical Astronomy*, pp. 111–12.

Anomalistic year: see under Year.

Aphelion: The point on the orbit of the Earth or a planet at which it is furthest from the Sun; *perihelion*, the point at which it is nearest the Sun.

Apogee: the point in the Moon's orbit which is furthest from the Earth; *perigee*, the point in the orbit which is nearest the Earth.

Appulse: the near approach of one celestial body to another.

Apses (apsides): the extremities of the major axis of the elliptical orbit.

Colure: see p. 346, note (6).

Conjunction of Sun and Moon: this occurs when the longitudes of the Sun and the Moon are equal; opposition when their longitudes differ by 180°. When their longitudes differ by 90°, the Sun and Moon are said to be in *quadrature*. The points midway between the quadratures are called the *octants*. The two positions, conjunction and opposition, when spoken of jointly, are called *syzygies*.

Culmination: a celestial body *culminates* when it reaches its highest point above the horizon, i.e. when it crosses the meridian.

Eccentricity: see p. 31, note (4).

Ephemeris (pl. *ephemerides*): a table giving the position of the Sun, Moon, planets, etc., for given dates at regular intervals.

Equation: see p. 8, note (3).

Equation of the centre: an inequality in the motion of the Moon due to the eccentricity of her orbit.

Equation of the Sun's centre: see p. 6, note (9).

Equation of time: see p. 73, note (5). It is the difference between apparent and mean time at any instant. See under *Time*.

Evection. Owing to the variation in the eccentricity of the Moon's orbit, there is a corresponding variation in the equation of the centre, between 5°.3′ and 7°.31. See Spencer Jones, p. 118, and H. Godfray, *Elementary Treatise on the Lunar Theory* (1855), pp. 92–3.

First point of Aries: see Introductory Note on the Lunar Theory, p. xxvii.

Heliacal rising: A star is said to have a heliacal rising when it is first seen in the morning twilight.

Horizontal parallax: see under *Parallax*.

Hour angle: The angular distance of a celestial body from the observer's meridian measured along the equator in hours, minutes and seconds. It is the angle which the declination circle makes with the meridian.

Kepler's equation: see p. 44, note (8).

Kepler's problem: see p. 21, note (3).

Lunation: see p. 25, note (8).

Meridian: The great circle passing through the zenith and the poles. It meets the horizon at the north and south points.

Month: the *sidereal month* is the interval given by the Moon's complete circuit of the stars as seen from the Earth. Its mean value is 27·3217 solar days. The *anomalistic month* is the interval during which the Moon moves in her path round the Earth from perigee to perigee. Its magnitude is 27·5546 mean solar days. The *synodical*, or *lunar* month is the period from conjunction to conjunction, or from opposition to opposition. Its value is 29·5306 mean solar days. (W. M. Smart, *Spherical Astronomy*, p. 133.)

Nodes: the two points in which the Moon's orbit crosses the ecliptic.

Nonagesimal point: see p. 156, note (2).

Nutation: the term applied to various periodic fluctuations in the mean precessional motion of the celestial poles.

Obliquity of the ecliptic: the great circle, the plane of which contains the Sun's (apparent) yearly path is called the ecliptic, and the angle it makes with the celestial equator is called the obliquity of the ecliptic.

Occultation: This takes place when the Moon or a planet passes in front of a celestial body shutting it out from view.

Octants: see *Conjunction of Sun and Moon*.

548

Opposition: see *Conjunction of Sun and Moon*.

Parallactic inequality: this arises from the sensible difference in the disturbing influence exerted by the Sun on the Moon according as the latter is in that part of the orbit nearest to, or most removed from, the Sun.

Parallax: the apparent change in the position of a body on the celestial sphere due to a change in the observer's position. The *annual parallax* of a star is the angle subtended at the star by the radius of the Earth's orbit. The *horizontal parallax* of the Moon (or a planet) is the angle subtended at that body by the Earth's equatorial radius. *Menstrual parallax*: 'The centre of gravity of the Earth and Moon always lies within the surface of the Earth, so that the monthly orbit described by the Earth's centre about the common centre of gravity is comprehended within a space less than the size of the Earth itself. The effect is, nevertheless, sensible in producing an apparent monthly displacement of the Sun in longitude, of a parallactic kind, which is called the *menstrual equation*, whose greatest amount is, however, less than the Sun's horizontal parallax, or about 8".6.' (Sir John F. W. Herschel, *Outlines of Astronomy* (1875), pp. 354–5.)

Perigee: see *Apogee*.

Perihelion: see *Aphelion*.

Precession: the axis of the Earth is continually changing its direction in space without changing its inclination to the plane of the orbit. This spinning motion is called precession. In consequence, the vernal equinox is not stationary; it moves along the ecliptic very slowly, its mean rate being 52".2 per year, the direction of motion being opposite to that of the Sun. This motion causes the vernal equinox to recur earlier than otherwise it would, hence the term 'precession'.

Prosthaphæresis: see p. 99, note (3). A derivation of this word, alternative to that there quoted, is given in Liddell and Scott, πρόσθε = 'previously', and ἀφαίρεσις = 'subtraction'.

Quadrature: see *Conjunction of Sun and Moon*.

Syzygies: see *Conjunction of Sun and Moon*.

Time: *Solar time* at any instant is the hour angle of the Sun's centre, reckoned westward from 0 h. to 24 h. This is called apparent solar time, and is the time indicated by a sun-dial. A *solar day* is the interval between two successive transits of the Sun's centre over the meridian. The *mean solar day* is the interval between two successive transits of the mean Sun across the observer's meridian. The *mean Sun* is assumed to move on the celestial equator at a uniform rate round the Earth, this rate being such that the mean Sun completes a revolution in the same time as that required by the Sun for a complete circuit of the ecliptic. (See W. M. Smart, *op. cit.* p. 42.) *Sidereal time* at any instant is the number of sidereal hours, minutes, seconds, since the last preceding transit of the First Point of Aries across the observer's meridian. A *sidereal day* is the interval between two consecutive transits of the First Point of Aries across the meridian.

Tychonic equation: see p. 44, note (3).

Variation: an inequality due to the variation in the magnitude of the residual solar attraction on the Earth-Moon system during a synodic month; its magnitude has been found to be $39' \sin 2E$, where E is the angle subtended at the Earth by the directions to the Moon and the Sun. See p. 31, note (3), and Spencer Jones, p. 120.

Year: the sidereal year is the interval between two successive passages of the Sun through any fixed point on the ecliptic. The tropical year is the average interval between two successive passages of the Sun through the vernal equinox (which is no longer regarded as fixed, owing to precession and nutation). The *anomalistic year* is the period between two successive passages of the Earth through perihelion. Since the apse line has an annual motion of $11°\cdot25$, the anomalistic year is slightly longer than the sidereal. The magnitudes of the different types of year are:

Tropical	365·242199 days
Sidereal	365·256360 days
Anomalistic	365·259641 days

(Spencer Jones, p. 62)

Zenith: the point on the celestial sphere vertically overhead. *Nadir* is the diametrically opposite point.

550

INDEX

Bold figures refer to Letter numbers

curvilinear figures of the second kind and *Opticks*, 414

Danzig Observatory, its position, 322
DASHWOOD and Redhead, 398, 399
DAWES, SIR WILLIAM, 440 n. 5
the declination of the magnetic needle and Halley's theory, 10, 11 n. 9
declination of the Sun, 75
DEMEE and Redhead, 399
DENMARK, KING OF, and his library, 482, 483
DENMARK, PRINCE GEORGE OF, 423 n. 10
    LETTER from the Referees: (23 January 1704/5), **685**, 436;
    and his Treasurer, LETTER from the Referees: (28 November 1706), **713**, 480
    a Fellow of the Royal Society, 431 n. 5
    and the *Historia*, 422, 482, 483
    and the Referees, 431
    and Römer and Tycho, 484 n. 3
DERBY, THE TENTH EARL, 520 n. 2
Derby and observations made there, 420, 449
De Re Metallica, 258 nn. 6 and 7
DESCARTES, RENÉ, and *Discours de la Méthode*, 102 n. 5
    and *Analysis Speciosa*, 237, 238
    and the nature of light, 508 n. 7
    and the rainbow, 266, 267, 268 n. 3
Descriptio Regni Japoniæ, 412 n. 3
designs for Queen Anne's Bounty medal, 429
DICKINSON, EDMUND, 198 n. 3
    and Jodochus a Rhe, 196
    and *Edmundus Dickinson, Medicus Regius, Et Theodorus Mundanus, Philosophus Adeptus, de Quintessentia Philosophorum . . .*, 198 n. 3
differential calculus and Leibniz, 15, 17, 18 n. 4
diatonic system, 275
ditone, 274
Doctrine of the Sphere and Flamsteed, 14 n. 8
    and eccentricity, 32 n. 6
    and eclipse of 1678, 110
    and the equation of apparent time, 85 n. 5
Dog Tavern and dinner for pyx trial jury, 371
DE DOMINIS, MARCO ANTONIO, 268 n. 1
DRUMMOND, JOHN, 528 n. 1
    LETTER to Newton: (20 July 1708), **748**, 528
DRUMMOND, WILLIAM, 523 n. 1
    LETTER to Newton: (12 July 1708), **745**, 522-3

DRUMMOND and the price of tin, 479
    and the sale of tin, 466
DUBLIN, ARCHBISHOP OF, and the Royal Society, 448
ducat, Holland, 282
DUMFORD, WILLIAM, and the Mint, 356
Dutch ambassador, his chaplain and the Royal Society, 543
duty mark on paper and parchment, 413

Earth, its axis, and the plane of the ecliptic, 333
    its orbit, and time, 71
    its shadow and Eclipse of 1678, 123
East India Company and coining, 317
Easter and its determination, 330 n. 2
ECCLESTON, CHARLES, 211, 213 n. 1
Eclipse of the Moon in 1678, 72, 110, 111, 123
    in 1682, 112
Eclipse of the Sun and Flamsteed, 112
écu and value in 1701, 374
Egham and the Holloways, 262
ELIZABETH I and 20s. piece, 283 n. 11
elliptical hypothesis and Curtz, 32 n. 5
ELLIS, CHRISTOPHER, additional clerk to Newton, 215
ELLIS, JOHN, 391 n. 9
    and Redhead, 401
    and the tin in the Mint, 411
ELLIS, JOHN, Vice-Chancellor of Cambridge University, his knighthood, 440 n. 9, 444
English and Wallis ('let those who desire to read it [*Opticks*], learn English'), 100
engravers at the Mint, 416-17
Enumeratio Linearum Tertii Ordinis by Newton, 278 n. 5
    and *Opticks*, 415 n. 6
Epistola Posterior and *Opticks*, 415 n. 5
    and Wallis, 184 n. 3
    and *Epistola Prior*, 129, 140-1
Epistola Prior and Wallis, 184 n. 3
equation, 8 n. 3
equation of right ascension and Flamsteed, 84
    of the Sun's centre and Hipparchus, 6 n. 9
    of time, 73 n. 5, 74, 328, 329
equations of the apogee and Newton, 291
    and eccentricities, 120, 137
    of the parallax, 120

555

Garways (Garraways), 285 n. 4, 435
  and meeting of Bentley, Flamsteed and Newton, 468
GASCOIGNE, WILLIAM, 423 n. 2
  and Crabtree, 32 n. 5
  and Witty, 438
GASTRELL, FRANCIS, 428 n. 10
Gate House prison and Newton, 212, 213 n. 1
GAYER (ANNESLEY), LADY ELIZABETH, 476 n. 2
geocentric system and Prolemy, 5 n. 2
geodetic measurements at present inaccurate, 105
GEORGE I, his currency engraved by Croker, 352 n. 3
German steel and making dies and puncheons, 487
GLOUCESTER, DUKE OF, 253 n. 2, 296 n. 9
  and Flamsteed, 294, 298
  and Gregory, 253
GODOLPHIN, FRANCIS, 440 n. 8
  and the 1705 election, 440, 445 n. 4
GODOLPHIN, SIDNEY, EARL OF, 386 n. 1
  LETTERS
    from David Gregory: (13 December 1707), **732**, 503–4
    from the Mint: (7 July 1702), **650**, 388–90; (January 1702/3), **657**, 396–7; (23 August 1704), **674**, 416–17; (13 September 1704), **675**, 418; (5 April 1705), **691**, 442–3; (23 September 1706), **712**, 478–9; (12 April 1707), **717**, 485–7; (26 May 1707), **720**, 489–90; (31 May 1707), **722**, 491–3; (July 1797), **726**, 496; (25 June 1708), **743**, 520
    from Newton: (15 April 1702), **647**, 385–6; (16 October 1702), **652**, 392; (21 December 1702), **653**, 393; (?1702), **655**, 394–5; (1702), **656**, 395–6; (13 July 1703), **666**, 408–9; (30 October 1703), **667**, 409–10; (?1703), **668**, 411; (9 January 1703/4), **669**, 411; (15 May 1704), **673**, 415–16; (12 October 1704), **676**, 419–20; (24 June 1707), **724**, 494–5; (14 November 1707), **729**, 500; (14 April 1708), **740**, 515–16; (April 1708), **741**, 516–18
    from Newton and Gregory: (30 June 1708, 4 August 1708), **744**, 521–2

Golden Square Tabernacle accounts, 377–80
gold, rate per guinea too high, 389
gold-silver ratio, is altered in France, 373
  Newton's attempt to establish, 390 n. 3
Goldsmiths' Company and the pyx, 371, 495–6
GOWER, SIR JOHN, 440 n. 11
graces, Newton's, 18 n. 1
grain tin, 517–18, 518 n. 2
Grantham, Soke of, 319
gravitation, and celestial phenomena, 1, 3
  and Fatio's theory, 11 n. 7
  laws of, and Flamsteed's observations, 332
  and Newton's explanation, 266, 267
  theory of confirmed by the satellites of Jupiter and Saturn, 62, 63
  theory of, and the Moon, 7 n. 2, 87
GRAY, ABBOM, and Halley, 254
Great Plague of 1665 and Newton, 142 n. 10
Greenwich Observatory, and Flamsteed's assistants, 474 n. 3
  and observations taken there, 420
  and its position, 78, 322
  Newton visits Flamsteed there, 1 September 1694, 7
GREGORY, DAVID
  LETTERS
    to Godolphin: (13 December 1707), **732**, 503–4; (30 June 1708), **744**, 521–2
    to Newton: (24 September 1694), **472**, 20; (23 December 1697), **577**, 253; (30 September 1702), **651**, 391–2; (12 August 1707), **727**, 497–8; (9 October 1707), **728**, 498–9; (28 December 1707), **733**, 505–6
GREGORY, DAVID
  LETTERS
    from Newton: (1707), **731**, 502–3
  MANUSCRIPT (15 April 1707), **718**, 487–8
  MEMORANDA (1 September 1694), **468**, 7; (7 September 1694), **471**, 15–18; (3 February 1694/5), **492**, 82; (20 February 1697/8), **584**, 265–7; (?July 1698), **589**, 276–7; (21 May 1701), **634**, 354–5; (March 1702/3), **622**, 402–3
  and appointed Savilian Professor of Astronomy, 73
  and *Astronomiæ Physicæ et Geometricæ Elementa*, 327 n. 8

NEWTON, ISAAC (*cont.*)

from Chaloner: (20–24 January 1698/9), **606**, 305; (*c.* 6 March 1698/9), **608**, 307–8

from Chamberlayne: (2 February 1703/4), **670**, 412

from Chrisloe: (6 June 1695), **515**, 131

from George Clark: (11 December 1704), **680**, 430

from John Drummond: (20 July 1708), **748**, 528

from William Drummond: (12 July 1708), **745**, 522–3

from Flamsteed: (7 September 1694), **470**, 12–13; (11 October 1694), **474**, 26–30; (25 October 1694), **476**, 36–7; (29 October 1694), **477**, 38–41; (3 November 1694), **479**, 44–6; (27 November 1694), **481**, 50–1; (6 December 1694), **483**, 53–5; (10 December 1694), **484**, 57–60; (31 December 1694), **486**, 63–6; (18 January 1694/5), **488**, 70–2; (29 January 1694/5), **490**, 77–9; (7 February 1694/5), **493**, 83; (2 March 1694/5), **495**, 89–92; (21 March 1694/5), **497**, 98–9; (20 April 1695), **499**, 103–4; (27 April 1695), **501**, 110–13; (6 May 1695), **506**, 121–3; (2 July 1695), **517**, 137–8; (13 July 1695), **521**, 144–8; (18 July 1695), **522**, 150–1; (23 July 1695), **524**, 152–5; (4 August 1695), **526**, 157–8; (6 August 1695), **527**, 158–63; (19 September 1695), **531**, 170–1; (11 January 1695/6), **543**, 191–2; (4 September 1697), **574** 249–50; (10 December 1697), **576**, 261–2; (10 October 1698), **595**, 286; (29 December 1689), **599**, 290–1; (2 January 1698/9), **600**, 292–5; (10 January 1698/9), **604**, 302–3; (18 June 1700), **627**, 333–46; (2 January 1704/5), **683** 433; (5 January 1704/5), **684**, 434–5; (5 April 1705), **690**, 441–2; (25 October 1705), **699**, 449–51; (26 February 1705/6), **707**, 473–4; (14 September 1706), **711**, 476–7; (10 April 1708), **739**, 515

from David Gregory: (24 September 1694), **472**, 20; (23 December 1697), **577**, 253; (30 September 1702), **651**,

391–2; (12 August 1707), **727**, 497–8; (9 October 1707), **728**, 498–9; (28 December 1707), **733**, 505–6

from Halifax: (17 March 1704/5), **688**, 439; (5 May 1705), **693**, 445

from Halley: (7 September 1695), **528**, 165; (28 September 1695), **532**, 171–2; (7 October 1695), **533**, 173–5; (15 October 1695), **534**, 176–9; (21 October 1695), **536**, 182–3; (1695/6), **542**, 190; (28 November 1696), **555**, 213–14; (13 February 1696/7), **563**, 230–1; (2 August 1697), **570**, 246; (30 December 1697), **578**, 254

from Harington: (22 May 1698), **587**, 272–3

from Hockett: (14 September 1699), **616**, 315–16

from Kemp: (15 February 1700/1), **632**, 353

from Lauderdale: (10 August 1709), **759**, 542

from Le Neve: (24 November 1705), **703**, 460

from de Monmort: (16 February 1708/9), **752**, 533–4

from Montague: (19 March 1695/6), **545**, 195

from Morland: (2 May 1695), **504**, 119

from Noah Neal: (15 August 1697), **572**, 248

from Henry Newton: (December 1707), **734**, 506–7

from Sir John Newton: (5 February 1697/8), **583**, 265

from Pepys: (13 May 1695), **507**, 126

from the Referees: (26 March 1708), **738**, 514

from Robinson: (5 February 1697/8), **582**, 263–4

from Seafield: (28 June 1709), **757**, 540

from Patrick Scott: (31 January 1707/8), **736**, 510–12

from Benjamin Smith: (18 November 1695), **540**, 187

from Stacy: (?1699), **619**, 318

from Tily: (24 November 1707), **730**, 501

from the Treasury: (12 August 1697), **571**, 247; (22 March 1702/3), **661**, 401

NEWTON, SIR JOHN (*cont.*)
and relationship to Newton, 461 n. 2, 489 n. 1
proposed for post at Christ's Hospital, 93
and Newton, 103, 132–3
and Pepys, 126
and Flamsteed, 98, 112
Noah's ark, ratios in its dimensions, 272
nodes, 42, 548
nonagesimary table, 138, 154–5, 156, 156 n. 2
non-jurors (*non jurats*), 293
NORRIS, SIR WILLIAM, 350 n. 4
North Witham, 319
Norwich Mint, 313, 396 n. 3
NORWOOD, RICHARD, 252 n. 4
notation of Barrow and Leibniz, 10 n. 2
Nugæ Antiquæ, by Sir Johh Harington, 273 n. 3
nutation, 333, 346 n. 4, 548

Oates, Newton's proposed visit, 406
obliquity of the ecliptic, 78, 326, 548
obol, 314 n. 3
observations by Flamsteed, 77, 156, 157, 291, 474
observations and weather, 77, 89
OLDENBURG, HENRY, and Newton concerning papers on light, 415 n. 2
and Newton's 'two letters', 186
and Slusius and his method of tangents, 141
Opera, Wallis's, letter from Leibniz, 239 n. 1
and *Philosophical Transactions*, 114–15
Opera Miscellanea, by Roger Cotes, 472 n. 7
Opera Posthuma and Horrocks, 72
Opticks, 117–18 n. 2, 414
a copy in Latin delivered to the College of Physicians, Edinburgh, 528
a copy sent to Rome by Henry Newton, 506, 507 n. 4, 508 n. 8
disputes and delay in printing, 414
and Flamsteed, 424, 463, 465 n. 7
and Gregory, 355
and Wallis, 100, 115, 116, 130
ORIGEN, 403 n. 3
ORMOND, DUKE OF, and Vigo Bay, 404 n. 2
OUGHTRED, W., *Clavis Mathematicæ*, 114
his method of writing decimal fractions, 252 n. 2
OVERTON, BENJAMIN, 189 n. 4, 196 n. 2
succeeded by Newton, 195

OVIEDO, 266, 267
the Oxford carrier, Matthews, 114
Oyer and Terminer, Commission of, 545 n. 2

PAGET, EDWARD, 38 n. 2
and Christ's Hospital, 37, 93, 126
PAMPHILION, EDWARD, and coinage offences, 218
paper-makers and counterfeiting, 416
parallactic equation, 34, 46, 143
and Flamsteed, 35, 137
and Halley, 13
parallax, horizontal, 67
parallax of the Pole Star, 279 n. 2
and Flamsteed, 83, 86 n. 6, 124, 369
and Wallis, 280, 300
parallaxes of six stars, 281 n. 3
The Paramour, and Halley, 190–1 n. 1, 230 n. 2, 311 n. 5
Paris Observatory, 112, 119, 322
PARKER, JOHN, and his almanac, 69, 70 n. 10
Parliamentary election, 383 n. 1
PARRY, SIR FRANCIS, and coinage of halfpence and farthings, 409 n. 2
PARSONS, COL. WILLIAM, 416, 417
PASCAL, BLAISE, and Bernoulli 221, 224
PATRICK, SIMON, 441 n. 2
PECK, and Chaloner, 307
PEERS, JOHN, 305, 544
PELHAM, SIR THOMAS (BARON PELHAM), 387 n. 7
PEMBROKE, EARL OF, 447 n. 3
introduces members of the Royal Society to Prince George, 449
and Psalmanazar, 413 n. 3
penny, use of copper recommended, 390
pennyweights and carats, number of grains in, 282
PEPYS, SAMUEL
LETTER to Newton: (13 May 1695), **508**, 126
and Flamsteed, 114 n. 12
and Halley, 99
and Paget, 126
and Wallis, 116 n. 9, 279 n. 2
perigee, 120, 547
Peruvian piece, 288
PETER THE GREAT, 265 n. 3
PETRONIUS, GAIUS, 435 n. 2
pewterers, and sale of tin, 466

571